NanoScience and Technology

The series NanoScience and Technology is focused on the fascinating nano-world, mesoscopic physics, analysis with atomic resolution, nano and quantum-effect devices, nanomechanics and atomic-scale processes. All the basic aspects and technology-oriented developments in this emerging discipline are covered by comprehensive and timely books. The series constitutes a survey of the relevant special topics, which are presented by leading experts in the field. These books will appeal to researchers, engineers, and advanced students.

More information about this series at http://www.springer.com/series/3705

Umberto Celano
Editor

Electrical Atomic Force Microscopy for Nanoelectronics

Springer

Editor
Umberto Celano
IMEC
Leuven, Belgium

ISSN 1434-4904 ISSN 2197-7127 (electronic)
NanoScience and Technology
ISBN 978-3-030-15611-4 ISBN 978-3-030-15612-1 (eBook)
https://doi.org/10.1007/978-3-030-15612-1

This Springer imprint is published by the registered company Springer Nature Switzerland AG
The registered company address is: Gewerbestrasse 11, 6330 Cham, Switzerland

Contents

Contributors

Oskar Amster PrimeNano, Inc., Santa Clara, CA, USA;
KLA, One Technology Drive, Milpitas, CA, USA

Nicolas Baboux Institut des Nanotechnologies de Lyon, INSA de Lyon, Université de Lyon, UMR CNRS 5270, Villeurbanne, France

Laurent Boudou LAPLACE, Université de Toulouse, CNRS, UPS, INPT, Toulouse, France

Stefano Brivio CNR-IMM, Unit of Agrate Brianza, Agrate Brianza, Italy

Adrian Carretero-Genevrier Institut d'Electronique et des Systemes (IES), CNRS, Universite Montpellier, Montpellier, France

Umberto Celano IMEC, Leuven, Belgium

Thierry Conard IMEC, Leuven, Belgium

Marco Corbetta NanoScan AG, Dübendorf, Switzerland

Graham Cross Adama Innovations, CRANN, Trinity College Dublin, Dublin, Ireland;
Trinity College, School of Physics, Dublin 2, Ireland

Pierre Eyben IMEC, Leuven, Belgium

Jacopo Frascaroli STMicroelectronics, Agrate Brianza, Italy

Brice Gautier Institut des Nanotechnologies de Lyon, INSA de Lyon, Université de Lyon, UMR CNRS 5270, Villeurbanne, France

Filippo Giannazzo Consiglio Nazionale delle Ricerche, Institute for Microelectronics and Microsystems (CNR-IMM), Catania, Italy

Martí Gich Institut de Ciència de Materials de Barcelona (ICMAB-CSIC), Bellaterra, Spain

Andres Gomez Institut de Ciència de Materials de Barcelona (ICMAB-CSIC), Bellaterra, Spain

Giuseppe Greco Consiglio Nazionale delle Ricerche, Institute for Microelectronics and Microsystems (CNR-IMM), Catania, Italy

Alexei Gruverman Department of Physics and Astronomy, University of Nebraska-Lincoln, Lincoln, NE, USA

Thomas Hantschel IMEC, Leuven, Belgium

Atsufumi Hirohata Department of Electronic Engineering, University of York, Heslington, York, UK

Jason Kilpatrick Conway Institute of Biomedical and Biomolecular Research, University College Dublin, Dublin, Ireland;
Adama Innovations, CRANN, Trinity College Dublin, Dublin, Ireland

Armin Wolfgang Knoll IBM Research Laboratory Zurich, Zürich, Switzerland

Mario Lanza Institute of Functional Nano & Soft Materials, Collaborative Innovation Center of the Ministry of Education, Soochow University, Suzhou, China

Min Hwan Lee Department of Mechanical Engineering, University of California —Merced, Merced, USA

Chandreswar Mahata Institute of Functional Nano & Soft Materials, Collaborative Innovation Center of the Ministry of Education, Soochow University, Suzhou, China

Kremena Makasheva LAPLACE, Université de Toulouse, CNRS, UPS, INPT, Toulouse, France

Simon Martin Institut des Nanotechnologies de Lyon, INSA de Lyon, Université de Lyon, UMR CNRS 5270, Villeurbanne, France

Shashank Misra Sandia National Laboratories, Albuquerque, NM, USA

Jay Mody GLOBALFOUNDRIES, Malta, NY, USA

Jochonia Nxumalo GLOBALFOUNDRIES, Malta, NY, USA

Kristof Paredis IMEC, Leuven, Belgium

Fabrizio Roccaforte Consiglio Nazionale delle Ricerche, Institute for Microelectronics and Microsystems (CNR-IMM), Catania, Italy

Christian Rodenbücher Forschungszentrum Jülich GmbH, Peter Grünberg Institute (PGI-7) and JARA-FIT, Jülich, Germany;
Forschungszentrum Jülich GmbH, Institute of Energy and Climate Research (IEK-3), Jülich, Germany

Kurt A. Rubin PrimeNano, Inc., Santa Clara, CA, USA;
KLA, One Technology Drive, Milpitas, CA, USA

Yu Kyoung Ryu Instituto de Ciencia de Materiales de Madrid, Consejo Superior de Investigaciones Científicas (CSIC), Madrid, Spain

Marjan Samiepour Department of Electronic Engineering, University of York, Heslington, York, UK

Andreas Schulze IMEC, Leuven, Belgium;
Applied Materials, Santa Clara, CA, USA

David A. Scrymgeour Sandia National Laboratories, Albuquerque, NM, USA

Kristof Szot Forschungszentrum Jülich GmbH, Peter Grünberg Institute (PGI-7) and JARA-FIT, Jülich, Germany;
University of Silesia, A. Chełkowski Institute of Physics, Katowice, Poland

Gilbert Teyssedre LAPLACE, Université de Toulouse, CNRS, UPS, INPT, Toulouse, France

Wilfried Vandervorst IMEC, Leuven, Belgium;
Department of Physics and Astronomy, KU Leuven, Leuven, Belgium

Christina Villeneuve-Faure LAPLACE, Université de Toulouse, CNRS, UPS, INPT, Toulouse, France

Marcin Wojtyniak University of Silesia, A. Chełkowski Institute of Physics, Katowice, Poland

Lennaert Wouters IMEC, Leuven, Belgium

Yongliang Yang PrimeNano, Inc., Santa Clara, CA, USA

Acronyms

AFM	Atomic Force Microscopy
AFM-IR	AFM-Based Infrared Spectroscopy
AFP	Atomic Force Prober
ALD	Atomic Layer Deposition
AM-KPFM	Amplitude Modulation KPFM
ASTC	Advanced Storage Technology Consortium
BD	Dielectric Breakdown
BE-PFM	Band-Excitation PFM
C-AFM	Conductive Atomic Force Microscopy
CBRAM	Conductive Bridge Random Access Memory
CDT	Coated Diamond Tip
CFs	Conductive Filaments
CGP	Contact Gate Pitch
CMOS	Complementary Metal-Oxide Semiconductor
CP-AFM	Conductive Probe AFM
CPE	Converse Piezoelectric Effect
C-SFM	Conductive Scanning Force Microscopy
C-SPM	Conductive Scanning Probe Microscopy
CVD	Chemical Vapour Deposition
DLC	Diamond-Like Carbon
DPFM	Direct Piezoelectric Force Microscopy
DRAM	Dynamic Random Access Memory
EFM	Electrostatic Force Microscopy
EOT	Equivalent Oxide Thickness
FA	Failure Analysis
FCVA	Filtered Cathodic Vacuum Arc
FDSOI	Fully Depleted Silicon on Insulator
FDT	Full Diamond Tip
FEM	Finite Element Model
FFM	Friction Force Microscopy

FFT-SSRM	Fast Fourier Transform SSRM
FIB	Focused Ion Beam
FMEA	Failure Mode and Effects Analysis
FM-KPFM	Frequency Modulated KPFM
FWHM	Full-Width at Half Maximum
GAA	Gate-all-Around
GBs	Grain Boundaries
HAMR	Heat-Assisted Magnetic Recording
HDC	High-Density Carbon
HET	Hot Electron Transistor
HRS	High Resistive State
HS-AFM	High-Speed AFM
ICs	Integrated Circuits
IDEMA	International Disk Drive Equipment and Materials Association
IFQW	Implant-Free Quantum Well
IoT	Internet of Things
IRDS	International Roadmap for Devices and Systems
KPFM	Kelvin Probe Force Microscopy
LCA	Liquid Crystal Analysis
LC-AFM	Local-Conductivity AFM
LIA	Lock-In Amplifier
LPFM	Lateral PFM
LRS	Low Resistive State
LSB	Least Significant Bit
MAMR	Microwave-Assisted Magnetic Recording
MBE	Molecular Beam Epitaxy
MCD	Magnetic Circular Dichroism
MEMS	Microelectromechanical Systems
MFM	Magnetic Force Microscopy
MIM	Metal–Insulator–Metal
MIM	Microwave Impedance Microscopy
MOKE	Magneto-Optical Kerr Effect
MOSFET	Metal-Oxide Field-Effect Transistor
MOVPE	Metalorganic Vapour Phase Epitaxy
MQWs	Multiple Quantum Wells
MUT	Material Under Test
NC	Negative Capacitance
NC-AFM	Non-Contact AFM
NDR	Negative Differential Resistance
NDSM	Non-Linear Dielectric Microscopy
NSMM	Near-Field Scanning Microwave Microscopy
NV	Nitrogen-Vacancy
ODT	Overcoated Diamond Tips
PCM	Phase Change Materials
PEM	Photoemission Analysis

PE-SC-DDP	Plasma Etched Single Crystal Doped Diamond Probes
PF-KPFM	Peak Force-KPFM
PFM	Piezoresponse Force Microscopy
PF-QNM	Peak Force Quantitative Nanomechanical
PSTM	Photon Scanning Tunneling Microscope
PUND	Positive Up Negative Down
PVD	Physical Vapour Deposition
QD	Quantum Dot
Q-factor	Quality Factor
QFSEG	Quasi-Free-Standing Epitaxial Graphene
QPC	Quantum Point Contact
RIE	Reactive Ion Etching
RRAM	Resistive Switching Random Access Memory
RS	Resistive Switching
SAD	Selective Area Deposition
SBH	Schottky Barrier Height
SCM	Scanning Capacitance Microscopy
SEM	Scanning Electron Microscopy
SEMPA	Scanning Electron Microscopy with Polarization Analysis
SHPM	Scanning Hall Probe Microscopy
SILC	Stress-Induced Leakage Currents
SKPFM	Scanning Kelvin Probe Microscopy
sMIM	Scanning Microwave Impedance Microscopy
SMM	Scanning Microwave Microscopy
SMR	Shingled Magnetic Recording
SNOM	Near-Field Optical Microscopy
SOI	Silicon on Insulator
SPL	Scanning Probe Lithography
SPM	Scanning Probe Microscopy
SP-STM	Spin-Polarized STM
SQUID	Superconducting Quantum Interference Devices
SRP	Spreading Resistance Probe
s-SNOM	Scattering Near-Field Optics Microcopy
SSRM	Scanning Spreading Resistance Microscopy
SThM	Scanning Thermal Microscopy
STM	Scanning Tunneling Microscopy
TAT	Trap-Assisted Tunneling
TCAD	Technology Computer-Aided Design
TDMR	Two-Dimensional Magnetic Recording
TEM	Transmission Electron Microscopy
TIs	Topological Insulators
TMDs	Transitions Metals Dichalcogenides
TOF-SIMS	Time-of-Flight Secondary Ion-Mass Spectrometry
TUNA	Tunneling AFM
UNCD	Ultra-Nanocrystalline Diamond

VPFM	Vertical PFM
WKB	Wentzel–Kramers–Brillouin
XMCD	X-ray Magnetic Circular Dichroism
XPEEM	X-ray Photoemission Electron Microscopy

Chapter 1
The Atomic Force Microscopy for Nanoelectronics

Umberto Celano

Abstract The invention of scanning tunneling microscopy (STM), rapidly followed by atomic force microscopy (AFM), occurred at the time when extensive research on sub-μm metal oxide field-effect transistors (MOSFET) was beginning. Apparently uncorrelated, these events have positively influenced one another. In fact, ultra-scaled semiconductor devices required nanometer control of the surface quality, and the newborn microscopy techniques provided unprecedented sensing capability at the atomic scale. This alliance opened new horizons for materials characterization and continues to this day, with AFM representing one of the most popular analysis techniques in nanoelectronics. This book discusses how the introduction of new devices benefited from AFM, while driving the analysis and sensing capabilities in novel directions. Here, the goal is to introduce the major electrical AFM methods, going through the journey that has seen our life changed by the advent of ubiquitous nanoelectronics devices, and has extended our capability to sense matter on a scale previously inaccessible.

1.1 Introduction

The atomic force microscope (AFM) is a special type of microscope using a mechanical sampling method to form images of surfaces at the nanoscale [1]. This is achieved by scanning an atomically sharp tip on the sample surface, at a controlled load force, while recording the tip-sample interactions. The origin of modern AFM can be traced back to the pioneering attempts to use an optical method to probe the deflection of a sharp stylus [2–5]. This demonstrated an effective alternative to previous implementations which involved optical interferometry or the measurement of tunneling currents between the stylus (i.e., a tip in contact with the surface) and a second fixed electrically conductive cantilever beam. Compared to its predecessor, the scanning tunneling microscope (STM) [6], the AFM based on the optical method enabled the use of functionalized probes that were easy to replace, and this turned out to be a

U. Celano (✉)
IMEC, Kapeldreef 75, 3001 Leuven, Belgium
e-mail: celano@imec.be

U. Celano (ed.), *Electrical Atomic Force Microscopy for Nanoelectronics*,
NanoScience and Technology, https://doi.org/10.1007/978-3-030-15612-1_1

disruptive innovation leading to a massive adoption of AFM to measure virtually all types of materials and related physical quantities. As a consequence, over the past thirty years, AFM has become one of the most important analysis technique for a multitude of disciplines including physics, chemistry, biology, material science, and nanotechnology [7–11]. Thus, it is not surprising that a large number of books and review articles have covered this analysis method [7–14].

Today, the most important application is considered to be the detection of relevant physical quantities such as structural, electrical, chemical, and mechanical information with a spatial resolution ranging from 0.1 to 100 nm. The fields of micro- and nanoelectronics have benefited greatly from the developments in AFM. This is true also for the industrial production of integrated circuits (ICs), where AFM techniques are used intensively at various stages of the chip manufacturing process. This book tries to introduce to students and researchers to the vast field of electrical AFMs, emphasizing the role that these techniques have had, and are still having, on the development of devices in modern nanoelectronics. In each chapter, one AFM technique is described, combining the explanation of the operation principles with a specific area of application in nanoelectronics, without overlooking the necessary underlying nanoscience. We introduce some of the requirements for the next generation of semiconductor devices by addressing the AFM techniques that are contributing to their development now and will do so in the future. In addition, our ambition is to put in context both the physical limitations encountered by a specific device when scaled, and the associated challenges for metrology when measuring a certain physical property. Each chapter will emphasize the must-meet metrology challenges that various AFM modes are trying to overcome to support the development of emerging technologies. This chapter gives a broad overview of the use of AFM in nanoelectronics, describes the basic principles of the technique, and explains the metrology challenges associated with future chip technology. Chapters 2–4 have been organized to cover contact mode techniques and their evolution for sensing the fundamental electrical properties such as electrical current flow, resistance, and capacitance. Chapter 5 addresses the possible use of AFM for nanoscale patterning, including electrochemical and thermomechanical lithography. The study of piezoelectric materials and resistive memories are described in Chaps. 6 and 7. Chapter 8 discusses the possible application of AFM for the analysis of magnetic fields for data storage. Photoconductive probing methods and 2D materials are discussed in Chaps. 9 and 10, respectively. For their pivotal role in enabling high lateral resolution in electrical modes, Chap. 11 describes the latest technologies for the fabrication of ultra-sharp conductive diamond probes. Finally, the last chapter is devoted to microwave-based methods which extend the current capability of electrical AFMs and is becoming increasingly important in the study of single-impurity devices and quantum bits.

1.2 Atomic Force Microscopy: The Swiss-Knife of Nanoelectronics

The central role of electrical AFMs for the development of micro- and nanoelectronics is shown schematically shown in Fig. 1.1. The correlation between various techniques and the introduction of new devices clearly illustrates their side-by-side evolution over the last few decades. Far from being complete, the list of AFM methods shown in the figure have supported the development of all the major integrated microelectronic devices, with an active contribution during (1) the research stage, and (2) the production phase (e.g., in-line metrology, failure analysis, etc.). This strong partnership is even more important today, when the industry is struggling to answer questions about the evolution of IC technology in the coming decades [15, 16].

Soon after the introduction of AFM, additional modes were rapidly conceived. Opportunities to sense various physical quantities were offered by the wide variety of forces detected during the tip-sample interaction. These include attractive and repulsive forces induced by electrostatic, magnetic, and chemical coupling, opening up multiple pathways for specialized electrical modes optimized to sense, among others, resistance, capacitance, and electric and magnetic fields [2, 18, 19]. Developments were favored by the inherent flexibility of using dedicated tips, replacing them, and applying/sensing customized electrical stimuli. For example, a probe can be vibrated (in non-contact) and the forces acting on the tip recorded by tracking the changes in the tip frequency while approaching the surface. Since long range forces affecting the frequency shift are electrostatic or magnetic, this allows for the direct probing of electric fields and work function differences as in electrostatic force microscopy (EFM) [20] and Kelvin probe force microscopy (KPFM), respectively [21]. Similarly, magnetic force microscopy (MFM) senses magnetic fields applying the same concept and using tips with magnetic coatings [22]. Developed at the end of the 1980s and the beginning of the 1990s, to date, lateral resolutions of c.a. 10–20 nm have been demonstrated for these techniques, while limitations exist owing to fringing fields induced by the tip-sample distance (i.e., affecting resolution), as well as tip-apex asymmetry, which can introduce artifacts in the results [23, 24].

Two-dimensional quantitative dopant and carrier profiling has been traditionally one of the most important areas of application for electrical AFMs. The first techniques used for this purpose were KPFM, scanning capacitance microscopy (SCM), and scanning spreading resistance microscopy (SSRM) [18, 25, 26]. In SCM the probe creates a movable nanosized metal-oxide-semiconductor structure whose capacitance changes with the local carrier concentration. Alternatively, one can measure the spreading resistance at the tip-semiconductor junction, relating the local conductivity with carrier concentration by a calibration procedure as done by SSRM. Not surprisingly, interest in quantitative carrier profiling started at the end of the 1980s when the transistor count on commercial chips had grown to over one million and the minimum feature size was reduced below 1 μm (Fig. 1.1). Applications for IC process control and failure analysis continue even today. However, the

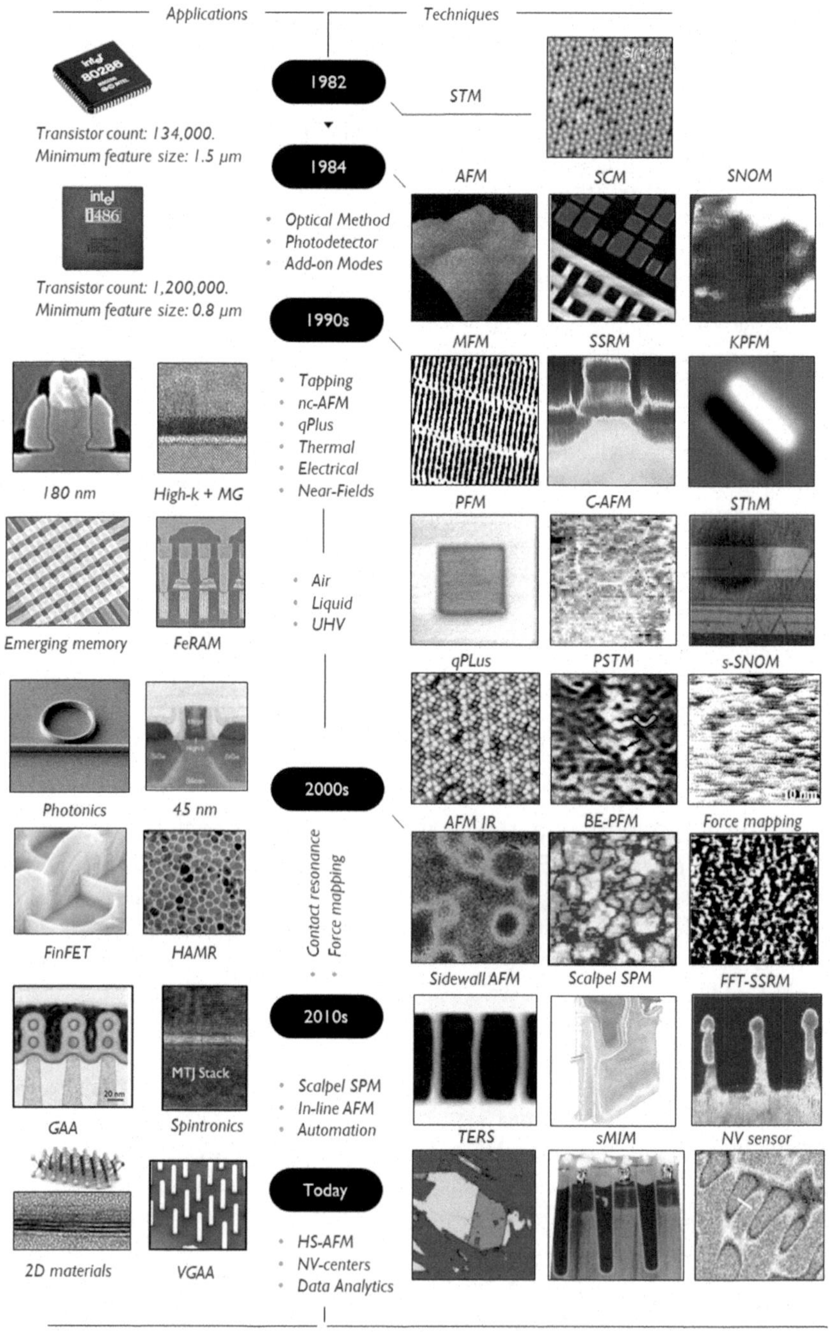

Fig. 1.1 Schematic illustration of nanoelectronics evolution side-by-side with development of AFM techniques over the last four decades. Adapted with permission from a collection of references as reported in [17]

current increase in complexity (i.e., sub-10 nm feature size, transistor counts in the billions) has required the introduction of advanced methods such as two- and three-dimensional Fast Fourier transform SSRM (FFT-SSRM), and scanning microwave impedance microscopy (sMIM) [27–30]. In the former, an additional force modulation is used to decouple the spreading resistance from parasitic series resistance components, while in sMIM the probe is a GHz emitter-receiver antenna used to measure the local impedance of the sample through high frequency sensing electronics [30].

During the 1990s when high-k dielectrics and metal gates received great attention to replace conventional SiO_2/Poly-Si gate stacks, conductive atomic force microscopy (C-AFM) and KPFM were used, respectively, to study leakage and localized states (traps) in thin oxides [23, 31, 32]. In particular, the C-AFM probe acts as a scaled movable electrode, sensing local changes in the electrical current flowing in the tip-sample system while in contact. Due to its ease of operation, C-AFM rapidly became widespread beyond thin oxides for the characterization of organic materials, nanowires and photovoltaics, to name but a few [31].

Interestingly, when in contact, the tip-induced electrical stimulation of the sample can also induce mechanical modification of the surface with practical applications. For example, the local oxidation of the sample can be used for nanopatterning, or the converse piezoelectric effect can be used to study piezo- and ferroelectric phenomena. This is the case for piezoresponse force microscopy (PFM) [33, 34], which has found many applications in the development of ferroelectric random access memory (FeRAM), sensors, and ferroelectric tunnel junctions. More recently, PFM has been improved by the introduction of band-excitation PFM (BE-PFM), which enhances the signal-to-noise ratio by working at the resonant frequency of the tip-sample mechanical system and allows for higher lateral resolution and hyperspectral data acquisition [35]. The evolution of mass production memory including volatile dynamic RAM (DRAM) and non-volatile floating gate (i.e., charge-trap flash) required extensive use of EFM, KPFM, SCM, and SSRM to assess device electrostatics. On the other hand, emerging memory concepts such as phase change materials (PCM), and filamentary- and interfacial-resistive switching devices, generated interest in thermo-chemical mapping and three-dimensional analysis of electrical transport in confined volumes [36–38]. The development of magnetic recording for non-volatile memory has been continuously supported by MFM [22]. However, at present, the considerable research on heat-assisted magnetic recording (HAMR) has renewed interest in techniques capable of studying radiative heat transfer, such as scanning thermal microscopy (SThM) in combination with MFM [39].

For exploratory research such as chemical self-assembly, 2D materials- and nanotubes-based devices, controlled environments (e.g., low temperature, ultrahigh vacuum, or liquid) are now of the utmost importance to enable atomic resolution. Of particular importance in this area is the advent of quartz tuning fork cantilevers (i.e., qPlus sensors). These have high spring constants (~kN) allowing oscillation amplitudes of a few tens of pm and have given birth to a new branch of atomically-resolved non-contact AFM techniques [5, 40–43]. To resolve features smaller than the diffraction limit of the applied radiation, and visualize local field enhancement induced by plasmonic resonances, near-field optics techniques such as near-field optical microscopy (SNOM) [44, 45], photon scanning tunneling microscope (PSTM) [46, 47], and AFM-based infrared spectroscopy (AFM-IR) have received much attention for the development of integrated optics, high speed interconnections, plasmonic waveguide and photonic crystals [48, 49]. Along with the evolution of IC process technology, a growing number of new materials have been introduced, creating the need for accurate characterization of specific mechanical properties such as adhesion, stiffness, friction, and wear. To this end, force mapping methods have been introduced to collect several force-distance spectroscopy curves in the area of interest, thereby imaging the local variation in mechanical properties of the surface along with its topography. The combination of force mapping and pre-existing modes (i.e., electrical, chemical and optical) often boosted the resolution of the existing techniques, adding mechanical information.

Recently, the introduction of three-dimensional architectures like FinFETs and vertical memory concepts has determined another inflection point for AFMs, now faced with the challenge of an additional spatial dimension to probe. A range of ingenious solutions have emerged. For example, specialized tips with T-shaped ends are used to access between fins [50]. Alternatively, tilting the scanning head eliminates certain tip-shape distortions and can be used to probe the side wall of adjacent fins [51]. Both solutions allow for the extraction of side-walls properties in fins (e.g., line edge roughness, side-wall angle, and side-wall roughness). On an industrial level, these wafer-compatible high-yield metrology tools are in the front line for FinFETs and memory fabrication. However, at the moment, FinFETs are already showing scaling issues which are mitigated by manufacturing increasingly taller fins of reduced pitch, thus introducing new challenges for AFMs. On the contrary, emerging concepts in logic and memory, such as spintronics, 2D materials and topological insulators (TIs) offer a unique playground for the introduction of new physics in AFM sensing [52, 53]. The use of nitrogen-vacancy (NV) centers in diamond, could offer a pathway for analyzing strong spin-orbit coupling, spin textures, and topology that would be useful for spintronics in general [54, 55]. Combined force mapping techniques and near-field optics methods are opening new areas of research on 2D materials e.g., tip-enhanced Raman and scattering-near field optics microcopy (s-SNOM) [56, 57], where chemical mapping and optical properties can be explored with nm-precision. Similarly, force spectroscopy and advanced imaging in liquids are extending the possibility for single-molecule analysis [58]. Despite great progress, AFMs remain relatively slow among scanning instruments. Recent work on high speed AFM (HS-AFM), which combines fast scanners and detectors with small cantilevers, has achieved a 10–20 frames/s sampling speed [59, 60]. Interesting new areas of development are currently offered by advanced data analytics, which enables AFM to use machine-learning methods and deep learning to automate measurements and the off-line treatment process [61, 62]. Finally, it is worth mentioning that the number of AFM modes available greatly exceeds the list reported in this book. Although Fig. 1.1 shows over 20 different techniques, this list is far from exhausting all the variations reported in the literature. Besides, there are at least as many dedicated techniques used by other communities such as bio-chemistry and geology, to name but two. Here we limit ourselves to modes of operation with a well-established track record in nanoelectronics.

1.3 Introduction to Atomic Force Microscopy

The original idea of STM to detect the tunneling current between a conductive tip and a sample while changing the position of the scanning probe as a function of the current intensity, uncloaked forever the manipulation and sensing of matter at the atomic scale [12]. However, the need for electric current flow between tip and sample hampered the application of STM in non-conductive materials, thus representing a major limitation. Very soon after the invention of STM, AFM exceeded this limit by changing the sensing technique from tunneling current to mechanical deflection as experienced by the tip in direct- or intermittent-contact with the sample surface. In AFM, a laser is shone on the rear side of the cantilever and the reflected light is used to record the force acting on the tip-sample system. In this way, the high conductivity of the sample was no longer a stringent requirement, nor the need for ultra-high vacuum. Rapidly, this method established itself in a multitude of disciplines including physics, chemistry, biology, materials science, and nanotechnology. This section describes the basic principles of operation of AFM in modern machines, addressing the fundamental mechanisms of contrast formation for various contact and non-contact modes.

1.3.1 Basic Operating Principles

Atomic force microscopes sense the force between a nanosized tip and the sample. In AFM, precise tip positioning and scanning is enabled by piezoelectric actuators. Piezoelectric materials can in principle deliver picometer scale motions in response to electrical stimuli [8]. In essence, if properly cut, quartz crystals can be used as transducers that convert electrical potential into precise mechanical motion. Therefore, an atomically sharp probe is mounted at the end of a microfabricated cantilever which is characterized by its spring constant k, measured in N/m. The same microfabrication techniques used for ICs are used for tip production. The length of the cantilevers can range from a few to hundreds of μm, with thickness generally inferior to 1 μm. The tip shape and cone angle are two variable parameters, while the apex is typically less than 10 nm in radius. Significant advances have been achieved in the area of probe development, with hundreds of possible alternatives available for the user in terms of tip size, geometry, materials, and spring constant, to name but a few [63]. In Chap. 11, we review the state-of-the-art for diamond tip fabrication, as they are widely used in electrical AFM modes. In modern microscopes, the deflection of the cantilever is detected with a laser beam that is reflected by the rear side of the tip

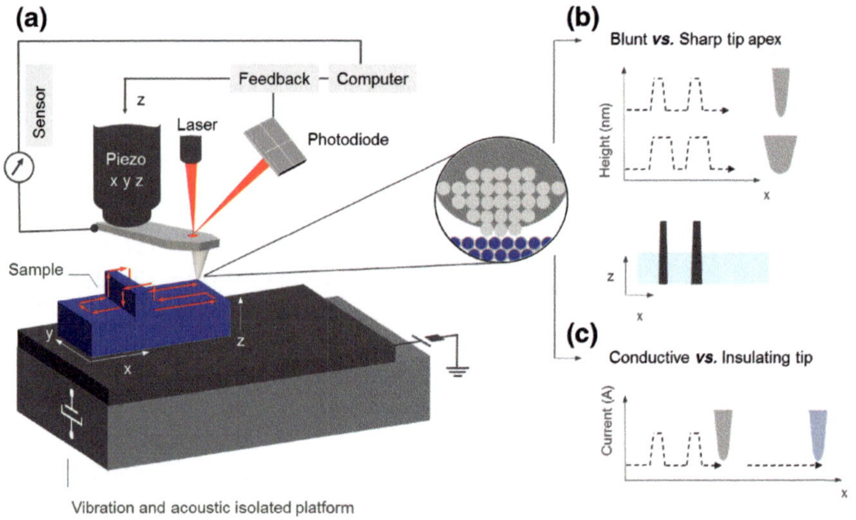

Fig. 1.2 **a**, Basic setup for electrical AFM. In this design the sample is held fixed and the probe can move in *x*, *y*, *z*. Tip-sample geometrical deconvolution and resolution limit imposed by the tip radius for **b** morphology and **c** current. We assume here that the two fins are conductive, connected to a bottom plane and embedded in an insulating material. The biased tip scans in contact and a current amplifier is used as the sensor for the secondary AFM channel

and collected on a four-quadrant photodiode (Fig. 1.2). As the cantilever raster-scans the surface, the physical tip-sample interactions can be recorded by the motion of the laser beam reflection changing the output of the photodiode. When scanning at constant force, an electronic feedback is used to hold the amplitude of the changes on the photodiode constant, so that the tip will scan the surface at constant load force. In this way, topographic information about the sample can be acquired in *x*, *y*, *z* coordinates and subsequently processed by a computer for image generation. Two main configurations exist, one design using a movable stage scanning the sample (in *x* and *y*) while the tip is fixed (i.e., tracking only *z*-changes), and a second design where the sample is held fixed and the probe can move in *x*, *y*, *z*. A clear engineering challenge is represented by the mechanical stability and noise isolation of the AFM stage. For electrical modes, the tip requires a direct connection which can be used to apply an external voltage or to connect a sensor. For a more specialized description of the hardware, the reader is referred to various excellent overviews [8, 10, 64]. In simple terms a microscope requires (Fig. 1.2):

1. A vibration and acoustic isolation platform
2. A tip connected to piezo transducers
3. A laser-based force detection system
4. A vibration isolated sample support
5. Control electronics (i.e., feedback) to sense and react to changes in the forces

6. A computer with software to control the parameters and output the results (e.g., surface map or secondary channels).

Despite the relatively simple instrumentation, several conditions must be met in order to obtain a good measurement. A comprehensive description of all possible tuning issues for each AFM mode would go beyond the scope of this chapter. However, three main points can be identified qualitatively for a successful measurement. First, understanding of the tip-sample system and selection of the right probe. Second, the selection and optimization of analysis conditions (scan speed, load force, etc.). And third, comprehension and mitigation of possible measurement-related artefacts. These may originate from a specific property of the sample, such as adhesion, hydrophilicity, etc., or by an error in the selection of the analysis conditions. Independently of the source, measurements artefacts are extremely common in AFM and a good practice requires the user to repeat the same experiment under a range of conditions, e.g., probe type, environment, and load force, to discard erroneous results.

The optimal tip selection is always relative to the sample and the specific feature to be sensed. This is due to the AFM image formation process. As an example, the probed surface morphology is determined by the geometrical tip-sample deconvolution (Fig. 1.2b). Significant asymmetry in the tip apex will introduce morphology artefacts, limiting the quantitative interpretation of the results. Therefore, the tip must be as sharp as possible, with the proper rigidity and stability to maximize probing while avoiding damages to the surface. It is generally important to know the mechanical properties of the surface, for example the pressure condition for elastic and plastic deformation. For electrical modes, the geometrical tip-sample deconvolution still applies. However, the tip must now be conductive, and the deconvolution effect must be considered between the electrical contact area of the probe and the electrical features. A simple example for the case of a biased tip sensing current is shown in Fig. 1.2c. It goes without saying that asymmetry or modifications of the tip apex will affect the electrical contact area, thus inducing among other things, variable current densities in the area of interest or partial loss of conduction when the tip is contaminated or not conductive. In general, a comprehensive understanding of the physical and electrical tip-sample contact is required for a good measurement.

Second, analysis conditions such as vacuum, controlled humidity, or an inert gas environment must be selected depending on the chemical reactivity of the sample. It is well known that hydrophilicity and surface termination can be used to modify the surface morphology. For example, in the presence of absorbed humidity on the surface, a biased tip can activate dissociative processes in the water, providing the chemical environment for tip-induced oxidation often used to generate tip-induced nanopatterning (Chap. 5) [65]. However, the same process can introduce artefacts in the measurement, by altering the morphological results and adding a highly resistive layer under the tip (e.g., in electrical measurements). For these reasons, it is common to enclose the AFM system in environmental chambers and connect them to vacuum systems to mitigate measurements artifacts while enhancing the sensing capability. Finally, the third and most difficult condition to realize is the prevention of tip wear and contamination, which can in turn engender the appearance of various measurement artefacts. If we consider a contaminated tip, the latter will affect the geometrical deconvolution, and hence also the morphological features, and depending on the contaminant's electrical properties, it could abruptly cut off the conductivity or modify the magnitude of the observed electrical features. For this type of issue, predicting and counteracting severe contamination is a major challenge, often requiring long experience and perseverance on the part of the experimenter. Through the remaining chapters of this book, the reader will become familiar with many details for establishing optimal measurement conditions in different analysis modes, from sample preparation to tip selection.

1.3.2 To Touch, or Not To Touch, That Is the Question

The way in which the tip apex interacts with the surface is so important that defines the entire experiment, including the name of the mode. Independently from the application of interest and the information to sense, there exist three main types of tip-sample contact. These determine three possible classes of operations for AFM, namely *contact*, *non-contact* and *intermittent*. Each method is characterized by a distinctive way to probe the surface morphology. In general, when the static deflection of the cantilever is measured in response to a physical contact with the surface, we refer to this as contact mode, often referred to as static (Fig. 1.3). In contrast, when dynamic oscillations of the cantilever are used, and no physical contact exists between tip and surface, we refer to this as non-contact mode (Fig. 1.3). Finally, if the morphology is reconstructed using dynamic oscillations of the cantilever, but the tip apex has some degree of physical contact with the surface, we refer to this as intermittent mode, often referred to also as dynamic (Fig. 1.3). While the contact method can be intuitively understood as the tip experiencing the normal reaction force from the sample surface, for the other two modes, we must consider the cantilever as oscillating (at or near its resonance frequency), with a small gap between tip and sample. Any possible interaction between tip and surface will affect the amplitude, phase, or frequency of the oscillating cantilever. When the gap between the tip and the sample is gradually reduced, the apex starts to interact with the force field of the sample, thus damping the cantilever oscillation. Since the cantilever is excited by an additional piezo transducer, the periodic forcing signal is known, and the oscillation monitored by the laser photodiode can be compared to the original to sense the force interaction. As for the static deflection signal of contact mode, a certain reduction in the oscillation amplitude or shift in frequency can be used here as input to the feedback element to adjust the z-motion during the x, y scanning (Fig. 1.3). Now, it can be understood that when the oscillation of the apex is damped without any physical contact with the surface, the mode is said to be *non-contact*. On the other hand, if during the oscillation the cantilever touches (taps) the surface mechanically, the mode is said to be *intermittent*, and this is often referred to as tapping mode [66]. Here, the existence of a real physical contact guarantees penetration through possible

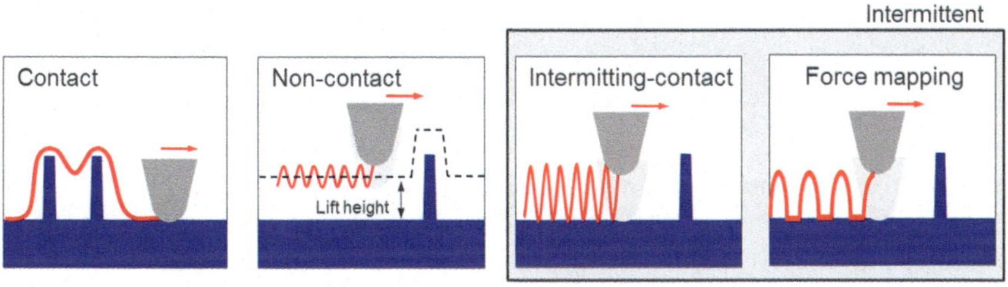

Fig. 1.3 Possible tip-sample interactions used by different AFM modes

contamination layers, but it increases the possibility of wear or contamination of the tip apex.

It is important to mention that the tip is oscillated at the natural resonance frequency (or higher harmonics), so the frequency response is primarily defined by the cantilever's mechanical properties (e.g., length, thickness, shape, stiffness, and material). In common with all resonators, an oscillating tip has a quality factor (Q-factor), which is a measure of the strength of the damping of its oscillations. Thanks to the great variety of tips, a wide range of resonance frequencies (50–2000 kHz), Q-factors and stiffnesses are available for various applications. However, once made to oscillate, the tip is raster-scanned across the sample with no control over its individual intermittent taps. The solution to this problem comes in the form of a hybrid methodology called force mapping modes (Fig. 1.3) [67, 68]. In essence, the idea is to oscillate the tip in non-resonant conditions, that is lower frequencies than the natural resonance. In general, the range of operation is 1–10 kHz, so that fast responding electronics (i.e., high-bandwidth force sensors) can track individual oscillations of the tip on each tap during the imaging process. This represents a considerable advantage, as it is equivalent to acquiring thousands of individual force curves while scanning adjacent pixels. Since the local mechanical properties of the sample can be obtained from force curves, this mode can be used to map adhesion, modulus, dissipation, and deformation with nm lateral resolution. Even more importantly, the force control on each tap offers a pathway to minimize the tip-sample interaction, i.e., limiting the contact area and reducing tip damage, while maintaining a short-controlled period in contact to probe electrical properties. It will be clear to the reader that force mapping modes have required recent technological advances for their implementation, making them relatively new compared with the original methods. However, they may be considered as belonging to the family of intermittent modes, provided that the advantage offered by the force control capability is understood.

1.3.3 Mechanisms of Contrast Formation

Independently of the type of information to be probed, virtually all AFM techniques rely on one or more of the three methods for surface sensing. It is important to understand the advantages and disadvantages of each approach if one is to select the correct technique based on one's needs. The tip-surface interaction contains simultaneous contributions from different forces, and these act in accordance with the measurement conditions and tip-sample separations (Fig. 1.4a). Dominant forces include chemical forces (i.e., atomic bonding forces modelled by a Lennard-Jones potential), van der Waals forces (i.e., when interaction occurs between dipoles), and magnetic, mechanical, and electrostatic force (i.e., assuming the tip-sample system to be a capacitor). The latter is important when an external voltage is applied to the tip-sample system, as an additional attractive force component must be considered. It is thus critical to comprehend the contrast formation mechanisms for each technique if one wants to maximize information extraction and avoid artefacts [69, 70]. Here, we briefly review general guidelines for understanding the contrast formation mechanisms of the electrical modes described in this book (Fig. 1.5).

When sensing morphological features, the AFM is a real-space imaging technique, so the collected data always represent the geometrical convolution between the tip and sample. Therefore, sharp probes with high aspect are generally preferred, and their results are considered to provide a more accurate representation of the sample surface (Fig. 1.2b). Similarly, better results can be obtained by minimizing the interaction between the apex and the surface, as for non-contact versus contact modes

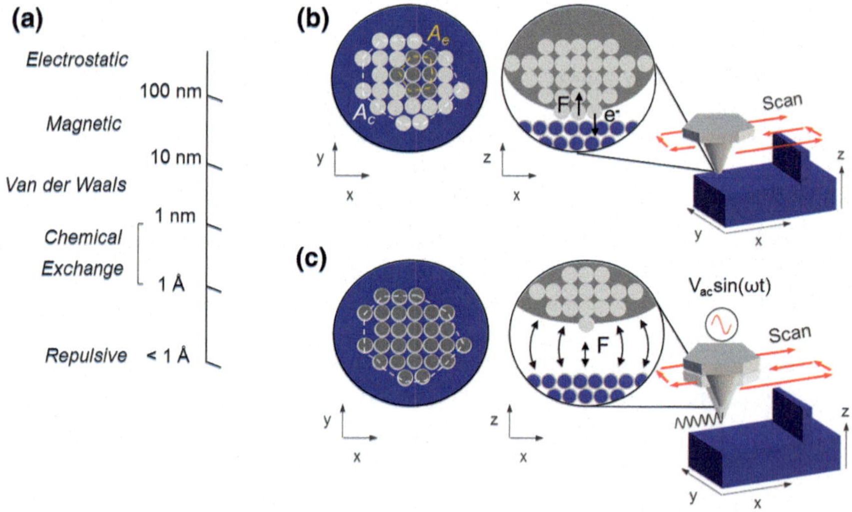

Fig. 1.4 **a** Forces acting on the tip at different heights from the surface. Scanning operation and tip-sample contact for **b** contact and **c** non-contact modes. Note the difference between the electrical and physical contact areas when the tip is in contact

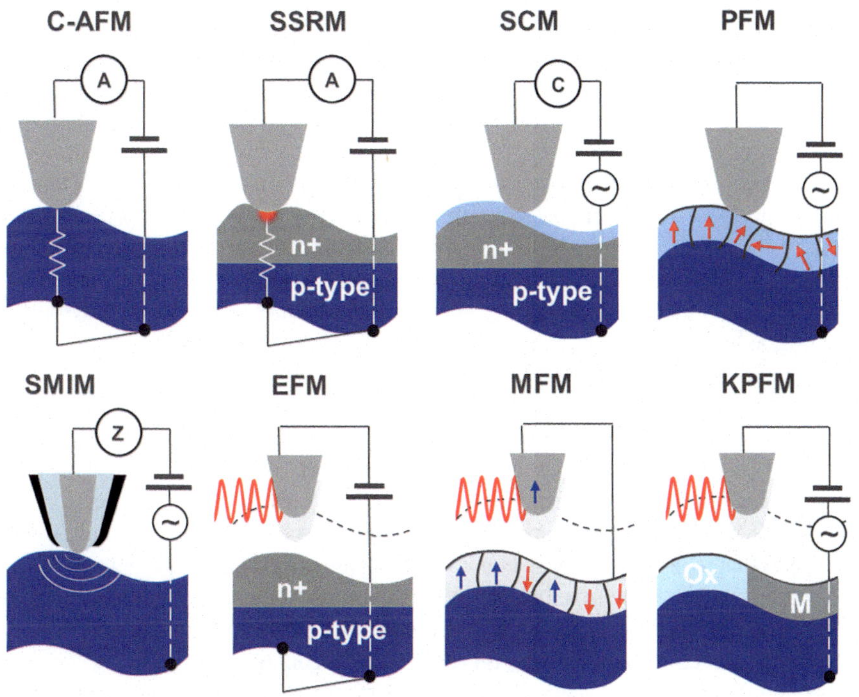

Fig. 1.5 Electrical AFM modes described in this book

(Fig. 1.3). Today, for topography measurements, sub-nm resolution has been demonstrated (<1 nm lateral and <0.1 nm vertical), with true atomic resolution achievable under proper conditions. For a detailed analysis of topography contrast formation in the different modes, the reader is referred to one of the reviews available in literature [8, 13, 64, 69].

Besides topography, the secondary channel of electrical AFMs (e.g., dopant/carrier, electrical current, resistance, capacitance, or work function, to name but a few) has a more complex interpretation. Here, instead of the physical tip-sample convolution between contacting objects, we must consider the electrical convolution between tip and surface as electrical objects, i.e., contacts. In other words, depending on the tip-surface interaction (i.e., contact, intermittent, or non-contact), the main question to answer is where the charge flow and how does it flow (contact), also what the surface potential and capacitive coupling of the system that is formed by the sample and tip apex (especially in non-contact).

The contact mode is one of the most widely used platforms for electrical methods as shown in Fig. 1.5. Conductive atomic force microscopy, scanning spreading resistance, scanning capacitance, piezoresponse force microscopy, and scanning microwave impedance microcopy are all based on contact mode AFM. In the contact regime, the repulsive force on the tip dominates, and physical contact exists between tip and sample, so this method is not ideal for high-resolution morphology imaging. Depending on the materials involved, the strong repulsive force could result in uncontrolled (or controlled) plastic deformation of the tip apex or sample surface. If the force range is safely away from the values of tip-sample surface modification, the force applied to the surface by the probe follows Hooke's law $F = -k * z$, where F is the force in (N), k the spring constant (N/m) and z the tip displacement (m). Mechanical models exist to predict the size of the contact area (Ac) at a given force. Among these, the Hertz, Derjaguin-Muller-Toporov and Johnson-Kendall-Roberts models are most widely used [71, 72]. The estimated contact areas are in good agreement with the experimental values defining the minimum feature size that can be resolved in topography.

As mentioned previously, it would be a mistake to consider the value of the contact area as defining the lateral resolution of the secondary channel. Indeed, for techniques relying on a flow of charge, such as C-AFM or SSRM, the lateral resolution of the electrical signal is generally much smaller than what is predicted by the physical contact area. This incongruence can be explained by the tendency of electrons to flow in the path of least resistance, and this does not necessarily coincide with the entire physical contact area. An electrical contact area (A_e) can thus be defined as the sub-region that poses the least resistive path for the flow of electric current. The main consequence of this is that the electrical contact area will set the minimum feature size that is electrically resolvable. As the electrical contact area can be as small as a few square nanometers, lateral resolution down to 1 nm has been demonstrated for these techniques, with atomic resolution obtained under optimal conditions for C-AFM [73–77].

Interestingly, even though SSRM and C-AFM have comparable resolution, the contrast formation mechanisms differ in the two cases. In C-AFM, a dc or ac electrical signal is applied to the sample and the current detected as the tip scans the sample. The contrast formation mechanism for the C-AFM current map is the passive sensing of local changes in the tip-sample contact resistance [70]. Particularly for thin oxides or highly resistive samples, C-AFM can be considered as STM performed without a physical tunneling gap, with the tip acting like an ultra-scale, movable electrode. Compared to C-AFM, SSRM is performed at relatively high load force (μN), on samples of larger resistance range. At high pressures, Si undergoes a phase transition from semiconducting diamond face-centered cubic (fcc) to conductive tetragonal β-Sn [76, 78]. Thus, the injection of the electrical current at the tip is mediated by a small metallic well (β-Sn pocket) which forms in the semiconductor as a result of the high pressure. The β-Sn pocket is formed in the area of maximum pressure (>10 GPa), thus in a small volume compared to the contact area. Therefore, a local metal-semiconductor junction is constantly present between the probe and the sample, defining the lateral resolution of SSRM which effectively senses the spreading current around the confined Schottky junction [79]. Another implication of flowing charge, is that the series of resistances present in the tip-sample-contact chain determines the effective voltage drop in the tip-sample junction as in a resistor voltage divider (Chap. 3).

Although based on contact mode, some techniques do not require the flow of charge to sense the electrical properties of the sample. This is the case for scanning capacitance, piezoresponse force microscopy and scanning microwave impedance microscopy. Scanning capacitance microscopy leverages the presence of a thin oxide to separate the tip from the sample, thus avoiding exchange of charge between the two. Here the bias is modulated by an ac component, while scanning to induce changes in the capacitance formed at the metal-oxide-semiconductor junction under the probe. The tip-induced field effect induces local changes in the carrier concentration which are sensed by the capacitance sensor (i.e., ultra-high frequency resonator). The technique is nondestructive and allows for quantitative 2D carrier profiling with excellent spatial resolution (nm), high dopant gradient resolution (3 nm/dec), and wide dynamic range (10^{15}–10^{20} cm^{-3}). The thin oxide that separates the tip and the sample must be sufficiently thin not to introduce strong field fringing that would deteriorate the lateral resolution of SCM. For this reason, in the case of SCM, sample preparation plays a crucial role in the final measurement (Chap. 4). Scanning microwave impedance microcopy is a natural evolution of SCM in which ac biasing is replaced by microwaves. These are applied to the sample by a specialized tip (i.e., mimicking a coaxial cable), and the reflected microwave signal is used to measure the sample permittivity and conductivity extracted by the impedance [80]. One of the main advantages of this approach is that there is no need for a complicated sample preparation and back-contact, because the tip is acting as a microwave generator and sensor. However, the tip is still a relatively complicated part of the setup and its dimensions often limit the lateral resolution of the technique. On the other hand, this technique has demonstrated the highest sensitivity of capacitance so far reported by an AFM method, and as a relatively young technique future improvements can be expected (Chap. 12).

Piezoresponse force microscopy, used for the analysis of ferroelectric domains, has probably the most peculiar mechanism of contrast formation. This method operates in contact mode with an additional ac bias applied between tip and sample. In piezoelectric materials, the converse piezoelectric effect induces voltage-induced deformation of the sample on the tip passage. Thanks to the small tip radius, the electric field is highly confined and decays rapidly, thus sensing only a small volume under the tip. The volume changes (i.e., expansion and contraction) are transmitted to the tip and sensed by a lock-in amplifier. In samples with multi domains, PFM can directly probe the domain size, polarization orientation (i.e., in-plane and out-of-plane), and electromechanical coupling coefficients (1–100 pm/V). Using accurate calibration procedures, the PFM allows for the quantitative analysis of piezoelectric coefficients [81, 82]. In PFM the tip acts as a mechanical transducer, therefore the geometrical and mechanical properties of the tip, including the contact area, are of the utmost importance. However, the converse piezoelectric effect is triggered by applying the ac bias, suggesting a major role for the electrical contact area (Chap. 6). Indeed, the contrast formation mechanism of PFM is strongly dependent on both the physical and electrical components of the tip-sample contact. The lateral resolution of the technique has been demonstrated in range 10 nm, while the use of contact resonant modes pushed this limit below 10 nm, exploiting techniques such as band-excitation PFM [83].

When in non-contact regime, an oscillating probe is brought close to the sample surface, while fine control over the tip-sample distance is enabled by tracking the amplitude, phase, or frequency of the oscillation. In these cases, for each oscillation the probe experiences first the attractive regime and then the repulsive regime. When a gap exists between tip and sample, long range attractive forces generally dominate the tip-sample interaction (Fig. 1.4a, c). Holding the frequency shift constant, the tip can follow the morphology of the surface at distances that can be lowered to minimum tip-sample distance, with great benefit for the lateral resolution. Today, thanks to the use of low temperature, ultrahigh vacuum, and quartz tuning fork cantilevers, these methods have achieved a high degree of accuracy, and in conjunction with STM atomic resolution, this has enabled single-atoms manipulation and molecular imaging at tip-sample distances of a few angstroms [84]. However, thanks to their ease of operation, non-contact and intermittent modes operated at room temperature still remain very popular for sensing electric and magnetic fields, work function differences, and chemical properties of the surface (in air, vacuum, and liquids).

When a biased tip scans above the surface, the electrostatic coupling dominates the interaction (Fig. 1.4c). The force acting on the tip varies nonlinearly with the distance from the surface, with a quadratic dependence on the applied voltage. In the absence of an electric current flow, the force is $F = \frac{1}{2}\frac{dC}{dz}V^2$, where V is the applied voltage, C the capacitance, and z the distance between the tip and sample. Considering this relationship, at a fixed dc bias, the variations in capacitance between the metallic tip apex and the sample surface will induce measurable changes in the resonant frequency (i.e., amplitude and phase) [85]. This is a common contrast mechanism for electrical non-contact modes. For example, it can be used to sense capacitance and electric fields as in electrostatic force microscopy. Besides electrostatic forces, this method can measure magnetic nanodomains by replacing the coating of the metal probe with a magnetic material. As a consequence, the frequency shift experienced by the tip on a magnetic material will be induced by the presence and distribution of its magnetic domains [22]. A careful selection of the tip and sample is required. For example, objects with a small magnetic moment can represent a challenge for detection, and for soft magnetic materials, the tip can alter the domain structure (Chap. 8). Finally, it should be noted that even though the electrostatic force is not used here as a contrast mechanism, its presence can introduce artefacts in MFM, particularly when measuring biased samples of materials with high contact potential differences. Therefore, although based on a simple physical mechanism for contrast generation, the quantitative interpretation of EFM and magnetic force microscopy remains a challenge [86].

Another good example of the non-contact regime is KPFM, often referred to as scanning Kelvin probe microscopy (SKPM), which is capable of probing the contact potential difference (i.e., the difference in work function of the tip and sample) with high spatial resolution [21]. Here, the contrast mechanism is based on the fact that materials with the same surface potential have no electrostatic force between them. On the contrary, an electrostatic force proportional to the contact potential difference is established between different materials. In KPFM, an oscillating [e.g., $V_{ac}\sin(\omega t)$] and dc bias are applied to the tip, while the feedback is used to nullify the frequency shift by adjusting the dc bias. The contact potential difference causes a shift in the mechanical vibration of the cantilever that is nullified by the application of the dc bias. By recording the value of the dc bias on the surface, the equivalent surface potential map can be obtained. KPFM has been widely used to study space charge regions and junction potentials for semiconductors and photovoltaics. Trapped charges, charge injection, and dielectric constant can now be measured by a standard non-contact AFM equipped with a conductive tip and a biased sample [87, 88]. Clearly, for doped samples or in the presence of trapped charges or adsorbates on the surface, the interpretation of KPFM becomes complicated due to the contribution of these terms [89–91]. As explained in Chap. 9, the extreme surface sensitivity of KPFM has enabled its use for dopant concentrations, photoinduced charge generation, ionic motion, and electronic band bending, to name but a few. To summarize, as these techniques are sensitive to electrostatic and magnetic forces, both amplitude- and frequency-sensitive modulation methods can be used. However, to identify the ultimate mechanism of contrast formation, it is important to understand the long-range interactions between a biased tip and the sample. In general, when measuring metals, due to the pronounced charge screening effect, the tip radius determines the force configuration, while in the case of dielectrics, the force will depend on the overall tip geometry, as described in the literature [87]. Moreover, the lateral resolution of these techniques is largely affected by the tip-sample distance (i.e., due to field fringing, as in the inset of Fig. 1.4c), tip-apex asymmetry, and surface potential.

In the case of force mapping (Fig. 1.3), with must once again consider the combination of the different contrast formation mechanisms valid for contact and non-contact modes. Force mapping methods have a short and controlled time in contact with the surface, during which most of the electrical properties can still be sensed. At the same time, on each tap, the probe also experiences a prolonged time when it is not in contact, and this can be used to probe properties based on long-range forces. It should come as no surprise that equivalents of C-AFM, SSRM, KPFM, and SCM have already been demonstrated using the force mapping method as imaging technique. A clear advantage is the minimization of the tip-sample interaction enabled by the force control on each tap. The latter also reduces potential tip damage. It should be noted, however, that a severe tradeoff exists between sampling force and electrical contact quality, especially for modes involving charge flow. In other words, an extremely weak contact combined with the absence of shear force (typical of tip dragging), can introduce an unacceptably high series resistance. This generally makes necessary to compensate by performing the measurements at higher load force to achieve the same level of electrical contact, thus reducing the advantages. However, without doubt, force mapping represents an interesting area of development for electrical AFMs.

1.3.4 Effective Voltage Drop, Phantom Force, and Biasing Schemes

An important aspect of electrical modes is the distribution of the voltage in the virtual circuit formed by the tip-sample system. For techniques that rely on the flow of charge, such as C-AFM or SSRM, a good approximation of the virtual circuit can be a series of resistors, each one describing an individual element (e.g., the cantilever, back contact, substrate, wires, and tip-sample junction) as in Fig. 1.6a. Generally, a simple voltage divider can be considered in which the tip-sample junction represents the larger resistance term (R_{TS}), so that the final applied bias will be $V_{eff} = V^{*}_{dc}(R_{TS})/(R_{TS} + R_s) + \Delta\Phi$, where V_{eff} is the effective voltage drop, V_{dc} the applied bias, R_s the remaining series resistance (including all sample and probe components), and $\Delta\Phi$ the difference in work function. A special case is represented by thin oxides with interfacial layers or a bilayer stack. Here, the effective voltage at the tip-sample junction is determined by the thickness of the oxide and the relative dielectric constant.

In simple terms, in the case of a bilayer oxide the effective voltage drop can be described by $V_{TS} = V_{dc}/\left(\left(\varepsilon^{*}_{ox1}d_{ox2}\right)/\left(\varepsilon^{*}_{ox2}d_{ox1}\right) + 1\right)$ where ε and d are, respectively, the dielectric constant and thickness of the two oxide layers. It is worth mentioning that most AFM systems generate the V_{dc} bias internally (often limited to ± 10 V). The voltage is therefore supplied to the stage while the conductive tip is grounded through the sensor for the mode of choice. Owing to the presence of a protection resistor between the stage and the voltage generator, a small dc bias offset is always present on the sample stage, even when the value selected by the experimenter is supposedly zero. Experimental values for this undesired voltage offset lie in the range 1–100 mV (depending on the tool maintenance and the degradation of the internal resistor). This may seem a small value, and in general it is negligible compared to the applied bias. However, when doing low- or zero-bias experiments (ca. V_{dc} <

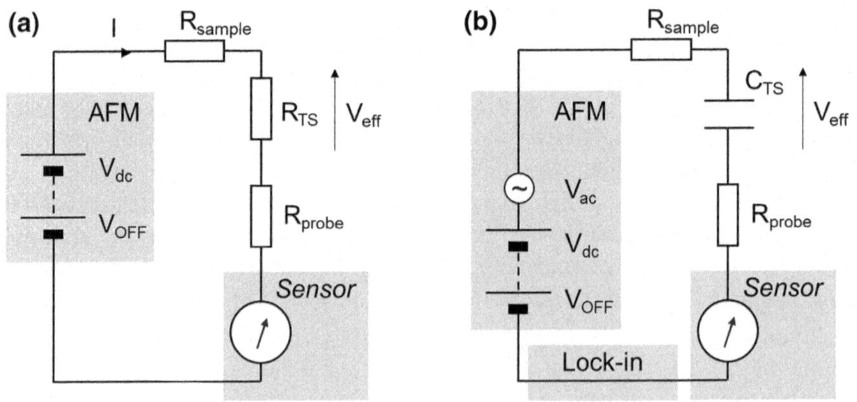

Fig. 1.6 Virtual circuit of electrical AFM modes in **a** contact and **b** non-contact methods

100 mV), the presence of the offset can dramatically affect the interpretation of the results.

Finally, in the absence of an electric current flowing between tip and sample, such as in SCM, KPFM, EFM and others, the virtual circuit can be modified as in Fig. 1.6b. In general, an ac voltage is supplied to the sample with a modulation in the kHz range, and a lock-in amplifier is used to sense variation of the signal at the modulation frequency. V_{eff} can still be used in the expression for the force, this time or the case of non-contact. This leads to the form $F = \frac{1}{2}\frac{dC}{dz}\left(V^{*}_{dc}(R_{TS})/(R_{TS} + R_s) + \Delta\Phi\right)^2$, which includes a component depending on the sample resistance. Weymouth and co-authors have shown that this effect, often referred to as phantom force, can have an impact on the interpretation of quantitative information in non-contact mode techniques [92].

1.4 Emerging Nanoelectronics Devices and Metrology Challenges

For over 50 years microelectronic devices have been steadily miniaturized to reduce the size of the final ICs, with great advantage for cost and performance. What became known as "Moore's law" describes this dramatic decrease in the size of components, by fixing a clear timeline associated with the pace of miniaturization [93]. The meticulous accomplishment of this roadmap has led to the broad ecosystem of electronics products that surround us today. Today, this translates into devices that are smaller in dimensions and more complex in shape, while making use of non-conventional materials like never before in the 50-year history of micro- and nanoelectronics. However, in the next 10–15 years the size of IC components is expected to reach the point where they can no longer get smaller due to physical atomic limitation, or because their cost and reliability will simply make their fabrication uneconomical. In addition, the last ten years have already seen a growing demand for application-specific microelectronics products such as mobiles and wearables, translating various design needs into new measurement methods.

The present era of transition is also associated with a massive effort to extend current devices beyond their limits, while exploring new solutions. The aim of this research goes well beyond mere device engineering and involves the exploration of new physics to enable novel technologies. However, to translate new physics into technologically relevant applications, we must be able to accurately measure all the parameters in the systems at the level of the material. It is the task of physical characterization and metrology techniques to sense the properties of the desired material at the relevant time-scale, with high accuracy and reproducibility. In general, analysis methods are highly specialized to sense individual material properties, thus requiring different analysis tools to enable one to measure multiple physical properties. As introduced in this chapter and described later in the book, AFM is a special case of a characterization technique for which many specialized modes provide a wide range of material properties thanks to the inherent flexibility of the method. This adaptability is attested by the continued appearance of new methodologies, developed in response to emerging materials requirements. Today we can easily count over twenty AFM techniques which are routinely used at industrial level to translate physical observations into technological applications. The number grows rapidly to over fifty methods if we consider also fundamental research and academic development.

A good example of the adaptability of AFM is provided by the evolution of the main technique used to measure local magnetic fields (Chap. 8). Magnetic force microscopy uses magnetic probes to sense the magnetic stray field originating at the surface of magnetic layers, for example, directly probing the encoded bits in the case of magnetic storage in the form of hard drives [22]. Here, the active component to store information is a ferromagnetic layer in which 0 and 1 s are encoded as small domains with opposite magnetization direction. With its high lateral resolution (ca. 10 nm), MFM has been used for years to study, control, and boost storage capacity, while bits were scaled down from hundreds of μm^2 to a few nm^2. This technique has supported the development of magnetic layers with a dramatic impact on commercial mass data storage right up the present time. However, manufacturers are currently exploring new concepts to overcome the scaling limitation of current bits and exceed the 1 Tbits/in.^{2} density. One solution currently under investigation is represented by magnetic skyrmions, which are quasiparticles existing in thin magnetic films. The recent stable observation at room-temperature has increased interest in their application to future data storage [94]. With skyrmions, the bits can be encoded in spin textures that are topologically different from other states with a theoretical size of 1 nm [95]. This represents a challenge for the existing magnetic metrology, including AFM-based techniques. While standard MFM cannot sense skyrmions directly, a sequence of images with variable externally applied magnetic field can be used to stabilize isolated skyrmions during the film re-magnetization phase around the saturation field [96]. This allows some degree of observation but does not provide information on the topology. In order to access this information as well, MFM must be reinvented, this time exploring the physics of nitrogen-vacancy centers in diamond, in order to probe the spin textures directly using a scanning magnetometry concept that allows for spin reconstruction [97]. While magnetic recording such as conventional hard drives are a successful commercial application of past and present, skyrmions and more generally spintronics are considered as a future disruptive technology for memory and logic. This brief example shows the intrinsic capability of AFM techniques to re-adapt to present needs for sensing different material properties in response to an emerging technological driver.

1.4.1 Device Scaling: An Increasingly Difficult Miniaturization Landscape

Today's main logic device for chip design is the FinFET, and projections of the international roadmap for devices and systems (IRDS) suggest that it will be extended to another two or three technology nodes (7, 5 and possibly 3 nm). It may be useful here to provide the numbers for a generic 10 nm FinFET technology node, currently based on a fin width < 10 nm, gate length < 20 nm, gate half-pitch (i.e., half the distance between two identical components) ca. 20 nm, and metal half-pitch < 30 nm. What comes next is still unclear, although concepts such as gate-all-around (GAA) using nanowires or nanosheets are considered candidates at node 3 nm, due to their improved electrostatic. In combination, memory-in-logic integration schemes, heterogeneous materials, and new vertical architectures are expected to continue to advance chip design in the near future. However, even now, IC scaling is an increasingly difficult task because, as the reduction in geometrical dimensions continues, the insurgence of parasitics, higher manufacturing costs, and ultimately the granularity of matter pose dramatic limitations.

It is worth noting that the same physical limitations that exist for operation of scaled devices, will also limit the techniques for nanoscale metrology, imposing clear boundaries on existing analysis methods. Measurements are required to sense near atomic scale dimensions, probing multiple physical properties that often deviate from their classical description as the quantum confinement limit (<10 nm) is approached. As devices have moved to 3D architectures, 3D metrology has become one of the most important topics addressed by the community. This extends to improvements of existing techniques to achieve 3D sensing capability, as well as the introduction of new methods. Ideally, metrology must also offer high speed and yield, low noise and reduced sources of error. Therefore, alongside chip manufacture, IC metrology is facing important challenges. The book discusses various challenges and opportunities for AFM techniques to provide information at the required length scales in various emerging devices, ranging from modern logic and memories to the discussion of new materials.

For each AFM method we will answer the following questions

- Regarding the principle of operation, how does it work and what is it used for?
- Where are the existing limitations?
- What is the physics behind such limitations (resolution, sensitivity, repeatability, and scalability)?
- Regarding possible improvements, can we extend capability?
- To what devices can it be applied and how far can it take us in the roadmap?

1.4.2 New Devices Based on New Physics

While the path toward the end of the roadmap seems well defined for charge-based devices, a few disruptive technologies are arising with different operating principles and great promise for performance. Devices based on 2D materials, single-atoms switches, and multiferroic-based devices are an emerging area of research where AFM finds wide application. In many cases, dedicated needs for physical analysis drive the development of new methods. A good example is provided by 2D materials.

Graphene, transition metal dichalcogenides (TMDs) as well as synthetic layers like silicene and germanene are providing the building blocks for a new set of atomically-thin devices in all areas (Chap. 10) [42, 98]. For these materials many fundamental processes occur differently compared to their bulk counterpart due to the different surface/bulk ratio. Here carrier profiling, strain, electrical, and chemical properties are substantially different from anything else, as their bulk counterpart is not relevant when considering their applications. Various extraordinary properties of these materials will involve unlocking surface and interface engineering. Similarly, as most of the atoms are on the surface, the functionalization of these layers can be achieved by atom substitution. Layer-dependent properties of TMDs such as band gap tunability, optoelectronics, or spin and valley coupling can be directly probed using AFMs. The latter also allows direct surface engineering to modify the electronic properties of heterostructures. Alternatively, using the AFM tip as an optical antenna, 2D materials offer an important platform for light-matter interaction due to their long propagation of plasmon polaritons [99].

In the direction of novel computing, devices to process quantum information and quantum materials are attracting the attention of the community [100, 101]. Si and Ge have demonstrated long electron spin coherence making them appealing for quantum devices [102]. Among the possible devices currently under investigation, arrays of dopant atoms in a silicon host are often considered as CMOS-compatible building blocks for quantum computers. As for 2D materials, the research effort in atomic fabrication for quantum computing opens a new playground to study the quantum effects of highly confined electrons. Interestingly, these devices require the use of atomic-precise STM lithography for their fabrication. The precise doping is accomplished by a selective de-passivation of the hydrogen terminated Si surface using the STM tip, followed by exposure to phosphine molecules. Here, SCM has proven to be a valuable technique for the non-destructive characterization of buried phosphorus nanostructures [103]. Alternatively, sMIM using impedance-matched microwave signals (at 3 GHz) can be applied to these structures to sense the tip-sample impedance (Chap. 12).

1.5 Present Status and Future Applications

Despite the successful half a century of development and scaling, the pace of innovation of IC devices is clearly slowing. Therefore, this is the perfect time to explore new physics and alternative computation paradigms. However, enabling large scale production of new devices means first the development of accurate and repeatable measurement solutions. Already today, as devices shrink in size and become more three-dimensional in geometry, flexibility of analysis techniques can make all the difference. Future trends in product diversification including Internet-of-things and artificial-intelligence will dictate new challenges and requirements. Hence, improved instruments and novel techniques are required. AFM remains a solid methodology to probe emerging devices in the single digit nm-range. The role of electrical AFMs cannot be overlooked, as these techniques are continuing to support and evolve in response to the requirements of future nanoelectronics, as they have done already in the past decades.

Looking forward, many emerging technologies will support the field in its continued evolution. Chemical analysis and compositional mapping are now accessible with nanoscale resolution, and many interesting options are currently exploring spectroscopic capability below diffraction limit [48, 104–106]. Thanks to recent developments in near-field microscopy and microwave technology, sophisticated techniques are flourishing for nanoscale sensing of the temporal or spectral evolution of propagating light fields, thermal properties, and impedance [30, 65, 107–109].

High-speed AFM (i.e., tens to hundreds of images per second) is rapidly evolving thanks to new phase detectors and aggressive tip engineering, thus extending the probing capability of dynamic processes and molecular imaging for physical chemistry and biology [40, 59, 62, 110, 111]. In this context, non-raster scanning AFM imaging is also showing great potential to achieve high speed and throughput across large areas [112].

Multi-frequency AFM, band-excitation and force mapping methods are extending our capability to extract previously inaccessible materials properties with increasingly high accuracy [83, 113]. As a consequence, large, complex and high-dimensional datasets are now available, stimulating the community to design new solutions for storage, interpretation and presentation [61, 62]. Three-dimensional probing concepts are emerging (Chap. 3). The combination of tip-induced material removal and sensing has enabled the analysis of confined volumes (<100 nm^3) with accurate removal control (<5 Å). A dedicated design consisting in a multi-head system for 3D tomography holds the potential for sensing structural, chemical, magnetic, and electrical properties of confined volumes with unprecedented resolution [70, 114].

Increasing attention to probe fabrication has led to a continued improvement in performance (Chap. 11). Examples are high aspect ratio full diamond tips, crucial for high lateral resolution in electrical contact modes. Thermally conductive probes for SThM act as nanoscale thermocouple or microbalance tips (i.e., mechanical transducers) to perform biomolecular recognition. Finally, the compact size and reduced costs of AFM equipment are driving the rapid integration of AFM with other analysis techniques. The combined use of AFM in situ with beam analysis techniques benefits the quality of information acquired (e.g., for imaging and spectroscopy), enabling increased functionality. On a longer timescale, artificial intelligence will assist automatic measurements and parameter optimization, while advanced data analytics will help to extract and process only those data needed for critical decisions. Physical modelling and atomistic simulation in combination with process information will help to extract further physical quantities from 3D images with high temporal resolution. These are only a few of the exciting advances currently being explored by the community in industry and academia.

References

1. G. Binnig, C.F. Quate, C. Gerber, Atomic force microscope. Phys. Rev. Lett. **56**, 930–933 (1986)
2. Y. Martin, C.C. Williams, H.K. Wickramasinghe, Atomic force microscope–force mapping and profiling on a sub 100-Å scale. J. Appl. Phys. **61**, 4723–4729 (1987)
3. G.M. McClelland, R. Erlandsson, S. Chiang, Review of progress in quantitative non-destructive evaluation. Rev. Prog. Quant. Non-Destructive Eval. **6 B**, 1307–1314 (1987)
4. G. Meyer, N.M. Amer, Novel optical approach to atomic force microscopy. Appl. Phys. Lett. **53**, 1045–1047 (1988)
5. F.J. Giessibl, Advances in atomic force microscopy. Rev. Mod. Phys. **75**, 949–983 (2003)
6. G. Binnig, R. Heinrich, W. Pauli, The scanning tunneling microscope. Sci. Am. **253**, 50–56 (1985)
7. B. Voigtlaender, *Scanning Probe Microscopy* (Springer, Berlin, 2015)
8. P. Eaton, P. West, *Atomic Force Microscopy* (Oxford University Press, Oxford, 2010)
9. S.K. Kulkarni, *Nanotechnology : Principles and Practices* (Springer, Berlin, 2014)
10. G. Haugstad, *Atomic Force Microscopy: Understanding Basic Modes and Advanced Applications* (Wiley, New York, 2012)
11. A. Foster, W.A. Hofer, *Scanning Probe Microscopy* (2006). https://doi.org/10.1007/0-387-37231-8
12. G. Binnig, H. Rohrer, In touch with atoms. Rev. Mod. Phys. **71**, S324–S330 (1999)
13. O. Custance, R. Perez, S. Morita, Atomic force microscopy as a tool for atom manipulation. Nat. Nanotechnol. **4**, 803–810 (2009)
14. J.Y. Park, S. Maier, B. Hendriksen, M. Salmeron, Sensing current and forces with SPM. Mater. Today **13**, 38–45 (2010)
15. S. Salahuddin, K. Ni, S. Datta, The era of hyper-scaling in electronics. Nat. Electron. **1**, 442–450 (2018)
16. N.G. Orji et al., Metrology for the next generation of semiconductor devices. Nat. Electron. **1**, 532–547 (2018)
17. For Fig. 1 the author acknowledges various sources including reprints from [U. Dürig, D.W. Pohl, F. Rohner, J. Appl. Phys. **59**, 10] and, [R.C. Reddick, R.J. Warmack, D.W. Chilcott et al., Rev. Sci. Instrum. **61**, 12] and, [Meyer, G. & Amer, N. M, Appl. Phys. Lett. **53**, 1045] and [P. Maletinsky, S. Hong, M.S. Grinolds, B. Hausmann, M.D. Lukin et al., Nat. Nanotechnol. **7**, 5]. In addition, web galleries including, R. Feenstra, Physics Dept. at Carnegie Mellon, Renishaw TERS examples (https://www.renishaw.com/en/carbon-2d-materials-and-nanotechnology–8269), Chipworks (h), and cntech (https://www.cntech.co.uk/contacts/)
18. Y. Martin, D.W. Abraham, H.K. Wickramasinghe, High-resolution capacitance measurement and potentiometry by force microscopy. Appl. Phys. Lett. **52**, 1103–1105 (1988)
19. H.J. Mamin, D. Rugar, J.E. Stern, B.D. Terris, S.E. Lambert, Force microscopy of magnetization patterns in longitudinal recording media. Appl. Phys. Lett. **53**, 1563–1565 (1988)
20. S. Xu, M.F. Arnsdorf, Electrostatic force microscope for probing surface charges in aqueous solutions. Proc. Natl. Acad. Sci. **92**, 10384–10388 (1995)
21. M. Nonnenmacher, M. O'Boyle, H.K. Wickramasinghe, Surface investigations with a Kelvin probe force microscope. Ultramicroscopy **42–44**, 268–273 (1992)
22. Hartmann, U. Magnetic force microscopy. Annu. Rev. Mater. Sci. **29**, 53–87 (1999)
23. W. Melitz, J. Shen, A.C. Kummel, S. Lee, Kelvin probe force microscopy and its application. Surf. Sci. Rep. **66**, 1–27 (2011)
24. X. Li et al., Quantitatively probing the magnetic behavior of individual nanoparticles by an AC field-modulated magnetic force microscopy. Sci. Rep. **6**, 22467 (2016)
25. W. Vandervorst, M. Meuris, Method for resistance measurements on a semiconductor element with controlled probe pressure, US5369372A (1994)
26. J.R. Matey, J. Blanc, Scanning capacitance microscopy. J. Appl. Phys. **57**, 1437–1444 (1985)
27. A. Schulze, R. Cao, P. Eyben, T. Hantschel, W. Vandervorst, Outwitting the series resistance in scanning spreading resistance microscopy. Ultramicroscopy **161**, 59–65 (2016)

28. P. Eyben, P. Bisiaux, A. Schulze, A. Nazir, W. Vandervorst, Fast Fourier transform scanning spreading resistance microscopy: a novel technique to overcome the limitations of classical conductive AFM techniques. Nanotechnology **26**, 355702 (2015)
29. R. Shao, S.V. Kalinin, D.A. Bonnell, Local impedance imaging and spectroscopy of polycrystalline ZnO using contact atomic force microscopy. Appl. Phys. Lett. **82**, 1869–1871 (2003)
30. O. Amster et al., Practical quantitative scanning microwave impedance microscopy of semiconductor devices, in *2017 IEEE 24th International Symposium on the Physical and Failure Analysis of Integrated Circuits (IPFA)* 1–4 (IEEE, New York, 2017). https://doi.org/10.1109/ipfa.2017.8060202
31. *Conductive Atomic Force Microscopy: Applications in Nanomaterials* (Wiley-VCH Verlag GmbH & Co. KGaA, 2017). https://doi.org/10.1002/9783527699773
32. M.P. Murrell et al., Spatially resolved electrical measurements of SiO2 gate oxides using atomic force microscopy. Appl. Phys. Lett. **62**, 786 (1993)
33. P. Güthner, K. Dransfeld, Local poling of ferroelectric polymers by scanning force microscopy. Appl. Phys. Lett. **61**, 1137–1139 (1992)
34. C. Barth, C.R. Henry, Related content: Scanning force microscopy studies of domain structure in $BaTiO_3$ single crystals (microscopy study) (1997)
35. S. Jesse, S.V. Kalinin, Band excitation in scanning probe microscopy: sines of change. J. Phys. D Appl. Phys. **44**, 464006 (2011)
36. S. Gomès, A. Assy, P.-O. Chapuis, Scanning thermal microscopy: a review. Phys. Status Solidi **212**, 477–494 (2015)
37. Y. Yang, R. Huang, Probing memristive switching in nanoionic devices. Nat. Electron. **1**, 274–287 (2018)
38. U. Celano et al., Imaging the three-dimensional conductive channel in filamentary-based oxide resistive switching memory. Nano Lett. **15**, 7970–7975 (2015)
39. R.B. Dinwiddie, R.J. Pylkki, P.E. West, Thermal conductivity contrast imaging with a scanning thermal microscope. Therm. Conduct. **22**, 668–677 (1994)
40. L. Gross, Scanning probe microscopy. Nat. Chem. **3**, 273–279 (2011)
41. S.A. Mari et al., Gating of the MlotiK1 potassium channel involves large rearrangements of the cyclic nucleotide-binding domains. Proc. Natl. Acad. Sci. **108**, 20802–20807 (2011)
42. S. Manzeli, D. Ovchinnikov, D. Pasquier, O.V. Yazyev, A. Kis, 2D transition metal dichalcogenides. Nat. Rev. Mater. **2**, 17033 (2017)
43. A.L. Weisenhorn, P. Maivald, H.-J. Butt, P.K. Hansma, Measuring adhesion, attraction, and repulsion between surfaces in liquids with an atomic-force microscope. Phys. Rev. B **45**, 11226–11232 (1992)
44. E. Betzig, A. Lewis, A. Harootunian, M. Isaacson, E. Kratschmer, Near field scanning optical microscopy (NSOM). Biophys. J. **49**, 269–279 (1986)
45. U. Dürig, D.W. Pohl, F. Rohner, Near-field optical-scanning microscopy. J. Appl. Phys. **59**, 3318–3327 (1986)
46. N.F. van Hulst, F.B. Segerink, F. Achten, B. Bölger, Evanescent-field optical microscopy: effects of polarization, tip shape and radiative waves. Ultramicroscopy **42–44**, 416–421 (1992)
47. R.C. Reddick, R.J. Warmack, D.W. Chilcott, S.L. Sharp, T.L. Ferrell, Photon scanning tunneling microscopy. Rev. Sci. Instrum. **61**, 3669–3677 (1990)
48. L. Novotny, N. Van Hulst, Antennas for light. Nat. Photonics **5**, 83–90 (2011)
49. S. Zhu, G.Q. Lo, D.L. Kwong, Performance of ultracompact copper-capped silicon hybrid plasmonic waveguide-ring resonators at telecom wavelengths. Opt. Express **20**, 15232 (2012)
50. L. Liu et al., Three-dimensional atomic force microscopy for sidewall imaging using torsional resonance mode. Scanning **2018**, 1–8 (2018)
51. T.-G. Kim et al., In-line critical dimension and sidewall roughness metrology study for compound nanostructure process control by in-line 3D atomic force microscope. ECS Trans. **75**, 761–767 (2016)
52. Y. Liu et al., Tuning dirac states by strain in the topological insulator Bi_2Se_3. Nat. Phys. **10**, 294–299 (2014)

53. P. Roushan et al., Topological surface states protected from backscattering by chiral spin texture. Nature **460**, 1106–1109 (2009)
54. Y. Dovzhenko et al., Imaging the spin texture of a skyrmion under ambient conditions using an atomic-sized sensor (2016). Available at: http://arxiv.org/abs/1611.00673
55. P. Maletinsky et al., A robust scanning diamond sensor for nanoscale imaging with single nitrogen-vacancy centres. Nat. Nanotechnol. **7**, 320–324 (2012)
56. F. Zenhausern, Y. Martin, H.K. Wickramasinghe, Scanning interferometric apertureless microscopy: optical imaging at 10 Angstrom resolution. Science (80) **269**, 1083–1085 (1995)
57. F. Zenhausern, M.P. O'Boyle, H.K. Wickramasinghe, Apertureless near-field optical microscope. Appl. Phys. Lett. **65**, 1623–1625 (1994)
58. F. Long, B. Cao, A. Khanal, S. Fang, R. Shahbazian-Yassar, Modification of a single-molecule AFM probe with highly defined surface functionality. Beilstein J. Nanotechnol. **5**, 2122–2128 (2014)
59. T. Ando, High-speed atomic force microscopy. J. Electron. Microsc. (Tokyo) **62**, 81–93 (2013)
60. G.R. Heath, S. Scheuring, High-speed AFM height spectroscopy reveals μs-dynamics of unlabeled biomolecules. Nat. Commun. **9**, 4983 (2018)
61. S.V. Kalinin et al., Big, deep, and smart data in scanning probe microscopy. ACS Nano (2016). https://doi.org/10.1021/acsnano.6b04212
62. X. Li et al., High-veracity functional imaging in scanning probe microscopy via Graph-Bootstrapping. Nat. Commun. **9**, 2428 (2018)
63. K.-H. Chung, Wear characteristics of atomic force microscopy tips: A reivew. Int. J. Precis. Eng. Manuf. **15**, 2219–2230 (2014)
64. D. Sarid, R. Coratger, F. Ajustron, J. Beauvillain, Scanning force microscopy—with applications to electric, magnetic and atomic forces. Microsc. Microanal. Microstruct. **2**, 649 (1991)
65. R. Garcia, A.W. Knoll, E. Riedo, Advanced scanning probe lithography. Nat. Nanotechnol. **9**, 577–587 (2014)
66. D. Sarid, Tapping-mode scanning force microscopy: metallic tips and samples. Comput. Mater. Sci. **5** (1996)
67. M. Stark, R.W. Stark, W.M. Heckl, R. Guckenberger, Inverting dynamic force microscopy: from signals to time-resolved interaction forces. Proc. Natl. Acad. Sci. **99**, 8473–8478 (2002)
68. T.J. Young et al., The use of the PeakForce™ quantitative nanomechanical mapping AFM-based method for high-resolution Young's modulus measurement of polymers. Meas. Sci. Technol. **22**, 125703 (2011)
69. E. Meyer, H.J. Hug, R. Bennewitz, *Scanning Probe Microscopy: The Lab on a Tip* (Springer, Berlin, 2003)
70. U. Celano, *Metrology and Physical Mechanisms in New Generation Ionic Devices* (Springer International Publishing, 2016). https://doi.org/10.1007/978-3-319-39531-9
71. J. Drelich, G.W. Tormoen, E.R. Beach, Determination of solid surface tension from particle-substrate pull-off forces measured with the atomic force microscope. J. Colloid Interface Sci. **280**, 484–497 (2004)
72. D.S. Grierson, E.E. Flater, R.W. Carpick, Accounting for the JKR–DMT transition in adhesion and friction measurements with atomic force microscopy. J. Adhes. Sci. Technol. **19**, 291–311 (2005)
73. U. Celano et al., Evaluation of the electrical contact area in contact-mode scanning probe microscopy. J. Appl. Phys. **117**, 214305 (2015)
74. C. Rodenbücher et al., Local surface conductivity of transition metal oxides mapped with true atomic resolution. Nanoscale (2018). https://doi.org/10.1039/c8nr02562b
75. T. Hantschel et al., Conductive diamond tips with sub-nanometer electrical resolution for characterization of nanoelectronics device structures. Phys. Status Solidi **206**, 2077–2081 (2009)
76. P. Eyben et al., Subnanometer characterization of nanoelectronic devices, in *Fundamentals of Picoscience,* ed. by D.S. Klaus (CRC Press, Boca Raton 2013), pp. 677–704. https://doi.org/10.1201/b15523-46

77. M. Enachescu, D. Schleef, D. Ogletree, M. Salmeron, Integration of point-contact microscopy and atomic-force microscopy: application to characterization of graphite/Pt(111). Phys. Rev. B **60**, 16913–16919 (1999)
78. P. Eyben, T. Janssens, W. Vandervorst, Scanning spreading resistance microscopy (SSRM) 2d carrier profiling for ultra-shallow junction characterization in deep-submicron technologies. Mater. Sci. Eng., B **124–125**, 45–53 (2005)
79. K. Mylvaganam, L.C. Zhang, P. Eyben, J. Mody, W. Vandervorst, Evolution of metastable phases in silicon during nanoindentation: mechanism analysis and experimental verification. Nanotechnology **20**, 305705 (2009)
80. S. Friedman, O. Amster, Y. Yang, Recent advances in scanning Microwave Impedance Microscopy (sMIM) for nano-scale measurements and industrial applications, ed. by M.T. Postek (2014), 917308. https://doi.org/10.1117/12.2063138
81. A. Gruverman, S.V. Kalinin, Piezoresponse force microscopy and recent advances in nanoscale studies of ferroelectrics. Front. Ferroelectr. A Spec. Issue J. Mater. Sci. **1**, 107–116 (2007)
82. A. Kholkin, S. Kalinin, A. Roelofs, A. Gruverman, Review of ferroelectric domain imaging by piezoresponse force microscopy. Scanning Probe Microsc. Electr. Electromech. Phenom. Nanoscale **1**, 173–214 (2007)
83. S. Jesse, S.V. Kalinin, R. Proksch, A.P. Baddorf, B.J. Rodriguez, The band excitation method in scanning probe microscopy for rapid mapping of energy dissipation on the nanoscale. Nanotechnology **18** (2007)
84. S. Morita, F.J. Giessibl, E. Meyer, R. Wiesendanger, *Noncontact Atomic Force Microscopy* (2015)
85. P.E. Hillner, S. Manne, A.J. Gratz, P.K. Hansma, AFM images of dissolution and growth on a calcite crystal. Ultramicroscopy **42–44**, 1387–1393 (1992)
86. L. Angeloni, D. Passeri, M. Reggente, D. Mantovani, M. Rossi, Removal of electrostatic artifacts in magnetic force microscopy by controlled magnetization of the tip: Application to superparamagnetic nanoparticles. Sci. Rep. **6**, 1–14 (2016)
87. S. Gómez-Moñivas, L.S. Froufe-Pérez, A.J. Caamaño, J.J. Sáenz, Electrostatic forces between sharp tips and metallic and dielectric samples. Appl. Phys. Lett. **79**, 4048 (2001)
88. L. Fumagalli, D. Esteban-Ferrer, A. Cuervo, J.L. Carrascosa, G. Gomila, Label-free identification of single dielectric nanoparticles and viruses with ultraweak polarization forces. Nat. Mater. **11**, 808–816 (2012)
89. A.K. Henning et al., Two-dimensional surface dopant profiling in silicon using scanning Kelvin probe microscopy. J. Appl. Phys. **77**, 1888–1896 (1995)
90. C. Maragliano, D. Heskes, M. Stefancich, M. Chiesa, T. Souier, Dynamic electrostatic force microscopy technique for the study of electrical properties with improved spatial resolution. Nanotechnology **24**, 225703 (2013)
91. D.A. Bonnell, J. Garra, Scanning probe microscopy of oxide surfaces: atomic structure and properties. Reports Prog. Phys. **71**, 044501 (2008)
92. A.J. Weymouth, T. Wutscher, J. Welker, T. Hofmann, F.J. Giessibl, Phantom force induced by tunneling current: a characterization on Si(111). Phys. Rev. Lett. **106**, 226801 (2011)
93. G.E. Moore, Cramming more components onto integrated circuits. Reprinted from Electronics **38**(8), 114 ff. (1965, Apr 19). IEEE Solid-State Circuits Soc. Newsl. **11**, 33–35 (2006)
94. S. Woo et al., Observation of room-temperature magnetic skyrmions and their current-driven dynamics in ultrathin metallic ferromagnets. Nat. Mater. **15**, 501–506 (2016)
95. N. Nagaosa, Y. Tokura, Topological properties and dynamics of magnetic skyrmions. Nat. Nanotechnol. **8**, 899–911 (2013)
96. A. Hrabec et al., Current-induced skyrmion generation and dynamics in symmetric bilayers. Nat. Commun. **8**, 1–6 (2017)
97. P. Maletinsky et al., A robust, scanning quantum system for nanoscale sensing and imaging. 1–11 (2011). https://doi.org/10.1038/nnano.2012.50
98. C. Musumeci, Advanced scanning probe microscopy of graphene and other 2D materials. Crystals **7**, 216 (2017)

99. G.X. Ni et al., Fundamental limits to graphene plasmonics. Nature **557**, 530–533 (2018)
100. G. Scappucci et al., A complete fabrication route for atomic-scale, donor-based devices in single-crystal germanium. Nano Lett. **11**, 2272–2279 (2011)
101. L. Oberbeck et al., Imaging of buried phosphorus nanostructures in silicon using scanning tunneling microscopy. Appl. Phys. Lett. **104**, 1–6 (2014)
102. R. Vrijen et al., Electron-spin-resonance transistors for quantum computing in silicon-germanium heterostructures. Phys. Rev. A **62**, 012306 (2000)
103. E. Bussmann et al., Scanning capacitance microscopy registration of buried atomic-precision donor devices. Nanotechnology **26**, 085701 (2015)
104. A. Dazzi, R. Prazeres, F. Glotin, J.M. Ortega, Local infrared microspectroscopy with sub-wavelength spatial resolution with an atomic force microscope tip used as a photothermal sensor. Opt. Lett. **30**, 2388 (2005)
105. C. Chen, N. Hayazawa, S. Kawata, A 1.7 nm resolution chemical analysis of carbon nanotubes by tip-enhanced Raman imaging in the ambient. Nat. Commun. **5**, 1–5 (2014)
106. A.J. Huber, F. Keilmann, J. Wittborn, J. Aizpurua, R. Hillenbrand, Terahertz near-field nanoscopy of nanodevices. Nano Lett. **8**, 3766–3770 (2008)
107. N. Rotenberg, L. Kuipers, Mapping nanoscale light fields. Nat. Photonics **8**, 919–926 (2014)
108. A. Soudi, R.D. Dawson, Y. Gu, Quantitative heat dissipation characteristics in current-carrying GaN nanowires probed by combining scanning thermal microscopy and spatially resolved Raman spectroscopy. ACS Nano **5**, 255–262 (2011)
109. E. Yalon et al., Spatially resolved thermometry of resistive memory devices. Sci. Rep. **7**, 15360 (2017)
110. T. Ando, High-speed atomic force microscopy and its future prospects. Biophys. Rev. **10**, 285–292 (2018)
111. F. Mohn, L. Gross, N. Moll, G. Meyer, Imaging the charge distribution within a single molecule. Nat. Nanotechnol. **7**, 227–231 (2012)
112. A. Ulčinas, Vaitekonis, Rotational scanning atomic force microscopy. Nanotechnology **28** (2017)
113. R. Garcia, E.T. Herruzo, The emergence of multifrequency force microscopy. Nat. Nanotechnol. **7**, 217–226 (2012)
114. U. Celano et al., Mesoscopic physical removal of material using sliding nano-diamond contacts. Sci. Rep. **8**, 2994 (2018)

Chapter 2
Conductive AFM for Nanoscale Analysis of High-k Dielectric Metal Oxides

Christian Rodenbücher, Marcin Wojtyniak and Kristof Szot

Abstract Conductive atomic force microscopy has become a valuable tool for investigation of electronic transport properties with utmost lateral resolution. In this chapter, we prevent an overview about C-AFM applications to high-k semiconductors, which are key materials for future energy-efficient information technology.

The ongoing miniaturization in nanotechnology has presented new challenges that must be overcome by advanced material technologies in order to build up energy-efficient logic and memory devices. One of the most relevant issues is the suppression of electron tunnelling in semiconductor transistors, which is present if gate thickness is reduced to the nanometre scale. Here, the introduction of high-k metal oxides has been proven to be beneficial allowing for an increase in the gate thickness without lowering the capacitance. However, metal oxides often exhibit nanoscale conduction paths related to extended defects, such as grain boundaries or dislocations that cause enhanced leakage through the dielectric. In order to detect those leakage paths and understand the fundamental electronic transport properties of metal oxides, C-AFM, with its conducting tip acting as a mobile nanoelectrode, is the technique of choice for providing a direct insight into nanoscale conductivity with maximum lateral resolution. In this chapter, we present a comprehensive introduction of the C-AFM technique and the basics of high-k dielectrics before illustrating the capabilities of C-AFM by examples of local conductivity investigations of metal oxide surfaces.

C. Rodenbücher (✉) · K. Szot
Forschungszentrum Jülich GmbH, Peter Grünberg Institute (PGI-7)
and JARA-FIT, 52425 Jülich, Germany
e-mail: c.rodenbuecher@fz-juelich.de

K. Szot
e-mail: k.szot@fz-juelich.de

C. Rodenbücher
Forschungszentrum Jülich GmbH, Institute of Energy and Climate Research (IEK-3), 52425 Jülich, Germany

M. Wojtyniak · K. Szot
University of Silesia, A. Chełkowski Institute of Physics, 40-007 Katowice, Poland
e-mail: marcin.wojtyniak@us.edu.pl

U. Celano (ed.), *Electrical Atomic Force Microscopy for Nanoelectronics*,
NanoScience and Technology, https://doi.org/10.1007/978-3-030-15612-1_2

In particular, we focus on the role of extended defects in current channelling and present possibilities on how to manipulate the conductivity of metal oxides, such as by creating 2D electron gases or tip-induced nano-patterning.

2.1 The C-AFM Technique

2.1.1 AFM in Contact Mode with Conducting Tips

Nanoscale technology has grown rapidly in recent years, demanding the electrical characterization of surfaces on the nanometre scale, which in turn leads to the development of a number of current or voltage sensing scanning probe microscopy (SPM) based techniques. In this chapter, we focus on conductive atomic force microscopy (C-AFM), which is also called local-conductivity AFM (LC-AFM), conductive probe AFM (CP-AFM), conductive scanning probe microscopy (C-SPM) or conductive scanning force microscopy (C-SFM). It was introduced in the mid-1990s, when it was used for local electrical characterization of thin silicon gate dielectrics on the nanometre scale [1]. Since the C-AFM enables the simultaneous detection of topography and the current, it can be applied to insulating systems (unlike the STM). Nowadays, it is used in many applications [2], such as modern semiconductor device design [3], molecular electronics [4] or even single-molecule biosensors [5]. Furthermore, a number of other systems were successfully investigated by the C-AFM, such as nanoparticles [6], nanowires [7], two-dimensional materials [8], local oxidation [9], topological insulators [10] and photoelectricity [11]. Aside from those examples, extensive research activities emerged in the field of redox processes on high-k transition metal oxides [12] connected with resistive switching phenomena [13, 14]. In these investigations, C-AFM has proven to be especially suited for the investigation of localized filamentary leakage paths and the influence of extended defects on the redox properties, as will be illustrated in detail in this chapter after giving an overview of the measurement principle.

The conventional setup for the C-AFM is shown in Fig. 2.1 and consists of an AFM operated in contact mode equipped with a conducting tip and current detection system. This setup allows for the simultaneous detection of the current flowing through the sample with nanoscale spatial resolution and surface topography. The working principle is identical to the standard contact-mode AFM, where the force between the sample and tip is detected and used for the feedback-loop on the piezo scanner in the *Z* direction (constant force mode). Typically, this force is calculated from the tip deflection that can be detected by the optical lever method. In this method, the light from a light source (laser) illuminates the AFM cantilever and is reflected into a 4-sector photodiode. In addition, a voltage is applied between the sample and tip. Thus, the conducting tip serves as a mobile top electrode connected to a sensitive current-to-voltage amplifier, allowing for the detection of currents in the pico- or even femto-ampere range [15].

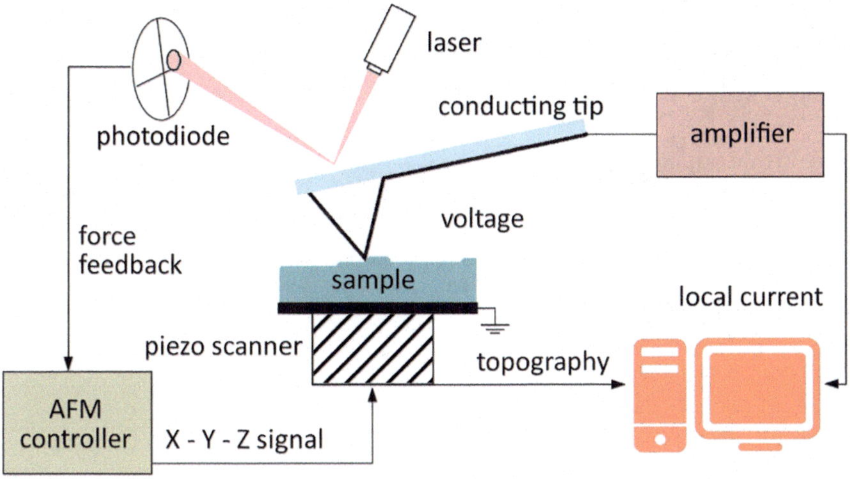

Fig. 2.1 The schematic of the C-AFM experimental setup

It is worth mentioning that as with other SPM techniques, the state and condition of the conducting tip have a substantial influence on the quality of the measurements. The tip sharpness will influence the resolution, both in the topography and local current map. Furthermore, depending on the construction of the conducing tip (PtIr, TiN, doped diamond-like carbon (DLC), TiO, bulk metal), various mechanical and thermal stressors can influence the properties of the conducting tip, thus changing its electrical properties. While the mechanical damage is relatively easy to detect (indicated by a sudden change in the current map and the existence of artefacts), the slow-changing tip modifications are more difficult to evaluate. For example, one of the most popular types of conducting tip—the Pt/Ir covered silicon AFM tip—experiences drastic changes in its mechanical and electrical properties under the influence of the elevated temperatures [16], as illustrated by TEM images of the tip apex after annealing under different atmospheres in Fig. 2.2. These changes include an increased tip apex radius, reduced resolution and increased resistivity (decrease in the local current measured with the use of such a tip). The changes originate from the formation of new phases (such as the Pt_2Si under UHV conditions, which has 60 times higher resistivity than pure Pt) or oxidation (interface SiO_2 formation under ambient conditions, which increased the resistivity of the tip by several orders of magnitude). While the C-AFM measurements at elevated temperatures are not that common, one must also include the increase in the temperature of the tip by means of Joule heating. Thus, the temperature of the tip apex can substantially increase, even with relatively low power output [17]. Naturally, the power depends strongly on the tip (or tip coating) material; however, for the standard Si tip in contact with the gold substrate, the 10 V and 17 mW of power was enough to raise the temperature of the tip above 700 °C [17].

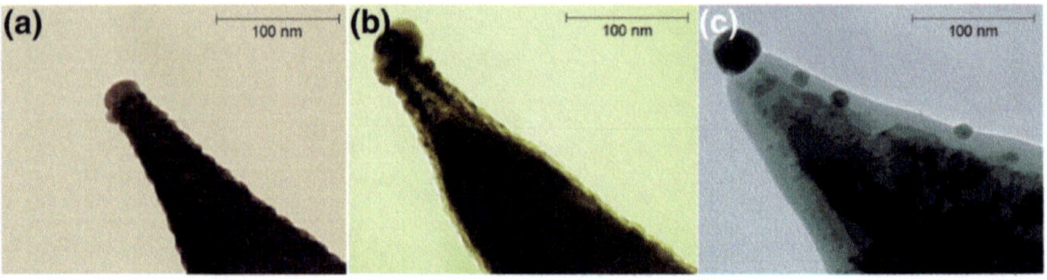

Fig. 2.2 Influence of the elevated temperature of the conducting tip measured by TEM: **a** reference tip (not used); **b** under the reducing conditions; **c** under the oxidizing conditions. Image adapted from Wojtyniak et al. [16]

2.1.2 C-AFM Modes

The local current measurement in the contact mode C-AFM can be obtained in several regimes, leading to the measurement of different physical quantities.

- Constant DC voltage mode can be implemented, giving the local current value as a response to the applied voltage (electrical resistance: $R = \frac{dV}{dI}$). In the general case of a nonlinear system, the entire current-voltage characteristics obtained by ramping the bias voltage must be analysed. Moreover, 2D arrays or any arbitrary paths of the tip can be implemented. However, the absolute value of the electrical resistance of the investigated material is difficult to obtain, as the C-AFM is a two-point method, meaning that all contributions to the current from the experimental setup and tip-sample geometry must be included [18].
- Constant DC current mode is typically used for the AFM-based lithography, as the C-AFM setup is capable of realizing arbitrary patterns for the tip movement. Thus, by applying constant current, the properties of a material under the tip can be changed. There are several mechanisms that can do that, and amongst others can be identified electrochemical degradation [19], tip-induced resistive switching [20] and local oxidation [21].
- AC mode can be implemented, giving the equivalent of the resistance as a function of frequency (electrical impedance: $Z(\omega) = R(\omega) + iX(\omega)$), where R is the AC resistance and X is called reactance. Typically, a Nyquist plot (impedance vs. frequency) can be obtained, which allows for the construction of the electrical equivalent of investigated material. Often, to improve the signal to noise ratio, a lock-in detection system is used.
- Potentiometry is typically measured using a null-force approach [22], which is based on the applied AC and DC voltage simultaneously. Under such conditions, the tip will oscillate due to electrostatic forces. The amplitude of oscillation at the tip-resonant frequency will depend on the product of the DC and AC voltages. Thus, by changing the DC voltage, the oscillation amplitude can be decreased to zero. This means that the set voltage is equal to the voltage under the tip.

Concerning all of the modes mentioned above, by varying the contact force, additional information can be obtained on: charge generation and injection or contact resistance. Naturally, the biggest advantage of the C-AFM is the ability to measure local current maps simultaneously with topographic imaging. Thus, it is very useful for the comparison of the local structure or morphology of the surface with local electrical properties such as conductivity, charge distribution or even the carrier concentration in semiconductors.

2.1.3 Electronics

- Current-to-voltage converter

The core of the electronic circuit for C-AFM measurements is a transimpedance transducer, a so-called current-to-voltage converter (I/V). This is essentially an ideal, virtually grounded ammeter with an internal resistance approaching zero [23–25]. The construction of an I/V converter can be realized on the basis of an electrometric operational amplifier (op-amp) with its reference resistor in a negative feedback configuration (see Fig. 2.3). Conventionally, the input of the I/V converter serving as the virtual ground is connected to the conducting C-AFM tip and the voltage applied to the sample via either the bottom electrode of a selected point or area on the surface. Mapping the conductivity on the macro- or nanoscale of high-k dielectrics with large band gaps, such as BO_2 or ABO_3 oxides, represents a major challenge due to the extremely high resistance of those materials in the stoichiometric state. Hence, the currents of only a fraction of pA must be detected. This can be realized using a modern electrometric amplifier such as an AD 4530 (see the datasheet of Analog Devices) or INA 116 (see datasheet of Texas Instruments). As these op-amps have a high current sensitivity of a few fA (in the range of 1–20 fA) and a very high input resistance (10^{14-15} Ω with a capacitance of 1–8 pF), they allow for the building of an I/V converter with a feedback resistor of 10^9 to 10^{10} Ω. In such a system, amplifications of 10 pA/V can be achieved. Because the resolution of a standard analog-to-digital converter, which is connected to the output of the I/V converter (and further with the controller of the C-AFM system to store the data), is four decades, the current sensitivity of the C-AFM system can be very high (up to 10 fA for the last significant bit (LSB) e.g., of 1 mV). For the optimal performance of an I/V converter, the resistor at the non-inverting input (R_+ in Fig. 2.3) should be adjusted to that in feedback and the resistance of the sample. A further important point for measuring currents with high sensitivity is the compensation of the thermal drift of the op-amp (~1 μV/°C) and the time drift of the offset voltage, which would lead to the shift of the "0" point of the I/V characteristics. Those compensations can be obtained manually or by means of an automatic controlling system (e.g., the ping-pong technique, chopper offset-stabilized amplifiers, etc. [26]). It should also not be forgotten that the active compounds (here the op-amp) and resistor of the I/V converter are, per se, the source of different kinds of the electronic noises (typically

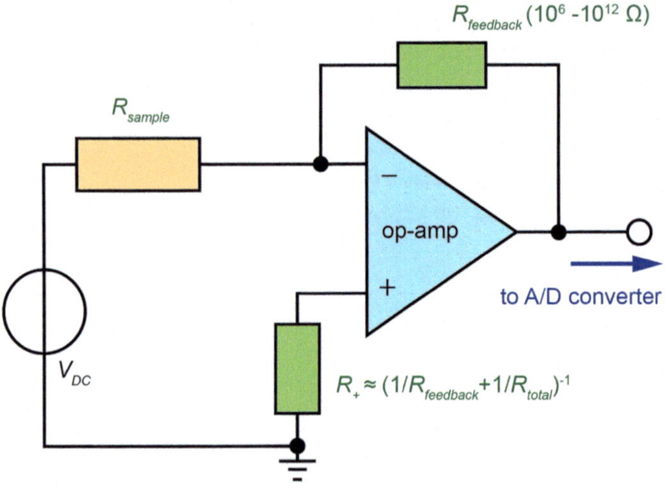

Fig. 2.3 Schematic illustration of a current-to-voltage converter

white noise), which can overlap with the measured signal. Therefore, it is essential to calculate the bandwidth of the system, in which a measurement with higher sensitivity than the level of the noise is possible. For an I/V converter employing a typical op-amp with a noise of 10–80 nV/(Hz)$^{0.5}$, the noise level is relatively low (about 1–9 μV for a bandwidth of 10 kHz). Also, the Johnson noise of the feedback resistor must be taken into account, e.g., for a resistor of 10 GΩ, the noise calculated with the formula $V = (4{\cdot}k_B{\cdot}T{\cdot}R{\cdot}\Delta f)^{0.5}$, ($k_B$: Boltzmann constant, T: temperature, R: resistance, Δf: frequency bandwidth) at a bandwidth of 10 kHz is of 1.2 mV, which means only approximately 1 LSB. Hence, these parameters do not limit the applicability of the I/V converter for the very sensitive current measurements on the sample with the extremal high resistance of many TΩ.

The most critical point for C-AFM measurements with a highly sensitive I/V converter is connected to the parasitic capacitance between the feedback resistor and surrounding (i.e., the casing). Assuming a parasitic capacitance of only 1 pF between a 10 GΩ feedback resistor and the rest of the electronic system, the bandwidth (so-called 3 dB band) of the system is reduced to only 300 Hz, as illustrated by SPICE simulations (Fig. 2.4a). Here, the feedback resistor was modelled as a series connection of five resistors, R_f, and the parasitic capacitance was represented by four capacitors C_p in between. Note that with the "perfect" amplification of "1" (meaning without non-linearity), it can only be measured up to a few dozen Hz.

This issue can be overcome using appropriate active compensation of the potential distribution along the feedback resistor by surrounding it with an additional resistor connected to ground [15]. Thereby, the bandwidth of the I/V converter can be increased by two or three orders of magnitude compared to a system without compensation (Fig. 2.4b). Hence, a C-AFM map can be obtained with 256 × 256 pixels in an acceptable time of a few dozen seconds.

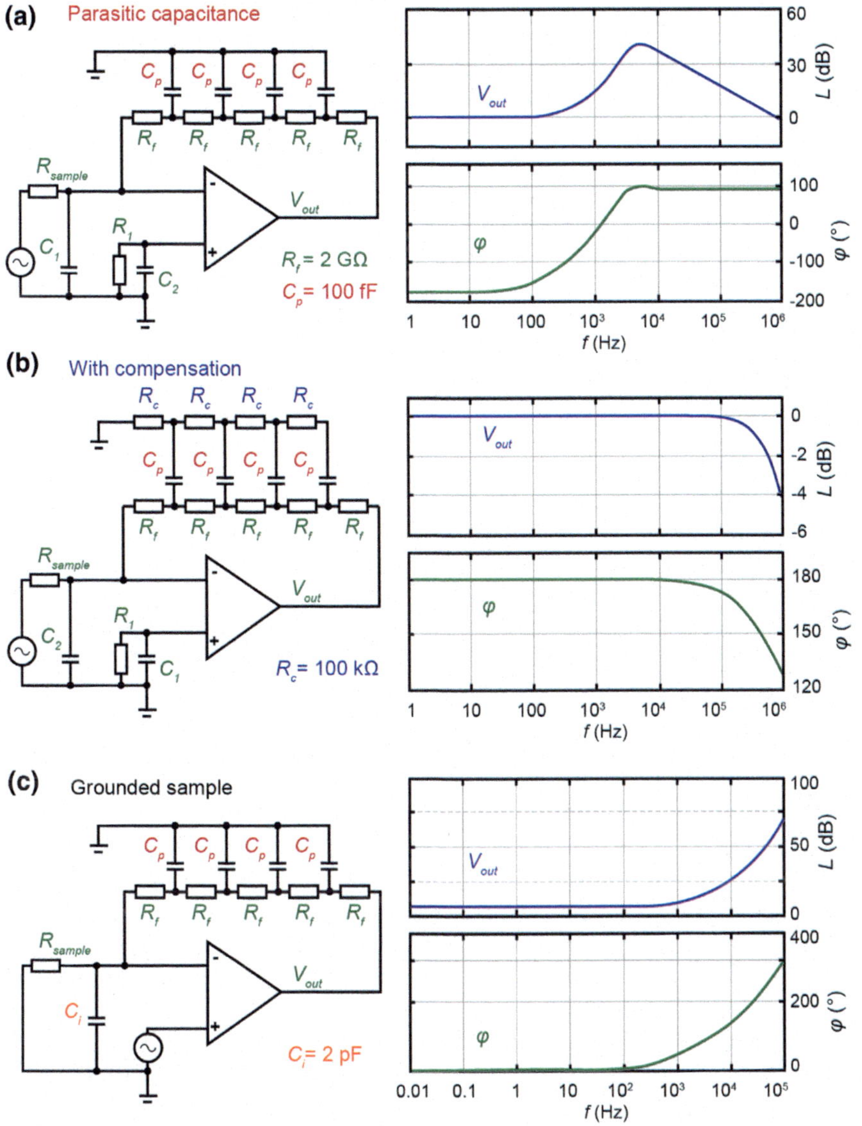

Fig. 2.4 SPICE modelling of a highly sensitive I/V converter using a feedback resistor of 10 GΩ: **a** Limitation of the bandwidth by parasitic capacitance $C_P = 4 \cdot C_p$; **b** active compensation by adjusting the potential distribution of the high-ohmic feedback resistor $R_F = 5 \cdot R_f$ by surrounding it with a low-ohmic resistor $R_C = 4 \cdot R_c$; **c** grounded sample configuration. Here, the bandwidth is limited by the input capacitance, C_i

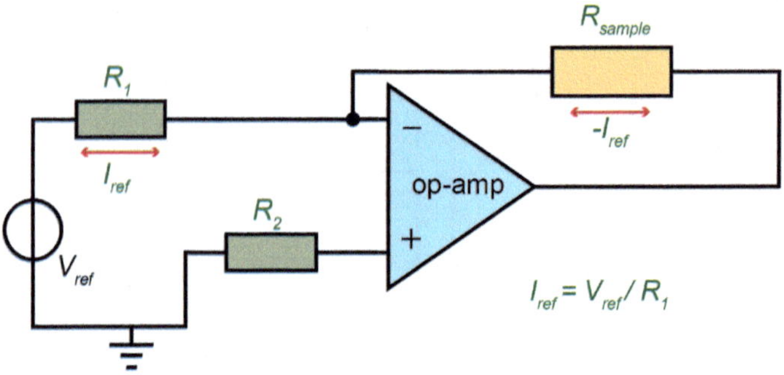

Fig. 2.5 Schematic illustration of a constant current source, which can be realized by simply interchanging the position of the feedback resistor (here R_1) and the sample relatively to the configuration of the I/V converter presented in Fig. 2.3

As the above-described configuration of the I/V converter can only be applied for a non-grounded sample, a simple modification is necessary for the electrical analysis of a grounded sample. This can be realized by connecting the voltage source to the non-inverting input of the amplifier. However, the drawback of this solution is that the bandwidth using a high feedback resistor ($R_f > 100\ \mathrm{M\Omega}$) is dramatically reduced (see Fig. 2.4c). Here, not only the parasitic capacitance between resistor and casing, but also the input capacitance C_i between both of the inputs of the op-amp play a decisive role.

When performing C-AFM measurements not in constant voltage mode, but in constant current mode, a precise current source is needed. Such a source can be constructed based on an I/V converter by simply changing the connections of the sample with the reference resistance (Fig. 2.5). The current of this source can be tuned by selecting the appropriate reference resistance and polarization voltage, and so it can be applied for the C-AFM measurements of non-grounded samples (Note: the control of the constant current flow is limited by the compliance voltage). Thereby, measurements in constant current mode can be obtained in the range of 10 μA to 10 pA (with the resolution of a few fA in the pA-range). For the highly sensitive current measurements, it is necessary to control the parasitic capacity of the sample relative to the ambient. Of course, such a constant current source can also be modified for grounded samples.

- Electrometer

Different processes observable in high-k oxides often require more precise electrical measurements than standard C-AFM. For example, the potential distribution in a polarized sample gives additional information about the homogeneity and current distribution of the investigated sample. This is especially important for subtle processes, such as electrodegradation, resistive switching and ion transport. Thus, a precise electrometer must be implemented for the C-AFM setup to be able to precisely detect the potential distribution on the sample surface. The design of an electrometer can be realized analogously to the I/V converter discussed above by employing the

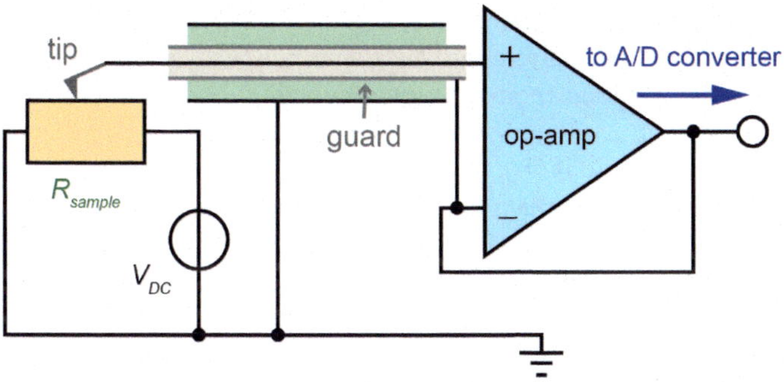

Fig. 2.6 Illustration of the measurement of the potential distribution on a polarized sample using an electrometer (voltage follower). Here, a triaxial shielding of the input connection is necessary as otherwise, e.g., for a sample resistance of 10 GΩ, the typical input capacitance of about 2 pF reduces the bandwidth of the system to a few Hz

same type of operational amplifier (Fig. 2.6). When attempting to map the potential distribution of the sample with a moderate resistance of, e.g., 10 GΩ, the input resistance of the op-amp (about 1 PΩ) is not the limiting factor of such an investigation. Only the input capacitance of the electrometric op-amp can lead to a noticeable reduction of the bandwidth of the system. Therefore, the position of the amplifier should be close to the sample, or triaxial screening of the connections should be applied. It should be noted that this consideration of input capacitance is not only relevant for electrometric measurements but also for conductivity detection via an I/V converter or constant current sources. In particular, when performing measurements under UHV conditions, an elaborate system design employing hermetical housing with appropriate electrical feedthroughs is of high importance.

2.1.4 Analysis of the Tip Sample Interaction: Energy Barriers

The electrical contact of a C-AFM tip contacting a typical semiconducting high-k metal oxide sample can be approached by analysing a metal semiconductor junction (the Schottky barrier junction) [27]. At the contact, due to differences in the work functions of the two materials (W_m for metal and W_s for semiconductor), a charge transfer occurs. Thus, the potential barrier emerges from the charge separation at the contact barrier. The charge separation leads to the formation of a high resistance region in the semiconductor. In the simple case of a metal-n-type-semiconductor junction at equilibrium, we can observe two possible cases. First, if W_m is larger than W_s, the electrons from the semiconductor are transferred to the metal. Second, if W_m is smaller than W_s, an electron accumulation layer is formed in the semiconductor near the junction. In the case of C-AFM, a small voltage is applied between the tip and sample, and so we must consider the influence of the voltage across the junction.

This can be discussed in the frame of voltage-current characteristics. Therefore, if W_m is larger than W_s, the current-voltage characteristic is asymmetric (rectifying). However, we obtain ohmic (non-rectifying) contact when W_m is smaller than W_s [28]. For the case of a p-type semiconductor, this behaviour is reversed. In that case, the contact is ohmic if W_m is larger than W_s and rectifying if W_m is smaller than W_s. Additionally, adsorbate layers or interface states may also influence the energy barrier between the tip and sample (the so-called Bardeen barrier [29]). Hence, when measuring the local I/V characteristics of metal oxides by C-AFM, imperfect Schottky-type curves are typically obtained, indicating complex electronic transport mechanisms.

The contact area between the tip and surface will play a significant role in C-AFM measurements and interpretation. It is possible to evaluate this by measuring the tunnel current when contacting a thin dielectric material [30]. It was also shown that the mechanical (physical) contact does not entirely contribute to the electrical transport, but an effective electrical contact exists that is approximately 10 times smaller than the physical contact. Additionally, the measured C-AFM current was linearly dependent on the force load, which is both tip- and sample-dependent. Thus, for Pt/Ir and diamond-like covered conducting tips with relatively small load forces (below 100 nN), contact areas of 1 nm^2 area or less can be expected.

Due to the small point contact between the tip and sample, the potential distribution is strongly localised below the tip. The potential distribution under the tip can be calculated, for instance by means of finite element calculations. This has been performed for the relatively simple case of the ideal dielectric or semiconductor with specific conductivity σ being at one point in contact with the tip of the cantilever. The potential distribution is hemispherical in shape and can be written as: $\varphi(r) = \varphi_0 \cdot \frac{r_0}{r}$, where r_0 is the tip radius, φ_0 is the potential of the tip. This means that 75% of the potential drop occurs in the radius of the contact tip-surface. Calculations performed for two types of common conducting tips (silicon covered with DLC and Pt/Ir) [18] show that the current density and potential distribution depends on the tip size and shape. The finite element calculations illustrate that most of the potential drop is localized very close to the tip's apex and the contact point as depicted in Fig. 2.7.

Since in a standard C-AFM experiment the current measurement is performed using a 2-point technique, the different contributions of the resistance present in the experimental setup must be taken into consideration. These contributions can be illustrated as a series of resistors presented in Fig. 2.8. Moreover, the $R_{contact}$ consists of spreading resistances of the tip and the sample. These spreading resistances are the nanocontact resistances of the material, which depend on the contact materials and geometry [31]. To calculate the spreading resistance, classical calculations using the Maxwell formula can be utilised [32]:

$$R_{spreading} = \frac{\rho_{tip}}{4a} + \frac{\rho_{sample}}{4a}$$

However, this formula is only valid for the cases of relatively large radii of the tip (large compared to the mean free path of electrons or holes). For cases when the contact size is extremely small (only a few atoms), the electron transport becomes ballistic and the conductance quantified. In this case, the Sharvin formula can be used [33, 34]:

$$R_{sharvin} = \frac{h}{2e^2}\frac{\lambda_F^2}{\pi^2 a^2}$$

where λ_F is the Fermi wavelength. This equation also allows for the calculation of the minimal resistance of the one-atom electrical contact in the form of $1/G_0 = h/2e^2 = 12.9\ \mathrm{k}\Omega$. This becomes particularly important in ultra-high resolutions C-AFM with atomic or sub-atomic resolution.

2.1.5 *Requirements for Sample Preparation—The Role of Adsorbates*

Typically, the first step of an AFM experiment is the sample preparation, which entails polishing, fracturing, heating, or ion bombardment. These steps are espe-

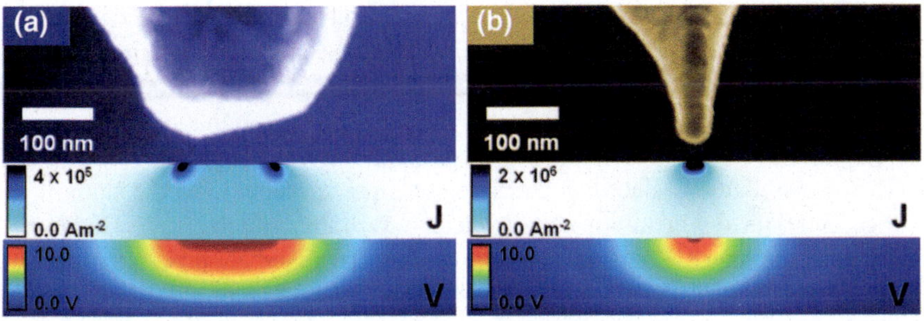

Fig. 2.7 Diamond-like (**a**) and platinum-coated (**b**) conducting AFM tips, along with the finite element calculations of current density and potential distribution in the case of a contact with the semiconducting surface. Adapted from Reid et al. [18]

Fig. 2.8 The resistance of the different components contributing to the total resistance measured in a typical C-AFM experiment

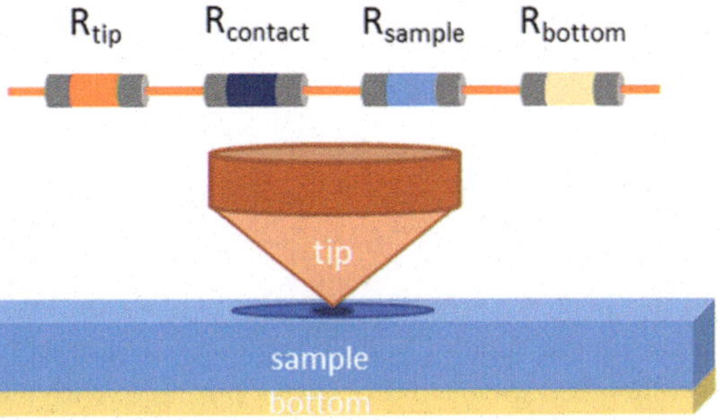

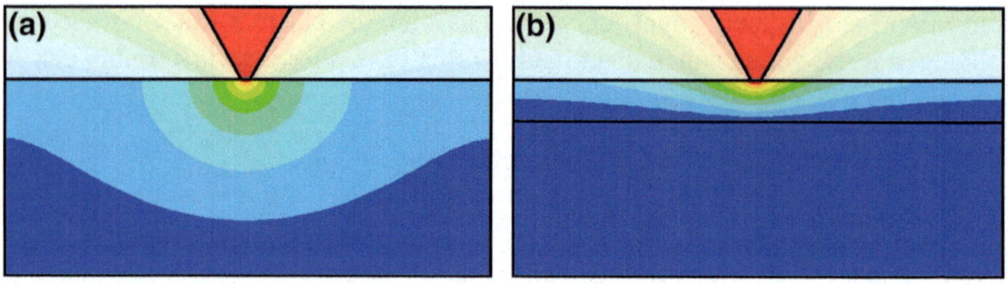

Fig. 2.9 FEM simulation of the influence of adsorbates on the potential distribution between the C-AFM tip and sample: **a** Sample without adsorbates; **b** sample with an additional adsorbate layer with lower permittivity

cially important in the case of UHV measurements and/or high resolution, where the state of the sample surface greatly influences the obtained data. In the case of electrical measurements, such as by C-AFM, PFM or SCM, the requirements concerning the surface quality are even higher. This originates from the fact that most contaminations, such as adsorbates, will influence the electrical measurements in a significant way. This can be seen from finite element calculations showing that the protrusion of the potential drop in the sample is significantly hindered if an adsorbate layer is present (Fig. 2.9).

A good example of the influence of the adsorbate layer on the electrical properties are measurements of piezo-response performed on the surface of single crystals of $BiTiO_3$. The data shows [35] that there is a detrimental effect of the adsorbate layer on the electrical signal measured from the sample. The contamination decreases the PFM signal due to the fact that most of the voltage drops in the contamination layer.

This shows that the electrical measurements are susceptible to the state of the surface, and typical preparation steps to clean it must therefore be made. However, a heat treatment under UHV conditions can remove the adsorbate layer (or part of it). As is illustrated in Fig. 2.10, showing the C-AFM measurement of a Pt surface, the conductivity of the surface can be significantly enhanced by annealing under vacuum conditions due to the removal of adsorbates from the Pt grains. A further method for obtaining an adsorbate-free surface is cleaving under reducing conditions, such as UHV or gas mixtures of Ar and H_2. The concept of a cleaving C-AFM setup is presented in Fig. 2.6b) and is particularly beneficial for obtaining a kind of 3D map of the local conductivity of the reduced oxides (see Fig. 2.10).

2.1.6 C-AFM with Atomic Resolution

As has been shown, the electrical contact between the C-AFM tip and sample surface is much smaller than the mechanical contact. Hence, the detection of localized conductance spots smaller than 1 nm^2 and, with atomic resolution, should eventually be possible. The first reports that this is indeed the case were published in the 1990s

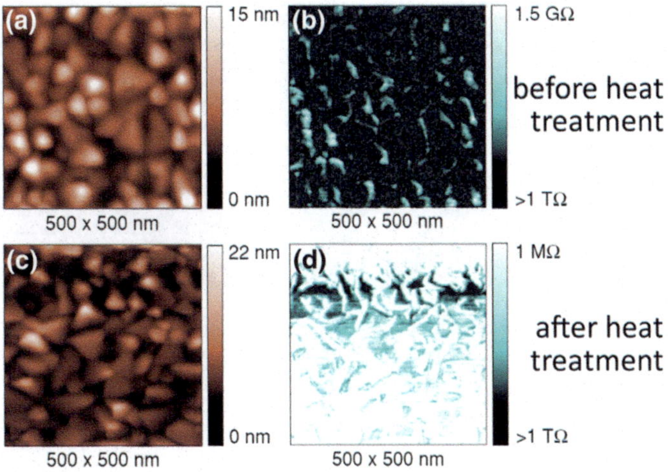

Fig. 2.10 Topography and local resistance map for platinum surface before (**a**, **b**) and after (**c**, **d**) heat treatment. The results show the influence of the adsorbate layer on the measured resistance. Adapted from Szot et al. [35]

by Enachescu et al., demonstrating atomic resolution on HOPG [36]. An example of such a C-AFM investigation on HOPG is shown in Fig. 2.11a, b, where the hexagonal lattice structure of graphite can be identified in the friction force microscopy (FFM), as well as in the simultaneously recorded C-AFM map [37]. However, there has been debate as to whether this is true atomic resolution or, rather, relates to a collective effect such as the Pethica effect, which was assumed to also be present in STM measurements of graphite. In contrast, Fig. 2.11c provides an indication that it is possible to map the current by C-AFM with true atomic resolution. Therefore, at first, a defective region in the HOPG was created by applying a pulse of several volts locally through the tip, resulting in a distorted area on the atomic length scale and in the subsequently recorded current map ($V = 15$ mV); even single surface point defects can be identified. As a consequence of the ongoing miniaturization in electronics, there has recently been an increased interest in high-resolution C-AFM and atomically resolved current maps of $Mo_xW_{1-x}Se_2$, $SrTiO_3$, and TiO_2 (see below) have been published [37, 38].

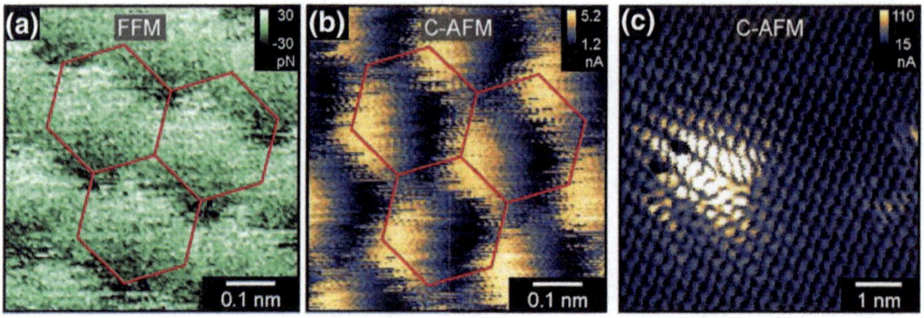

Fig. 2.11 LC-AFM measurement of HOPG. **a** Maps of friction force and current ($U = 15$ mV); **b** current map of an inhomogeneous region on HOPG demonstrating true atomic resolution. Adapted from Rodenbücher et al. [37]

2.2 Basics of High-k Dielectrics

2.2.1 Scaling Limits and Challenges in Semiconductor Technology

Ever since the development of semiconductor electronics, there has been an ongoing miniaturization of logic circuits. As predicted by G. Moore in 1965, the complexity in terms of the number of components of an integrated circuit doubles every one or two years [39]. This trend is directly connected with the feature size of single electronic components such as transistors exponentially decreasing over the course of time, posing new challenges related to physical scaling limits [40, 41] regarding, for example, a CMOS (complementary metal-oxide semiconductor) field effect transistor (FET), where the silicon gate is separated from the channel by a thin layer of silicon oxide (Fig. 2.12).

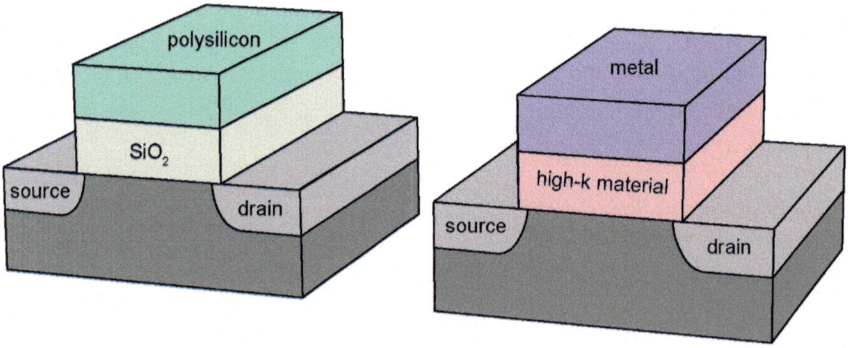

Fig. 2.12 Illustration of a standard transistor: (a) with SiO_2 as a gate oxide and of a transistor employing high-k materials

Table 2.1 Relative permittivity and band gap of selected technologically important metal oxides [42, 49]. The permittivity values of the ferroelectric materials PZT and $BaTiO_3$ show a wide spread and depend on parameters such as the grain size of the ceramics, doping and composition

Material	Relative permittivity	Band gap (eV)
SiO_2	3.9	9
Si_3N_4	7	5.3
Al_2O_3	9	8.8
Ta_2O_5	22	4.4
HfO_2	25	5.8
TiO_2	80	3.5
$SrTiO_3$	300	3.2
$Pb[Zr_xTi_{1-x}]O_3$ (PZT)	up to 1700	3.9
$BaTiO_3$	300–3000	3.2

When scaling down the feature size and, in turn, the thickness of the SiO_2 gate oxide, the leakage current increases due to direct electron tunnelling through the oxide, causing the dissipation of the power of the integrated circuit to increase drastically, to unacceptable values. Hence, new concepts must be developed that allow for a further downscaling of CMOS technology without increasing power consumption. Here, it is practical that the FET is a capacitance-operated device with a source-drain current depending on the gate capacitance:

$$C = \varepsilon_0 k \frac{A}{t}$$

where ε_0 is the vacuum permittivity, k is the relative permittivity, A is the gate area and t is the thickness of the gate oxide. Hence, the use of a material with a higher dielectric constant k would create the opportunity to increase the oxide thickness t in order to prevent leakage without decreasing the capacitance [42].

2.2.2 Technologically Relevant Metal Oxides with High k

Aside from the requirement that a potential gate oxide with reduced leakage current should have a higher k value than the 3.9 of SiO_2, it should also be thermodynamically stable and compatible with the standard CMOS production processes. Moreover, it should also be an insulator with a large enough band gap. As is illustrated in Table 2.1, which gives an overview of the most relevant high-k oxides under investigation; there is, however, a general trend that materials with a higher k value than SiO_2 also possess a lower band gap demanding for a well-considered trade-off when choosing the gate oxide [43–47].

Furthermore, materials with too high a k value are also unsuitable, as they would cause unwanted fringing fields at the source and drain electrode [48]. Hence, mate-

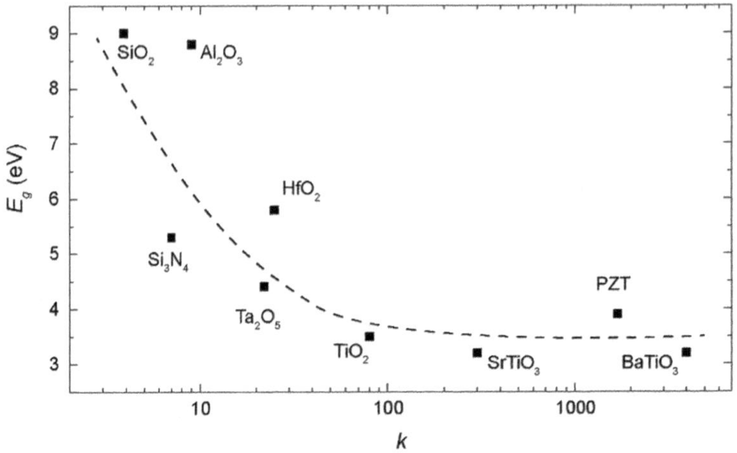

Fig. 2.13 Band gap as a function of dielectric constant for selected metal oxides

rials such as the ferroelectric $BaTiO_3$ providing k values up to 4000 are not frequently under consideration for use with CMOS technology and Al_2O_3 and HfO_2 have become the most prominent representatives of the class of high-k materials (Fig. 2.13).

2.2.3 Synthesis of Metal Oxides

As the properties of a metal oxide in particular in terms of leakage current are not only determined by its thickness and band gap, but are also strongly influenced by the presence of defects such as clusters or dislocations that evolved during oxide growth, it is essential to analyse the synthesis methods applied.

At first, we will discuss the growth of single crystals, which are frequently used for the analysis of basic material properties, in particular with the C-AFM technique, and are often implemented as a substrate for the deposition of functional oxide layers. Most of the metal oxide crystals available, such as TiO_2, $SrTiO_3$, or $BaTiO_3$, are produced by the Verneuil technique. In this technique, the raw materials are fed into a tube as powders where they are melted in a flame before falling onto a support rod, where the crystallization takes place. After cooling, the crystal is cut into slices (typical thickness: 0.5 mm) along the crystal axes and the surfaces are polished. Compared to other crystal growth techniques, such as the Czochralski method or the top-seeded solution growth where the crystal is pulled out of the melt or solution, respectively, the Verneuil method is very fast and cost-efficient, although this comes at the expense of crystal quality. Crystallographic perfection and homogeneity is particularly important when single crystals such as Nb-doped $SrTiO_3$ are used as substrates for the growth of thin films. Here, it has been found that the homogeneity

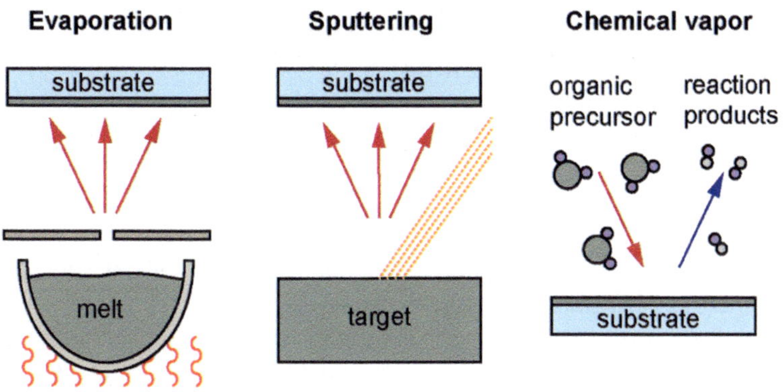

Fig. 2.14 Illustration of the basic concepts of metal oxide thin film deposition

of donor distribution depends strongly on the growth conditions. While in carefully prepared crystals a homogeneous Nb distribution on the macro- and nanoscales has been found [50], inhomogeneities related to growth defects and segregation can be present as artefacts of excessively rapid crystal growth [51]. Hence, careful analysis is recommended before using such crystals as substrates.

In order to prepare metal oxides as thin films, as is needed in CMOS technology, various methods, primarily based on gas phase deposition, have been applied. Here, a distinction can be made between physical vapour deposition (PVD) involving a net transfer of material from a target on an appropriate substrate and chemical vapour deposition (CVD) related to chemical reactions within the gas phase above the substrate [52]. PVD comprises evaporation techniques, such as molecular beam epitaxy (MBE), sputter deposition and pulsed laser deposition (PLD).

In CVD, organic precursors that react within the deposition chamber are frequently used, leaving behind layers of metal oxides, as in metalorganic vapour phase epitaxy (MOVPE) or atomic layer deposition (ALD), where layer-by-layer growth is induced by a dedicated sequence of pulse and purge steps involving reactions with different gases, allowing for a conformal coating of the substrate by the metal oxide film. An illustration of the main concepts in metal oxide thin film deposition is shown in Fig. 2.14. In contrast to the use of single crystals, thin film deposition methods allow for the easy tuning of the deposition parameters, and in turn the thin film properties, in particular the conductivity. This can be achieved by varying the chemical composition when adjusting the oxygen partial pressure in the deposition chamber during growth, or by modifying the local crystallographic structure, e.g., by choosing a substrate with a slightly different lattice constant than the film, leading to mismatch and strain.

2.3 Local Analysis of Electronic Transport Properties in Metal Oxide Thin Films

2.3.1 Variation of Nanoscale Conductivity Seen by Nanoelectrodes

By decreasing the feature size of nanoelectronic devices, local variations in electronic properties on the nanoscale gain ever more importance. The presence of these variations can be readily detected when depositing an array of planar nanoelectrodes of identical size on a metal oxide film and addressing each nanoelectrode individually to measure the conductance. This was exemplarily conducted for a PZT thin film deposited on an $SrTiO_3$ substrate [53]. At first, on each nanoelectrode, a DC polarization voltage of -2.1 V was applied via a C-AFM tip and, subsequently, an area of 7×7 μm was mapped in a conventional C-AFM scan with a bias of 0.9 V applied to the tip [35]. As is illustrated in Fig. 2.15, a huge variation in conductance between the individual electrodes can be identified. While some electrodes show a current of several microamperes, many electrodes only show a current of a few nanoamperes, although an identical voltage was applied. This unexpected observation poses a major challenge to the design and production of reliable metal oxide devices with reproducible electronic properties and therefore needs to be investigated in detail, with a particular focus on the microscopic origin of the observed variations. C-AFM is the method of choice for this endeavour, as the conducting AFM tip consisting of Pt-coated Si, for example, can be regarded as a mobile nanoelectrode that reflects the situation of a local metal-metal oxide contact without the need to deposit electrodes in a technologically demanding process. Furthermore, the contact area between a C-AFM tip as a mobile electrode and the oxide is much smaller than that of any deposited electrode and information about the local electronic properties can be obtained with a high degree of local resolution.

2.3.2 Localized Nature of Leakage Current

In recent years, various high-k metal oxide surfaces have been investigated by C-AFM, and it has often been shown that leakage and dielectric breakdown phenomena are related to localized conducting paths on the nanoscale. For the prototypical high-k material HfO_2, it has been revealed that on polycrystalline HfO_2/SiO_2 stacks, grain boundaries (GBs), for instance, are the primarily conduction paths through the dielectric [54]. Electrical stress-induced breakdown sites were also found to be located at the GBs, suggesting that the polycrystalline phase of the gate dielectric may impair reliability. This is illustrated in Fig. 2.16, which presents the topography and corresponding C-AFM map obtained at a bias of 6.5 V. The line profiles obtained across and along the grain boundary (Fig. 2.16) reveal not only that the leakage

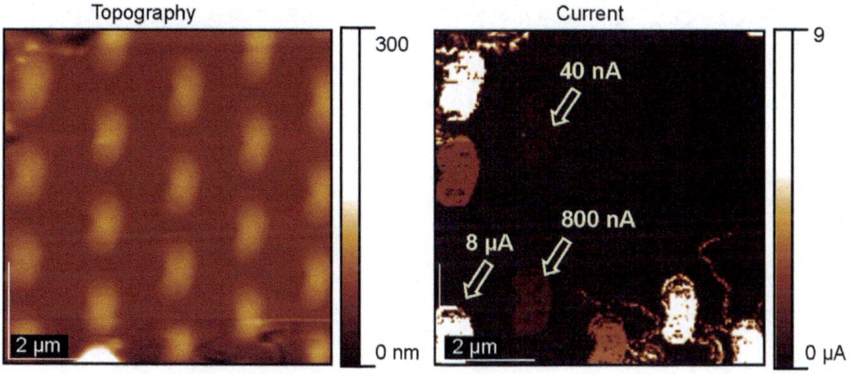

Fig. 2.15 C-AFM investigation of an array of nanoelectrodes deposited on a 4 nm PZT film after DC polarization of each nanoelectrode with −2.1 V. Adapted from Szot et al. [35]

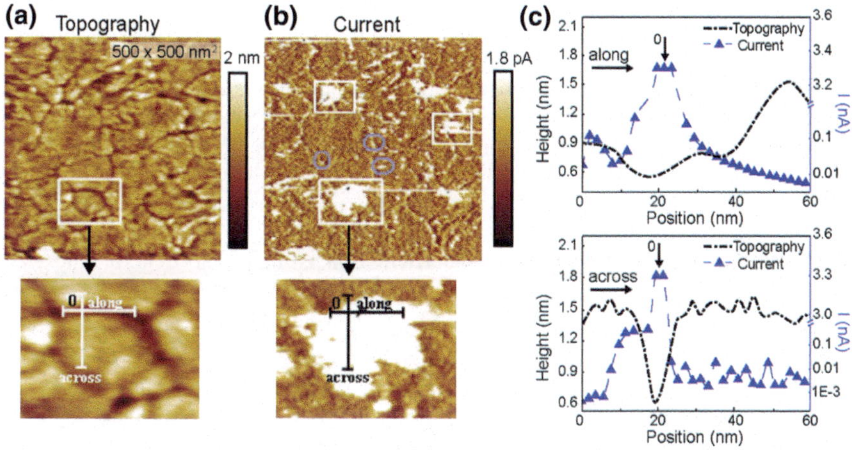

Fig. 2.16 **a** Topographical; and **b** the current images (500 × 500 nm^2) of an HfO_2/SiO_2 stack obtained at −6.5 V (applied to the substrate, tip grounded) in UHV. The circles and squares represent leakage sites and breakdown spots, respectively. **c** Topographical and current profiles obtained across and along the grain boundary, respectively, at the site of the largest breakdown spot (point 0 is the position of maximum current). Adapted from Iglesias et al. [54]

current through the high-k dielectric stack preferentially flows through the GBs, but also that the breakdown spots are located at the GBs, which in turn propagate into the surrounding regions, preferentially along the GBs.

Another example of the importance of grain boundaries for conductance in high-k dielectrics is polycrystalline Ta_2O_5 [55]. Here, it was found that upon thermal treatment, grain boundaries develop into small gaps that create a network of insulating material across the entire surface (Fig. 2.17). Although the grain boundaries here were found to be insulating, in contrast to the HfO_2 films discussed above, it was

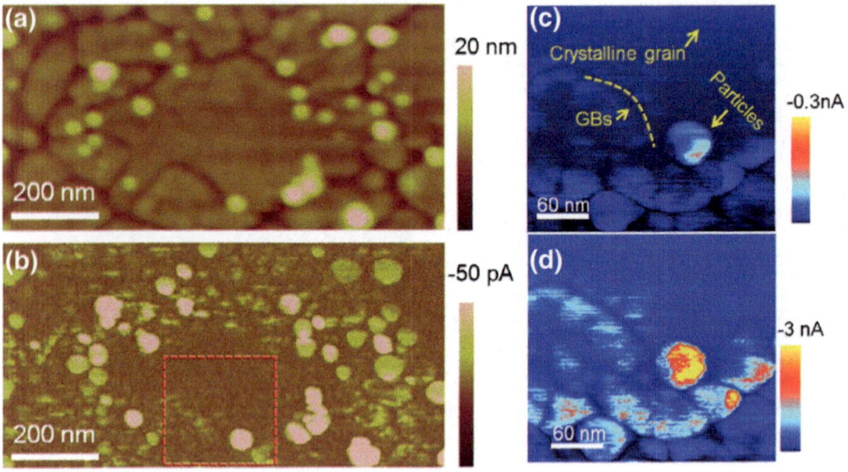

Fig. 2.17 **a** Topography image of polycrystalline Ta_2O_5 showing crystalline islands surrounded by grain boundaries; **b** current image as measured by C-AFM (−0.5 V bias applied to the sample); **c** C-AFM magnification of the area marked in (**b**). Three main features are visible: (1) a highly conductive rounded particle in the centre; (2) crystal grain showing good insulating properties at the top right; and (3) insulating GBs running through the entire sample surface. Adapted from Weaver [55]

also found that some highly conductive particles consisting of TaN decorate the grain boundaries and thus provide increased leakage through the dielectric (Fig. 2.17c, d).

An illustrative example of how C-AFM can detect localized conductance effects on metal oxide is the investigation of $NdNiO_3$, a prototype-correlated metal oxide [56]. The dielectric shows a reversible temperature-dependent insulator-metal transition, making it a promising candidate for memristive applications. As is shown in Fig. 2.18a, at low temperatures the oxide has a high resistance, which turns into metallic behaviour upon being heated above 100 K. This transition was followed by obtaining C-AFM maps (Fig. 2.18c) in a warm-up run for the different temperatures of 4, 78, 100, 150 and 300 K. Far below (4 K) and above (300 K), the MIT of the resistive maps look roughly homogeneous, with average resistance values differing by 2–3 orders of magnitude (around 10^{10}–10^{11} Ω and 10^8–10^9 Ω at 4 and 300 K, respectively). Upon warming (see the image at 78 K), low-resistance regions appear in a high-resistance background. At 100 K, more of these regions nucleate, grow in size and begin to percolate. At 150 K, the density of these low-resistance regions is even higher, with the high-resistance ones having almost vanished. Figure 2.18b plots the relative fraction of insulating regions deduced from the images (symbols), along with the same quantity extracted from the macroscopic resistivity data (lines), which shows that the transition is not perfectly abrupt, but shows a kink at a fraction of about 0.5, the percolation threshold, and reflects the transition between the nonpercolating and percolating regimes.

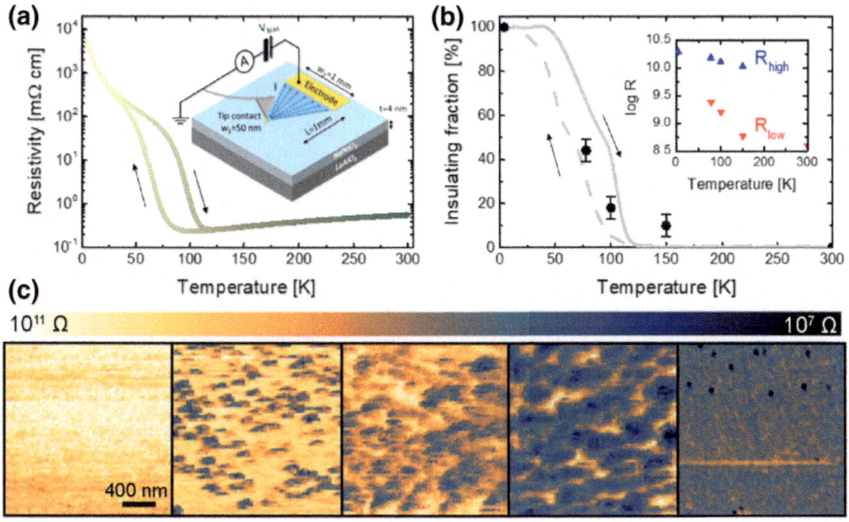

Fig. 2.18 C-AFM investigation of the $NdNiO_3/LaAlO_3$ interface: **a** Resistivity as a function of temperature; **b** insulating fraction as obtained from the resistivity data (lines) and the C-AFM images (symbols). The inset shows the average resistance in the insulating and metallic regions as a function of temperature. **c** Resistance maps acquired at several temperatures. Adapted from Preziosi et al. [56]

2.3.3 *Correlation of the Localized Conductivity with the Surface Potential*

By combining C-AFM investigations with Kelvin probe force microscopy, a deeper understanding of the electronic surface properties of high-k materials can be obtained, simultaneously providing information about local conductivity and surface potential on the nanoscale. In particular, when the same cantilever is used for both measurements, the same sample area can be scanned. This can sometimes be challenging, as for contact mode, long cantilevers with a low resonance frequency of around 10 kHz are normally used and are difficult to operate in non-contact mode under vacuum conditions, where the cantilever frequency is the input parameter of the feedback loop. Hence, for a stable oscillation, not the resonance oscillation itself, but a higher harmonic, has been used.

In this way, the TiO_2 (110) surface, for example, reduced at 750 °C under UHV conditions, was investigated in situ [57]. The C-AFM measurement presented in Fig. 2.19 shows an inhomogeneous, cloud-like structure typical for slightly reduced metal oxide. In addition, the CPD map obtained at the same scan area revealed local variations in surface potential. These variations can be visualized by calculating histograms (Fig. 2.19, right). The conductance histogram can be simulated by a log-normal distribution, reflecting its inhomogeneous nature, while the CPD histogram is close to a normal distribution. This difference in the variations of the current and

CPD maps can be understood by taking into account that the current between the tip and sample is determined by the energy barrier (of Schottky or Bardeen type) at the surface. As the normal-distributed variations in CPD can be regarded as a measure for variations in surface barrier height, they will be amplified to a lognormal distribution in current due to a non-linear relation such as that caused by an activated transport mechanism. This effect has also been observed on $SrTiO_3$, indicating that local variations in electronic properties on the nanoscale are a common effect on slightly reduced metal oxide surfaces [58]. An inhomogeneous oxygen vacancy distribution induced upon thermal reduction in the surface layer has been identified as the origin of the variations in local barrier height and conductivity [57, 59].

The combined analysis by C-AFM and KPFM is particularly useful to understanding the leakage mechanisms on materials with intrinsic polycrystalline morphology such as annealed HfO_2 systems. There has been intense discussion of whether grain boundaries can act as leakage paths. With the help of C-AFM and KPFM, this issue was tackled and now there is a general consensus that grain boundaries in HfO_2 are leakier due to the higher concentration of positive charges [54]. This is illustrated in Fig. 2.20, which shows C-AFM and KPFM measurements of a polycrystalline 2.5 nm-HfO_2 film on 1 nm-SiO_2/Si stacks (crystallization was induced by annealing at 1000 °C). It can be clearly identified that conducting spots are located in between the crystallites at the grain boundary and that there, the surface potential is several tens of mV higher than on the grain. This result is highly relevant for resistive switching applications based on HfO_2, indicating that the growth of the switchable filament during an initial electroforming step would preferentially occur at grain boundaries and could therefore offer a potential perspective on how to control the filament position by means of material engineering.

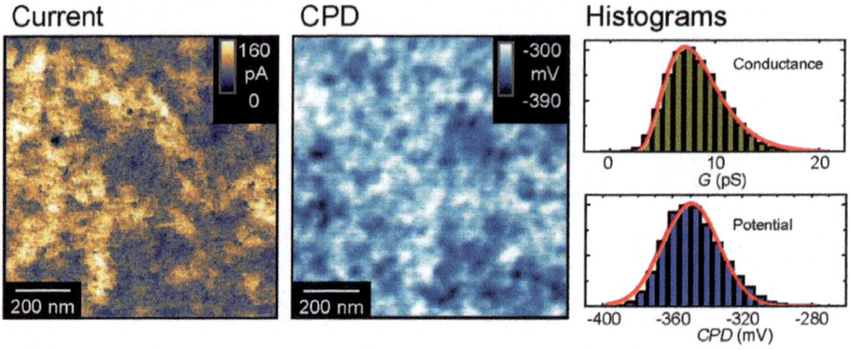

Fig. 2.19 In situ C-AFM ($U = 0.1$ V) and KPFM investigation of the TiO_2 surface annealed under UHV conditions at 750 °C and recorded at the same area. The histograms of conductance and contact potential difference were extracted from the maps. Adapted from Rodenbücher et al. [57]

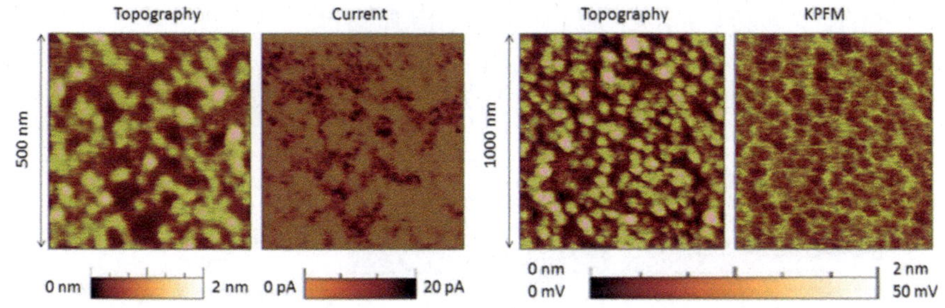

Fig. 2.20 Simultaneously collected topography-current (left) and topography-CPD images (right) on the surface of a 2.5 nm-HfO_2/1 nm-SiO_2/Si stack annealed at 1000 °C. Adapted from Lanza [60]

2.3.4 Tuning the Conductivity by Thermal Annealing

Metal oxides provide a unique opportunity to tune the conductivity by changing the oxygen stoichiometry, making them promising candidates for future memristive applications. This effect is mainly related to the excorporation of oxygen from the crystal lattice, leaving behind oxygen vacancies that are compensated by electrons, leading to a valence change of the transition metal oxide [13]. Hence, such a change in conductivity cannot only be induced by gradients of the electric potential such as in a memristive cell, but also by thermal annealing under reducing conditions. Using single crystals of TiO_2 and $SrTiO_3$, the local nature of the reduction upon annealing has been analysed in detail [61]. As an example, a C-AFM analysis of the TiO_2 (110) surface doped with 0.5 wt% Nb is shown. The epi-polished surface was annealed at different temperatures under UHV conditions and, after each annealing step, the local conductivity was mapped, in situ, at room temperature [62]. The average surface conductivity increases nearly exponentially with increasing annealing temperature, as can be expected given that the excorporation of oxygen is a temperature-activated process. With increasing surface conductivity, interesting changes in the local conductivity can also be seen. After annealing at lower temperatures (700–800 °C), a high level of inhomogeneity can be seen in the current maps, revealing a cluster-like conductivity on the nanoscale. It is noteworthy that this inhomogeneity is not related to the presence of adsorbates, as even chemisorbed organic components are released from the surface at 700 °C, as has been confirmed by XPS investigations (Fig. 2.21).

At higher reduction temperatures above 900 °C, however, a distinct change occurs and the surface conductivity becomes more homogeneous and, eventually, atomically-flat steps evolve that can be easily identified in topography, as well as in the current maps as having a slightly lower conductivity at the step edges. It has been argued that those changes in surface conductivity often relate to changes in reconstruction: in the case of TiO_2:Nb, a transition from (1×1) to (4×2) has been detected, while in $SrTiO_3$ a transition from (1×1) to $(\sqrt{5} \times \sqrt{5})$ R 26.6° had a similar impact on the conductivity [59]. This indicates that the first stage of ther-

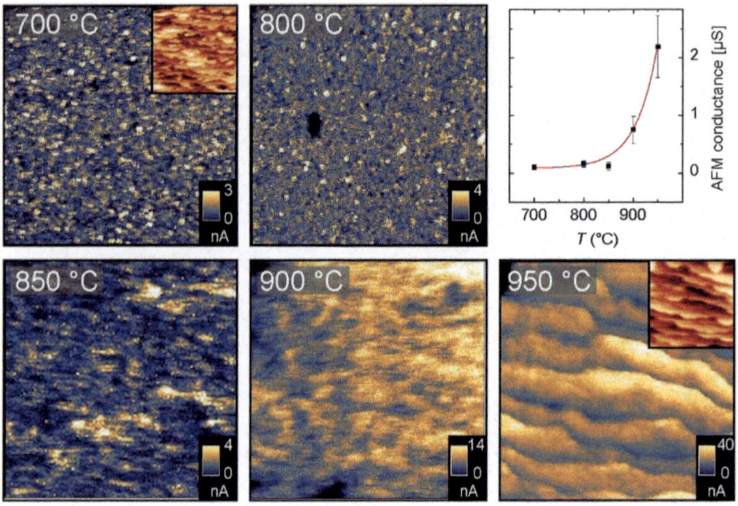

Fig. 2.21 C-AFM investigation of Nb-doped TiO_2 after thermal reduction under UHV conditions at different temperatures. Adapted from Wrana et al. [62]

mal reduction of a metal oxide relates to an inhomogeneous distribution of oxygen vacancies, e.g., related to the presence of dislocations, where the formation enthalpy is significantly lower than in the bulk, thus providing easy reduction sites where channels of high conductivity are formed (see below) [63]. After prolonged reduction, the vacancy generation within the matrix also becomes activated, resulting in more homogeneous conductivity.

Using high-resolution C-AFM, the changes in surface conductivity upon thermal annealing have been mapped with atomic resolution for the prototype metal oxide TiO_2. Similar to $SrTiO_3$, the surface conductivity of the TiO_2 (110) surface changes from highly inhomogeneous to quasi-homogeneous in the course of reduction, which also relates to a change in reconstruction from (1×1) to (2×1). Using friction force microscopy (FFM), in combination with C-AFM, these changes could be tracked on the atomic scale [37]. After a slight reduction at 700 °C in UHV, the surface shows a flat but fairly unstructured morphology, while the local conductivity was found to be confined to conducting spots with diameters of several tens of nanometres embedded within an insulating matrix (Fig. 2.22a). Remarkably, C-AFM with atomic resolution reveals that the transition from highly conducting to insulating is spatially confined to only a few lattice constants, indicating a highly localized arrangement of oxygen vacancies in linear structures or along dislocations. After a high-temperature reduction at 1000 °C, the well-known (2×1) reconstruction could be identified by STM and FFM. The corresponding C-AFM maps show that the conductivity of the reconstructed surface is also more homogeneous than that of the slightly reduced surface and periodic arrangement of the atoms can be identified (Fig. 2.22b). However, this arrangement was found to be superposed by a further local conductivity

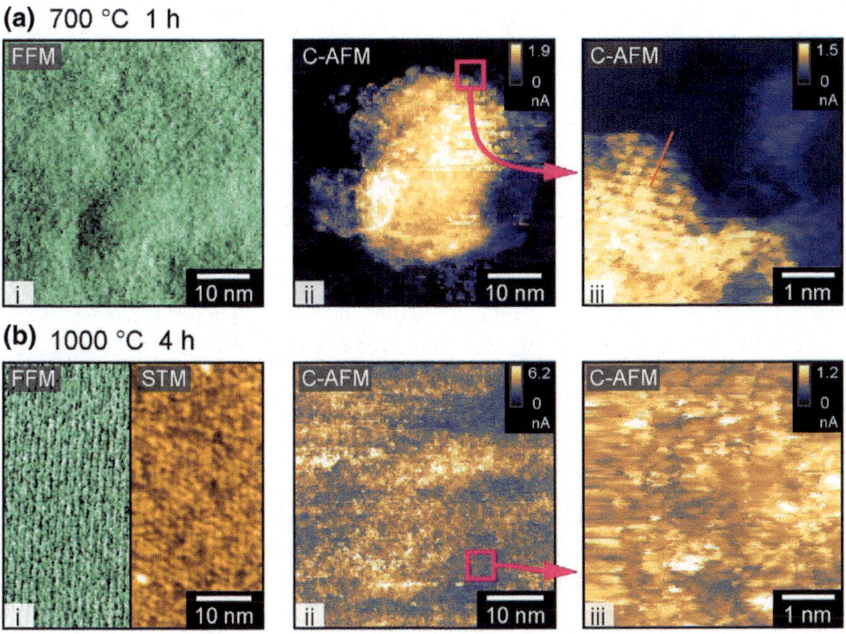

Fig. 2.22 Combined high-resolution friction force and local conductivity measurements on the TiO_2 (110) surface after reduction at **a** 700 °C and **b** 1000 °C. (i) FFM and (ii-iii) C-AFM current maps ($U = 20$ mV). The comparison between FFM and STM in (**b**, i) confirms the periodic arrangement of surface atoms in 2×1 reconstruction. Adapted from [37]

variation, indicating that the surface conductivity of metal oxides is a highly complex phenomenon related not only to extended defects, such as dislocations, but also to local variations in the oxygen vacancy density on the surface as well as in the subsurface region.

2.3.5 Confinement of Conductivity at Surfaces and Interfaces

In recent years, the opportunity to confine the conductivity at surfaces and interfaces has attracted enormous attention, allowing for the generation of 2D electron gases that exhibit a variety of extraordinary properties, such as magnetism and superconductivity [64]. A prominent example is an interface between $LaAlO_3$ and $SrTiO_3$ (LAO/STO interface) that can be easily produced by thin film technology, using $SrTiO_3$ as a substrate. While this phenomenon itself has been known for several years, it has always been a question of whether the electron gas is actually confined to the interface as predicted or has a much broader spatial extension [65]. Using C-AFM, this question could be solved by Basletic et al. [8], as illustrated in Fig. 2.23. In order to investigate the thickness of the electron gas, LAO/STO structures grown

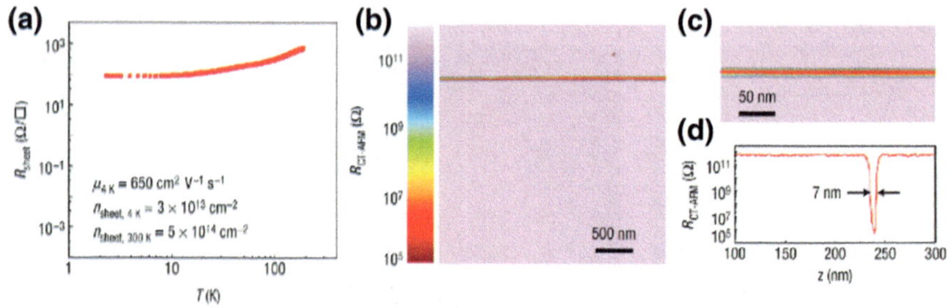

Fig. 2.23 C-AFM investigation of the $LaAlO_3/SrTiO_3$ interface: **a** Temperature dependence of the sheet resistance R_{sheet}; **b** C-AFM resistance mapping; **c** CT-AFM resistance image in high-resolution mode; **d** resistance profile across the LAO/STO interface extracted from (**c**). Adapted from Basletic et al. [8]

by PLD were cut and polished and C-AFM mapping was performed on the sectional plane. It has been shown that the spatial extension of the region with high conductivity depends strongly on the growth and annealing process and can be varied from hundreds of micrometres to a few nanometres next to the interface. Additionally, those experiments provided direct evidence for the presence of a metallic electron gas with a carrier density of around 10^{21} cm^{-3} confined at room temperature within a few nanometres next to the $LaAlO_3/SrTiO_3$ interface in annealed samples, being in agreement with a proposed mechanism based on charge transfer as the origin of this metallic state.

A further method to generate 2D electron gases is the modification of crystal surfaces [67]. It has been shown that even simple vacuum cleaving and the subsequent measuring of photoemission spectra is sufficient to induce metallic conductivity on the $SrTiO_3$ surface [68]. Similarly, using bombardment by electrons or ions, a valence change of the transition metal ion, leading to metallic behaviour in confined surface regions of nominally insulating metal oxides, can be induced. Figure 2.24 shows an example of Ar-sputtered TiO_2 investigated by means of C-AFM after cleaving the sample perpendicular to the sputtered surface [66]. A well-conducting quasi-2D layer can be observed as a bright (highly conductive) stripe in the centre of the C-AFM map. The thickness of the well-conducting layer was estimated to be about 30 nm using the cross sections shown in Fig. 2.24b. It is noteworthy that the sputtered surface showing metallic conductivity also revealed a certain level of inhomogeneity when mapped by C-AFM, related to a grainy surface morphology and illustrating that defects in metal oxide surfaces tend to agglomerate and cluster under the influence of external driving forces.

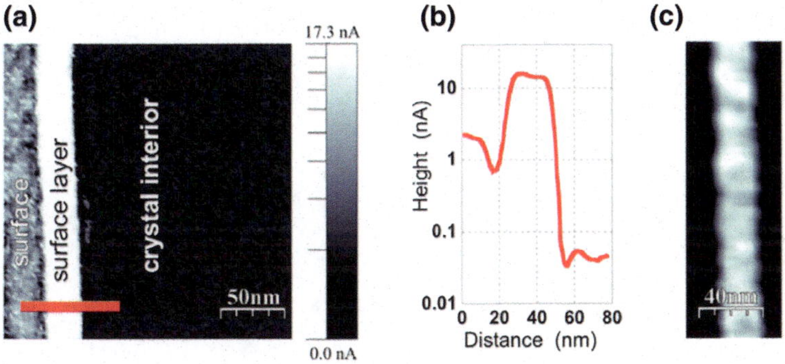

Fig. 2.24 C-AFM measurements in the cleavage plane of previously Ar-sputtered TiO_2: **a** local conductivity image (225 × 225 nm^2; sample-tip polarization 3.0 V); **b** cross sections along the marked line; **c** magnification of the region of high-conductivity surface layer (60 × 180 nm^2; polarization 2.0 V). Adapted from Rogala et al. [66]

2.4 Influence of Extended Defects on the Local Conductivity of Single Crystals

2.4.1 Current Channelling Along Dislocations

As a prototype material for transition metal oxides, $SrTiO_3$ has been intensively studied, experimentally as well as by atomistic simulations, indicating that dislocations play a crucial role for redox processes. It was calculated that the oxygen vacancy formation energy in $SrTiO_3$ is lower at sites close to the dislocation core by as much as 2 eV compared to that in the bulk, making dislocations preferential reduction sites during the first stage of thermal annealing under reducing conditions [63]. This effect may also be a reasonable explanation for the presence of inhomogeneous C-AFM maps on slightly reduced metal oxides, as discussed above. In order to elucidate the influence of dislocations on the local electronic transport in more detail, a system with a controlled arrangement of dislocations must be investigated. Hence, synthetic bicrystals of $SrTiO_3$ are the material of choice for such an investigation. Typically, they are produced by hot pressing two crystal pieces whose surfaces have been polished with a dedicated miscut angle, resulting in a periodic array of dislocations aligned along the bicrystal boundary. It must be noted that the quality of available bicrystals varies and they are often not only a defined array of dislocations, but rather a broad band of irregular agglomerations and bundles are present close to the boundary [25]. In Fig. 2.25, a C-AFM scan of the 36.8° bicrystal boundary of $SrTiO_3$ is shown. The measurement was performed under vacuum conditions at an elevated temperature in order to remove adsorbates from the surface. The line profile in the inset extracted from the C-AFM scan clearly shows that the local conductivity at the boundary is significantly increased, supporting the assumption that dislocations act as conducting paths for the current flow. This is in line with the C-AFM measure-

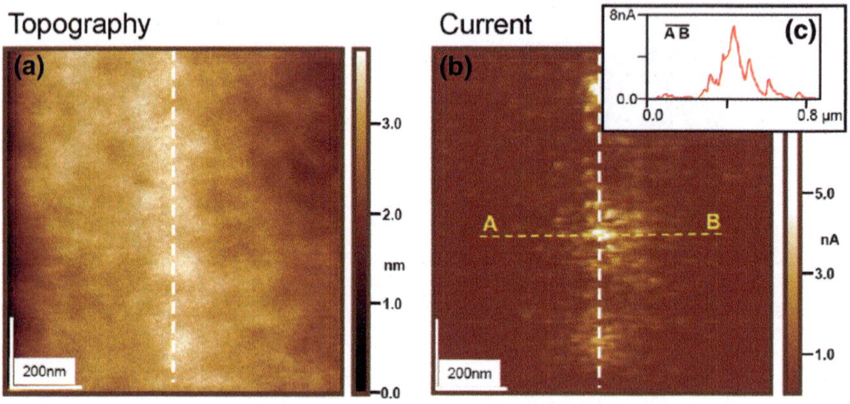

Fig. 2.25 **a** AFM topography; and **b** C-AFM map of the bicrystalline boundary (36.8°) of $SrTiO_3$ (obtained under HV conditions at elevated temperature). The cross-section (inset) shows an increased conductivity in a broad region perpendicular to the bicrystalline boundary. Adapted from Szot et al. [25]

ments of the local conductivity of single or bundles of dislocations marked by etch pits, which show an increased conductance at the centre of the etch pit [25].

2.4.2 *Analysis of Bulk Conductivity by the Cleaving Method*

Having seen that the surface conductivity of slightly reduced metal oxides such as $SrTiO_3$ reveals heterogeneity, which can be potentially related to the exits of dislocation acting as current channels, C-AFM was applied in order to analyse the conductivity in deeper parts of the sample towards the bulk. This was done by cleaving reduced crystals and analysing the cleaving plane, analogously to the investigation of 2D electron gases discussed above. As an example, such an investigation of reduced $SrTiO_3$ is shown in Fig. 2.26. In order to prevent a reoxidation and contamination of the cleaved face, cleaving was performed inside the AFM under reducing Ar/H_2 atmosphere by means of a dedicated cleaving system. The C-AFM maps reveal that there is decreasing local conductivity from the original surface towards the bulk. Within the first few micrometres from the surface, a highly conducting region can be identified in the map, indicating that from a macroscopic point of view a quasi-2D conductance within the surface layer is present in slightly reduced $SrTiO_3$. In deeper parts of the sample, towards the original bulk, it can be seen that single conductive spots are present and exhibit a nearly exponential decay in density as a function of the distance from the surface layer. This observation can be regarded as further indication for channelling the current along the dislocations, as TEM investigations have revealed that the dislocation density also decays exponentially from the surface to the bulk [69]. This is due to mechanical polishing of the epi-ready surface normally

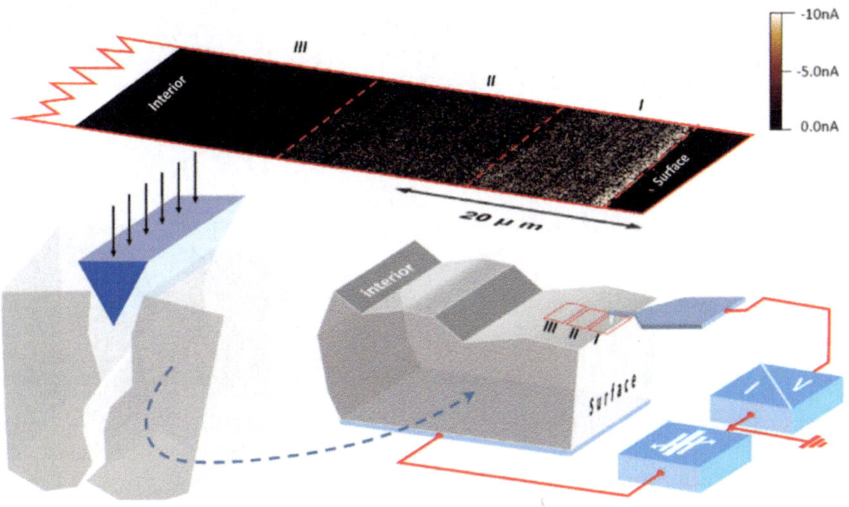

Fig. 2.26 Distribution of the conducting filaments in the surface region of a reduced $SrTiO_3$ crystal revealed by C-AFM mapping. Adapted from Szot et al. [25]

done by the manufacturer of crystal substrates, thereby inducing a high density in the dislocations in the surface layer. Hence, a dislocation network with the shape of a hierarchical tree is present and is transformed by slight thermal reduction into a network of metallic nanowires due to the localized formation of oxygen vacancies and the related charge compensation by electrons [70], changing the valence of the Ti atoms within the core of the dislocations from +4 to +3.

2.4.3 Ferroelectric Domain Walls as Conducting Paths

Amongst the class of high-k materials, ferroelectrics provide a variety of interesting electronic properties. Due to the broken crystallographic inversion symmetry, the spontaneous electric polarization of these materials can be easily switched by a small external electric field that is reflected by a dielectric constant that can reach a value of several thousand. Similar to ferromagnets, ferroelectrics can form a domain structure on the micro- and nanoscale, consisting of areas with differently orientated spontaneous polarization that are separated by a domain wall. Although it had already been predicted in 1973 that domain walls in insulating ferroelectrics can have metallic conductivity due to the movement of the compensating electrons or holes along the wall [72], it took until the availability of reliable C-AFM apparatus for it to be possible to investigate this phenomenon in detail. These investigations can easily be combined with piezo force microscopy (PFM), which is needed to detect the domain structure and orientations, as both measurements are conducted in contact mode

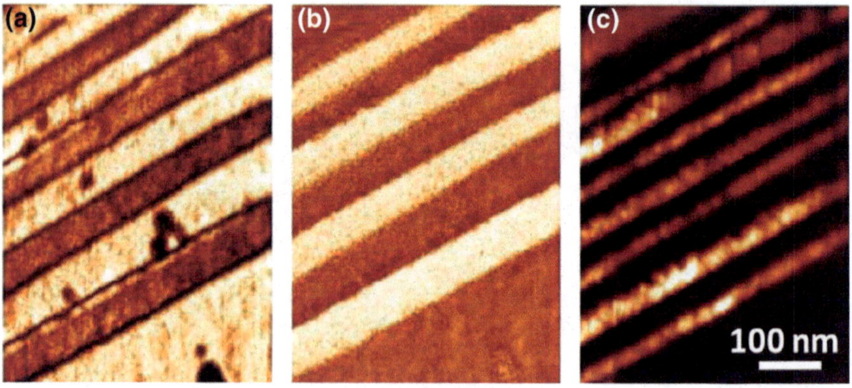

Fig. 2.27 **a** PFM amplitude; **b** PFM phase images of a $BiFeO_3$ sample with 109° stripe domains; **c** simultaneously acquired C-AFM image of the same area showing that each 109° domain wall is electrically conductive. Adapted from Seidel et al. [71]

and hence the same tip can be used. Figure 2.27 shows an example of a $BiFeO_3$ film doped with 10% La that was grown by PLD on $SrRuO_3/DyScO_3$ substrates [71]. The amplitude and phase signal mapped by PFM clearly reveal a striped domain pattern with a domain width of several 100 nm. The domains were separated by 109° walls. A comparison between the PFM images and C-AFM map reveals a clear correlation between the domain wall position and local conductivity indicated by high C-AFM (Fig. 2.27c). The resulting conductive stripes are about 20 nm wide.

With regard to applications, the discovery of enhanced conductivity at domain walls is highly promising, as the domain structure of ferroelectrics and, in turn, the geometric structure of the domain walls can be easily manipulated by applying an electrical field. Using C-AFM, this can be done via the conducting tip on the nanoscale with high precision, as is shown for a $Pb(Zr_{0.2}Ti_{0.8})O_3$ thin film [73] deposited on $SrRuO_3/SrTiO_3$ in Fig. 2.28. Here, a conducting domain wall has been electrically written in the pristine film and can be directly observed by C-AFM. The domain walls can form a complex network on the nanoscale, as demonstrated in Fig. 2.28b, c presenting a C-AFM and PFM investigation of the PZT thin film after annealing under reducing conditions, which resulted in a complex domain structure.

Recently, it has also been proven that the control of conducting domain walls is possible in single crystals using, e.g., MgO-doped $LiNbO_3$, which is of particular interest due to its unusual optical properties [74]. Following the established method of calligraphic poling, where a macroscopic tip is moved across the surface while a voltage is applied, domains were oriented in a controlled way. As is shown in Fig. 2.29a, enhanced conductivity at the domain walls could be detected even when performing a linescan using the macroscopic tip used for poling. A more detailed analysis by C-AFM and PFM (Fig. 2.29b, c) revealed a direct correlation between the position of the domain wall and localized conductivity, also in this material. Due to the potential of $LiNbO_3$ for photonic applications, the observed effect can pave the

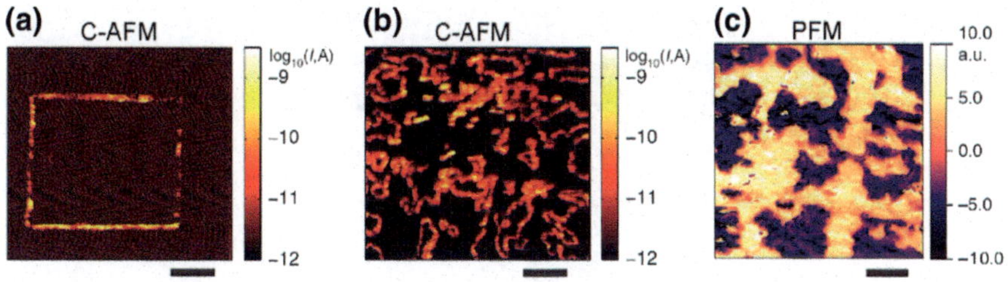

Fig. 2.28 C-AFM measurement of PZT films ($U = 1.9$ V). The image in (**a**) reveals a conducting domain wall written in the pristine film; the image in (**b**) shows a spontaneously formed network of conducting domain walls in the annealed film; while **c** depicts an out-of-plane PFM image from the annealed film. The spots on the film, where images in **b**–**c** were recorded, are different, but located close to each other. Scale bars: 280 nm (**a**); 400 nm (**b**–**c**). Adapted from Tselev et al. [73]

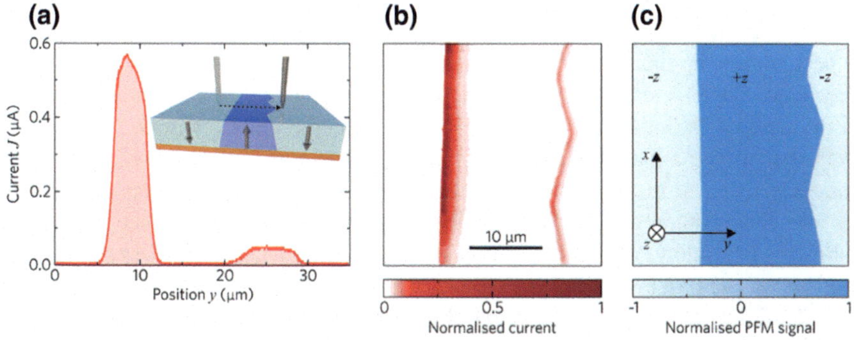

Fig. 2.29 **a** Current J versus tip position y when scanning across a domain line in a MgO-doped $LiNbO_3$ single crystal with a macroscopic tip ($U = 50$ V); **b** spatial distribution $J(x, y)$ obtained with a C-AFM probe; **c** corresponding PFM image identifying the $\pm z$ domains. Adapted from Werner et al. [74]

way to novel integrated electronic-optical devices employing calligraphically-written charged domain walls as electrodes inside the bulk, providing electrically-controlled local phase modulation.

It should be noted that, depending on the properties of the surface layer of the ferroelectric material, the evolution of domains can occur when being poled by applying a high voltage locally, either by a macroscopic tip as in calligraphic poling or by a C-AFM tip. Figure 2.30 shows an example of a PZT single crystal where, at elevated temperatures of 190 °C in vacuum conditions, an area of $4 \times 4\ \mu m^2$ has been scanned beforehand while a poling voltage was applied. Subsequently, a $10 \times 10\ \mu m^2$ scan was obtained in the same area in PFM, as well as in C-AFM mode. The topography map (Fig. 2.30a) shows that the initial crystal surface was relatively rough, with single grains or particles having heights of more than 100 nm. The poled area, however, showed a flat surface without any significant particles, indicating that during the poling scan, a transformation and removal of the material of the surface layer had taken place. The corresponding PFM out-of-plane amplitude map ($U_{ac} =$

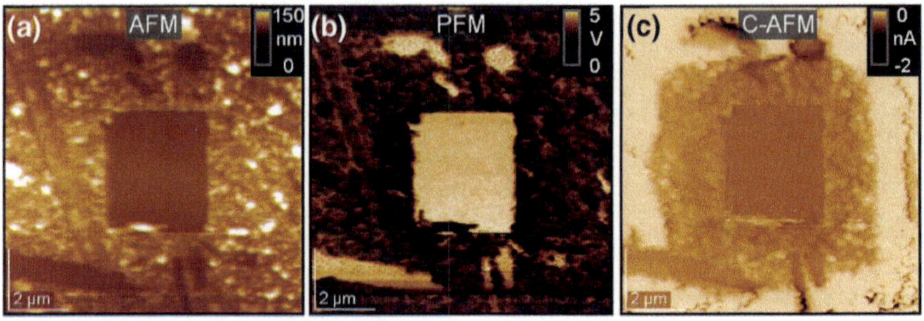

Fig. 2.30 Investigation of tip-induced switching on the surface of a PZT single crystal. **a** Topography; **b** PFM amplitude; and **c** C-AFM map obtained with a bias of −40 V

5 V, $f = 6$ kHz) proves that the poled area had changed its polarization state, as expected upon application of a high voltage (Fig. 2.30b). Surprisingly, the C-AFM map (Fig. 2.30c) does not reveal an increased conductivity at the domain walls, as in the case of the thin film (Fig. 2.28), but across the entire switched region and even the surface areas, several micrometres next to this region showed increased conductivity. Within the switched area, the conductivity is homogeneous, but in the previously unscanned region, an inhomogeneous distribution of conductivity following the grainy surface structure is present. This indicates that by applying high voltage via the C-AFM tip, distinct surface modifications can be induced, allowing for material processing on the nanoscale. Furthermore, the simultaneous presence of polarization and conductance switching could open up applications, combining ferroelectric and resistive switching for stable non-volatile memory applications.

2.5 Manipulation of the Conductivity by C-AFM

2.5.1 Tip-Induced Memristive Switching of Single Dislocations in $SrTiO_3$

As the local resistivity of high-k transition metal oxides can be easily manipulated by gradients of the chemical and electrical potential, they are promising candidates for non-volatile memory and logic applications based on the resistive switching effect. Using C-AFM, this effect has been intensively investigated on prototypic memristive cells, as well as on single crystals, offering insight into the basic mechanism of this effect. In general, it is now widely accepted that resistive switching is related to the generation and movement of oxygen vacancies within locally-confined regions of the transition metal oxide, thus forming switchable filaments. In order to investigate whether this effect is scalable, as is needed for building up highly integrated circuits, the localized nature of the resistive switching effect has been investigated on $SrTiO_3$

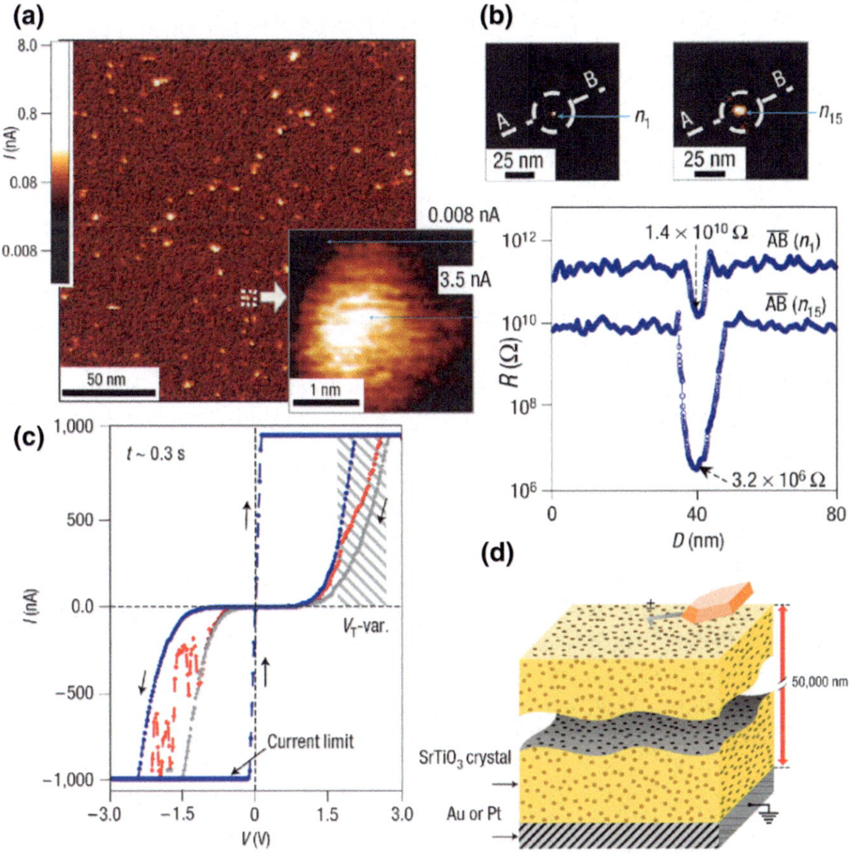

Fig. 2.31 **a** A conductivity map of the surface of a reduced and re-oxidized $SrTiO_3$ single crystal, as recorded by C-AFM. Inset: spot with a dimension of 1–2 nm corresponding to the core of a typical edge-type dislocation. **b** Conductivity maps and line scans across a selected spot (D denoting distance along AB) before and after application of a negative tip voltage bias; **c** I/V characteristics for different dislocations crossing the surface of a single crystal; **d** schematic illustration of the three-dimensional network of filaments present in the $SrTiO_3$ single crystal after thermal treatment. Adapted from Szot et al. [76]

single crystals as the traditional model material of functional electronic oxides [75]. Considering that dislocations in $SrTiO_3$ act as preferential reduction sites due to the lowered formation enthalpy of oxygen vacancies, as discussed above, it is unsurprising that dislocations also play a vital role in resistive switching phenomena. This has been confirmed by using the conducting tip of the C-AFM as a mobile nanoelectrode, as shown in Fig. 2.31.

2.5.2 *Creation of Conducting Nanowires on LAO/STO Structures*

As was already seen in Fig. 2.23, the heterostructures of polar $LaAlO_3$ and non-polar $SrTiO_3$ exhibit fascinating properties related to the evolution of a quasi-two-dimensional electron gas at the interface. It has been elaborated that such an electron gas is only present if the $LaAlO_3$ exceeds a critical thickness of several unit cells. However, if the $LaAlO_3$ layer is thinner (in the range of 3 unit cells), the electron gas becomes metastable, allowing for the manipulation of the conductance by the application of an electric field probably related to the generation of oxygen vacancies at the interface and a bi-stable insulator-metal transition can be tuned. Using C-AFM, this can be done at the nanoscale and a dedicated metallic pattern can be written within an insulating surrounding [78]. The schematic setup of such an experiment [77] is presented in Fig. 2.32. On top of an LAO/STO structure, two electrodes were deposited within a distance of several micrometres. Then, a conducting nanowire was written by C-AFM that connected the two electrodes. It can be seen that the conductance measured between the electrodes shows a steep increase as soon as the connection between the electrodes by the nanowire is established (Fig. 2.32a). This process can also be reversed by cutting the nanowire using opposite polarization and scanning with the C-AFM tip perpendicular to the nanowire. Here, the conductance reveals a steep drop as soon as the nanowire is cut (Fig. 2.32b).

By combining C-AFM writing with electric force microscopy (EFM), the conducting nanowires have been directly visualized [79]. In this experiment, shown in Fig. 2.33, two electrodes were written ($V_{\mathrm{write}} = +8$ V) on a $d_{\mathrm{LAO}} = 3$ uc. sample connecting to Al electrodes and an external circuit (Fig. 2.33a). Then, the electrodes were connected ($V_{\mathrm{write}} = +6$ V, Fig. 2.33b), resulting in a sharp conductance increase (label 1, Fig. 2.33e). Next, the conductive path was cut (Fig. 2.33c) by writing perpendicular to the wire ($V_{\mathrm{write}} = -3$ V), giving a sharp conductance decrease (label 2, Fig. 2.33e). This process is reproducible, and a reconnection was made writing vertically ($V_{\mathrm{write}} = +6$ V, label 3, Fig. 2.33e). Using the EFM phase images, the surface fields, corresponding to the conductive paths (dark contrast, reading tip voltage $V_{\mathrm{read}} = -2$ V) and the insulating paths (bright contrast), which directly correlate to the resistance changes, were imaged. These results indicate that the conductivity changes induced by C-AFM are due to the writing surface charge [80]. Writing metallic nanowires on thin LAO/STO structures reversibly is highly promising in order to 'sketch' electronic circuits with extremely small dimensions using only a C-AFM tip as an electrical pencil. For example, it has been demonstrated that single-electron transistors, in which single electrons tunnel resonantly between the source and drain electrodes through a conducting oxide island with a diameter of ~1.5 nm, can be realized providing insight into novel device functionalities based on quantum effects [81].

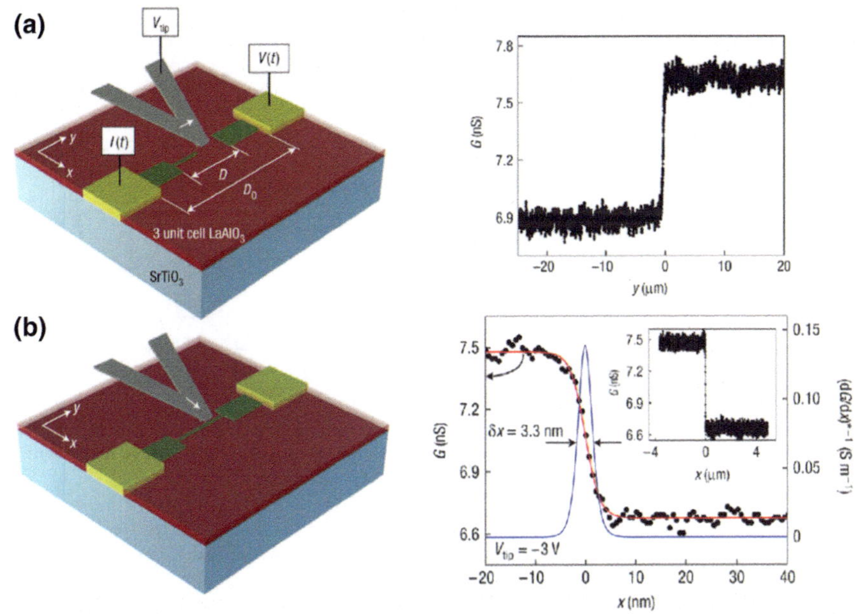

Fig. 2.32 **a** Schematic diagram of the experimental set-up for writing a conducting wire. A voltage-biased AFM tip is scanned from one electrode towards a second in contact mode and the conductance between the two electrodes measured as a function of the tip position while writing a conducting wire with 3 V bias applied to the tip; **b** schematic diagram of the experimental set-up for cutting a conducting wire by a negatively bias C-AFM tip and conductance between the two electrodes measured as a function of the tip position across the wire while cutting at −3 V. The inset shows the conductance measured over the entire 8 µm scan length. The deconvolved differential conductance $(dG/dx)^{*-1}$ shows a full-width at half-maximum $x = 3.3$ nm that is also plotted. Adapted from Cen et al. [77]

2.5.3 Electrical Nanopatterning of Oxide Surfaces

The ability of high-k metal oxides to exhibit the resistive switching effect related to local redox processes has triggered numerous investigations employing the C-AFM tip as mobile nanoelectrode in order to investigate the fundamental nature of the valence change mechanism. So far, many experiments have been performed on the prototype memristive material TiO_2, revealing the possibility of switching the local conductance state of the surface areas in the nanometer range [82–85]. Similar results have been obtained on other binary oxides, such as NiO [86]. In these experiments, at first, an area of the surface is scanned with a higher writing voltage applied in order to generate and move oxygen vacancies. Subsequently, the modified area is mapped with a lower reading voltage by a standard C-AFM scan. By combining these investigations by EFM or KPFM, information about the changes of the local surface potential induced by the C-AFM writing can be analysed. Figure 2.34 shows an example of the slightly reduced TiO_2 (110) single crystal surface [57] that had

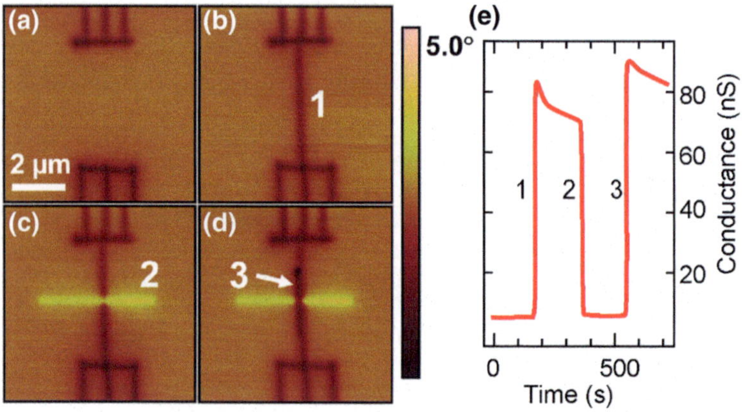

Fig. 2.33 Writing and erasing of conductive paths for d_{LAO} ~ 3 uc. EFM phase images with V_{read} = −2 V. **a** Dark contrast shows the electrodes connected to the outer Al wiring, V_{write} = +8 V; **b** the written wire connects the electrodes, V_{write} = +6 V, corresponding to (1) in (**e**); **c** the wire is cut (bright phase contrast) with V_{write} = −3 V, corresponding to (2) in (**e**); **d** electrodes reconnected with a vertical line using V_{write} = +6 V, corresponding to (3) in (**e**); **e** two-point conductance between the Al electrodes. Adapted from Xie et al. [79]

been reduced at 750 °C under UHV conditions before the switching experiment in order to remove contaminations and induce oxygen vacancies in the surface layer. As an illustration of the resistive switching capability of this oxide, a pattern consisting of 12 squares, having sizes of 200 × 200 nm^2, was written while a different tip voltage between −6 V and +6 V was applied to each square (Fig. 2.34a). Subsequently, the local conductivity and contact potential differences were measured by C-AFM and KPFM, respectively (Fig. 2.34a, b). It can be seen that upon writing with a positive tip voltage, the surface conductivity was increased by more than three orders of magnitude, while writing with a negative voltage resulted in a significant decrease in conductivity. The CPD measurements obtained from the same modified area may offer an explanation for this behaviour. Here, a writing with positive bias resulted in a decrease of CPD and vice versa, showing that by tuning the writing voltage, different resistance states can be induced, which illustrates the property of TiO_2 to allow for multilevel resistive switching, which is a prerequisite for neuromorphic applications.

It has also been demonstrated that the electrical nanopatterning is reversible by first switching an area to the ON state by scanning with +4 V applied and subsequently performing a smaller scan with −4 V applied in the center of the previous scan (Fig. 2.34d). To investigate the scaling limits of electrical nanopatterning, a binary checkerboard-like pattern with decreasing feature size was written where white color corresponds to a switching voltage of +4 V and black color to zero bias. A readout was performed at 100 mV, resulting in the current map shown on the bottom of Fig. 2.34e. It can be seen that while the 50 nm pattern can be unambiguously identified (line profile in Fig. 2.34f), an overlap of the ON and OFF areas occurs below a feature

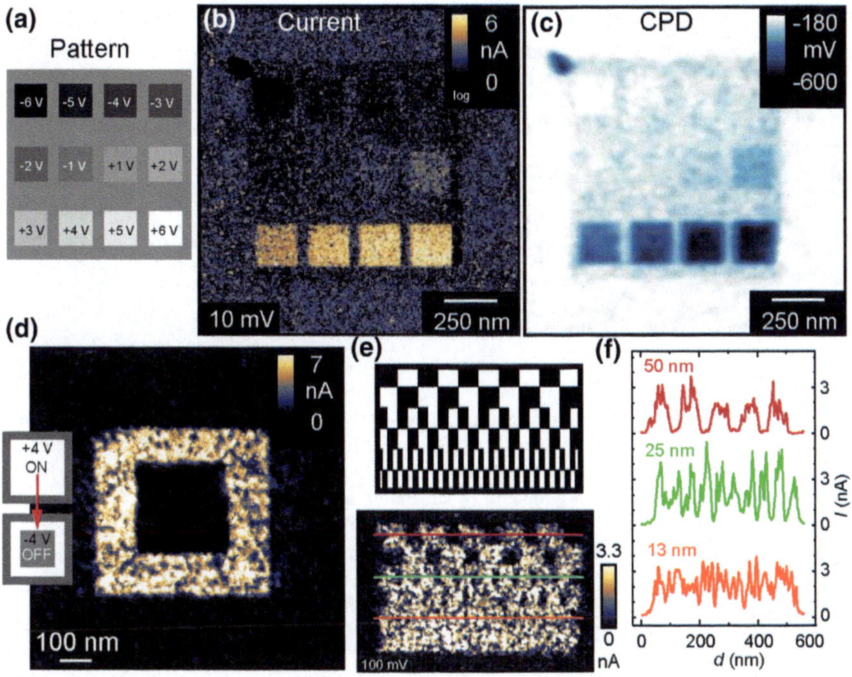

Fig. 2.34 Electrical tip-induced nanopatterning of the slightly reduced TiO_2 surface. A pattern (**a**) of squares with different voltages between −6 and +6 V was written and subsequently read by scanning with 10 mV bias; **b** the corresponding CPD map; **c** display changes in the surface potential upon nanopatterning; **d** current map (100 mV) after writing a 500 × 500 nm^2 ON pattern with +4 V and subsequently a 250 × 250 nm^2 OFF pattern with −4 V; **e** current map (bottom) of a region where a nanopattern (top) had been written with +4 V (white rectangles); the line profiles (**f**) were extracted from regions with three different feature sizes of the nanopattern. Adapted from Rodenbücher et al. [57]

size of 20 nm, which has been related to surface inhomogeneities such as clustered vacancies [57].

Finally, to illustrate that a nanopattern with a controlled shape can be written, an example from the reduced TiO_2 surface is shown in Fig. 2.35. Here, it has been reported that annealing in UHV at 1100 °C, followed by room temperature oxidation, results in the evolution of a quasi-homogeneous distribution of switchable nanofilaments related to linear arrangements of oxygen vacancies, interacting with initially present dislocations [20]. Hence, such a surface structure allows for writing conductive nanopatterns with high spatial resolution. It must be noted that the tip-induced nanopatterning is only an electrical modification, without changing the morphology of the crystal as provided by the topography map presented on the right of Fig. 2.35, which promises to use this effect for reliable data storage applications.

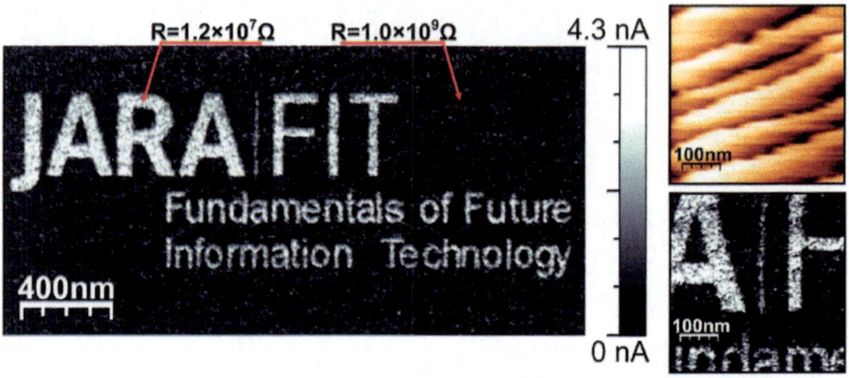

Fig. 2.35 LC-AFM conductivity map (sample tip polarization 50 mV) of TiO_2 previously patterned by selectively applying a low voltage ($U < 4$ V) in the scanned area. An enlarged section of the centre of the pattern is shown on the right, together with the corresponding topography map. Before patterning, the TiO_2 crystal was prepared by annealing at 1100 °C in UHV and further surface reoxidation by dosing 10^6 L of oxygen at room temperature. Adapted from Rogala et al. [20]

References

1. M.P. Murrell, M.E. Welland, S.J. O'Shea, T.M.H. Wong, J.R. Barnes, A.W. McKinnon, M. Heyns, S. Verhaverbeke, Spatially resolved electrical measurements of SiO_2 gate oxides using atomic force microscopy. Appl. Phys. Lett. **62**, 786–788 (1993). https://doi.org/10.1063/1.108579
2. M. Lanza, *Conductive Atomic Force Microscopy : Applications in Nanomaterials* (Wiley-VCH, 2017)
3. M. Deen, F. Pascal, Electrical characterization of semiconductor materials and devices—review. J. Mater. Sci.: Mater. Electron. **17**, 549–575 (2006). https://doi.org/10.1007/s10854-006-0001-8
4. C. Joachim, J.K. Gimzewski, A. Aviram, Electronics using hybrid-molecular and mono-molecular devices (2000)
5. D. Porath, A. Bezryadin, S. De Vries, C. Dekker, Direct measurement of electrical transport through DNA molecules. Nature **403**, 635–638 (2000). https://doi.org/10.1038/35001029
6. J.Y. Sze, B.K. Tay, C.I. Pakes, D.N. Jamieson, S. Prawer, Conducting Ni nanoparticles in an ion-modified polymer. J. Appl. Phys. **98** (2005). https://doi.org/10.1063/1.2014938
7. Z.L. Wang, J. Song, Piezoelectric nanogenerators based on zinc oxide nanowire arrays. Science (80-.). **312**, 242–246 (2006). https://doi.org/10.1126/science.1124005
8. M. Basletic, J.-L. Maurice, C. Carrétéro, G. Herranz, O. Copie, M. Bibes, É. Jacquet, K. Bouzehouane, S. Fusil, A. Barthélémy, Mapping the spatial distribution of charge carriers in $LaAlO_3/SrTiO_3$ heterostructures. Nat. Mater. **7**, 621–625 (2008). https://doi.org/10.1038/nmat2223
9. M. Lanza, Y. Wang, T. Gao, A. Bayerl, M. Porti, M. Nafria, Y. Zhou, G. Jing, Y. Zhang, Z. Liu, D. Yu, H. Duan, Electrical and mechanical performance of graphene sheets exposed to oxidative environments. Nano Res. **6**, 485–495 (2013). https://doi.org/10.1007/s12274-013-0326-6
10. R. Rapacz, K. Balin, M. Wojtyniak, J. Szade, Morphology and local conductance of single crystalline Bi_2Te_3 thin films on mica. Nanoscale **7**, 16034–16038 (2015). https://doi.org/10.1039/C5NR02551F

11. S.L. Howell, S. Padalkar, K. Yoon, Q. Li, D.D. Koleske, J.J. Wierer, G.T. Wang, L.J. Lauhon, Spatial mapping of efficiency of GaN/InGaN nanowire array solar cells using scanning photocurrent microscopy. Nano Lett. **13**, 5123–5128 (2013). https://doi.org/10.1021/nl402331u
12. M. Lorenz, M.S. Ramachandra Rao, T. Venkatesan, E. Fortunato, P. Barquinha, R. Branquinho, D. Salgueiro, R. Martins, E. Carlos, A. Liu, F.K. Shan, M. Grundmann, H. Boschker, J. Mukherjee, M. Priyadarshini, N. Dasgupta, D.J. Rogers, F.H. Teherani, E.V. Sandana, P. Bove, K. Rietwyk, A. Zaban, A. Veziridis, A. Weidenkaff, M. Muralidhar, M. Murakami, S. Abel, J. Fompeyrine, J. Zuniga-Perez, R. Ramesh, N.A. Spaldin, S. Ostanin, V. Borisov, I. Mertig, V. Lazenka, G. Srinivasan, W. Prellier, M. Uchida, M. Kawasaki, R. Pentcheva, P. Gegenwart, F. Miletto Granozio, J. Fontcuberta, N. Pryds, The 2016 oxide electronic materials and oxide interfaces roadmap (2016)
13. R. Waser, R. Dittmann, G. Staikov, K. Szot, Redox-based resistive switching memories-nanoionic mechanisms, prospects, and challenges. Adv. Mater. **21**, 2632–2663 (2009). https://doi.org/10.1002/adma.200900375
14. D. Ielmini, R. Waser, *Resistive Switching: From Fundamentals of Nanoionic Redox Processes to Memristive Device Applications*
15. J. Herion, R. Otto, K. Szot, European Patent EP0789840B1: Current/voltage transformer for the detection of the electron current in a scanning tunnelling-electron microscope, (1999). https://patents.google.com/patent/EP0789840B1/en
16. M. Wojtyniak, K. Szot, R. Waser, The thermal stability of Pt/Ir coated AFM tips for resistive switching measurements. Appl. Surf. Sci. **257**, 7627–7632 (2011). https://doi.org/10.1016/J.APSUSC.2011.03.149
17. A. Parkhi, T.S. Gross, Impact of I-V behavior and estimated temperature rise on surface and tip modification of the nanocontact between a highly doped silicon scanning probe microscope tip and gold surface under ambient conditions. J. Appl. Phys. **109**, 014323 (2011). https://doi.org/10.1063/1.3531545
18. O.G. Reid, K. Munechika, D.S. Ginger, Space charge limited current measurements on conjugated polymer films using conductive atomic force microscopy. Nano Lett. **8**, 1602–1609 (2008). https://doi.org/10.1021/nl080155l
19. H. Sugimura, N. Nakagiri, AFM lithography in constant current mode. Nanotechnology **8**, A15–A18 (1997). https://doi.org/10.1088/0957-4484/8/3A/004
20. M. Rogala, G. Bihlmayer, W. Speier, Z. Klusek, C. Rodenbücher, K. Szot, Resistive switching of a quasi-homogeneous distribution of filaments generated at heat-treated TiO_2 (110)-surfaces. Adv. Funct. Mater. **25**, 6382–6389 (2015). https://doi.org/10.1002/adfm.201500855
21. H.C. Day, D.R. Allee, Selective area oxidation of silicon with a scanning force microscope. Appl. Phys. Lett. **62**, 2691–2693 (1993). https://doi.org/10.1063/1.109259
22. J.M.R. Weaver, D.W. Abraham, High resolution atomic force microscopy potentiometry. J. Vac. Sci. Technol. B Microelectron. Nanom. Struct. **9**, 1559 (1991). https://doi.org/10.1116/1.585423
23. Tektronix: Low Level Measurements Handbook—7th Edition. (2014)
24. L. Libioulle, A. Radenovic, E. Bystrenova, G. Dietler, Low noise current-to-voltage converter and vibration damping system for a low-temperature ultrahigh vacuum scanning tunneling microscope. Rev. Sci. Instrum. **74**, 1016–1021 (2003). https://doi.org/10.1063/1.1533100
25. K. Szot, C. Rodenbücher, G. Bihlmayer, W. Speier, R. Ishikawa, N. Shibata, Y. Ikuhara, Influence of dislocations in transition metal oxides on selected physical and chemical properties. Crystals **8**, 241 (2018). https://doi.org/10.3390/cryst8060241
26. J.F. Witte, K.A.A. Makinwa, J.H. Huijsing, *Dynamic Offset Compensated CMOS Amplifiers* (2009)
27. M.S. Tyagi, Physics of Schottky barrier junctions, in *Metal-Semiconductor Schottky Barrier Junctions and Their Applications*, (Springer US, Boston, MA, 1984), pp. 1–60
28. R. Dalven, Physics of metal-semiconductor and metal-insulator-semiconductor junctions, in *Introduction to Applied Solid State Physics*, (Springer US, Boston, MA, 1980), pp. 109–125
29. J. Bardeen, Surface states and rectification at a metal semi-conductor contact. Phys. Rev. **71**, 717–727 (1947). https://doi.org/10.1103/physrev.71.717

30. U. Celano, T. Hantschel, G. Giammaria, R.C. Chintala, T. Conard, H. Bender, W. Vandervorst, Evaluation of the electrical contact area in contact-mode scanning probe microscopy. J. Appl. Phys. **117**, 214305 (2015). https://doi.org/10.1063/1.4921878
31. S. Kalinin, A. Gruverman, *Scanning Probe Microscopy* (Springer, New York, NY, 2007)
32. R. Holm, E. Holm, *Electric Contacts : Theory and Application* (Springer-Verlag, 1967)
33. M. Brandbyge, J. Schiøtz, M.R. Sørensen, P. Stoltze, K.W. Jacobsen, J.K. Nørskov, L. Olesen, E. Laegsgaard, I. Stensgaard, F. Besenbacher, Quantized conductance in atom-sized wires between two metals. Phys. Rev. B. **52**, 8499–8514 (1995). https://doi.org/10.1103/PhysRevB.52.8499
34. M. Büttiker, Y. Imry, R. Landauer, S. Pinhas, Generalized many-channel conductance formula with application to small rings. Phys. Rev. B. **31**, 6207–6215 (1985). https://doi.org/10.1103/PhysRevB.31.6207
35. K. Szot, B. Reichenberg, F. Peter, R. Waser, S. Tiedke, Electrical characterization of perovskite nanostructures by SPM, in *Scanning Probe Microscopy*. (Springer, New York, NY, 2007), pp. 746–775
36. M. Enachescu, D. Schleef, D.F. Ogletree, M. Salmeron, Integration of point-contact microscopy and atomic-force microscopy: application to characterization of graphite/Pt (111). Phys. Rev. B - Condens. Matter Mater. Phys. **60**, 16913–16919 (1999). https://doi.org/10.1103/physrevb.60.16913
37. C. Rodenbücher, G. Bihlmayer, W. Speier, J. Kubacki, M. Wojtyniak, M. Rogala, D. Wrana, F. Krok, K. Szot, Local surface conductivity of transition metal oxides mapped with true atomic resolution. Nanoscale (2018). https://doi.org/10.1039/c8nr02562b
38. P. Bampoulis, K. Sotthewes, M.H. Siekman, H.J.W. Zandvliet, Local conduction in $Mo_xW_{1-x}Se_2$: the role of stacking faults, defects, and alloying. ACS Appl. Mater. Interfaces (2018). https://doi.org/10.1021/acsami.8b01506
39. G.E. Moore, Cramming more components onto integrated circuits. Proc. IEEE **86**, 82–85 (1998). https://doi.org/10.1109/JPROC.1998.658762
40. M.M. Waldrop, The chips are down for Moore's law. Nature **530**, 144–147 (2016). https://doi.org/10.1038/530144a
41. I.L. Markov, Limits on fundamental limits to computation. Nature **512**, 147–154 (2014). https://doi.org/10.1038/nature13570
42. J. Robertson, High dielectric constant oxides. Eur. Phys. J. Appl. Phys. **28**, 265–291 (2004). https://doi.org/10.1051/epjap:2004206
43. H.E. Weaver, Dielectric properties of single crystals of $SrTiO_3$ at low temperatures. J. Phys. Chem. Solids **11**, 274–277 (1959). https://doi.org/10.1016/0022-3697(59)90226-4
44. H. Lee, Y.S. Kang, S.-J. Cho, B. Xiao, H. Morkoç, T.D. Kang, G.S. Lee, J. Li, S.-H. Wei, P.G. Snyder, J.T. Evans, Dielectric functions and electronic band structure of lead zirconate titanate thin films. J. Appl. Phys. **98**, 094108 (2005). https://doi.org/10.1063/1.2128043
45. T.S. Low, W. Guo, Modeling of a three-layer piezoelectric bimorph beam with hysteresis. J. Microelectromech. Syst. **4**, 230–237 (1995). https://doi.org/10.1109/84.475550
46. K. Suzuki, K. Kijima, optical band gap of barium titanate nanoparticles prepared by RF-plasma chemical vapor deposition. Jpn. J. Appl. Phys. **44**, 2081–2082 (2005). https://doi.org/10.1143/JJAP.44.2081
47. C.A. Miller, Hysteresis loss and dielectric constant in barium titanate. Br. J. Appl. Phys. **18**, 1689–1697 (1967). https://doi.org/10.1088/0508-3443/18/12/303
48. J.D. Plummer, P.B. Griffin, Material and process limits in silicon VLSI technology. Proc. IEEE **89**, 240–258 (2001). https://doi.org/10.1109/5.915373
49. R. Waser, *Nanoelectronics and Information Technology* (2012)
50. C. Rodenbücher, M. Luysberg, A. Schwedt, V. Havel, F. Gunkel, J. Mayer, R. Waser, Homogeneity and variation of donor doping in Verneuil-grown SrTiO3: Nb single crystals. Sci. Rep. **6**, 32250 (2016)
51. C. Rodenbücher, T. Gensch, W. Speier, U. Breuer, M. Pilch, H. Hardtdegen, M. Mikulics, E. Zych, R. Waser, K. Szot, Inhomogeneity of donor doping in $SrTiO_3$ substrates studied by fluorescence-lifetime imaging microscopy. Appl. Phys. Lett. **103**, 162904 (2013)

52. A. Yanguas-Gil, Fundamentals of gas phase transport in nanostructured materials, in *Growth and Transport in Nanostructured Materials: Reactive Transport in PVD, CVD, and ALD* (2017), pp. 39–67
53. J. Rodríguez Contreras, H. Kohlstedt, A. Petraru, A. Gerber, B. Hermanns, H. Haselier, N. Nagarajan, J. Schubert, U. Poppe, C. Buchal, R. Waser, Improved $PbZr_{0.52}Ti_{0.48}O_3$ film quality on $SrRuO_3/SrTiO_3$ substrates. J. Cryst. Growth. **277**, 210–217 (2005). https://doi.org/10.1016/j.jcrysgro.2004.12.137
54. V. Iglesias, M. Porti, M. Nafría, X. Aymerich, P. Dudek, G. Bersuker, Dielectric breakdown in polycrystalline hafnium oxide gate dielectrics investigated by conductive atomic force microscopy. J. Vac. Sci. Technol. B, Nanotechnol. Microelectron. Mater. Process. Meas. Phenom. **29**, 01AB02 (2011). https://doi.org/10.1116/1.3532945
55. U. Celano, R.C. Chintala, C. Adelmann, O. Richard, W. Vandervorst, The unexpected effects of crystallization on Ta_2O_5 as studied by HRTEM and C-AFM. Microelectron. Eng. **109**, 318–321 (2013). https://doi.org/10.1016/j.mee.2013.03.147
56. D. Preziosi, L. Lopez-Mir, X. Li, T. Cornelissen, J.H. Lee, F. Trier, K. Bouzehouane, S. Valencia, A. Gloter, A. Barthélémy, M. Bibes, Direct mapping of phase separation across the metal-insulator transition of NdNiO3. Nano Lett. **18**, 2226–2232 (2018). https://doi.org/10.1021/acs.nanolett.7b04728
57. C. Rodenbücher, D. Wrana, P. Meuffels, M. Rogala, F. Krok, K. Szot, Electrical nanopatterning of TiO_2 single crystal surfaces in situ via local resistance and potential switching. APL Mater. (2018)
58. D.M. Schaadt, E.T. Yu, V. Vaithyanathan, D.G. Schlom, Nanoscale current transport in epitaxial $SrTiO_3$ on n + -Si investigated with conductive atomic force microscopy. J. Vac. Sci. Technol. B Microelectron. Nanom. Struct. **22**, 2030 (2004). https://doi.org/10.1116/1.1768529
59. D. Wrana, C. Rodenbücher, W. Bełza, K. Szot, F. Krok, In situ study of redox processes on the surface of $SrTiO_3$ single crystals. Appl. Surf. Sci. **432** (2018). https://doi.org/10.1016/j.apsusc.2017.06.272
60. M. Lanza, Mario: a review on resistive switching in high-*k* dielectrics: a nanoscale point of view using conductive atomic force microscope. Mater. Basel **7**, 2155–2182 (2014). https://doi.org/10.3390/ma7032155
61. K. Szot, G. Bihlmayer, W. Speier, Nature of the resistive switching phenomena in TiO_2 and $SrTiO_3$: origin of the reversible insulator–metal transition, in *Solid State Physics,* Vol 65, eds. by R.E. Camley, R.L. Stamps (Academic Press, 2014), pp. 353–559
62. D. Wrana, C. Rodenbücher, M. Krawiec, B.R. Jany, J. Rysz, M. Ermrich, K. Szot, F. Krok, Tuning the surface structure and conductivity of niobium-doped rutile TiO_2 single crystals via thermal reduction. Phys. Chem. Chem. Phys. **19**, 30339–30350 (2017). https://doi.org/10.1039/C7CP03136J
63. D. Marrocchelli, L. Sun, B. Yildiz, Dislocations in $SrTiO_3$: easy to reduce but not so fast for oxygen transport. J. Am. Chem. Soc. **137**, 4735–4748 (2015). https://doi.org/10.1021/ja513176u
64. A. Ohtomo, H.Y. Hwang, A high-mobility electron gas at the LAO/STO heterointerface. Nature **427**, 423–426 (2004). https://doi.org/10.1038/nature04773
65. G. Herranz, M. Basletić, M. Bibes, C. Carrétéro, E. Tafra, E. Jacquet, K. Bouzehouane, C. Deranlot, A. Hamzić, J.-M. Broto, A. Barthélémy, A. Fert, High Mobility in $LaAlO_3/SrTiO_3$ heterostructures: origin, dimensionality, and perspectives. Phys. Rev. Lett. **98**, 216803 (2007). https://doi.org/10.1103/physrevlett.98.216803
66. M. Rogala, Z. Klusek, C. Rodenbücher, R. Waser, K. Szot, Quasi-two-dimensional conducting layer on TiO_2 (110) introduced by sputtering as a template for resistive switching. Appl. Phys. Lett. **102**, 131604 (2013). https://doi.org/10.1063/1.4801437
67. E. Frantzeskakis, T.C. Rödel, F. Fortuna, A.F. Santander-Syro, 2D surprises at the surface of 3D materials: Confined electron systems in transition metal oxides. J. Electron. Spectros. Relat. Phenomena. **219**, 16–28 (2017). https://doi.org/10.1016/j.elspec.2016.10.001
68. A.F. Santander-Syro, O. Copie, T. Kondo, F. Fortuna, S. Pailhés, R. Weht, X.G. Qiu, F. Bertran, A. Nicolaou, A. Taleb-Ibrahimi, P. Le Évre, G. Herranz, M. Bibes, N. Reyren, Y. Apertet, P. Lecoeur, A. Barthélémy, M.J. Rozenberg, Two-dimensional electron gas with universal

subbands at the surface of $SrTiO_3$. Nature. **469**, 189–194 (2011). https://doi.org/10.1038/nature09720
69. R. Wang, S.M. Shapiro, Structural defects and the origin of the second length scale in $SrTiO_3$. Phys. Rev. Lett. **80**, 2370–2373 (1998). https://doi.org/10.1103/physrevlett.80.2370
70. P. Gao, R. Ishikawa, B. Feng, A. Kumamoto, N. Shibata, Y. Ikuhara, Atomic-scale structure relaxation, chemistry and charge distribution of dislocation cores in $SrTiO_3$. Ultramicroscopy. **184**, 217–224 (2018). https://doi.org/10.1016/j.ultramic.2017.09.006
71. J. Seidel, P. Maksymovych, Y. Batra, A. Katan, S.-Y. Yang, Q. He, A.P. Baddorf, S.V. Kalinin, C.-H. Yang, J.-C. Yang, Y.-H. Chu, E.K.H. Salje, H. Wormeester, M. Salmeron, R. Ramesh, Domain wall conductivity in La-Doped $BiFeO_3$. Phys. Rev. Lett. **105**, 197603 (2010). https://doi.org/10.1103/physrevlett.105.197603
72. B.M. Vul, G.M. Guro, I.I. Ivanchik, Encountering domains in ferroelectrics. Ferroelectrics **6**, 29–31 (1973). https://doi.org/10.1080/00150197308237691
73. A. Tselev, P. Yu, Y. Cao, L.R. Dedon, L.W. Martin, S.V. Kalinin, P. Maksymovych, Microwave a.c. conductivity of domain walls in ferroelectric thin films. Nat. Commun. **7**, 11630 (2016). https://doi.org/10.1038/ncomms11630
74. C.S. Werner, S.J. Herr, K. Buse, B. Sturman, E. Soergel, C. Razzaghi, I. Breunig, Large and accessible conductivity of charged domain walls in lithium niobate. Sci. Rep. **7**, 9862 (2017). https://doi.org/10.1038/s41598-017-09703-2
75. C. Rodenbücher, P. Meuffels, W. Speier, M. Ermrich, D. Wrana, F. Krok, K. Szot, Stability and Decomposition of Perovskite-Type Titanates upon High-Temperature Reduction. Phys. Status Solidi—Rapid Res. Lett. **11** (2017). https://doi.org/10.1002/pssr.201700222
76. K. Szot, W. Speier, G. Bihlmayer, R. Waser, Switching the electrical resistance of individual dislocations in single-crystalline $SrTiO_3$. Nat. Mater. **5**, 312–320 (2006). https://doi.org/10.1038/nmat1614
77. C. Cen, S. Thiel, G. Hammerl, C.W. Schneider, K.E. Andersen, C.S. Hellberg, J. Mannhart, J. Levy, Nanoscale control of an interfacial metal-insulator transition at room temperature. Nat. Mater. **7**, 298–302 (2008). https://doi.org/10.1038/nmat2136
78. M. Boselli, D. Li, W. Liu, A. Fête, S. Gariglio, J.M. Triscone, Characterization of atomic force microscopy written conducting nanowires at $LaAlO_3/SrTiO_3$interfaces. Appl. Phys. Lett. **108** (2016). https://doi.org/10.1063/1.4941817
79. Y. Xie, C. Bell, T. Yajima, Y. Hikita, H.Y. Hwang, Charge writing at the $LaAlO_3/SrTiO_3$ surface. Nano Lett. **10**, 2588–2591 (2010). https://doi.org/10.1021/nl1012695
80. Y. Xie, C. Bell, Y. Hikita, H.Y. Hwang, Tuning the electron gas at an oxide heterointerface via free surface charges. Adv. Mater. **23**, 1744–1747 (2011). https://doi.org/10.1002/adma.201004673
81. G. Cheng, P.F. Siles, F. Bi, C. Cen, D.F. Bogorin, C.W. Bark, C.M. Folkman, J.W. Park, C.B. Eom, G. Medeiros-Ribeiro, J. Levy, Sketched oxide single-electron transistor. Nat. Nanotechnol. **6**, 343–347 (2011). https://doi.org/10.1038/nnano.2011.56
82. M.H. Lee, C.S. Hwang, Resistive switching memory: observations with scanning probe microscopy. Nanoscale **3**, 490–502 (2011). https://doi.org/10.1039/C0NR00580K
83. Y. Du, A. Kumar, H. Pan, K. Zeng, S. Wang, P. Yang, A.T.S. Wee, The resistive switching in TiO_2 films studied by conductive atomic force microscopy and Kelvin probe force microscopy. AIP Adv. **3**, 1–8 (2013). https://doi.org/10.1063/1.4818119
84. P. Bousoulas, J. Giannopoulos, K. Giannakopoulos, P. Dimitrakis, D. Tsoukalas, Memory programming of TiO_{2-x} films by conductive atomic force microscopy evidencing filamentary resistive switching. Appl. Surf. Sci. **332**, 55–61 (2015). https://doi.org/10.1016/j.apsusc.2015.01.133
85. W. Lu, L.-M. Wong, S. Wang, K. Zeng, Probing electrochemically induced resistive switching of TiO_2 using SPM techniques. Phys. Chem. Chem. Phys. **19**, 31399–31409 (2017). https://doi.org/10.1039/C7CP06992H
86. M.H. Lee, S.J. Song, K.M. Kim, G.H. Kim, J.Y. Seok, J.H. Yoon, C.S. Hwang, Scanning probe based observation of bipolar resistive switching NiO films. Appl. Phys. Lett. **97** (2010). https://doi.org/10.1063/1.3479526

Chapter 3
Mapping Conductance and Carrier Distributions in Confined Three-Dimensional Transistor Structures

Andreas Schulze, Pierre Eyben, Jay Mody, Kristof Paredis, Lennaert Wouters, Umberto Celano and Wilfried Vandervorst

Abstract Probing the distribution of charge carriers in semiconductor device structures is of crucial importance to better understand semiconductor fabrication processes and how they affect the incorporation, diffusion and activation of dopants and hence the final device performance. Scanning spreading resistance microscopy (SSRM) has emerged as the most valuable technique for 2D and 3D carrier mapping in semiconductor device structures due to its excellent spatial resolution, sensitivity and ease of quantification. The present chapter first introduces the principles of the technique, thereby discussing the underlying physical mechanisms such as the nanometer-size probe-semiconductor contact. Faced with the stringent requirements imposed by advanced 3D device architectures, novel approaches and concepts such as 3D carrier profiling and fast Fourier transform-SSRM (FFT-SSRM) have been

A. Schulze (✉) · P. Eyben · K. Paredis · L. Wouters · U. Celano · W. Vandervorst
IMEC, Kapeldreef 75, 3001 Leuven, Belgium
e-mail: andreas_schulze@amat.com

P. Eyben
e-mail: pierre.eyben@imec.be

J. Mody
e-mail: jay.mody@globalfoundries.com

K. Paredis
e-mail: kristof.paredis@imec.be

L. Wouters
e-mail: lennaert.wouters@imec.be

U. Celano
e-mail: celano@imec.be

W. Vandervorst
Department of Physics and Astronomy, KU Leuven, 3001 Leuven, Belgium
e-mail: Wilfried.Vandervorst@imec.be

A. Schulze
Present address:
Applied Materials, 3340 Scott Boulevard, Santa Clara, CA 95054, USA

J. Mody
Present address:
Globalfoundries, 400 Stone Break Rd Extension, Malta 12020, USA

U. Celano (ed.), *Electrical Atomic Force Microscopy for Nanoelectronics*,
NanoScience and Technology, https://doi.org/10.1007/978-3-030-15612-1_3

developed in the recent years. These methods aid in extending conventional SSRM toward quantitative carrier profiling in aggressively scaled 3D device structures which is illustrated on the example of selected relevant applications such as FinFETs and nanowire-based transistors.

3.1 Introduction: The Fundamentals of SSRM

3.1.1 Basic Principles

Scanning spreading resistance microscopy (SSRM) is an atomic force microscopy (AFM)-based technique which is widely used by research institutes as well as the semiconductor industry to determine the distribution of charge carriers in semiconductor structures and devices in two and even three dimensions with unrivaled spatial resolution and sensitivity. SSRM has been conceived by Vandervorst et al. [1, 2] and was first implemented by De Wolf et al. [3, 4] at imec starting in 1994.

The measurement principle is based on the spreading resistance probe (SRP), which was used for one-dimensional carrier profiling for several decades [5]. SSRM extends the capabilities of the classical SRP thereby enabling the in-depth study of carrier distributions with excellent spatial resolution in all three dimensions. This leap forward was enabled largely by the invention of the atomic force microscope (AFM) in 1982 [6] which represents the basic building block of the technique. In SSRM, a conductive probe is scanned in contact mode across the cross-section of the device of interest while a DC bias is applied between the scanning probe and an electrical back-contact on the backside of the sample (Fig. 3.1).

Due to the small apex of the probe, the electrical current experiences a geometrical constriction and then a spreading when passing from the tiny probe into the (large) sample volume. This current spreading is associated with an electrical resistance typically termed *Spreading Resistance*. The spreading resistance for a flat circular contact of radius a can be derived by solving the Laplace equation for the electrostatic potential [7]

Fig. 3.1 SSRM setup and measurement principle. A conductive probe (A) is scanned in contact mode across the sample/device cross-section (B) while a bias is being applied between the scanning probe (A) and a sample back-contact (C). The resulting current is amplified by means of a logarithmic current amplifier

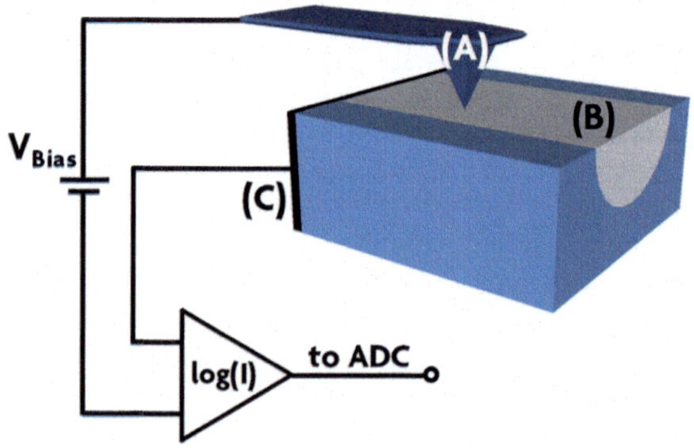

$$\nabla^2 V = \frac{\partial^2 V}{\partial r^2} + \frac{1}{r}\frac{\partial V}{\partial r} + \frac{\partial^2 V}{\partial z^2} = 0 \tag{3.1}$$

and yields

$$R_{\mathrm{sp}} = \frac{\rho}{4a} \tag{3.2}$$

where ρ corresponds to the local resistivity underneath the probe.

From (3.2) it becomes apparent that the spreading resistance is directly proportional to the local resistivity ρ with a sensitivity equal to one on a logarithmic scale, i.e. the spreading resistance varies one order of magnitude if the local resistivity changes by one order of magnitude. With other words, spreading resistance measurements and hence SSRM offer a unique sensitivity toward changes in carrier density and by using a logarithmic current amplifier it is possible to cover the entire carrier concentration range of interest (1e15..1e21) cm^{-3} providing simultaneously the required dynamic range. The local net carrier concentration $|n - p|$ can then be derived using dopant dependent mobility models [8, 9]:

$$|n - p| = \frac{1}{\rho\mu e}. \tag{3.3}$$

Due to the nanometer-scale dimensions of the electrical contact, the drop of the electrostatic potential occurs mainly in close proximity of the scanning probe. Hence, the electric field (Fig. 3.2) between probe and back-contact is not constant but drops sharply within the first few nanometers around the small contact. The spatial extent of this region characterized by the decay of the electric field defines the probing depth and lateral resolution of the technique. By establishing a small contact as indicted in Fig. 3.2a, measurements with a lateral resolution in the nanometer-range can be realized easily.

The unique combination of excellent sensitivity, dynamic range and spatial resolution represents the key differentiator of SSRM and is the reason why the technique outperforms all other carrier profiling methodologies such as scanning capacitance

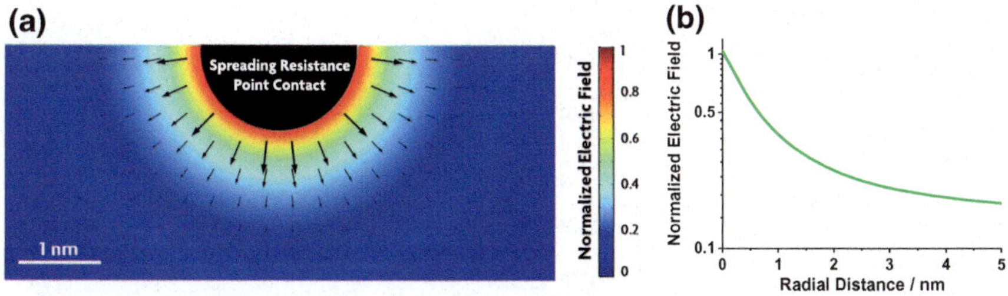

Fig. 3.2 **a** Two-dimensional distribution of the electric field in a Si sample upon applying a bias between the semi-hemispherical nanoscale contact (radius of 1 nm) and a back-contact placed at a distance of several micrometers. The arrows indicate the direction of the electric field and hence current flow. **b** One-dimensional section exemplifying the fast decay of the electric field in close proximity to the nanoscale contact

microscopy (also AFM-based) and electron holography (TEM-based) for most applications.

3.1.2 *Physics of the Nanoscale SSRM Contact*

The spreading resistance described by (3.2) holds only for a perfectly ohmic electrical contact between the scanning probe and the sample under test without any additional parasitic contact resistance at the interface. One can easily imagine, however, that this requirement is not fulfilled per se for a semiconductor substrate (potentially covered with a native insulating oxide layer) and a scanning probe made of doped diamond or metal. In order to limit the impact of parasitic contact resistances, SSRM measurements on Si and Ge samples are performed at a high contact force leading to a metallic-like phase underneath the AFM tip which acts as a virtual probe forming a Schottky-like contact with the semiconductor sample. More details on this phase transformation and the resulting Schottky contact are provided in the following two sections. Moreover, the electrical characteristics of the SSRM contact are influenced by surface states distorting the band structure and carrier density on the device cross-section which is being analyzed by the AFM tip. This phenomenon is discussed in more detail in Sect. 3.1.2.3.

3.1.2.1 High-Pressure Phase Transformation

The sensitivity of SSRM toward variations in the sample's net carrier concentration critically relies on the condition that the spreading resistance dominates any potential parasitic series resistance. At a low contact force, the interfacial contact resistance is dominant. Since the latter exhibits only a weak dependence on the underlying carrier concentration, the response function of the measurement (i.e. measured resistance vs. sample resistivity) remains rather flat (dashed line in Fig. 3.3).

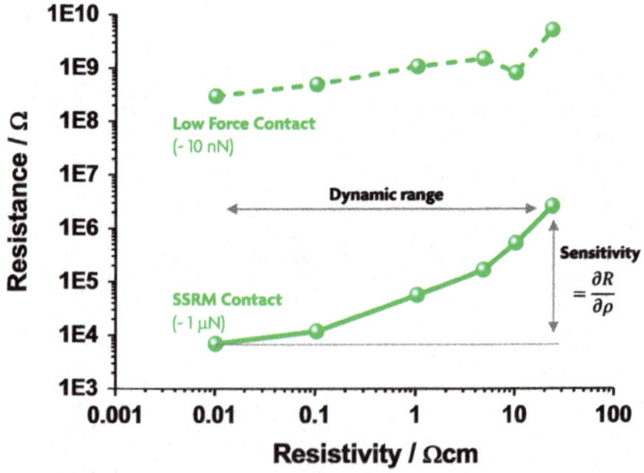

Fig. 3.3 Resistance measured for different sample resistivities (i.e. carrier concentrations) with low respectively high load force applied to the scanning probe. A sufficiently high load force triggers the formation of a metallic-like β-Sn phase at which point the spreading resistance dominates any interfacial contact resistances

The picture changes significantly once the pressure underneath the probe reaches the GPa regime. High pressure experiments have shown that Si undergoes a phase transformation from the diamond cubic structure to the β-Sn phase at moderately elevated pressures [10–13]. The β-Sn phase is metallic in nature and exhibits a body centered tetragonal configuration with a coordination number of six (inset Fig. 3.4). It is gradually formed due to the relative sliding between atoms along the compressive direction during indentation [14]. For such a phase transformation from the semiconducting to the metallic state a stress between 11.3 and 12.5 GPa (assuming pure hydrostatic conditions) is typically required [12]. However, shear stress components can decrease this value to about 8 GPa [10].

Similar to the Si case, Ge samples undergo a phase transformation toward a metallic-like β-Sn phase for sufficiently large pressures as well. Moreover, by means of molecular dynamics simulations and DFT calculations Lu et al. demonstrated that the pressure induced phase transformation leads to a zero bandgap region (β-Sn pocket) as well as a transition region of reduced bandgap between the β-Sn pocket and the Ge substrate [15].

The formation of such a metallic-like β-Sn phase is evident from the so-called resistance-force curves which reflect the resistance between probe and sample back-contact as a function of the force applied to the scanning probe. At a certain threshold force the resistance drops significantly (Fig. 3.4) which can be attributed to the formation of the β-Sn pocket underneath the probe. A further force increase (above 1 μN) leads to a slower continuous decay of the resistance, consistent with the assumed increase in contact area due to the larger indentation depth. SSRM measurements are typically carried out in the range between (1..2) μN which translates into the required pressure given the nanometer dimensions of the tip-sample contact area.

From Fig. 3.3 it becomes apparent that the sensitivity (i.e. slope of the response curve) is greatly enhanced when measuring at increased force due to the formation of

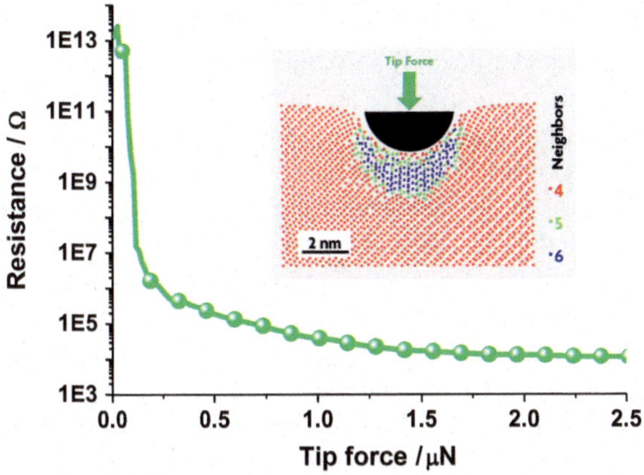

Fig. 3.4 Resistance as a function of force applied to the SSRM probe for a *p*-type (5e19 cm^{-3}) Si sample. The resistance drop is due to a local transformation of the diamond cubic crystal lattice toward the metallic-like β-Sn phase. A bias of +500 mV was applied between tip and sample. The inset shows in cross-sectional view the number of neighbors of each Si atom in proximity of the contact upon indentation as obtained by molecular dynamics simulations

the β-Sn pocket which aids in suppressing the (nearly) carrier density independent interfacial contact resistance. Nevertheless, the obtained response function clearly deviates from the linear relationship between measured resistance R and sample resistivity ρ one would expect based on (3.2). The deviation in the highly resistive regime is caused by distortions in the carrier density due to the Schottky-like nature of the β-Sn-semiconductor interface and the presence of surface states near the contact. The parasitic probe resistance on the other hand is responsible for the reduced slope in the low resistivity regime.

3.1.2.2 Schottky Contact

Given the existence of a metallic-like β-Sn pocket acting as a virtual probe and a transition region characterized by a reduced band gap, the overall SSRM contact can be understood as a series connection of the scanning probe, the metallic-like β-Sn pocket and a Schottky diode formed between the β-Sn pocket and the underlying semiconductor sample [7]. Through a systematic study of IV curves acquired on n- and p-type doped samples of different carrier concentration, the work function and thus the position of the Fermi level of the β-Sn pocket can be deduced qualitatively.

In the case of Si samples, Eyben et al. observed a strongly rectifying behavior for lowly doped p-type samples, and a weakly rectifying contact for lowly doped n-type samples [16, 17]. This means that the Fermi level of the β-Sn pocket must be located above mid-bandgap, leading to a situation where for any p-type doping level the work function of the β-Sn pocket is significantly smaller than the work function of the semiconductor, equivalent with a rectifying Schottky contact [18]. Given the weakly rectifying behavior observed on lowly n-type doped Si, Eyben et al. chose a β-Sn work function equal to 4.4 eV. In case of lowly n-type doped samples this leads to a situation where the semiconductor work function is slightly smaller than the work function of the β-Sn pocket. For highly doped samples (p- and n-type), charge carriers can tunnel efficiently through the Schottky tunneling barrier leading to a symmetric IV curve respectively an ohmic-like contact. The latter is in good agreement with experimental observations [16, 17].

For Ge samples, a comprehensive model of the electrical nanocontact has been established by Schulze et al. [19]. Experimentally obtained IV-curves for p- and n-type Ge of different doping levels are shown in Fig. 3.5.

The IV-curves for p-type Ge (purple) exhibit a symmetric behavior over a wide range of carrier concentrations, whereas in case of n-type Ge (green) a more rectifying behavior with reduced current for positive sample bias is identified at lower carrier concentrations. Although this observation is opposite to the Si case (see above), it confirms the Schottky-like nature of the nanocontact between the virtual probe (=β-Sn phase) and the Ge sample. For lowly doped p-type Ge, the symmetry of the IV-curve cannot be explained by tunneling through the Schottky barrier in reversed bias due to the weak band bending and hence wide tunneling barrier at the β-Sn - Ge interface. Hence, this symmetry indicates that the work function of the β-Sn phase must be larger than that of Ge at this particular doping level [18]. Given a

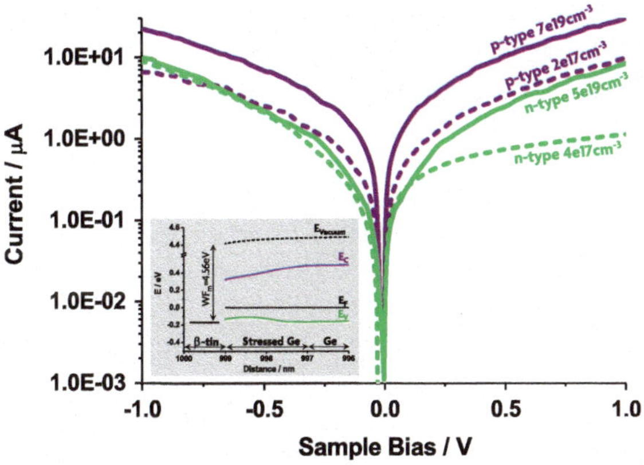

Fig. 3.5 Point contact IV-curves for p- and n-type Ge of different doping level recorded using full-diamond tip (FDT) probes. The inset shows the band structure along a radial section through the β-Sn pocket, Schottky diode into bulk Ge

location of the Fermi level in the semiconductor approximately 100 meV above the valence band edge E_V (for a carrier concentration of 2e17 cm^{-3}), the Fermi level of the β-Sn phase is expected in that range or closer to E_V. For n-type Ge (Fermi level always located above mid-band gap), this implies inevitably a situation where the work function of the β-Sn phase is larger than that of the semiconductor. Such a configuration represents a rectifying Schottky contact [18] which is in agreement with the experimentally obtained IV curve for lowly n-type doped Ge (Fig. 3.5). At high doping levels, charge carriers can tunnel efficiently through the Schottky barrier leading to a symmetric IV-curve. In order to fulfill the requirements put forward by the (non-)linearity of the n-type and p-type IV-curves respectively, Schulze et al. proposed a β-Sn work function of 4.56 eV, i.e. aligned to the valence band edge E_V of the compressed Ge region characterized by a reduced bandgap (inset Fig. 3.5) [19]. The experimentally observed steep onset of the IV-curve obtained on lowly doped n-type Ge (Fig. 3.5) has been explained by trap-assisted tunneling (TAT) via probe-generated defects [19]. TAT is a phenomenon whereby electrons cross a forbidden bandgap via tunneling and possibly thermal excitation processes via defects (traps creating discrete allowed energy levels inside the forbidden bandgap) [20]. TAT typically starts at a lower electric field than barrier tunneling and since the defect density is normally smaller than the atomic density, the total TAT current at high electric fields (high sample bias) remains smaller than the total barrier tunneling current. For n-type Ge, the electrons tunnel from the virtual probe (β-Sn) through the Schottky barrier into the semiconductor sample (vice versa for small negative bias).

3.1.2.3 Surface States

So far we did not consider the impact of dangling bonds and interface/surface states on the carrier concentration in the near-surface region and hence on the measurement. However, it is important to realize that SSRM measurements are typically carried out on cross-sections characterized by dangling bonds which are introduced during (1) sample preparation and (2) the high-force measurement itself. The dangling bonds create states in the band gap which can either accept or donate charges, depending on the position of their charge neutral level and the work function of the underlying semiconductor. For typical surface state densities, their impact on the underlying carrier profile can be neglected for high doping levels. On the other hand, surface states can cause a significant distortion of the carrier profile in the near-surface region for carrier densities below 1e18 cm^{-3}.

For Si, a charge neutral level of 280 meV above mid-bandgap and a surface state density of 1e13 cm^{-2} have been reported [16, 17]. Depending on the size of the β-Sn pocket, surface states present on the sample surface next to the actual electrical contact do impact on the region below the probe and hence influence the measurement [21]. This is of importance in particular when analyzing lowly doped *p*-type Si which will experience an inversion of the near-surface region upon alignment of the semiconductor Fermi level with the surface states charge neutral level. As a consequence, lowly doped *p*-type Si surface regions appear as *n*-type. Hence, due to the impact of surface states a cross-sectioned p^{-}-n junction in Si turns into an n^{-}-n configuration where no junction cusp is being observed [17].

For the case of Ge, Schulze et al. found a charge neutral level of 100 meV above the valence band edge [19], which is in agreement with earlier studies on the Ge surface [22] and with the earlier reported formation of a *p*-type inversion layer on lowly *n*-type doped Ge surfaces [23]. Moreover, the aforementioned charge neutral level causes an accumulation of holes near the sample surface for lowly *p*-type doped Ge thereby increasing the apparent hole concentration (compared to the bulk value). This leads to a saturation of the *p*-type calibration curve for high sample resistivities, i.e. a limited dynamic range of the measured resistance. By matching the onset and the degree of simulated and experimentally observed saturation of the *p*-type calibration curve in the lowly doped regime, Schulze et al. derived a surface state density of 1e13 cm^{-2} [19], which appears to be a reasonable value for Ge surfaces as shown earlier [22, 23].

3.1.3 Quantification

SSRM raw data reflect the total resistance between the conductive scanning probe and the sample back-contact which is given by

$$R_{\text{total}} = R_{\text{Sp}} + R_{\text{bulk}} + R_{\text{tip}} + R_{\text{BC}}, \tag{3.4}$$

where R_{Sp} is the spreading resistance (3.2), R_{bulk} the resistance experienced by the current on its trajectory toward the current collecting back-contact, R_{tip} the intrinsic resistance of the probe and R_{BC} the resistance associated with the back-contact. The spreading resistance R_{Sp} itself constitutes the only term in (3.4) which is directly linked to the *local* sample resistivity ρ and thus to the *local* net-carrier concentration one wants to determine. Hence, the local resistivity respectively net-carrier concentration can be extracted as long as the spreading resistance R_{Sp} dominates the sum of the parasitic resistance terms in (3.4) significantly [24]. However, simply applying (3.2) and (3.3) in order to derive accurate quantitative results is in practice not feasible as

- The exact value of the contact radius a is unknown. Furthermore, a varies with the dopant concentration [25] for which reason it is not sufficient to determine a using a single reference sample of known carrier concentration.
- Maxwell's spreading resistance (3.2) is valid only under the assumption of an ohmic contact. However, in reality the carrier concentration is distorted by surface states and the Schottky-like nature of the contact between β-Sn pocket and semiconductor sample as discussed in the previous sections.
- The high pressure applied during SSRM measurements leads to bandgap narrowing underneath the β-Sn pocket. Hence, the effective mass and mobility values are affected for which reason standard mobility curves cannot be used to convert resistivity into carrier concentration values (3.3).

Although the SSRM contact on Si [17] and Ge [19] has been modeled in detail, it is not trivial to comprise all of the above mentioned phenomena quantitatively to derive accurate information about the local net-carrier concentration from the measured resistance value. For this reason, calibration samples covering the entire carrier concentration range of interest are used to obtain an accurate quantification. To facilitate this process, a single sample consisting of several differently doped layers is typically used [26]. Acquiring a 2D resistance map across such a staircase sample (Fig. 3.6a) allows one to extract a 1D resistance profile across the different layers (Fig. 3.6b) by averaging over several lines which leads to a very precise profile with low noise level.

The chemical dopant concentration as well as the net-carrier concentration in the various layers have been determined by means of SIMS and SRP, respectively.

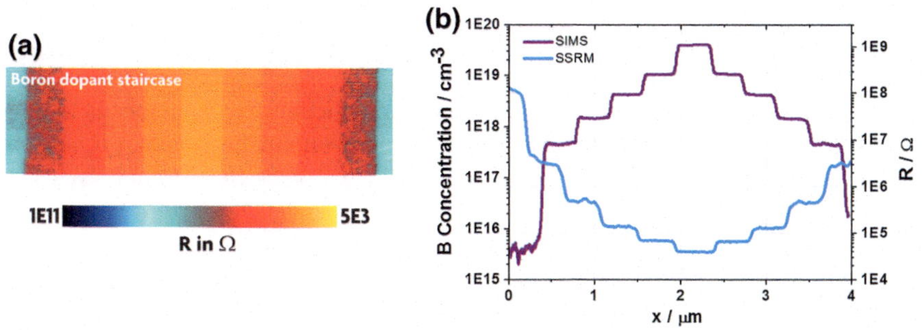

Fig. 3.6 **a** 2D SSRM map obtained on a dedicated dopant (Boron) staircase sample which facilitates the generation of a calibration curve with a single scan. **b** One-dimensional resistance profile (right vertical axis) and Boron concentration profile (left vertical axis) visualizing the individual differently doped layers

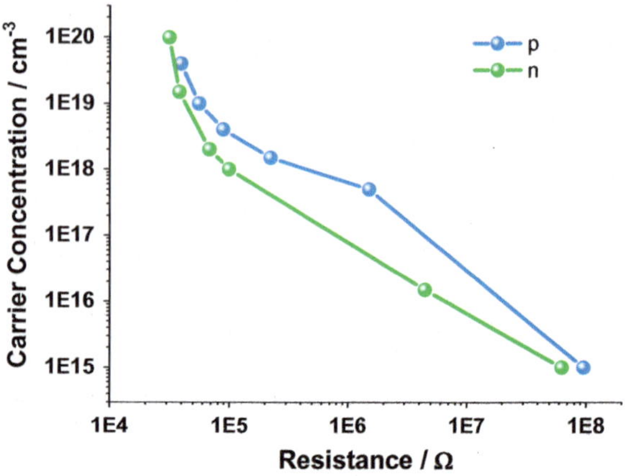

Fig. 3.7 SSRM calibration curves for Si (p- and n-type) extracted from measurements on dedicated dopant staircase samples. Such a calibration curve can be used to assign a carrier concentration value to a resistance value measured using the same probe and identical measurement conditions

Moreover, the carrier concentration was confirmed using sheet resistance and mobility measurements on individual layers grown under identical conditions. By linking the resistance measured with SSRM to the carrier concentration determined by SRP and R_s measurements, a calibration curve can be established. Calibration curves for both carrier types (Fig. 3.7) can be generated by repeating this procedure for an n-type and p-type calibration sample. It is then possible to quantify resistance values on unknown samples measured under the exact same conditions.

It is important to note that any quantification toward net carrier concentration values using this approach holds only under the assumption that the mobility in the reference sample and the actual sample under test are identical. This criterion, however, is not always fulfilled in case of advanced devices, where e.g. mechanical strain and crystalline defects can alter the charge carrier mobility. In such cases a quantification toward a local resistivity value is more meaningful and correct.

3.1.4 Practical Aspects

In the following sub-sections the most relevant practical aspects one needs to consider when targeting high-quality and artifact-free measurements shall be pointed out briefly. More extended discussions on the respective topics can be found elsewhere [7, 27, 28].

3.1.4.1 Sample Preparation

Sample preparation can typically be divided in two parts: (1) The device structure to be analyzed needs to be cross-sectioned and (2) a lowly resistive electrical back-contact to the device under test needs to be established in order to drain the current injected through the probing tip at the cross-section surface.

In order to map the carrier distribution within a device or structure fabricated on a substrate (i.e. wafer), a cross-section through the object of interest needs to be realized. Thereby the accurate positioning of the cross-section is not particularly crucial when analyzing a planar structure. However, in case of a confined volume such as a fin or nanowire-based device, the accurate positioning of the cross-section is of the utmost importance. Moreover, the surface quality (i.e. smoothness) directly affects the noise level of the electrical data and hence the precision of the measurement and therefore represents an additional criterion. The most frequently applied procedures for generating cross-sections for SSRM analysis are (micro)cleaving, (chemical-) mechanical polishing and focused ion beam (FIB) milling. Cleaving works nicely for cross-sectioning large structures but fails when targeting a specific nanometer-scale device. Although (successive) polishing combined with scanning electron microscope (SEM) inspection can in principle be used for sectioning confined 3D structures, the approach is time-consuming and cumbersome and hence less appealing for device failure analysis. Nowadays a focused ion beam (FIB) in combination with a SEM is mostly applied whereby e.g. Ga ions are used to mill a section through the device of interest. It has been demonstrated that by reducing the ion beam energy to 2 keV and by choosing an angle of incident parallel to the sample cross-section i.e. perpendicular to the sample surface, the amorphization of the cross-section surface can be avoided [29]. Moreover, considering the limited dimensions of state-of-the-art devices, it is often sufficient to use FIB and fabricate a cross-section a few nanometer next to the actual plane of interest and then to use the scanning probe to remove the remaining material. This approach will be discussed in more detail in Sect. 3.2.4.2.

As mentioned above, the actual spreading resistance has to dominate all parasitic series resistance components in (3.4) in order to ensure sensitivity toward variations in the local carrier concentration. By placing the electrical back-contact, one automatically defines the path of the current through the sample and hence the bulk series resistance R_{bulk}. On the other hand, the way the contact is realized defines the resistance of the back-contact R_{BC} itself (i.e. contact resistivity, contact area). While in case of planar structures due to the large cross-section the placement and fabrication of a back-contact is rather simple [16], the fabrication of the back-contact can become extremely challenging for confined nanoscale devices. This has mainly three reasons: First, the presence of electrical junctions requires contacts to all the differently doped regions to prevent the current from passing through these junctions as this would otherwise introduce an additional bias-dependent series resistance. Second, the cross-sectional area available for current flow toward the back-contact is reduced significantly leading to a dramatic increase of R_{bulk}. Third, due to the reduced dimensions a poor contact resistivity cannot be compensated for by an increased contact area thus inevitably leading to a large R_{BC}. In practice, the back-contact configuration depends on the particular device structure under test. Often FIB milling is used to establish a trench parallel to the cross-section plane. Subsequently, this trench is filled with a metal by means of a FIB-assisted CVD process. This approach, however, becomes difficult in case of aggressively scaled devices where only a very limited volume between device cross-section and a plane-parallel back-contact would be left. Alternatively, the actual device contact plugs established during fabrication (e.g. S/D and gate contacts) can be used directly to drain the current [30].

3.1.4.2 Tip Requirements

In SSRM one measures an electrical current flowing between the scanning probe and a stationary sample back-contact upon applying a bias voltage. As such, the probe itself is part of the electric circuit and the resistance of the probe contributes directly to the overall measured resistance (3.4). For this reason, SSRM probes need to be highly conductive in order to keep their impact as small as possible. Only if this is assured, the measured total resistance is dominated by the spreading resistance and hence the measurement output signal is sensitive toward variations in the local net carrier concentration. Additionally, SSRM probes need to be particularly wear resistant since the apex of an SSRM probe typically experiences giant mechanical forces during the measurement. Accordingly, the very first SSRM measurements were carried out using natural diamond tips implanted with boron ions in order to make them electrically conductive [4]. Over the last two decades, however, much effort has been put into the development of integrated probes dedicated for high resolution SSRM measurements. Nowadays, pyramidal full diamond tips (FDT) [31, 32] as well as coated diamond tips (CDT) [33] are most commonly used (Fig. 3.8). The fabrication process and characteristics of these probes is discussed in detail in a dedicated chapter within this book.

3.1.4.3 Environmental Aspects

Nowadays, SSRM measurements are typically carried out under controlled environment (glove box with <1 ppm moisture and oxygen content) or high-vacuum

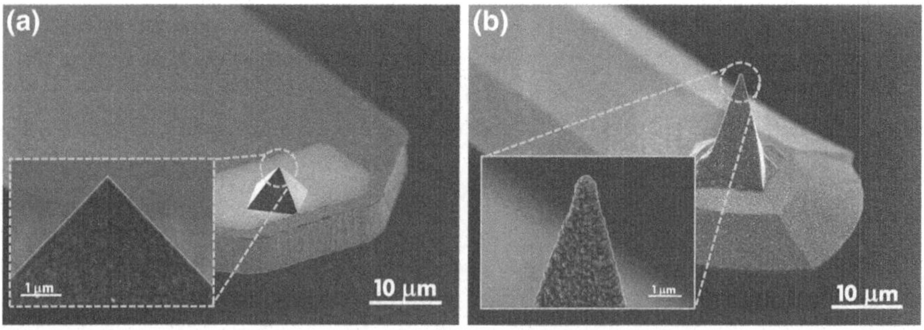

Fig. 3.8 SEM micrographs of **a** a full-diamond tip (FDT) developed and manufactured at imec and **b** a coated diamond tip (CDT)

conditions. This leads to three remarkable improvements as compared to SSRM measurements performed in ambient environment: (1) improved measurement precision, (2) reduced threshold force required to trigger the formation of a β-Sn pocket and (3) improved spatial resolution. The improved precision is directly linked to an increased signal-to-noise ratio and can be explained by the absence/reduction of contaminants and a water meniscus between sample and probe. Moreover, sample oxidation during the measurement is significantly reduced leading to a more stable and reproducible electrical contact. Concerning reduction of the required force, Tang et al. have shown that in the presence of water the formation of a β-Sn pocket was retarded which they explained by a change in the local stress distribution due to the water meniscus between probe and sample [34]. Moreover, Tang et al. demonstrated that in the presence of water the volume of the transformed Si is higher which can be explained by the fact that during the loading, water molecules behave as tiny moving surface indenters which penetrate into the sample surface and modify the stress field in their neighborhood, thereby increasing the volume of transformed Si leading to a reduced spatial resolution [25, 34].

3.1.5 Application to Planar Transistor Technologies

Due to its excellent characteristics in terms of sensitivity and spatial resolution, SSRM has been used extensively to study doping technologies and active dopant distributions in planar transistor structures. Representative, the two-dimensional spreading resistance map obtained on a planar *p*-MOSFET structure is shown in Fig. 3.9a. The various regions (S/D, extension regions, halo implants) as well as the gate electrode are clearly visible. Targeting better performance through the use of high mobility channel materials has led to the integration and exploration of SiGe channels into scaled *p*-MOSFET devices. As the entire structure becomes heterogeneous, understanding and predicting the finer details of dopant diffusion, deactivation, and defect interaction becomes challenging and extensive use of SSRM has been made to unravel some of them. For instance, in Fig. 3.9b, the 2D spreading resistance map is shown for the case of an implant-free quantum well (IFQW) device using a boron doped SiGe layer for S/D-extension formation. The SSRM results clearly indicate the diffusion of boron from the raised SiGe S/D regions into the SiGe channel, thereby facilitating the extension gate overlap (approx. 15 nm). Moreover, SSRM also evidenced that defects at the interface between the SiGe channel and the SiGe S/D regions reduce the degree of dopant activation [35].

3.2 SSRM Applied to 3D Transistor Architectures

The remainder of the chapter is devoted to 3D device architectures and their characterization. After motivating the need for confined and fully-depleted 3D channels

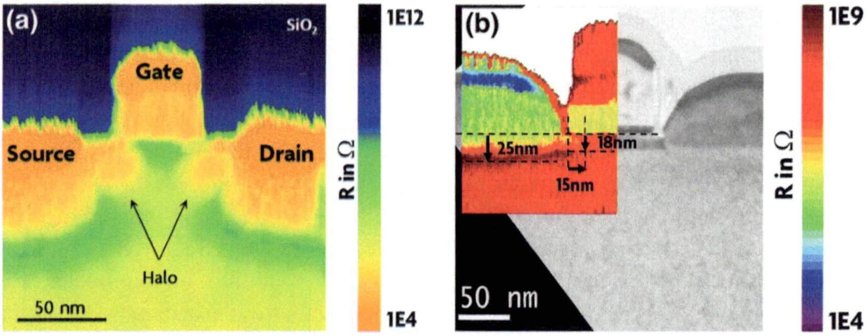

Fig. 3.9 **a** Two-dimensional spreading resistance map obtained on a planar MOSFET structure. The differently doped regions (S/D, extension regions, halo implants) as well as the gate electrode are clearly visible. **b** SSRM and TEM analysis of an IFQW device. The inset exemplifies the deactivation at the interface

such as fins and nanowires, we will discuss in detail the various challenges this imposes on SSRM as well as the developed solutions. The value of 2D and 3D SSRM will be demonstrated based on relevant examples.

3.2.1 CMOS Scaling and the Advent of 3D Devices

The tremendous and remarkable performance improvements related to the CMOS technology and integrated circuits was until the 130 nm technology node mainly driven by dimensional scaling and hence rather straightforward [36]. After that point, however, more disruptive materials and process innovations such as strained silicon [37] and high-κ metal gates [38] were required and had to be introduced in order to further improve the transistor performance (i.e. drive current). Interestingly, gate length scaling was nearly omitted beyond the 90 nm node in order to keep short channel effects and transistor off-currents under control. However, gate length scaling became relevant again to allow for a continued reduction of the contacted gate pitch (CGP) which is crucial for area density scaling. In order to limit short channel effects and off-currents at scaled channel lengths, improved electrostatics of the channel region controlled by the gate electrode are of the utmost importance. Depending on the level of acceptable short channel effects, the minimum channel length should be between 3λ and 7λ with λ being the characteristic length often used to describe the potential distribution along the channel [39]. λ is given by [40]

$$\lambda = \sqrt{\frac{\varepsilon_{\text{Si}}}{\varepsilon_{\text{ox}}} \cdot t_{\text{channel}} \cdot t_{\text{oxide}}} \tag{3.5}$$

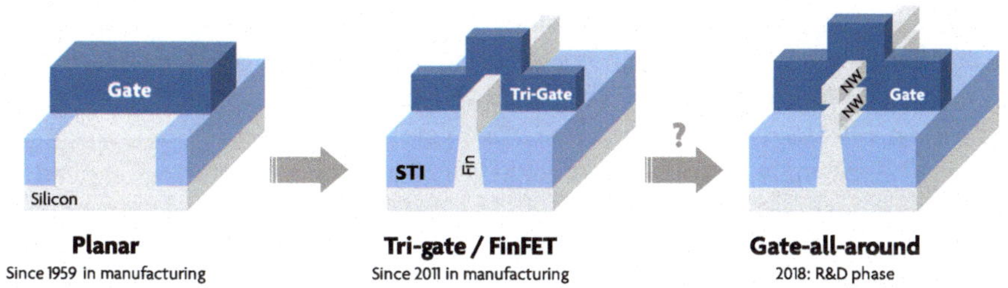

Fig. 3.10 Evolution of device architecture from planar transistors to confined 3D structures allowing for a better electrostatic control of the channel region

with ε_{Si} and ε_{ox} being the relative permittivity of Si and gate oxide, respectively. t_{ox} corresponds to the thickness of the gate oxide and $t_{channel}$ to the thickness of the conducting channel. In case of a planar MOSFET, $t_{channel}$ is determined by the junction depth within the channel region and can only be reduced by means of ultra-shallow junctions and increased channel doping. However, the latter causes a degradation of the carrier mobility and is therefore often not desired. One of the alternatives are transistors based on fully-depleted silicon-on-insulator (FDSOI) technology [41] whereby $t_{channel}$ is solely determined by the thickness of the SOI layer. However, FDSOI-based devices have not emerged in high-performance logic applications due to the relatively high cost of SOI substrates, challenges related to the SOI thickness control across the entire wafer as well as increased self-heating effects promoted by the thermally insulating buried oxide. On the other hand, 3D FinFET architectures such as dual- [42] or tri-gate transistors [43] with the channel placed on top of the substrate have demonstrated superior electrostatic behavior and as such could successfully extend gate length scaling from the 22 nm technology node onward. With the gate electrode covering at least two sides of the fin, the channel thickness is effectively reduced to $w_{fin}/2$ with w_{fin} representing the width of the fin. Horizontal gate-all-around transistors represent a promising option to further improve the electrostatic control by wrapping the gate electrode around the entire channel. The performance and feasibility of such devices is subject of extensive research at this point in time [44, 45]. Schematics of the planar, FinFET and gate-all-around architectures are shown in Fig. 3.10.

3.2.2 *Revisiting Dopant Metrology Requirements*

The characterization of junctions for planar transistors has been dominated for a very long time by two simple metrics: the sheet resistance R_s and the junction depth x_j. Both parameters have typically been extracted from blanket layers using (micro) four point probe (R_s) and SIMS (x_j) measurements. However, due to the advent of ever shallower junctions and shorter gate lengths as imposed by the Dennard Scaling [36], 1D metrology approaches were no longer sufficient to adequately characterize the complex carrier distributions resulting from 2D interactions such as the interplay between extension and halo implants and lateral dopant diffusion. Given this increased complexity for junction engineering, additional metrics such as the lateral junction depth (i.e. overlap between S/D and gate) and the lateral profile steepness became increasingly important and could only be extracted from real 2D carrier distribution maps as measured using SSRM. Dopant metrology requirements for 3D devices such as FinFETs deviate from the standard approaches for planar devices in the sense that the 3D geometry needs to be considered as well. One of the most important parameters to be measured is the degree of dopant conformality. The latter describes potential differences between the doping characteristics on the top and the sidewall of the fins. In principle, dopant conformality can be assessed on a cross-section prepared perpendicular to the fin by means of a 2D measurement (Sect. 3.2.3.1). However, due to the lateral dopant diffusion (i.e. along the fin) in case of a real device, the dopant distribution on such a 2D map strongly depends on the precise location of the cross-section plane in the direction along the fin with respect to the positions of S/D, spacer and gate. Hence, to obtain a complete picture of an actual device, a full 3D analysis is required in order to determine for instance the source/gate overlap on the sidewalls as well as the top of the fin.

3.2.3 Understanding Dopant Incorporation and Activation in 3D Structures

As mentioned in the previous paragraph, 2D measurements are still meaningful and can be used e.g. to determine dopant conformality in long fin (test) structures where variations along the fin due to the absence of lateral dopant diffusion can be neglected. In the following two sub-sections we illustrate how this approach can aid in understanding dopant incorporation in real 3D device architectures such as fins (Sect. 3.2.3.1) and vertical nanowires (Sect. 3.2.3.2).

3.2.3.1 Dopant and Carrier Conformality in Fins

Dopant (i.e. chemical) conformality in fins can be assessed using 1.5D SIMS [46] or atom probe tomography [47]. However, these techniques do not provide any insight in carrier conformality. For this reason, Mody et al. did a combined study using SIMS/APT and SSRM in order to investigate the degree of dopant and carrier conformality in Si fins doped by ion implantation (As and BF_2) whereby they analyzed in particular the influence of the implantation angle. 2D carrier concentration maps obtained using SSRM on As doped Si fins implanted under 10° (left) and 45° (right) tilt angle are shown in Fig. 3.11. For the 45° implant (right), the junction depth appears to be pretty conformal, i.e. the spatial extent of the doped region is similar on the top and the sidewall of the fin. In contrast, however, the 10° tilt angle (left) leads

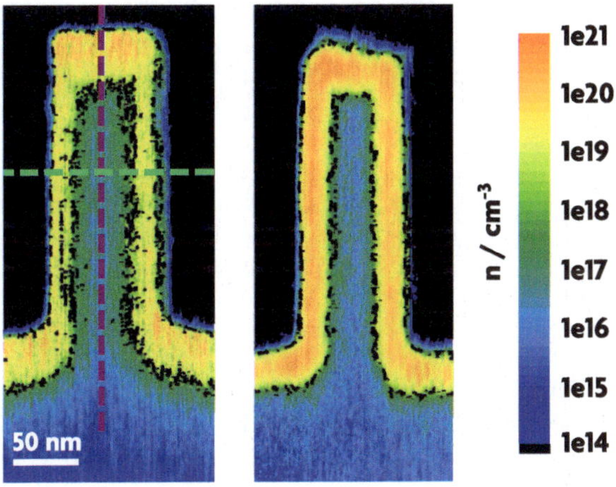

Fig. 3.11 Quantitative 2D carrier concentration maps obtained on ion implanted (As, 5 keV) Si fins using 10° (left) and 45° (right) tilt angle. The 1D profiles corresponding to the dashed lines in the left image are shown in Fig. 3.12

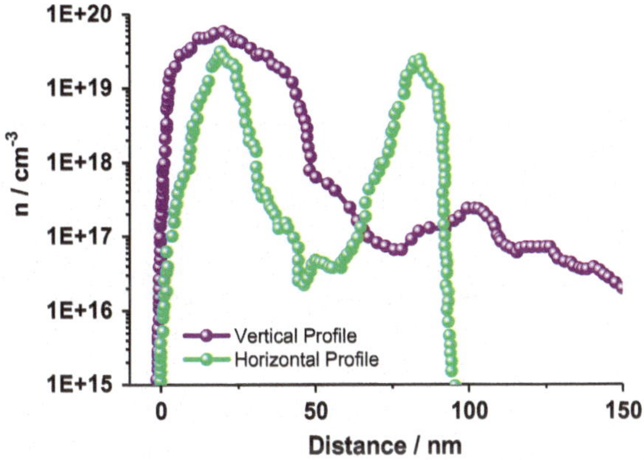

Fig. 3.12 Comparison of a vertical and horizontal 1D carrier concentration profile extracted from the As doped fin implanted under 10° tilt angle (Fig. 3.11)

to a shallower junction depth on the fin sidewall (i.e. horizontal junction depth) as compared to the top of the fin.

The latter observation is further exemplified by the 1D profiles (Fig. 3.12) extracted from the 2D carrier concentration map. From these sections the horizontal and vertical junction depths can be extracted straightforwardly. Since the implantation was carried out in two-quad mode, i.e. half of the dose from the left respectively right side, a higher carrier concentration on top of the fin as compared to the sidewalls can be observed.

In the case of 40 nm wide Boron doped Si fins, Mody et al. found a significant difference between the chemical and electrical conformality independent of the implantation angle. While essentially all dopants on the fin sidewall are electrically active (i.e. equal dopant and carrier concentration), only ~(30 ... 50)% of the incorporated Boron atoms (depending on the implantation angle) were found to be active on the top of the fin [48] (Table 3.1).

Table 3.1 Comparison of chemical (i.e. dopant) and electrical (i.e. carrier) conformality as determined by SIMS/APT and SSRM on 40 nm wide Boron doped Si fins (BF_2, 5keV, 1e14 cm^{-2}) using 10° and 45° tilt angle

	Conformality 10° tilt (Sidewall: Top) in %	Conformality 45° tilt (Sidewall: Top) in %
SIMS (Chemical)	9	36
Atom probe (Chemical)	12	39
SSRM (Carriers)	29	78

This explains the higher degree (~ 2 to 3x) of carrier conformality as compared to the dopant conformality. Indeed, even for the 10° tilt Mody et al. found a carrier conformality of 29% [48]. This is important information as the transistor performance eventually depends on the carrier rather than the dopant conformality.

3.2.3.2 Dopant Incorporation and Deactivation in Nanowires

SSRM has also been used successfully to study the distribution of charge carriers in true 3D structures such as nanowire-based transistors [24, 49, 50]. For example, Schulze et al. studied in detail the incorporation of boron by means of a tilted ion implantation as well as its role in an unexpected and size-dependent deactivation of As dopants. The investigated devices were fabricated by a top-down process which is summarized schematically in Fig. 3.13. The top junction of these transistors was formed by ion implantation (Fig. 3.13e) in four quadrant (4Q) mode at 45° tilt (3e15 ions/cm^2, 5 keV). The 4Q-mode is essential to dope the nanowires from all sides evenly. A nitride spacer and a Si cap layer have been added to realize the top source contact. The cap layer has then been doped using another ion implantation step (3e15 ions/cm^2, 5 keV, 45°) followed by a spike anneal at 1050 °C.

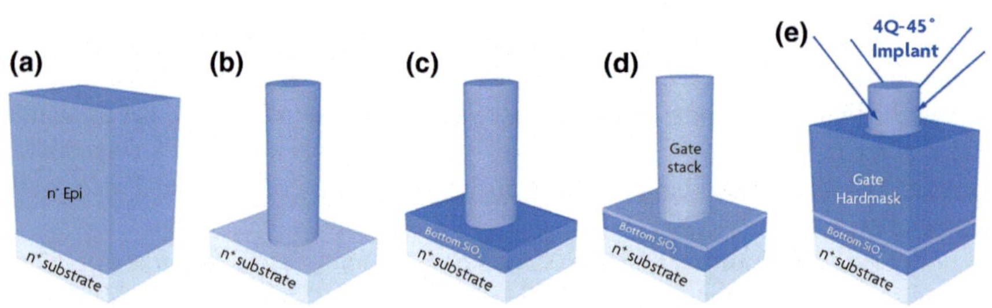

Fig. 3.13 Schematic process description of Si nanowire-based TFET with implanted top junction. **a** Epitaxial growth of lowly As doped layer on highly As doped substrate. **b** Nanowire patterning. **c** Bottom oxide to isolate drain from gate. **d** Deposition of high-κ metal gate stack. **e** Fabrication of gate hardmask and subsequent etch-back of gate stack to reveal the top of the nanowire (source) which is doped using a 4Q-45° ion implantation step followed by a spike anneal

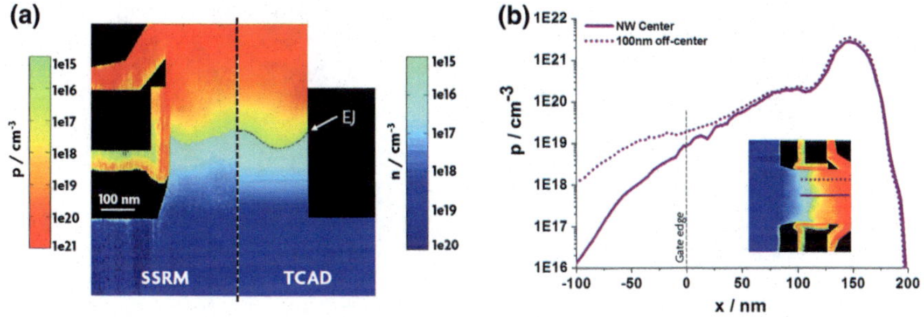

Fig. 3.14 **a** Comparison of the two-dimensional carrier distribution obtained by SSRM and process simulations on 400 nm wide Si-based TFET structure with implanted tunnel junction. **b** Vertical line sections (center and 100 nm off-center) visualizing the hole concentration relative to the position of the gate electrode

The obtained carrier distribution map for a 400 nm wide structure is shown in Fig. 3.14a. The various differently doped regions as well as the conductive metal gate wrapped around the structure can be clearly identified. For comparison, Schulze et al. furthermore performed process simulations using Sentaurus Process [51]. For this purpose the process parameters of the tilted ion implantation step and the subsequent spike anneal were used as input to predict the 2D active dopant profile. The simulated profile is shown next to the map obtained by SSRM in Fig. 3.14a. The simulation and measurement results are in excellent quantitative agreement and reveal both the formation of a dual-bump structure in the nanowire top section. Two vertical line sections, one in the center and one 100 nm off-center are shown in Fig. 3.14b. By comparing the two sections one can observe a steeper gradient in the center of the nanowire and a longer tail for the off-center profile which coincides with one of the implantation pockets. The formation of these pockets is directly linked to the tilted ion implantation step, which dopes the sidewall and the top of the nanowires simultaneously. The deeper junction (i.e. implantation pocket) is induced mainly by the sidewall doping, whereas the shallower junction depth (in between both pockets) is formed by doping through the top surface of the structure only.

By looking at the bottom profile of the nanowires and taking the position of the metal gates in Fig. 3.15 as a reference, it becomes apparent that the extent of the electron density profile from the substrate into the nanowire depends strongly on the nanowire diameter. As indicated by the arrows, a longer tail in the nanowire bottom (drain) could be detected by SSRM for wider structures. This is illustrated more quantitatively by means of vertical carrier concentration profiles shown in Fig. 3.16.

The electron concentration obtained on the substrate is ~1e19 carriers/cm^3 and the substrate surface after etching the nanowires is located at $x = 0$. A pronounced diffusion tail which extents further into the nanowire for larger nanowire diameter is evident (Fig. 3.16). According to the process flow of the investigated devices [52], this pronounced diffusion tail is due to the diffusion of As atoms from the substrate into the epitaxially grown layer during a LOCOS process (associated with a thermal

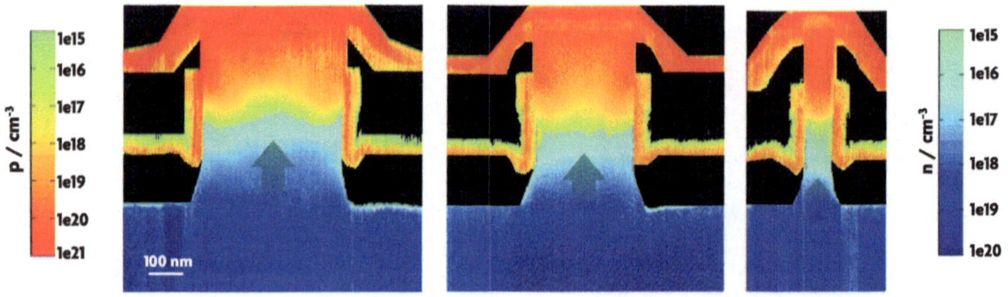

Fig. 3.15 Quantitative two-dimensional carrier distribution maps for nanowire diameters of **a** 400 nm, **b** 300 nm and **c** 100 nm. In case of the narrower devices, the electron density decays faster in the direction from the substrate into the nanowire

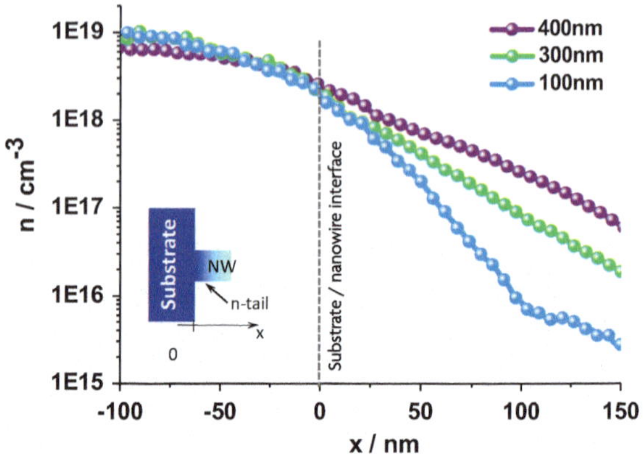

Fig. 3.16 Vertical section (cross-section average) through the substrate and the bottom part of the nanowires extracted from the carrier distribution maps in Fig. 3.15. A steeper gradient in the narrower devices is apparent

budget of 975 °C applied for 200 min) used to define the active respectively non-active areas on the wafer. However, this process step and its associated diffusion occurred before the nanowires were etched. Hence, one would expect a diameter independent gradient of the electron profile at the nanowire bottom. Assuming that the carrier profile is not affected by the nanowire patterning, the experimentally observed diameter dependence of the electron profile can then only be explained by a dopant deactivation process which is more efficient in narrower structures causing the steeper profile for nanowires with smaller diameter.

In order to verify this assumption, a SSRM measurement on a planar stack (∼ infinite nanowire diameter) grown under identical conditions and exposed to an equivalent thermal budget was carried out. The electron density profile obtained on that blanket sample was determined to be in agreement with the one measured on the largest structure (400 nm) [49]. This validates the assumption of a dopant deactivation mechanisms taking place in the narrower devices after nanowire patterning. Indeed, the deactivation of As atoms caused by ion implantation and subsequent annealing in case of blanket samples and high As concentrations has been reported before [53, 54]. In the present case, the bombardment of the nanowires with B ions during the implantation introduces Si self-interstitials and vacancies which diffuse during the subsequent annealing step toward the substrate. Here As-vacancy complexes can form which lead to a deactivation of As and hence a lower carrier (i.e. electron) density. The size dependence results from the increased proximity of surfaces for smaller structures. The surfaces act as a sink for interstitials and as such they reduce the vacancy-interstitial recombination probability which in turn leads to a promotion of As-vacancy agglomeration and thus a more efficient dopant deactivation in the case of narrower nanowires [49].

The observed size dependence is a clear indication of the presence of physical phenomena which are operative in small, three-dimensional structures but cannot be observed in simple blanket experiments. Only high resolution, quantitative measurements as done here with SSRM can provide evidence for their existence and aid in understanding size dependent device characteristics.

3.2.4 Tomographic Carrier Mapping

As already mentioned in Sect. 3.2.2, in order to fully characterize a 3D device like a FinFET, one needs to determine the carrier concentration in three dimensions. However, since SSRM is inherently a 2D technique, solution(s) extending the approach into the third dimension are required. In the following sections we will discuss two methodologies whereby the first one is based on using a dedicated test structure which facilitates the generation of a series of device cross-sections, each one placed a little further into the third dimension (Sect. 3.2.4.1). The second approach is based on the successive removal of material by means of the scanning probe itself which literally moves the cross-section plane through the device under test during the course of the measurement (Sect. 3.2.4.2).

3.2.4.1 Design for Metrology

For the first approach one generates a single cross-section through several nearby devices. If one arranges these nearby devices in a staggered fashion such that the position of each device is shifted by a certain distance in the direction perpendicular to the cross-section plane, one can easily create a series of cross-section planes whereby each plane is placed a little further into the device relative to the previous one. If one assumes now equivalent processing conditions and thus equivalent carrier profiles for these nearby devices, it is valid to combine the 2D carrier concentration maps obtained on the different devices into a 3D carrier distribution map by stacking the various 2D maps on top of each other and interpolating between them. The offset between two neighboring devices in the staggered array then determines the resolution in the third dimension [55–57].

In case of a FinFET, this approach can be implemented by fabricating an array of fins and then staggering the (dummy) gate such that it is shifted for each fin in the longitudinal direction by a certain distance (Fig. 3.17a). In case of a self-aligned gate,

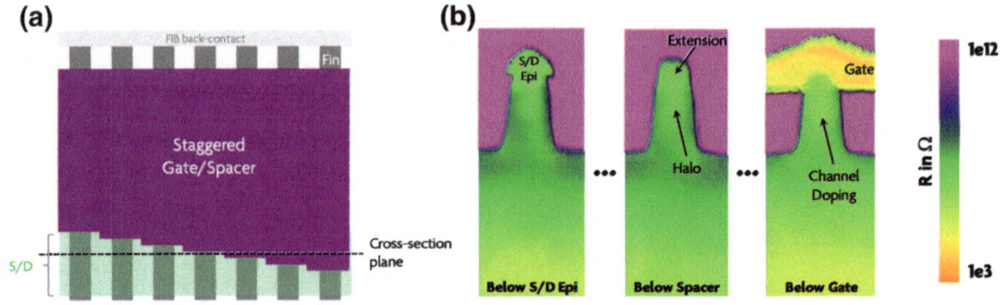

Fig. 3.17 **a** Schematic top-view illustrating the concept of a staggered gate. By establishing a single cross-section through such an array one actually generates a series of device cross-sections perpendicular to the fin, each located at a different position along the fin. **b** 2D Spreading resistance maps acquired on cross-section planes situated at S/D, spacer and gate, respectively

the aforementioned offset in gate position then leads to an identical shift of the entire dopant profile in the underlying fin. By cleaving through this array one produces a series of cross-sections moving along the fin from S/D toward the channel (i.e. below the gate).

Mody et al. used this methodology to study the 3D carrier distribution in n-type FinFETs whereby they used a gate stagger of 3 nm. 2D SSRM maps originating from different positions along the fin, i.e. below the gate (channel region), spacer and in the S/D region are shown in Fig. 3.17b. Analyzing the SSRM image below the gate (in the channel region) one can clearly see the channel implant as well as the shape of the conductive gate wrapped around the top of the fin. Below the spacer, the extension profile and the halo pocket can be seen. Eventually the epitaxially grown raised S/D diamonds can be observed [57].

By stacking these multiple 2D maps and interpolating between them, a detailed 3D carrier distribution map of a FinFET structure extending from S/D till the middle of the gate can be obtained (Fig. 3.18a). From this tomogram e.g. a 2D slice along the fin can be extracted (Fig. 3.18b). The latter nicely illustrates the details of the S/D (extension) carrier profile along the fin from which the gate overlap can be estimated [57].

3.2.4.2 Scalpel-SSRM

The methodology described in the previous section enables the acquisition of a 3D carrier distribution map by collecting and combining 2D profiles from nearby (thus different) devices. The procedure requires a dedicated test structure based on a (staggered) array of devices for which reason the method cannot be applied to a single device (e.g. for the purpose of failure analysis). Moreover, material removal during the acquisition of the individual 2D maps will affect the position of each map in the third dimension, thereby hampering a precise reconstruction leading to a (potentially) erroneous depth scale. To overcome these limitations, we developed

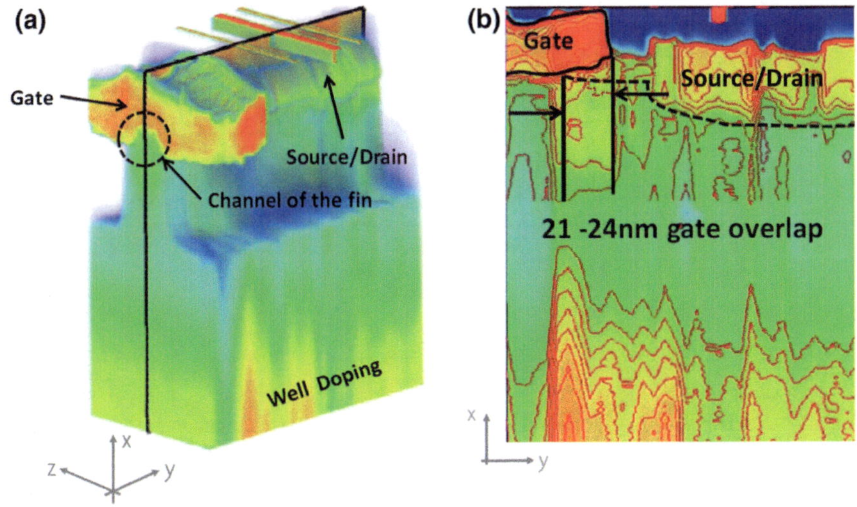

Fig. 3.18 **a** Reconstructed tomogram visualizing the distribution of carriers in a FinFET test structure in 3D. **b** 2D map along the fin extracted from the tomogram shown in (**a**). The gate-overlap can be extracted straightforwardly

a slice-and-view approach whereby the scanning probe itself is used to successively remove material which literally moves the cross-section plane through the device of interest during the course of the analysis. Individual maps originating from different depth are then reconstructed into a tomogram [58].

In practice, we first use micro-cleaving and FIB lapping in order to fabricate a starting cross-section plane as close as possible (typically less than 10 nm) to the structure of interest. The scalpel concept is then used to progress in the z-direction toward the center of the device (Fig. 3.19). In this example we chose a cross-section plane oriented along the fin (i.e. x-y-plane) which facilitates the analysis of the lateral and vertical dopant profile under the gate as any potential error during the reconstruction would only affect the z-scale [59].

Through continuous scanning one can remove a certain amount of material in a controlled manner depending on the force applied to the probe. The step size along the z-direction (i.e. the amount of material removed per scan) can be reduced down to below 1 nm [58, 60]. We applied this approach to study the carrier distribution in the S/D and extension regions of Si channel FinFETs. More specifically, the difference between devices with S/D regions doped by ion implantation (Device A) and in-situ doping (Si:P, Device B) were analyzed [59]. The process flow of the devices is depicted in Fig. 3.19b. 2D resistance maps extracted from the obtained tomograms at the center of the fin/channel are shown in Fig. 3.20b, c together with a TEM image (a).

By correlating the TEM and SSRM data, the various regions of the FinFET such as S/D and extension regions, metal gate and contacts can be clearly identified. The

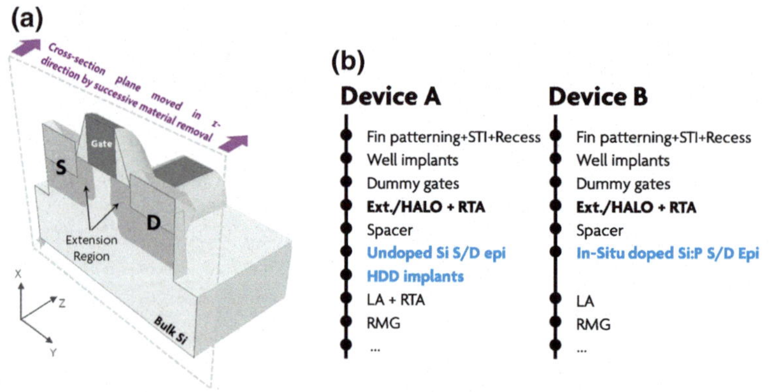

Fig. 3.19 **a** Schematic of a FinFET device cross-sectioned parallel to the channel. Through successive scanning material is being removed whereby the cross-section plane is moved in a plane-parallel manner through the device in the *z*-direction. **b** Process flow for Device 1 and 2. The different S/D doping strategies are highlighted

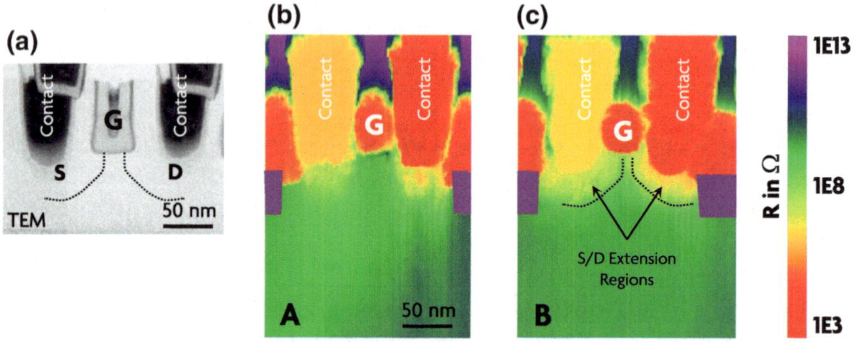

Fig. 3.20 **a** Cross-sectional TEM image indicating the position of the gate and the S/D contacts. 2D spreading resistance maps for devices A and B are shown in (**b**) and (**c**), respectively. The S/D extension regions are highlighted in (**c**)

source contact appears to be more resistive in both cases which is due to a parasitic series resistance between the source contact and the actual macroscopic back-contact. For this reason we focused our study on the analysis of the profile on the drain side. Resistance data were translated into carrier concentration values using the calibration procedure discussed in Sect. 3.1.3. Vertical and lateral carrier concentration profiles at the drain junction are shown in Figs. 3.21 and 3.22, respectively.

By comparing Device A and Device B, a higher active dopant concentration could be found in case of the device with the in-situ doped S/D region (Device B, Fig. 3.21). This higher carrier concentration results in a lower access resistance which in turn can be associated with a larger on current I_{on} and a reduced $R_{on,lin}$ determined during the electrical testing of these devices [59].

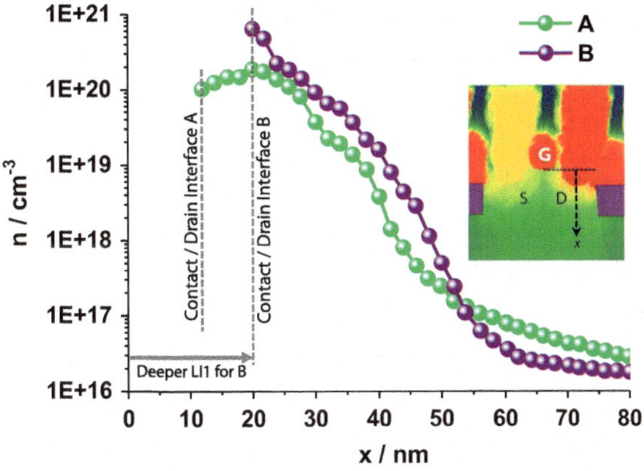

Fig. 3.21 Vertical carrier concentration profiles for devices A and B, respectively. The profiles are aligned w.r.t. the position of the gate oxide at $x = 0$. A higher active dopant concentration in case of device B is apparent. Furthermore, a 8 nm deeper local interconnect LI1 could be observed in case of device B. The inset illustrates the position from where the 1D profile has been extracted

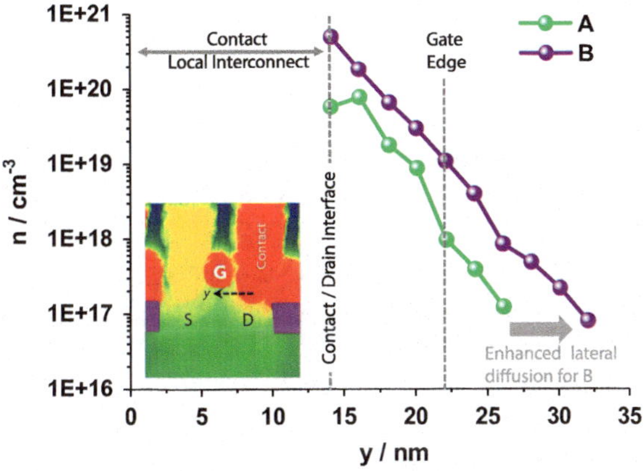

Fig. 3.22 Lateral carrier concentration profiles for devices A and B, respectively. A higher carrier concentration and increased lateral diffusion (6 nm at 1e17 cm^{-3}) can be observed for device B. The inset illustrates the position from where the 1D profile has been extracted

Similarly, a higher carrier concentration for Device B becomes apparent from the lateral profile (Fig. 3.22) as well. Moreover, a significantly increased lateral diffusion from the S/D regions under the gate could be observed in case of Device B. This inevitably leads to more pronounced short-channel effects. Indeed, devices fabricated using the in-situ doped Si:P S/D regions showed stronger V_t roll-off characteristics (Fig. 3.23a) as well as increased overlap capacitance C_{ov} (Fig. 3.23b).

3.2.5 *Outsmarting Parasitic Resistances: Fast Fourier Transform-SSRM*

Excellent sensitivity over a wide range of carrier concentrations represents the key characteristic of SSRM and the reason for its supremacy over other carrier profiling methods. However, this quality is only ensured as long as the spreading resis-

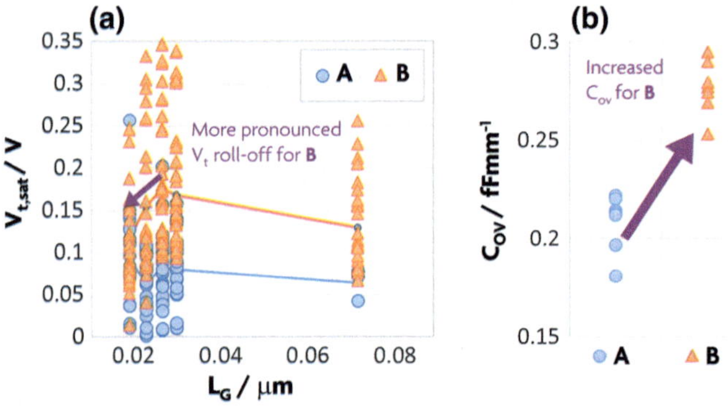

Fig. 3.23 **a** $V_{t,\mathrm{sat}}$ as a function of gate length for devices A and B respectively. A more pronounced V_t roll-off can be found in case of device B. **b** Overlap capacitance C_{ov} for devices A and B showing an increased C_{ov} for device B. Both performance observations can be explained by the difference in lateral diffusion detected by SSRM (see Fig. 3.22)

tance dominates the sum of all series resistance components as discussed in detail in Sect. 3.1.3. Fulfilling this requirement is challenging when applying SSRM to highly doped samples where the increasing impact of the tip series resistance (R_{tip}) limits the technique's dynamic range. Moreover, in case of confined volumes the bulk resistance (R_{bulk}) experienced by the current on its trajectory toward the current collecting back-contact as well as the resistance linked to the back-contact itself (R_{BC}) can increase significantly. This leads to a situation in which the parasitic series resistances (which are not related to the *local* sample resistivity) screen the spreading resistance (R_{Sp}) which then becomes unidentifiable.

To overcome this deficiency and to allow for quantitative carrier mapping in the presence of parasitic resistances, Schulze et al. conceived a methodology termed FFT-SSRM which essentially decouples the actual spreading resistance (containing the information about the local sample resistivity) from the parasitic series resistance components [28, 61, 62]. This is achieved by modulating the force applied to the scanning probe at angular frequency ω which leads to a modulation of the spatial extent of the $\beta - Sn$ pocket and hence of the electrical contact radius a and the spreading resistance R_{Sp}. Since the parasitic resistance components are force independent, information on the spreading resistance can be retrieved by extracting the component related to the modulation frequency out of the measured resistance signal (Fig. 3.24) [61].

The relation between the measured resistance and the force applied to the probe is shown for *p*-type doped Si of different dopant concentration in Fig. 3.25.

The resistance drop for forces <0.5 μN is caused by the formation of the β-Sn phase whereas the gradual resistance decrease for forces >0.5 μN is linked to the increasing electrical contact radius resulting from the larger indentation depth. Since the contact is more stable for forces >0.5 μN, an operating point in this regime is recommended although a larger force amplitude is required to achieve a detectable

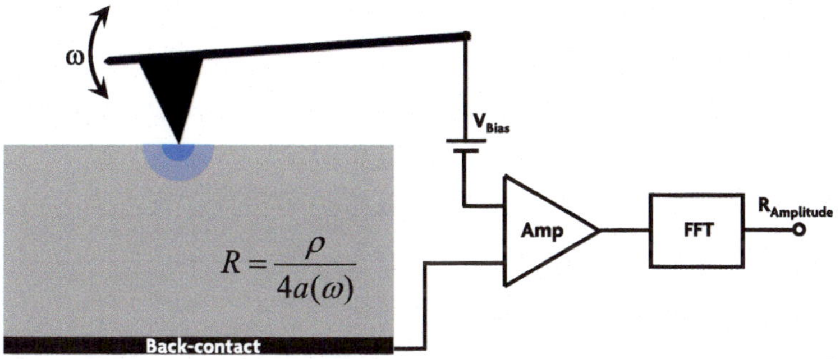

Fig. 3.24 Operation principle of FFT-SSRM. The spreading resistance is decoupled from parasitic series resistances by a sinusoidal modulation of the force applied to the cantilever, which causes a modulation of the contact size and thus a modulation of the spreading resistance. Series resistance components remain unaffected. Using FFT, the resistance amplitude at the modulation frequency is extracted

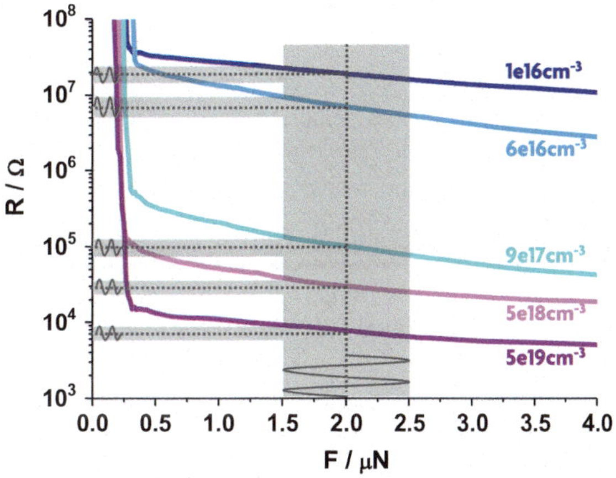

Fig. 3.25 Resistance-force curves obtained on p-type silicon for different doping levels. The modulation of the measured resistance signal due to a sinusoidal force variation is indicated

resistance modulation given the reduced slope of the resistance-force curve in this regime. The expected resistance variations for a force modulation with $F_0 = 2$ μN and a force amplitude of $A = 0.5$ μN) is indicated in Fig. 3.25. In order to determine the resistance amplitude in practice one needs to sample the resistance signal at high speed while scanning the probe under force modulation across the sample of interest. From the resistance signal the amplitude at the modulation frequency can be extracted using FFT or a lock-in amplifier. A typical frequency spectrum obtained by FFT is shown in Fig. 3.26a. The peak at $f = 0$ Hz represents the DC component of the resistance signal and corresponds to the resistance value one would measure at constant force, i.e. in conventional SSRM. Due to the periodic force modulation, a distinct resistance peak at the modulation frequency can be observed. Additional peaks at frequencies equal to multiples of the modulation frequency represent higher harmonics. Their existence can be explained by the non-linearity of the resistance-force

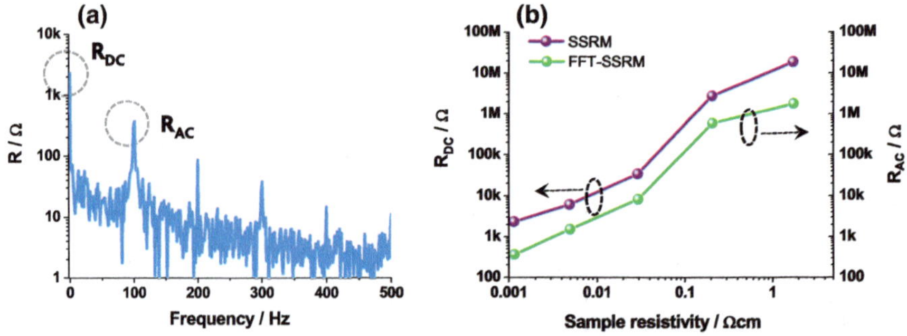

Fig. 3.26 **a** Typical frequency spectrum obtained by FFT from a measured resistance signal on a highly doped Si layer. **b** Calibration curves generated from a modulated resistance signal measured across a dopant staircase sample. The calibration curve for conventional SSRM has been obtained by extracting the constant component of the resistance (R_{DC} at $f = 0$ Hz), whereas the calibration curve for FFT-SSRM has been generated by extracting the resistance amplitude (R_{AC}) at the modulation frequency

relation. It must be noted that the amplitude of the first harmonic alone represents a good proxy for the amplitude of the overall resistance signal since the amplitudes of the higher harmonics are significantly reduced [61, 63].

By extracting the resistance amplitude R_{AC} at the modulation frequency (first harmonic) for different doping levels using e.g. a dopant staircase sample, a FFT-SSRM calibration curve can be established straightforwardly (Fig. 3.26b, right vertical axis). For comparison, a calibration curve based on the DC resistance component of the resistance signal ($f = 0$ Hz), corresponding to conventional SSRM, is shown (Fig. 3.26b, left vertical axis). The monotonic behavior of the FFT-SSRM calibration curve proves the possibility of quantitative carrier profiling by evaluating the resistance amplitude (R_{AC}) rather than the DC resistance component (R_{DC}) as done in conventional SSRM.

The influence of a parasitic series resistance on the SSRM calibration curve is illustrated in Fig. 3.27. The solid lines in this figure correspond to the calibration curve of regular SSRM (i.e. R_{DC}) whereas the dashed lines correspond to the FFT-SSRM calibration curves (i.e. R_{AC}). Looking at the regular calibration curves, a saturation of the DC resistance at low sample resistivities becomes apparent. This can be explained by the fact that the resistivity dependent spreading resistance component (R_{Sp}) is screened by the parasitic bulk series resistance which causes a loss of sensitivity for highly doped samples and hence limits the dynamic range of the technique. The onset and degree of saturation depend on the magnitude of the parasitic resistance with a more pronounced saturation for the higher parasitic resistance. On the other hand, the slopes of the FFT-SSRM calibration curves (R_{AC}) do not show any saturation at low sample resistivities, irrespectively of the parasitic resistance. This validates that the FFT-SSRM method successfully isolates information related to the spreading resistance and hence the local carrier concentration from undesired series

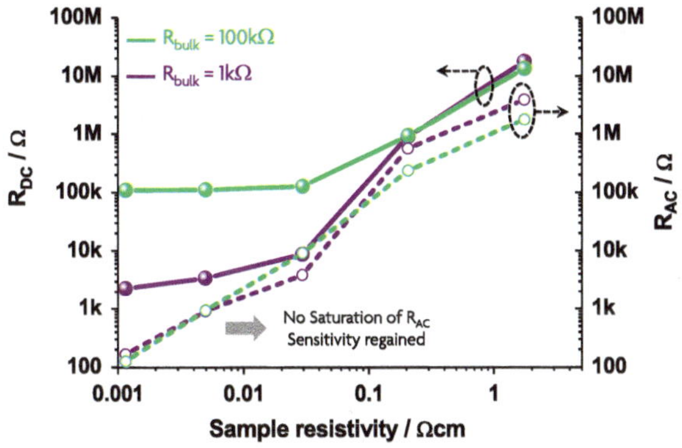

Fig. 3.27 Calibration curves for conventional SSRM (R_{DC}) and FFT-SSRM (R_{AC}) with a parasitic series resistance of 1 kΩ and 100 kΩ, respectively. As expected, the parasitic resistance screens the spreading resistance (R_{DC}) for lowly resistive samples leading to a saturation of the DC calibration curve (solid lines). The saturation and hence reduction of the dynamic range is more pronounced for larger parasitic resistances. In contrast, the calibration curve based on the extracted resistance amplitude (R_{AC}) resulting from the periodic force modulation does not show any sign of saturation at low sample resistivities (dashed lines). This verifies the capability of FFT-SSRM to eliminate undesired parasitic resistances, thereby regaining excellent sensitivity over a wide range of carrier concentrations

resistance components, thereby enabling excellent sensitivity over the entire carrier concentration range of interest.

The value of FFT-SSRM is furthermore exemplified by the example shown in Fig. 3.28 which illustrates the electrical characterization of in-situ doped Si:P S/D diamonds grown epitaxially on bulk Si fins. Given the confined nature of the fins and the S/D regions, the current experiences a significant resistance on its trajectory toward the back-contact which is plane-parallel to the cross-section plane. This parasitic series resistance screens the spreading resistance in case of highly-doped regions for which reason the conventional SSRM resistance (R_{DC}) saturates within the S/D region at approximately 7 MΩ.

Due to the dominance of the parasitic resistance, the true spreading resistance and its variation in the S/D region cannot be assessed. However, by applying the FFT-SSRM concept and extracting the resistance amplitude (R_{AC}) it is possible to retrieve information about the local resistivity. Since the spreading resistance is effectively decoupled from the parasitic resistances, even fine details such as a non-uniform carrier density profile with an increasing conductivity toward the top of the S/D diamond can be detected.

This application nicely illustrates that FFT-SSRM aids in extending the success story of spreading resistance-based carrier profiling toward nanometer-scale true 3D device structures such as fins and nanowires.

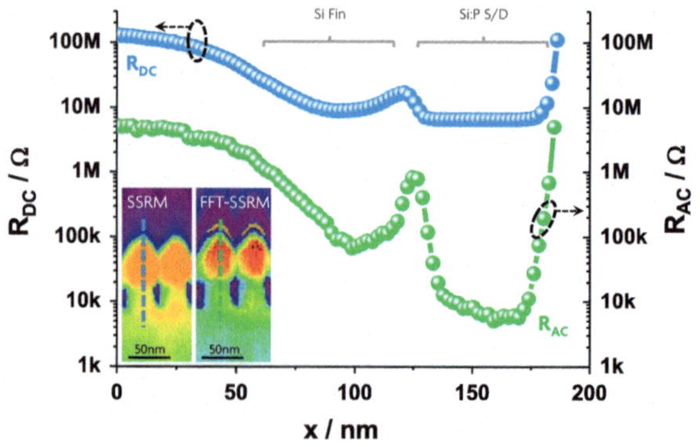

Fig. 3.28 Measurements on highly doped Si:P S/D diamonds grown on lowly doped Si fins. In case of regular SSRM (R_{DC}), a saturation of the resistance within the S/D region due to a parasitic series resistance caused by the confined nature of the structure can be observed. Details of the S/D carrier profile can therefore not be assessed. In the case of FFT-SSRM, however, the resistance amplitude clearly shows a non-uniform conductivity within the S/D region with an increased conductivity at the top of the S/D diamond. The insets show the two-dimensional distribution of R_{DC} (SSRM) and R_{AC} (FFT-SSRM), respectively

3.2.6 Toward Holistic Transistor Metrology: Combining TEM and SSRM

Although SSRM has experienced a number of significant improvements over the recent years, metrology requirements associated with complex device architectures often based on heterogeneous structures are steadily increasing, thereby challenging the state-of-the-art in materials characterization and diagnostics. For instance, SSRM is capable of analyzing the carrier profile in transistors in great detail, however, it does not provide chemical/compositional or structural information. As such, the characterization of carrier profiles at a semiconductor heterojunction (e.g. between Ge and SiGe) becomes difficult since the accurate position of the actual heterojunction (i.e. interface between the different materials) cannot be obtained. Moreover, using only SSRM one cannot attribute a measured variation in sample resistivity to a certain physical characteristic in the sample (e.g. crystalline defects, strain, dopants etc.). The correlation of SSRM with (S)TEM-based techniques on the other hand could aid in bridging this metrology gap, especially if SSRM and TEM analysis could be performed on the same device. Thereby SSRM and/or other AFM-based techniques would be applied to obtain high-resolution electrical profiles whereas (S)TEM related techniques would be used to obtain structural and chemical information [64].

The successful combination of SSRM and TEM characterization performed on the same device is discussed in the following paragraph. This result has been achieved by developing a dedicated sample preparation procedure that enables the nm-resolved 3D electrical and chemical analysis of Si- and Ge-based structures [65]. The main challenge is thereby imposed by the need for an analysis flow that meets the require-

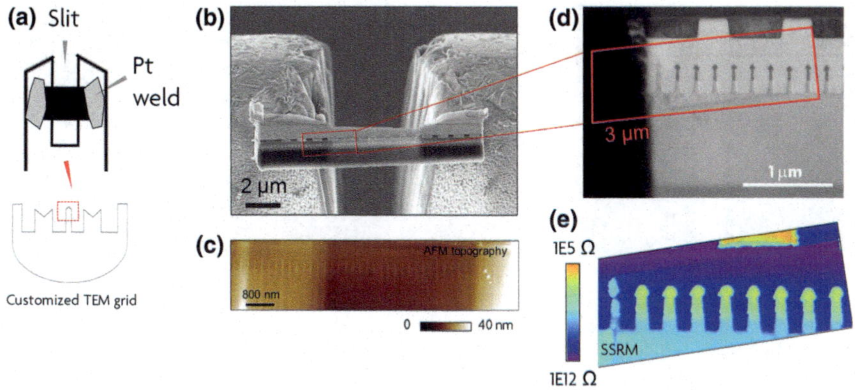

Fig. 3.29 **a** A standard Cu-grid TEM holder is modified by FIB to create a small slit of approx. 2 μm) in the central part of the middle pin. At this position the lamella is placed and anchored as shown in (**b**). AFM techniques can now be used on the area of interest with (**c**) showing the topography of the specimen. **d** and **e** show high-resolution TEM and SSRM images obtained on a series of nearby fin structures

ments of both techniques, namely (1) electron transparency (i.e. thin sample with a thickness well below 100 nm) for TEM, and (2) mechanical stability to withstand the forces acting on the specimen during contact-mode electrical AFM measurements such as SSRM. Figure 3.29a shows a possible solution based on a customized TEM-sample holder. Using focused ion beam (FIB), standard holders can be modified inserting a small slit in the central pin (inset Fig. 3.29a). The sample can be prepared using the standard process for TEM lamella extraction and thinning. Once released from the substrate, the lamella can be positioned in the central pin of the custom TEM holder in correspondence of the slit, and finally anchored e.g. by Pt welds (Fig. 3.29b). This approach ensures a certain degree of mechanical stability as well as electron transparency and as such facilitates subsequent AFM/SSRM (Fig. 3.29c, e) and TEM (Fig. 3.29d) analysis.

As an example, we report the in-depth analysis of the source/drain regions of strained Germanium FinFETs (Fig. 3.30a). The strained channel is obtained by growing Ge on top of a $Si_{0.3}Ge_{0.7}$ strain relaxed buffer (SRB) whereby boron-doped $Si_{0.25}Ge_{0.75}$ is used for the raised S/D regions. The specimen prepared for the analysis contains a series of nearby devices as can be seen from Fig. 3.29d, e. Structural, chemical and electrical information are now accessible for any device within the analysis area. Figure 3.30f–i show magnified views of one single fin probed by means of TEM, SSRM and EDX.

As described in the previous section, the use of FFT-SSRM is essential to obtain quantitative carriers profiles in such devices. From our analysis we could observe a resistive interfacial layer between strained Ge and boron doped SiGe S/D diamonds (Fig. 3.30c). This layer can be attributed to an insufficient pre-epitaxial clean treatment applied to the fin surface prior to the S/D epitaxial growth. Using STEM the

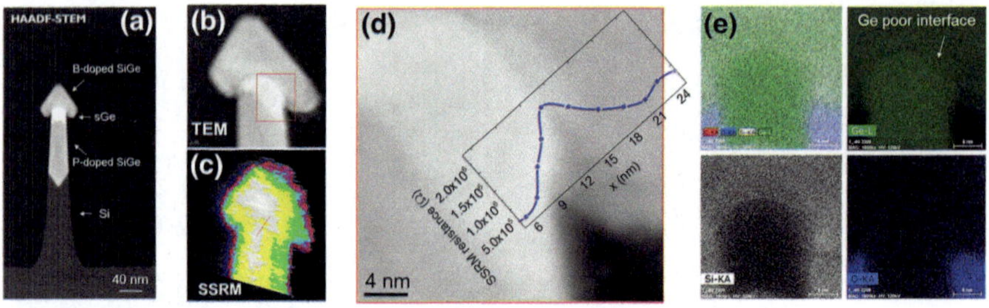

Fig. 3.30 **a** Boron doped SiGe S/D diamonds grown on strained Ge channels are used as test vehicles. **b–e** Site-specific structural, chemical and electrical information of the same device can be obtained by SSRM and (S)TEM. In this case, the chemical sensitivity of EDX (**e**) is combined with the carrier profile capability of SSRM (**c**). A correlation of structural and electrical data is shown in (**d**) whereby a thin interfacial layer of reduced conductivity between S/D and strained Ge could be revealed

thin interfacial oxide can be resolved accurately and its thickness be determined to approximately 2–3 nm. The correlation between STEM and resistance profile extracted from a SSRM map is shown in Fig. 3.30d. This highly resistive interfacial layer degrades the overall resistance in the raised S/D area, thereby leading to an increased access resistance and hence reduced drive current. Although the presence of oxygen could not be verified (i.e. below the detection limit), a reduction of the Ge intensity at the location of the interfacial layer became apparent from the EDX analysis shown in Fig. 3.30e.

In summary, we presented a promising path toward combined structural, chemical and electrical characterization of a single device by means of subsequent TEM and SSRM analysis on the same device. Given the ever increasing complexity of emerging devices, such a correlative approach appears indispensable as a single characterization technique is typically no longer sufficient to provide the required insights into novel materials and processes.

3.3 Summary

Thanks to the unrivaled combination of excellent spatial resolution and sensitivity, SSRM constitutes the method of choice for profiling the distribution of charge carriers in semiconductor devices, thereby providing invaluable information on dopant incorporation, diffusion and (de)activation mechanisms. Like any materials characterization technique, SSRM is continuously being challenged by the ever increasing complexity of emerging devices. However, breakthrough developments enabling a 3D profiling capability, the FFT-SSRM concept as well as the continued efforts leading to advanced probes successfully aided in moving the frontiers forward. SSRM is used to study various semiconductor materials and device structures and applications cover junction engineering, characterization of heterojunctions, failure analysis and TCAD calibration.

Acknowledgements The authors acknowledge C. Drijbooms, P. Favia and H. Bender for their contributions related to FIB and TEM work.

References

1. W. Vandervorst, M. Meuris, Method for resistance measurements on a semiconductor element with controlled probe pressure, EP-466274 (1992)
2. W. Vandervorst, M. Meuris, Method for resistance measurements on a semiconductor element with controlled probe pressure, US-5369372 (1994)
3. P. De Wolf, Two-dimensional carrier profiling of semiconductor structures with nanometer resolution. Ph.D. thesis (KU Leuven, Leuven, Belgium 1998)
4. P. De Wolf, J. Snauwaert, T. Clarysse, W. Vandervorst, L. Hellemans, Characterization of a point-contact on silicon using force microscopy-supported resistance measurements. Appl. Phys. Lett. **66**(12), 1530 (1995)
5. T. Clarysse, D. Vanhaeren, I. Hoflijk, W. Vandervorst, Characterization of electrically active dopant profiles with the spreading resistance probe. Mater. Sci. Eng.: R: Rep. **47**(5–6), 123–206 (2004)
6. G. Binnig, C.F. Quate, Ch. Gerber, Atomic force microscope. Phys. Rev. Lett. **56**, 930–933 (1986)
7. A. Schulze, P. Eyben, T. Hantschel, W. Vandervorst, Scanning spreading resistance microscopy (SSRM): high-resolution 2D and 3D carrier mapping of semiconductor nanostructures, in *Metrology and Diagnostic Techniques for Nanoelectronics*, pp. 377–445
8. W.R. Thurber, R.L. Mattis, Y.M. Liu, J.J. Filliben, Resistivity-dopant density relationship for boron-doped silicon. J. Electrochem. Soc. **127**(10), 2291–2294 (1980)
9. W.R. Thurber, R.L. Mattis, Y.M. Liu, J.J. Filliben, Resistivity-dopant density relationship for phosphorus-doped silicon. J. Electrochem. Soc. **127**(8), 1807–1812 (1980)
10. Mool C. Gupta, Arthur L. Ruoff, Static compression of silicon in the [100] and in the [111] directions. J. Appl. Phys. **51**(2), 1072–1075 (1980)
11. J.Z. Hu, L.D. Merkle, C.S. Menoni, I.L. Spain, Crystal data for high-pressure phases of silicon. Phys. Rev. B **34**(7), 4679–4684 (1986)
12. G.M. Pharr, W.C. Oliver, R.F. Cook, P.D. Kirchner, M.C. Kroll, T.R. Dinger, D.R. Clarke, Electrical resistance of metallic contacts on silicon and germanium during indentation. J. Mater. Res. **7**(4), 961–972 (1992)
13. K. Mylvaganam, L.C. Zhang, P. Eyben, J. Mody, W. Vandervorst, Evolution of metastable phases in silicon during nanoindentation: mechanism analysis and experimental verification. Nanotechnology **20**, 305705 (2009)
14. Y.-H. Lin, T.-C. Chen, P.-F. Yang, S.-R. Jian, Y.-S. Lai, Atomic-level simulations of nanoindentation-induced phase transformation in mono-crystalline silicon. Appl. Surf. Sci. **254**(5), 1415–1422 (2007)
15. A.K. Lu, Modeling of nano pressure contacts on Ge. Master thesis (University of Liege, Liege, Belgium, 2012)
16. P. Eyben, Scanning spreading resistance microscopy: high resolution two-dimensional carrier profiling of semiconductor structures. Ph.D. thesis (KU Leuven, Leuven, Belgium, 2004)
17. P. Eyben, S. Denis, T. Clarysse, W. Vandervorst, Progress towards a physical contact model for scanning spreading resistance microscopy. Mater. Sci. Eng. B **102**(1–3), 132–137 (2003)
18. B.G. Streetman, S.K. Banerjee, *Solid State Electronic Devices* (Pearson Prentice Hall, Upper Saddle River, 2010)

19. A. Schulze, A.S. Verhulst, A. Nazir, T. Hantschel, P. Eyben, W. Vandervorst, A comprehensive model for the electrical nanocontact on germanium for scanning spreading resistance microscopy applications. J. Appl. Phys. **113**, 114310 (2013)
20. E. Simoen, F. De Stefano, G. Eneman, B. De Jaeger, C. Claeys, F. Crupi, On the temperature and field dependence of trap-assisted tunneling current in Ge $p^{+}n$ junctions. IEEE Electron Device Lett. **30**(5), 562–564 (2009)
21. M. Xu, Application of scanning spreading resistance microscopy on InP-based materials and devices. Ph.D. thesis (KU Leuven, Leuven, Belgium, 2005)
22. D. Lin, G. Brammertz, S. Sioncke, C. Fleischmann, A. Delabie, K. Martens, H. Bender, T. Conard, W.H. Tseng, J.C. Lin, W.E. Wang, K. Temst, A. Vatomme, J. Mitard, M. Caymax, M. Meuris, M. Heyns, T. Hoffmann, Enabling the high-performance InGaAs, Ge CMOS: a common gate stack solution, in *Electron Devices Meeting*, IEDM (Technical Digest, IEEE International, 2010), p. 2009
23. T. Clarysse, P. Eyben, T. Janssens, I. Hoflijk, D. Vanhaeren, A. Satta, M. Meuris, W. Vandervorst, J. Bogdanowicz, G. Raskin, Active dopant characterization methodology for germanium. J. Vac. Sci. & Technol. B **24**(1), 381–389 (2006)
24. A. Schulze, T. Hantschel, P. Eyben, A. Vandooren, R. Rooyackers, J. Mody, A.S. Verhulst, W. Vandervorst, Quantitative two-dimensional carrier mapping in silicon nanowire-based tunnel-field effect transistors using scanning spreading resistance microscopy, in *Proceedings MRS Spring Meeting 2010*, vol. 1258 (San Francisco, USA, 2010)
25. P. Eyben, F. Clemente, K. Vanstreels, G. Pourtois, T. Clarysse, E. Duriau, T. Hantschel, K. Sankaran, J. Mody, W. Vandervorst, K. Mylvaganam, L. Zhang, Analysis and modeling of the high vacuum scanning spreading resistance microscopy nanocontact on silicon. J. Vac. Sci. & Technol. B: Microelectron. Nanometer Struct. **28**(2), 401–406 (2010)
26. T. Clarysse, M. Caymax, P. De Wolf, T. Trenkler, W. Vandervorst, J.S. McMurray, J. Kim, C.C. Williams, J.G. Clark, G. Neubauer, Epitaxial staircase structure for the calibration of electrical characterization techniques. J. Vac. Sci. & Technol. B **16**(1), 394–400 (1998)
27. P. Eyben, J. Mody, A. Nazir, A. Schulze, T. Clarysse, T. Hantschel, W. Vandervorst, Sub-nanometer characterization of nanoelectronic devices, in *Fundamentals of Picoscience*, pp. 677–705
28. A. Schulze, Two- and three-dimensional dopant and conductivity profiling in confined volumes. Ph.D. thesis (KU Leuven, Belgium, 2013)
29. P. Eyben, J. Mody, A. Nazir, A. Schulze, T. Hantschel, W. Vandervorst, *Development and Optimization of FIB-Based Sample Preparation for SSRM* (E-MRS Fall Meeting, Warsaw (Poland), 2010)
30. W. Vandervorst, C. Fleischmann, J. Bogdanowicz, A. Franquet, U. Celano, K. Paredis, A. Budrevich, Dopant, composition and carrier profiling for 3D structures. Mater. Sci. Semicond. Process. **62**, 31–48 (2017)
31. T. Hantschel, C. Demeulemeester, P. Eyben, V. Schulz, O. Richard, H. Bender, W. Vandervorst, Conductive diamond tips with sub-nanometer electrical resolution for characterization of nanoelectronics device structures. physica status solidi (a) **206**(9), 2077–2081 (2009)
32. T. Hantschel, M. Tsigkourakos, L. Zha, T. Nuytten, K. Paredis, B. Majeed, W. Vandervorst, Diamond scanning probes with sub-nanometer resolution for advanced nanoelectronics device characterization. Microelectron. Eng. **159** (2016)
33. P. Niedermann, W. Hanni, N. Blanc, R. Christoph, J. Burger, Chemical vapor deposition diamond for tips in nanoprobe experiments. J. Vac. Sci. & Technol. A **14**, 1233–1236 (1996)
34. C.Y. Tang, L.C. Zhang, A molecular dynamics analysis of the mechanical effect of water on the deformation of silicon monocrystals subjected to nano-indentation. Nanotechnology **16**(1), 15–20
35. T. Noda, J. Mitard, L. Witters, G. Hellings, C. Vrancken, P. Eyben, A. Thean, N. Horiguchi, W. Vandervorst, Analysis of dopant diffusion and defects in SiGe-channel implant free quantum well (IFQW) devices using an atomistic kinetic monte carlo approach (IEEE, New York), pp. 30.2.1–30.2.4

36. R.H. Dennard, F.H. Gaensslen, V.L. Rideout, E. Bassous, A.R. LeBlanc, Design of ion-implanted MOSFET's with very small physical dimensions. IEEE J. Solid-State Circuits **9**(5), 256–268 (1974)
37. T. Ghani, M. Armstrong, C. Auth, M. Bost, P. Charvat, G. Glass, T. Hoffmann, K. Johnson, C. Kenyon, J. Klaus, B. McIntyre, K. Mistry, A. Murthy, J. Sandford, M. Silberstein, S. Sivakumar, P. Smith, K. Zawadzki, S. Thompson, M. Bohr, A 90 nm high volume manufacturing logic technology featuring novel 45 nm gate length strained silicon CMOS transistors (IEEE, New York), pp. 11.6.1–11.6.3
38. K. Mistry, C. Allen, C. Auth, B. Beattie, D. Bergstrom, M. Bost, M. Brazier, M. Buehler, A. Cappellani, R. Chau, C.-H. Choi, G. Ding, K. Fischer, T. Ghani, R. Grover, W. Han, D. Hanken, M. Hattendorf, J. He, J. Hicks, R. Huessner, D. Ingerly, P. Jain, R. James, L. Jong, S. Joshi, C. Kenyon, K. Kuhn, K. Lee, H. Liu, J. Maiz, B. McIntyre, P. Moon, J. Neirynck, S. Pae, C. Parker, D. Parsons, C. Prasad, L. Pipes, M. Prince, P. Ranade, T. Reynolds, J. Sandford, L. Shifren, J. Sebastian, J. Seiple, D. Simon, S. Sivakumar, P. Smith, C. Thomas, T. Troeger, P. Vandervoorn, S. Williams, K. Zawadzki, A 45 nm logic technology with high-k+metal gate transistors, strained silicon, 9 Cu interconnect layers, 193 nm dry patterning, and 100% Pb-free packaging (IEEE, New York), pp. 247–250
39. S.-H. Oh, D. Monroe, J.M. Hergenrother, Analytic description of short-channel effects in fully-depleted double-gate and cylindrical, surrounding-gate MOSFETs. IEEE Electron Device Lett. **21**(9), 445–447
40. R.-H. Yan, A. Ourmazd, K.F. Lee, Scaling the Si MOSFET: from bulk to SOI to bulk. IEEE Trans. Electron Devices **39**(7), 1704–1710
41. N. Planes, O. Weber, V. Barral, S. Haendler, D. Noblet, D. Croain, M. Bocat, P.-O. Sassoulas, X. Federspiel, A. Cros, A. Bajolet, E. Richard, B. Dumont, P. Perreau, D. Petit, D. Golanski, C. Fenouillet-Beranger, N. Guillot, M. Rafik, V. Huard, S. Puget, X. Montagner, M.-A. Jaud, O. Rozeau, O. Saxod, F. Wacquant, F. Monsieur, D. Barge, L. Pinzelli, M. Mellier, F. Boeuf, F. Arnaud, M. Haond, 28 nm FDSOI technology platform for high-speed low-voltage digital applications (IEEE, New York), pp. 133–134
42. X. Huang, W.-C. Lee, C. Kuo, D. Hisamoto, L. Chang, J. Kedzierski, E. Anderson, H. Takeuchi, Y.-K. Choi, K. Asano, V. Subramanian, T.-J. King, J. Bokor, C. Hu, Sub 50-nm FinFET: PMOS (IEEE, New York), pp. 67–70
43. C.-H. Jan, U. Bhattacharya, R. Brain, S.-J. Choi, G. Curello, G. Gupta, W. Hafez, M. Jang, M. Kang, K. Komeyli, T. Leo, N. Nidhi, L. Pan, J. Park, K. Phoa, A. Rahman, C. Staus, H. Tashiro, C. Tsai, P. Vandervoorn, L. Yang, J.-Y. Yeh, P. Bai, A 22 nm SoC platform technology featuring 3-D tri-gate and high-k/metal gate, optimized for ultra low power, high performance and high density SoC applications (IEEE, New York), pp. 3.1.1–3.1.4
44. H. Mertens, R. Ritzenthaler, V. Pena, G. Santoro, K. Kenis, A. Schulze, E. D. Litta, S. A. Chew, K. Devriendt, r. Chiarella, S. Demuynck, D. Yakimets, D. Jang, A. Spessot, G. Eneman, A. Dangol, P. Lagrain, H. Bender, S. Sun, M. Korolik, D. Kioussis, M. Kim, K-.H. Bu, S. C. Chen, M. Cogorno, J. Devrajan, J. Machillot, N. Yoshida, N. Kim, K. Barla, D. Mocuta, N. Horiguchi, Vertically stacked gate-all-around Si nanowire transistors: key process optimizations and ring oscillator demonstration (IEEE, New York), pp. 37.4.1–37.4.4
45. L. Witters, F. Sebaai, A. Hikavyy, A. P. Milenin, R. Loo, A. De Keersgieter, G. Eneman, T. Schram, K. Wostyn, K. Devriendt, A. Schulze, R. Lieten, S. Bilodeau, E. Cooper, P. Storck, C. Vrancken, H. Arimura, P. Favia, E. Vancoille, J. Mitard, R. Langer, A. Opdebeeck, F. Holsteyns, N. Waldron, K. Barla, V. De Heyn, D. Mocuta, N. Collaert, Strained germanium gate-all-around PMOS device demonstration using selective wire release etch prior to replacement metal gate deposition (IEEE, New York), pp. T194–T195
46. A.A. Budrevich, W. Vandervorst, Sims analysis on the transistor scale: probing composition and dopants in nonplanar, confined 3D volumes, in *Metrology and Diagnostic Techniques for Nanoelectronics*, pp. 153–206
47. A.K. Kambham, G. Zschaetzsch, Y. Sasaki, M. Togo, N. Horiguchi, J. Mody, A. Florakis, D.R. Gajula, A. Kumar, M. Gilbert, W. Vandervorst, Atom probe tomography for 3D-dopant analysis in FinFET devices (IEEE, New York), pp. 77–78

48. J. Mody, A.K. Kambham, G. Zschatzsch, T. Chiarella, N. Collaert, L. Witters, P. Eyben, M. Gilbert, S. Kolling, A. Schulze, T.-Y. Hoffmann, W. Vandervorst, Dopant and carrier profiling for 3D-device architectures (IEEE, New York), pp. 108–113
49. A. Schulze, T. Hantschel, P. Eyben, A.S. Verhulst, R. Rooyackers, A. Vandooren, J. Mody, A. Nazir, D. Leonelli, W. Vandervorst, Observation of diameter dependent carrier distribution in nanowire-based transistors. Nanotechnology **22**(18) (2011)
50. A. Schulze, A. Florakis, T. Hantschel, P. Eyben, A.S. Verhulst, R. Rooyackers, A. Vandooren, W. Vandervorst, Diameter-dependent boron diffusion in silicon nanowire-based transistors. Appl. Phys. Lett. **102**(5), 052108
51. Sentaurus process, rel. f-2011.09, synopsys, 2011
52. A. Vandooren, R. Rooyackers, D. Leonelli, F. Iacopi, E Kunnen, D. Nguyen, M. Demand, P. Ong, L. Willie, J. Moonens, O. Richard, A.S. Verhulst, W.G. Vandenberghe, G. Groeseneken, S. De Gendt, M. Heyns, A 35 nm diameter vertical silicon nanowire short-gate tunnelFET with high-k/metal gate, in *Silicon Nanoelectronics Workshop* (Kyoto, 2009), p. 21
53. D.C. Mueller, E. Alonso, W. Fichtner, Arsenic deactivation in Si: electronic structure and charge states of vacancy-impurity clusters. Phys. Rev. B **68**(4), 045208 (2003)
54. D.W. Lawther, U. Myler, P.J. Simpson, P.M. Rousseau, P.B. Griffin, J.D. Plummer, Vacancy generation resulting from electrical deactivation of arsenic. Appl. Phys. Lett. **67**(24), 3575–3577 (1995)
55. A. Schulze, T. Hantschel, P. Eyben, A.S. Verhulst, R. Rooyackers, A. Vandooren, W. Vandervorst, Quantitative three-dimensional carrier mapping in nanowire-based transistors using scanning spreading resistance microscopy. Ultramicroscopy **125**, 18–23 (2013)
56. J. Mody, P. Eyben, W. Polspoel, J. Malgorzata, W. Vandervorst, Scanning spreading resistance microscopy for 3D-Carrier profiling in FinFET-Based structures (San Francisco, 2008)
57. J. Mody, G. Zschatzsch, S. Kolling, A. De Keersgieter, G. Eneman, A.K. Kambham, C. Drijbooms, A. Schulze, T. Chiarella, N. Horiguchi, T.-Y. Hoffmann, P. Eyben, W. Vandervorst, 3D-carrier profiling in FinFETs using scanning spreading resistance microscopy, in *2011 IEEE International Electron Devices Meeting (IEDM)* (2011)
58. A. Schulze, T. Hantschel, A. Dathe, P. Eyben, X. Ke, W. Vandervorst, Electrical tomography using atomic force microscopy and its application towards carbon nanotube-based interconnects. Nanotechnology **23**, 305707 (2012)
59. P. Eyben, T. Chiarella, S. Kubicek, H. Bender, O. Richard, J. Mitard, A. Mocuta, N. Horiguchi, A.V.-Y. Thean, Scalpel soft retrace scanning spreading resistance microscopy for 3D-carrier profiling in sub-10 nm WFIN FinFET (IEEE, New York), pp. 14.1.1–14.1.4
60. U. Celano, F.-C. Hsia, D. Vanhaeren, K. Paredis, T.E. M. Nordling, J.G. Buijnsters, T. Hantschel, W. Vandervorst, Mesoscopic physical removal of material using sliding nano-diamond contacts. Sci. Rep. **8**(1)
61. A. Schulze, R. Cao, P. Eyben, T. Hantschel, W. Vandervorst, Outwitting the series resistance in scanning spreading resistance microscopy. Ultramicroscopy **161**, 59–65
62. Method for determining local resistivity and carrier concentration using scanning spreading resistance measurement set-up
63. P. Eyben, P. Bisiaux, A. Schulze, A. Nazir, W. Vandervorst, Fast fourier transform scanning spreading resistance microscopy: a novel technique to overcome the limitations of classical conductive AFM techniques. Nanotechnology **26**(35), 355702
64. N.G. Orji, M. Badaroglu, B.M. Barnes, C. Beitia, B.D. Bunday, U. Celano, R.J. Kline, M. Neisser, Y. Obeng, A.E. Vladar, Metrology for the next generation of semiconductor devices. Nat. Electron. **1**, 532–547 (2018)
65. U. Celano, Individual device analysis using hybrid TEM-scalpel SSRM metrology, in *Frontiers for Materials Characterization at the Nanoscale (FMCN)*

Chapter 4
Scanning Capacitance Microscopy for Two-Dimensional Carrier Profiling of Semiconductor Devices

Jay Mody and Jochonia Nxumalo

Abstract As the semiconductor technology matures from research to development and eventually entering manufacturing, there is a consistent focus on reducing defects and yield detractors. This results in engineers utilizing the Failure Mode and Effects Analysis (FMEA) duplicate of integrated circuits. In failure analysis (FA) of integrated circuits, Scanning Capacitance Microscopy (SCM) has been used to identify failure mechanisms, such as regions of incorrect doping and electrical shorts, thereby indicating the appropriate corrective actions required to remedy the device. Because sample preparation and data interpretation are relatively straightforward, FA applications of SCM can be performed with quick turnaround and with few ambiguities that can arise in quantitative applications. In this chapter, we will focus on SCM applications, highlighting work performed at the state-of-the-art chip manufacturing facility of GLOBALFOUNDRIES.

4.1 Working Principle of Scanning Capacitance Microscopy

Scanning capacitance microscope (SCM) is a variation of scanning probe microscope (SPM) that extracts the two-dimensional profile in semiconductor devices by measuring the local capacitance with a tip [1–9] Fig. 4.1a. When a metallic tip is placed on a semiconductor surface with an oxide layer in between, as shown in Fig. 4.1a, a metal-oxide-semiconductor (MOS) capacitor is formed as shown in Fig. 4.1b.

In SCM, the tip is usually grounded and a bias voltage (V) is applied to the sample instead of a gate electrode as in semiconductor devices. The depletion region formed under the tip (see Fig. 4.1a) and the associated depletion capacitance (C_{dep}) is detected by measuring the capacitance (C) between the tip and the sample and the voltage dependence of the capacitance (C-V characteristics) is mapped in the

J. Mody (✉) · J. Nxumalo
GLOBALFOUNDRIES, 400 Stone Break Rd Extension, Malta, NY 12020, USA
e-mail: jay.mody@globalfoundries.com

J. Nxumalo
e-mail: jocho.nxumalo@globalfoundries.com

U. Celano (ed.), *Electrical Atomic Force Microscopy for Nanoelectronics*, NanoScience and Technology, https://doi.org/10.1007/978-3-030-15612-1_4

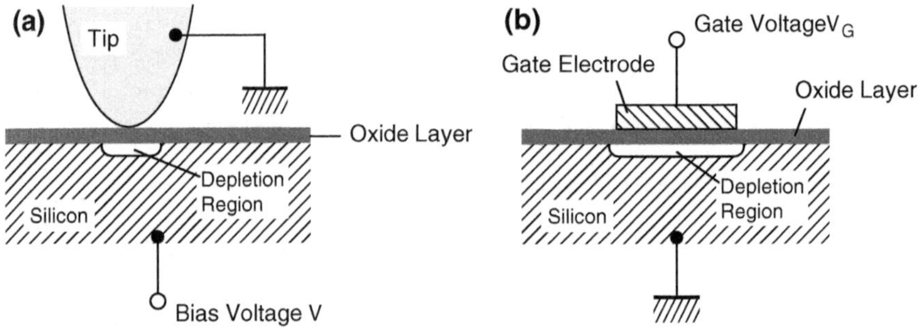

Fig. 4.1 **a** MOS structure of tip and sample in SCM observation scheme. **b** MOS structure in a semiconductor device

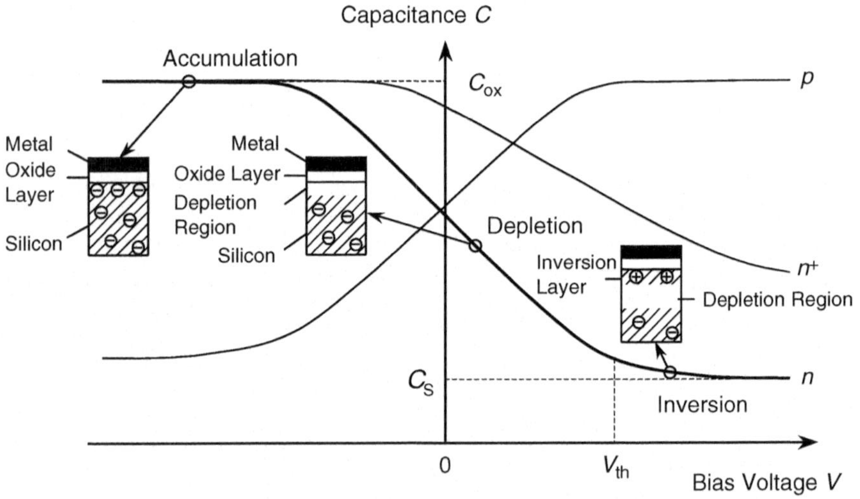

Fig. 4.2 *C-V* curves of one-dimensional MOS structure in SCM scheme

scanning area. As a practical matter, SCM measures a capacitance variation (ΔC) caused by a voltage modulation (ΔV) and we obtain dC/dV, because C is too small to be measured.

SCM signal is reviewed hereafter using a one-dimensional MOS structure model (Fig. 4.2) for simplicity. For an n-type semiconductor, a negative sample bias voltage causes majority carrier (electron) accumulation and then the MOS capacitance becomes equal to that of a capacitor formed by the oxide layer as a dielectric layer (C_{ox}). When a positive bias voltage is applied to the sample, a depletion region is induced just under the oxide layer and the effective MOS capacitance is decreased due to the series connection of C_{ox} and the capacitance of the depletion layer (C_{dep}). C decreases monotonically with positive V because of the increase of the thickness of the depletion layer. If V exceeds the threshold voltage V_{th}, the inversion occurs under the oxide and after that, the C converges to a specific value of inversion capacitance (C_s) because a higher bias voltage just accumulates the minority carrier (hole) in the inversion layer with a constant thickness (x_{dm}) of the depletion layer. (Here it is assumed that the carriers in the inversion layer follow only the bias voltage, not the UHF voltage for the capacitance measurement. However, in reality, the carriers in the inversion layer sometimes do not follow the AC bias voltage to form deep depletion or it follows the UHF voltage due to minority carrier generation by photoexcitation or minority carrier injection from p layer close to the measuring position. Higher dopant concentration raises V_{th} and lowers x_{dm} (i.e., increases C_s), resulting a lower dC/dV signal through a gentler slope in the C-V curve. In the case of p-type semiconductor, the majority carriers (holes) depletes at a negative sample bias, giving a dC/dV signal with the opposite polarity. Based on the principle mentioned above, we can distinguish the carrier type by the polarity of the dC/dV signal and evaluate the dopant concentration from the amplitude of the signal. In the real case, a dC/dV signal has a peak at a specific dopant concentration and drops with decreasing dopant concentration lower than the peak value because of a shift in the flat band voltage and/or a non-negligible spreading resistance in series with the capacitance in measurement. Thus, sufficient care needs to be taken while interpreting an SCM image for the non-unique dopant concentration related to the dC/dV signal obtained.

The instrumentation of SCM is a contact mode AFM combined with an electronic system that detects the dC/dV signal (Fig. 4.3). An RCA (Radio Corporation of America) sensor for a video disc pickup is used for capacitance detection. The sensor consists of a UHF (c.a. 1 GHz) oscillator, a resonator, and a detector. The capacitance between the sample and the tip is used as a part of the resonator. The capacitance variation conducts a resonance frequency shift of the resonator to change the amplitude of the UHF coming into the detector from the oscillator through the resonator. The dC/dV signal is obtained by lock-in detection of the detector output with an AC bias voltage (20–150 kHz) to the sample.

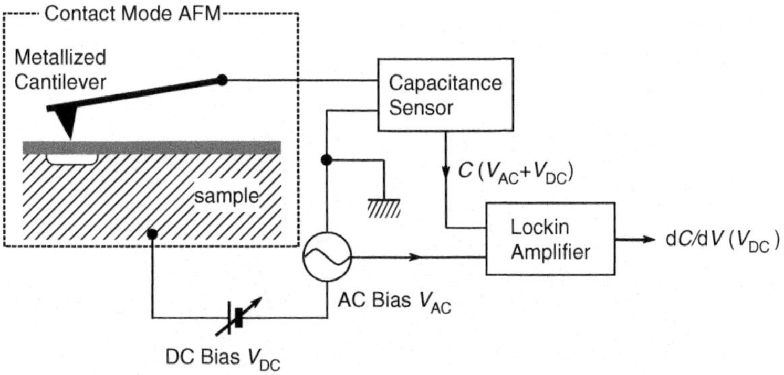

Fig. 4.3 Schematic illustration of SCM

4.2 Applications of Scanning Capacitance Microscopy

Failure analysis is a key component in the development and manufacturing of new semiconductor technologies. During the initial stages of process development and integration, feedback from FA studies is paramount to identifying root causes of device failures and providing the subsequent process, integration or design changes that are necessary to eliminate the observed failure mechanism and, during manufacturing failure analysis is utilized to eliminate the yield detractors. Failure analysis uses a wide array of analytical tools to isolate and then characterize the failure site physically, electrically, and chemically. Scanning capacitance microscopy has attracted increasing attention as a failure analysis tool because of its ability to provide 2D free carrier profiles in a semiconductor device, yielding information that, in many cases, cannot be obtained with other analysis techniques.

4.2.1 Effect of Hot Carrier Stress on Device Junctions

Device reliability is a key issue for semiconductor manufacturing and development. One of the main detractors for reliability is attributed to the hot carrier effect. During the hot carrier stress for NFET, a constant voltage is applied to the drain and the gate of a device, which causes device drive current, I, to degrade over time. The mechanism of hot carrier degradation has been widely discussed over the past few decades [10–12]. Multiple publications have attributed the degradation to hot electron injection into the gate oxide creating trapping states at the Si/SiO_2 interface, as depicted in Fig. 4.4.

In addition to this charge trapping mechanism, an alternative model of hydrogen passivation of dopant during hot carrier stress has been proposed for bipolar device [13], but has been overlooked by the reliability community over the years for MOS FET devices [10–12, 14, 15]. Wang et al. [16] discuss the direct observation of the junction profile and active carrier concentration for hot carrier stressed devices which are consistent with this alternative mechanism.

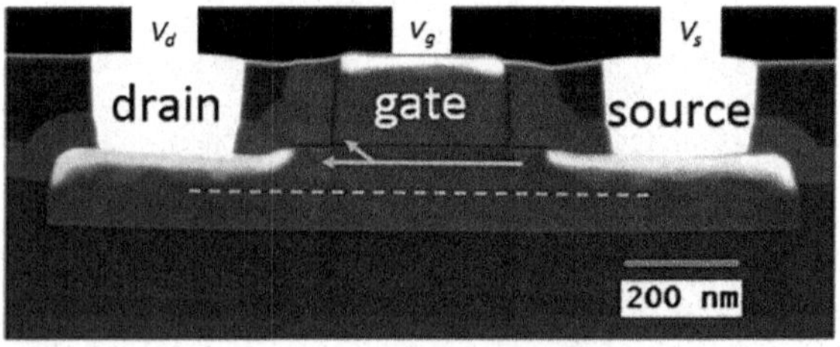

Fig. 4.4 Dark field STEM (Scanning Transmission Electron Microscopy) image of a device with hot carrier stressed testing. Arrow indicates the electron flowing from source to drain and injecting into the gate oxide, with grounded source (V_s of 0 V)

Electron holography is carried out on a dual lens holography system on a FEI Titan microscope. The holography fringe spacing is set to 1.6 nm, and microscope operating voltage is 200 kV. Samples are prepared by focused ion beam (FIB) method to about 400 nm thick. A full characterization of FIB prepared samples for junction profile measurement can be found in the article by Cooper et al. [17]. The junction profile is obtained by tilting the sample off of the [1 1 0] zone axis along the [1 $\bar{1}$ 0] axis by about 3°–4°, and along the [0 0 1] to less than 1° to minimize the projection and optimize the spatial resolution along the channel. The principle of electron holography is to use two coherent electron beams to form an interference hologram with one beam passing through vacuum as reference and other passing through the sample. Through data processing, amplitude and phase maps are derived from the hologram using the Holoworks plugin package running within Gatan's Digital Micrograph software. Under conditions of uniform sample thickness and low elastic scattering contrast, the recorded phase map is directly proportional to the electric potential map. The recorded maps in this study have a spatial resolution of about $\approx$ 5 nm (3x fringe spacing). For SCM samples, devices are delayered down to the CA (contact to active region) level. The CA contacts are connected to the ground. The sample is cross sectioned by polishing from one side.

In this study, three kinds of device were mapped by holography and SCM: they are listed in Table 4.1.

Device J1 (Junction 1) is an NFET silicon-on-insulator (SOI) device with gate length of 300 nm and with As implanted in the extension region. The extension region is defined as the region in Si under the nitride spacer near the gate, shown in Fig. 4.5. Device J2 (Junction 2) is a device with 280 nm gate length in which a low dose of P is additionally implanted into the extension region to broaden the junction and reduce the electric field there. Note, the difference in gate length between J1 and J2 does not influence the junction width under the extension region of the device. Device J3 (Junction 3) is device J2 that has undergone the hot carrier reliability stress test whereby the device was stressed by applying 3.1 V on the drain, 1.5 V on the gate and grounding the source. The key device parameters such as saturation current, $I_{d(sat)}$, and threshold voltage, V_t, were monitored periodically for the duration of the test. After 10,000 s stress, the device showed 7.3% degradation in the drive current.

Figure 4.6a–c are phase maps measured by dual lens electron holography for J1, J2, and J3, respectively. J1 and J2 in Fig. 4.6a, b show a symmetrical device: source (right) and drain (left) side exhibit similar junctions. Equipotential lines for J1 are

Table 4.1 Sample list for the process and stress condition

	Extension implant	Stress condition and result
Device J1	As	No stress
Device J2	As/P	No stress
Device J3	As/P	3.1 V(drain), 1.5 V (gate), 10^4 s, −7.3% ($\Delta I/I$)

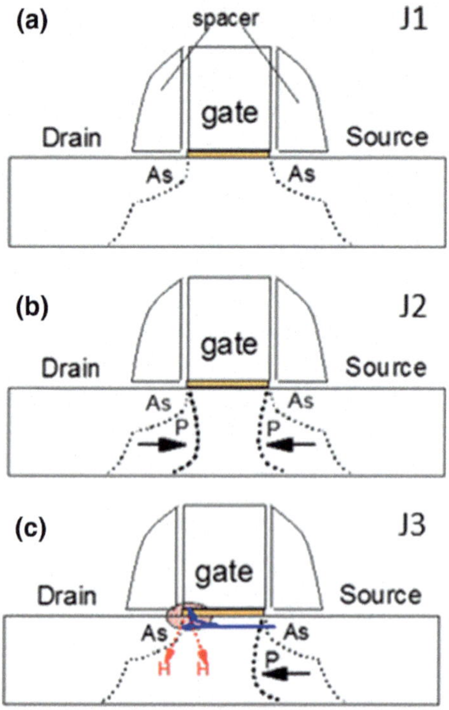

Fig. 4.5 **a** J1 with As implant in the extension; **b** J2 (As/P implant) without stress; and **c** J3 (As/P implant) after hot carrier stress testing, energetic electron (blue line) injects into drain side gate oxide, breaking Si-H bond in the gate oxide, generating free hydrogen to form an H-P (or H-Si-P) bond, and passivating P dopant in the extension region on the drain side

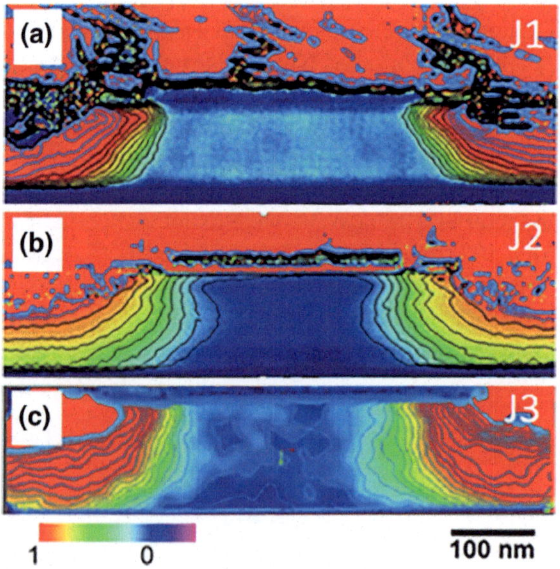

Fig. 4.6 Junction profile by dual lens electron holography: phase map for device J1, J2, and J3. The drain is on the left side and the source is on the right side

closely spaced on both source and drain sides, indicating abrupt junctions; while those for J2 are further apart indicating broader junctions. By contrast, J3 in Fig. 4.6c shows an asymmetric device, with close packed equipotential lines on drain side and less dense equipotential lines on the source side.

The intensity line profiles taken across the channel region of the electric potential maps from the three devices are shown in Fig. 4.7. The trajectory of the line profile is depicted as a dashed line in Fig. 4.4. The width of the junction on the source and drain side, W_s and W_d, respectively, are measured from the separation of the inflection points of these line profiles and the values are listed in Table 4.2. The junction widths of the source and drain for J1 are the same within experimental error. This is also true for J2. The junction widths for J2 are larger than those of J1 by approximately 46 ± 5 nm (average value). For the stressed device, J3, there is a junction width difference between source and drain of about 38 ± 5 nm with drain side being narrower. Furthermore, the reduced drain side junction width of J3 (161 nm) is closer to that of J1 (152 nm) than J2 (197 nm).

Figure 4.8a–c are the SCM active carrier maps for J1, J2, and J3, respectively. In these images, the dark region in Si indicates n-type carrier; while the bright region indicates p-type carrier. The line profiles near the top of channel region for these three devices are shown in Fig. 4.9 whereby a positive dC/dV value indicates p-type carrier while a negative one indicates n-type carrier. As expected, for all

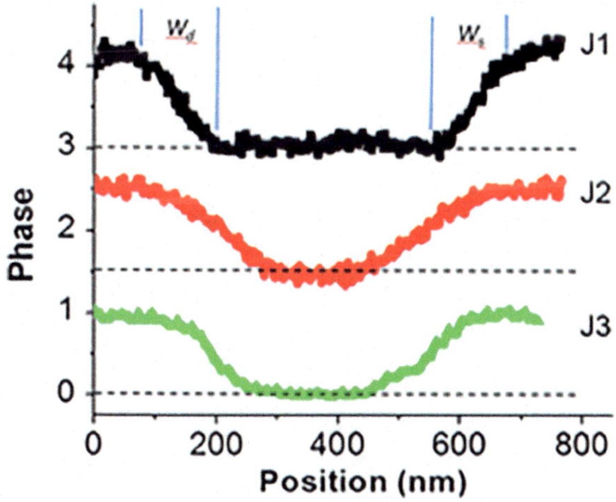

Fig. 4.7 Line profile of phase obtained from Fig. 4.6 in the middle of Si (indicated in Fig. 4.4 as the dashed line) for the three devices listed in Table 4.1

Table 4.2 Source and drain junction width for the three devices, with the measurement error of ±5 nm

	W_d (nm)	W_s (nm)
Device J1	153	150
Device J2	200	195
Device J3	161	199

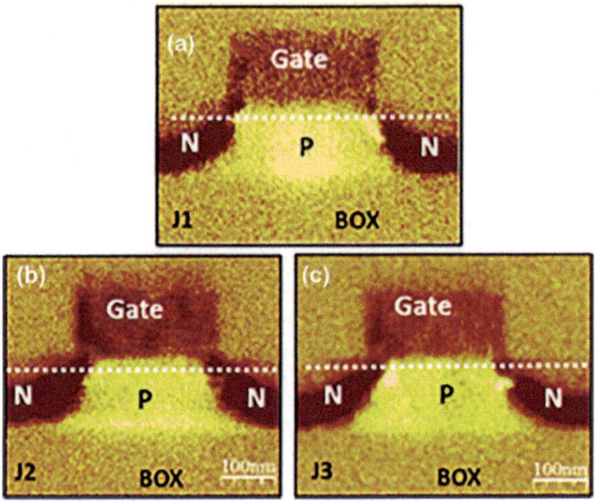

Fig. 4.8 SCM carrier density maps **a**, **b**, and **c** corresponding to three devices listed in Table 4.1: J1, J2, and J3, respectively. Left is the drain side, and right is the source side

devices, the n carrier concentration gets lower on approaching p-n junction from source and drain sides reflecting the effect of low dose extension doping. When approaching both junctions from the channel region for the device J1, with As dopant only, there is a positive increase in dC/dV indicating a hole (majority carrier) depletion on p-type Si in that region (indicated by the arrows). For device J2, with As/P co-implant, there is no increase in dC/dV near either junction approaching from the channel region, indicating no hole depletion on the p side of the junction. For stressed device J3, although having the same implant as J2, the drain side exhibits a hole depletion (indicated by the arrow) similar to the carrier depletion of the device J1, while the source side exhibits no hole depletion similar to J2.

We also performed TCAD (Technology Computer Aided Design) simulation shown in Fig. 4.10. Figure 4.10a depicts the case of J1, with As extension implant only, which shows abrupt junctions (narrow junction width) on both sides, while Fig. 4.10b depicts the case of J2, with As/P co-implant condition, which shows broad junctions on both sides. Figure 4.10c shows simulation of a hypothetical device proposed for the stressed device J3, where the source (right) side has As/P co-implant same as J2 and the drain (left) side has As only implant without P, same as J1. This model of asymmetric implant resembles the observed junction profile of J3.

The extension of n-type region under the gate (as indicated by the arrows in Fig. 4.10 is known as overlap capacitance (C_{OV}), and it is one of the parameters which determines the turn on voltage and drive current. The smaller the C_{OV} is, the less the drive current for the device. In this case, the C_{OV} shown in Fig. 4.10c is smaller on the left side comparing with the one on the right side of the device. The simulated trends in C_{OV} changes correlate well with the measured ones by holography (Fig. 4.6).

The direct observations of stressed device junction profiles present us an alternative mechanism to explain the device degradation due to hot carrier stress. Whilst previous studies have generally attributed the degradation to gate oxide damage near

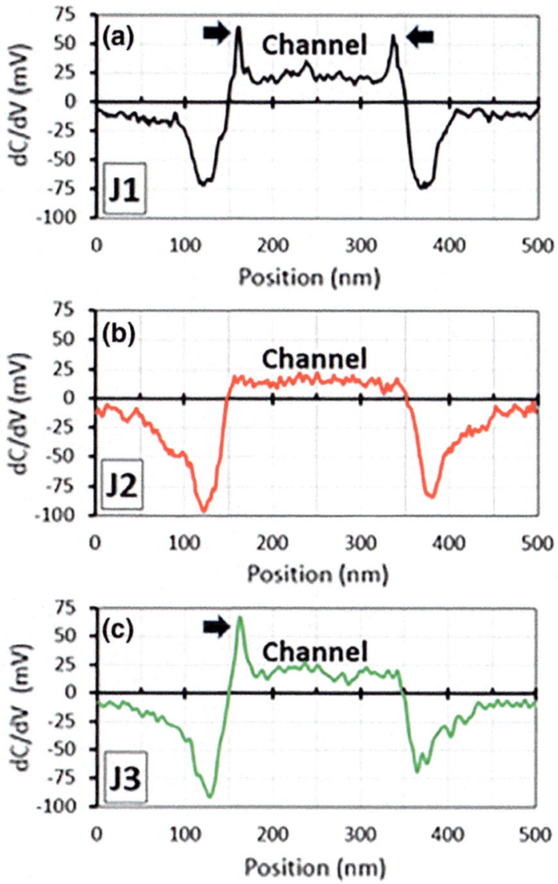

Fig. 4.9 SCM line profile **a**, **b**, and **c** near the top of channel region for three devices listed in Table 4.1: J1, J2, and J3, respectively. Left is the drain side, and right is the source side. Arrows indicate regions of hole depletion on p side

the drain side by the impact of the hot electrons, this study shows that the stress leads to an abrupt junction on the drain side of J3 which reduces C_{OV} and causes drive current degradation as electrically measured.

The observation for J1 and J2, experimental or simulation, is as expected, which is also depicted in Fig. 4.5a, b, respectively. The profiles of J1 and J2 can be understood as follows: the As implant alone for J1 would result in abrupt junctions due to the low diffusive nature of As; while, the addition of P for J2 broadens the junction due to high diffusivity of P in Si and causes the junction fronts to move towards the channel region as illustrated in Fig. 4.5. However, J3 exhibits an asymmetric device, both by holography and SCM. The source side of the stressed device looks like the unstressed As/P co-implanted device, whereas the drain side more closely resembles the unstressed device without P co-implant. It indicates that the junction of J3 on the source side is intact and not affected by the hot carrier stress test, while the junction on the drain side is changed to the one similar as that of J1, where P dopant is absent in the extension region. Thus, this leads us to attribute the observed change in the

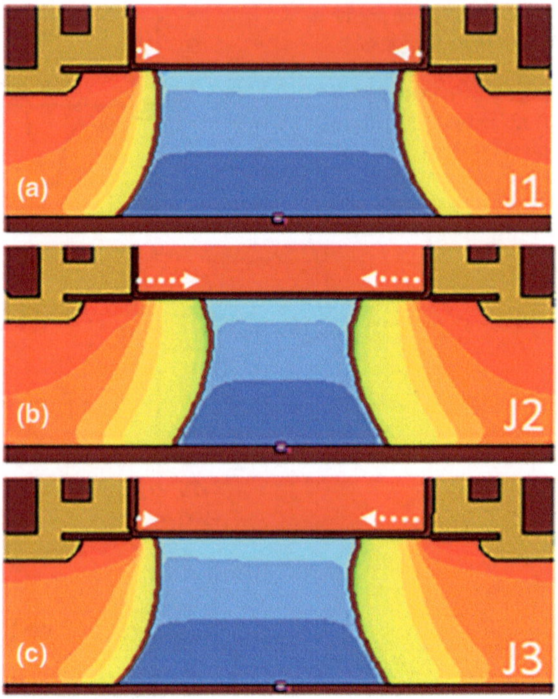

Fig. 4.10 Active dopant concentration simulation: **a** As only implant in extension; **b** As/P co-implant; and **c** As only implant on the left and As/P co-implant on the right

junction profile and carrier concentration at the drain side of J3 to the loss of active P in the extension region, as illustrated in Fig. 4.5c.

The loss of P dopant activity maybe due to displacement or out-diffusion of P from the Si lattice into an interstitial position or through passivation of P. Since the stress test was carried out at 85 °C, it is unlikely that P would be mobile during the stress testing. Accordingly, P probably remains in the Si lattice but is passivated on the drain side during the stress.

We further elaborate that P maybe passivated due to release of H resulting from electron impact of the dielectric in the gate oxide near the drain side, as illustrated in Fig. 4.5c. This hypothesis is in line with a theoretical model of hot carrier stress published by Hu et al. [11]. According to Hu et al., the hot electrons inject into gate oxide near the drain site, breaking Si-H bonds at the SiO_2/Si interface, creating trapped states and causing a change of the drive current. In this section, to explain our experimental data, we propose that the released hydrogen is free to capture an electron from P to form an H-P bond (or H-Si-P) thereby passivating P and causing the change of the drive current. Such a hydrogen passivation mechanism has been previously documented for blanket Si wafer [18–22]. It was suggested that free hydrogen can form B-H or H-Si-P (or P-H) bonds, thereby passivating dopant in Si. One recent experiment [19] shows that exposing P doped Si wafer to hydrogen plasma reduces phosphorus dopant activity in various dopant concentration. A hydrogen passivating model has also been proposed for hot carrier stress degradation in bipolar transistors

[13, 14] and as well as in the degradation of Si based photo-emitters for solar cell application [23–26].

One may argue that charge trapping may also influence the electric potential on the drain side. However, since the charge from such trapping would form a line defect in the gate oxide along the edge of the gate, the electric potential generated by this trapping charge would fall off quickly as $1/r$, which is not observed in these measurements. Therefore, such a mechanism cannot be used to explain the observed change of the junction profile on the drain side of the stressed device in this case.

Both scanning capacitance microscopy and electron holography show an asymmetrical junction profile for the hot carrier stressed device. Comparing this result with the ones for the unstressed devices with/without phosphorus extension implant and with simulation, we interpret that phosphorus in the extension region near the drain side of a hot carrier stressed device is likely to be inactive, which in turn leads to an abrupt junction and reduces the overlap capacitance near the drain side, which causes a low drive current. We further propose that the loss of active phosphorus dopant is due to hydrogen passivation, the hydrogen being released after energetic electrons are injected into the gate oxide during the hot carrier stress.

4.2.2 Root Cause Analysis for Pin Leakage

Pin leakage continues to be on the list of top yield detractors for microelectronics devices. As device geometry keeps shrinking, leakage issue is becoming more prominent. From the electrical point of view, pin leakage is simply manifested as elevated current with one pin or several pins during pin continuity test. For pin leakage issue, liquid crystal analysis (LCA) [27], photoemission analysis (PEM) [28] and laser stimulating techniques such as Optical Beam Induced Resistance Change (OBIRCH) or Thermally Induced Voltage Alteration [29–31], are effective techniques for globally localizing its fault. However, in terms of physical defects, it is not so simple. Pin leakage can be caused (1) by any type of defects, such as bridging type defects, open type defects in poly gate interconnect stack, dielectric issue and transistor junction issue etc., (2) any layers in the device, like front end of line, middle of line and back end of line, (3) any process step, such as deposition, lithography, etch, chemical mechanical polish and implantation etc. Thus, root cause analysis and identification for pin leakage are still very challenging with today's advanced failure analysis tools and techniques.

To understand the electrical signature of the failure, Song et al. [32] perform *I-V* curve tracing for a failing input pin and two output pin, comparing to several reference pins. The *I-V* curves, as shown in Fig. 4.11 for input pins and Fig. 4.12 for output pins, manifested that both failing input and output pins had nearly ohmic type leakage. As the input pin was routed to gate of input buffer and output pin was routed to drain of output buffer, but both input and output pins had Electrostatic Discharge (ESD) protection diode before their buffers, it suggested that a short type

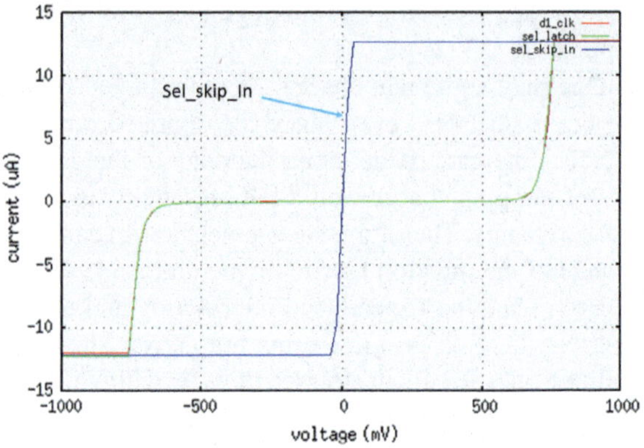

Fig. 4.11 *I-V* curves of a failing input pin (`Sel_skip_in`) with reference input pins (`d1_clk` and `sel_latch`)

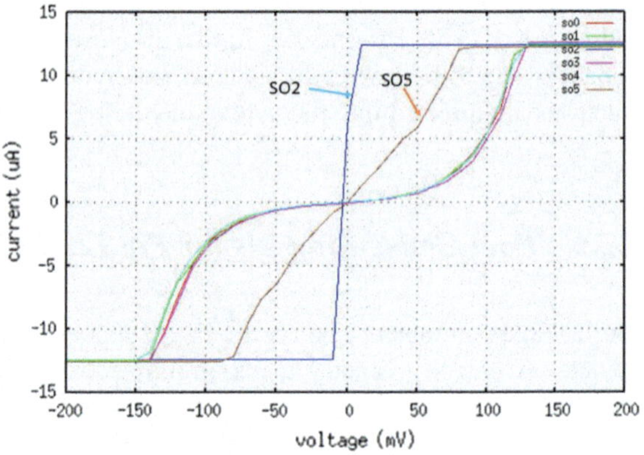

Fig. 4.12 *I-V* curves of two failing output pins (SO2 and SO5) with reference output pins (SO0, SO1, SO3 and SO4)

defect might present in the ESD protection diode for the failing I/O pins. Thus, the sample was further analyzed using OBIRCH.

OBIRCH analysis was performed on leaky input and output pins biased at a constant voltage of 0.1 V. For comparison, OBIRCH analysis was also performed on good input and output pins under the same bias condition. As expected, OBIRCH analysis showed abnormal hotspots in the ESD protection diode area for both failing input and output pins. Figure 4.13 shows the OBIRCH images.

The ESD protection diode includes many groups of diodes, connected in parallel. Each small rectangular shape in the OBIRCH image corresponds to a smallest group of diodes, which are connected in parallel too. Figure 4.14 is the layout view of the smallest group of diodes. It was interesting to notice that each failing input or output pin had multiple hotspots. Each hotspot was in oval shape and associated with a smallest group of diodes. Based on the OBIRCH response, as well as the ohmic leakage signature, the first hypothesis was a possible local bridging defect within the smallest group of diodes. Thus, the sample was subjected to physical failure analysis.

Since the sample had two similar OBIRCH spots, located at different pins. The following physical failure analysis strategy was adopted in order to allow inspection

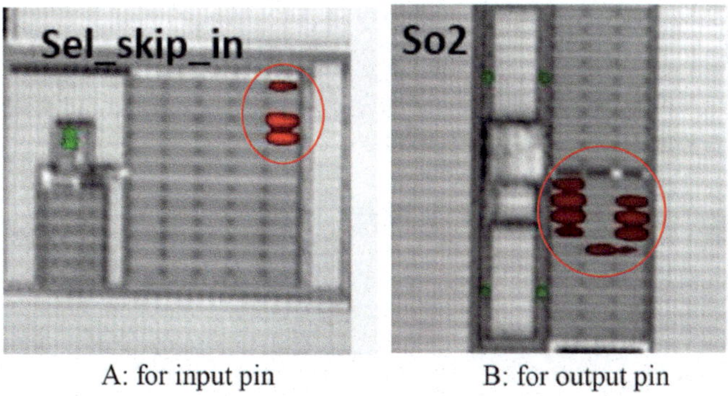

Fig. 4.13 OBIRCH image of a failing input pin and a failing output pin

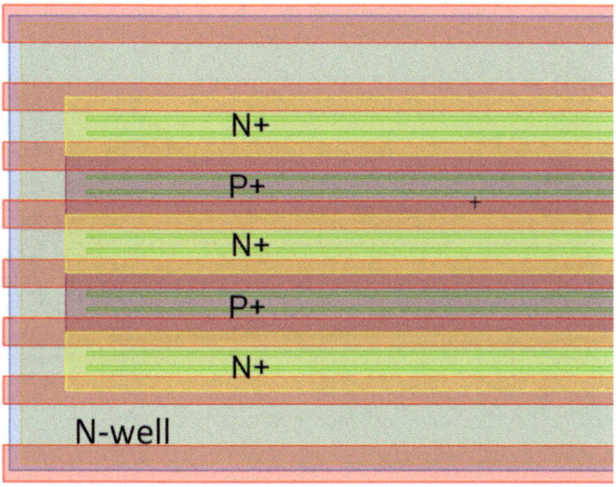

Fig. 4.14 The layout view of a smallest group of diodes in an ESD protection diode

to cover any possibilities of bridging defect within same layer or vertically between two adjacent layers. First, FIB cross-section was performed in perpendicular direction to the hotspots of the failing input pin: `Sel_Skip_In`. No abnormality was observed. Then top-down SEM inspection was performed layer by layer until contact level near OBIRCH area of the failing output pin: SO2. Again, no defect was discovered. Since the sample was still at contact level, it was worth trying electrical nano-probing to verify whether the defect was removed or still below contact. Electrical nano-probing was performed with Atomic Force Prober (AFP). Before AFP probing, it is necessary to understand the structure of the ESD protection diode. At contact level, all groups of diodes are separated from each other and each of the smallest group of diodes is divided into 4 diodes (see Fig. 4.14).

Figure 4.15 shows the cross-sectional view of a single diode based on the layout view in Fig. 4.14. AFP probing was performed on 3 diodes from 3 groups of diodes with hotspots, labeled as F1, F2 and F3 and also 3 diodes from 3 groups of diodes without hotspots, labelled as R1, R2 and R3. Figure 4.16 shows the AFP two-probe electrical probing results of the 3 diodes of F1, F2 and F3 and 3 diodes of R1, R2

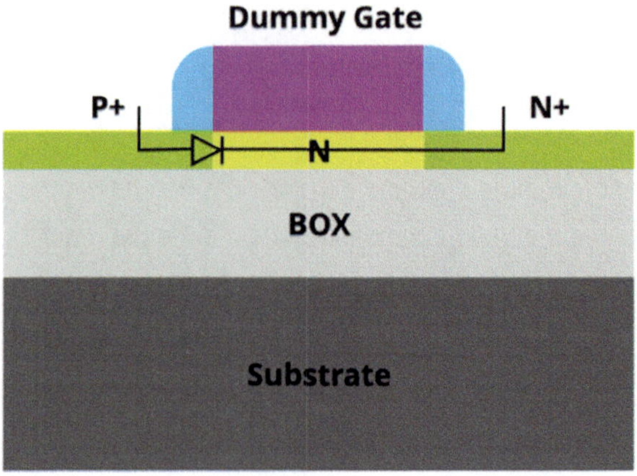

Fig. 4.15 Cross-sectional view of a single diode

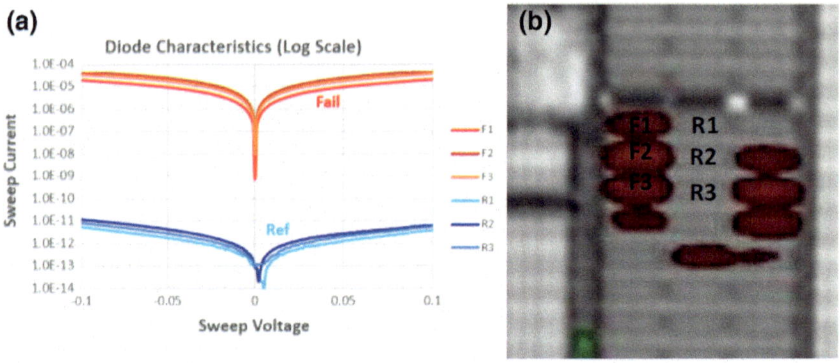

Fig. 4.16 **a** Semi-log scale I-V plots of 3 diodes with OBIRCH hotspot and 3 diodes without OBIRCH hotspot, which are labelled in (**b**)

and R3. The *I-V* plots with semi-log scale clearly shows that all 3 diodes of F1, F2 and F3 have elevated leakage >20 μA at 0.1 V, while the 3 reference diodes of R1, R2 and R3 measured ≈10 pA at 0.1 V. Although the diode characteristics might have been altered due to the 20 keV SEM inspection, the data clearly indicates that the defect causing the pin leakage is below contact level, and most likely implantation issue.

To verify the hypothesis of implantation issue, the sample was submitted for SCM analysis. For the first sample, SCM analysis was performed on a cross-sectional plane of diode structure. When the sample was mechanically polished to the right cross-sectional plane, which covered bad diodes as well as good diodes, chemical etch was performed to remove the tungsten (W) in gates and contacts in order to minimize the topography as well as avoid probe damage from imaging over W. Then, a very thin oxide was deposited on the cross-sectional plane. Figure 4.17 shows the SEM image of the cross-sectional plane analyzed by SCM.

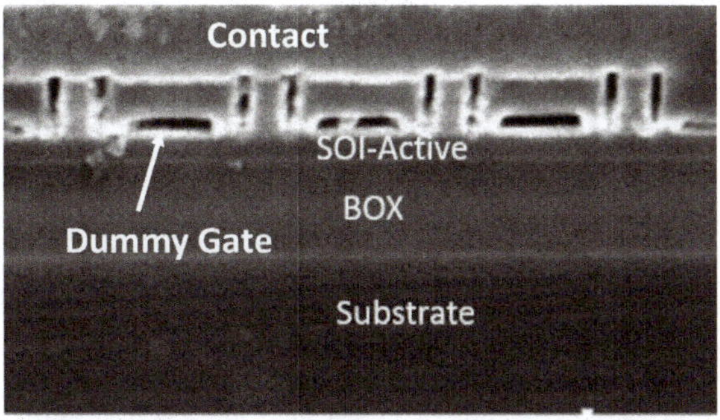

Fig. 4.17 The SEM image of the cross-sectional plane analyzed by SCM

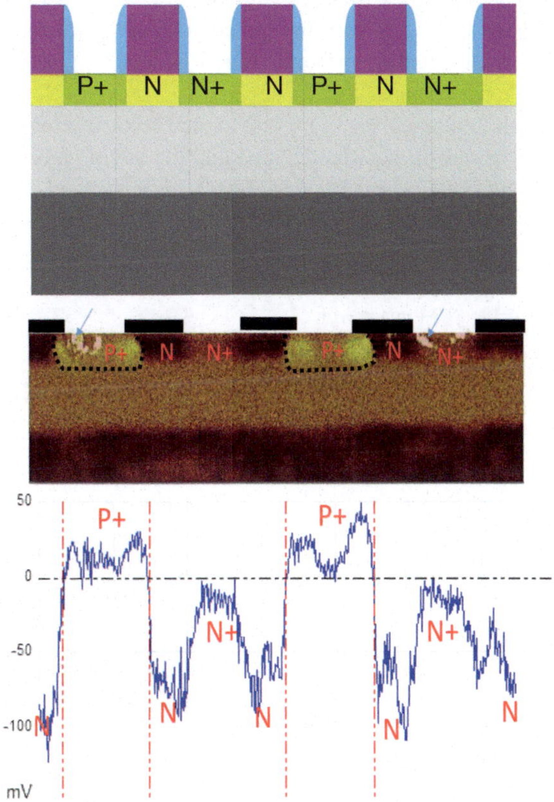

Fig. 4.18 SCM image with corresponding signal amplitudes for a good area

Figure 4.18 is the SCM image with corresponding SCM signal for a good area. SCM measured the amplitude of $\Delta C/\Delta V$ and the phase of $\Delta C/\Delta V$, which is dependent on the doping concentration and types. The amplitude of $\Delta C/\Delta V$ is higher for lower carrier concentration and lower for higher concentration. In the current measurement, the SCM is setup such that the phase of $\Delta C/\Delta V$ is positive for p-type doping and negative for n-type doping. Figure 4.18 clearly shows that p+, n+ and n-well under gate appears different contrast and different signal phases and amplitudes. p+ region shows positive low amplitude, n+ region shows negative low amplitude and n-well region shows negative high amplitude.

Figure 4.19 is the SCM image with corresponding signal amplitudes for the bad diode area. It shows that all signal are negative, indicating no p-type dopant at the supposed p+ area. Since p+ implantation was missing, refer to Fig. 4.15, the structure became n-well → n+, an equivalent resistor. This finding was consistent with the electrical signature of ohmic type leakage.

Cross-sectional SCM analysis had showed p-type dopant missing in p+ area, which explained the electrical signature of the leakage. To further understand the dimensions of the affected area and the lateral geometry where the p+ implantation was blocked, planar SCM analysis was performed on another sample after OBIRCH analysis identifying the bad diode area. For this sample, the SCM analysis was done

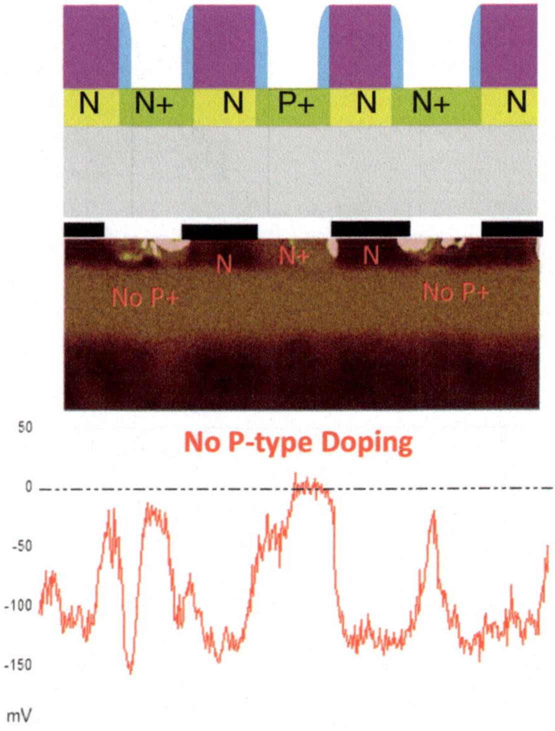

Fig. 4.19 SCM image with corresponding signal amplitudes for a bad area

from backside. For accessing to the backside surface of the active, the sample was prepared with the following procedures. First, the sample was parallel lapped to M1 level and attached to a piece of blank Si. Then, an opening window was created at the ESD protection diode area of the leaky pin with stop on BOX and subsequently, the backside surface of the SOI was exposed using diluted HF (DHF) etching away the BOX.

Figure 4.20 shows the cross-sectional view of the as-prepared sample and the diagram of the planar AFM and SCM analyses. First, normal AFM was performed to analyze the topography of the backside surface after DHF etch. The topographic image shown in Fig. 4.21 manifested that the p+ doped area was wide and brighter for the good diode area, while narrow and darker for the bad diode area, or totally missing for the worst case matching the OBIRCH hotspot size very well. Figure 4.22 shows the height profile of line scan crossing the good diode area and the bad diode area. It indicated that the wide and bright area had wide and tall height profile peak. From the layout, it was understood that the long wide and bright lines between the adjacent smallest groups of diodes are STI area, where the nitride layer was exposed after BOX was removed. Because nitride was not etched by DHF, it appeared as the tallest peaks in the height profile for good and bad diode area. For the good diode area, the p+ doping region also appeared as high peaks, see the blue line in the height profile, while the corresponding "supposed to be p+" region in the bad diode area appear short peaks or totally missing, see the red line and the peaks pointed by red arrow in the height profile. First, in the good diode area, p+ regions were implanted deep enough to reach the BOX. When the bottom side of SOI was exposed to HF, the p doped silicon was etched slower than surrounding N doped silicon. As a result, the p+ doped region remains a topographically higher than the surrounding n-type region. Differential etching of doped silicon in DHF was published in a study of the etching mechanism of silicon in HF [33, 34]. The authors concluded that etching of heavily doped silicon is driven by the availability of free electrons at the silicon surface, and therefore n-type silicon is etched faster than p-type silicon in DHF. Secondly, as shown in Fig. 4.19, P doping in the bad diode area are shallow or missing. This means that initially only n-type SOI is exposed to DHF in the bad diode area. Consequently, since both areas (good and bad diode) are etched simultaneously, partially blocked

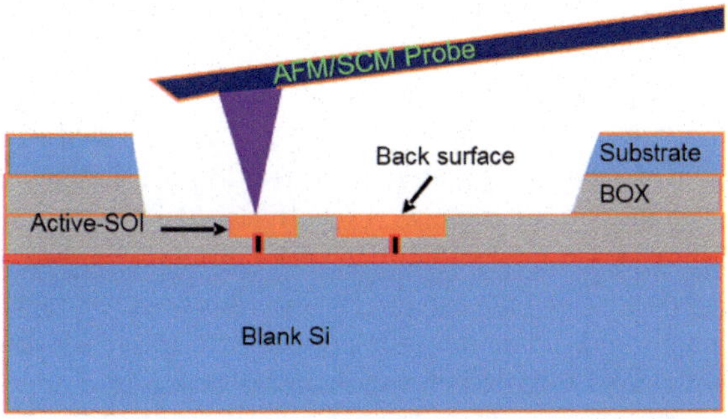

Fig. 4.20 The cross-sectional view of the sample and the diagram of the plan-view AFM and SCM analyses

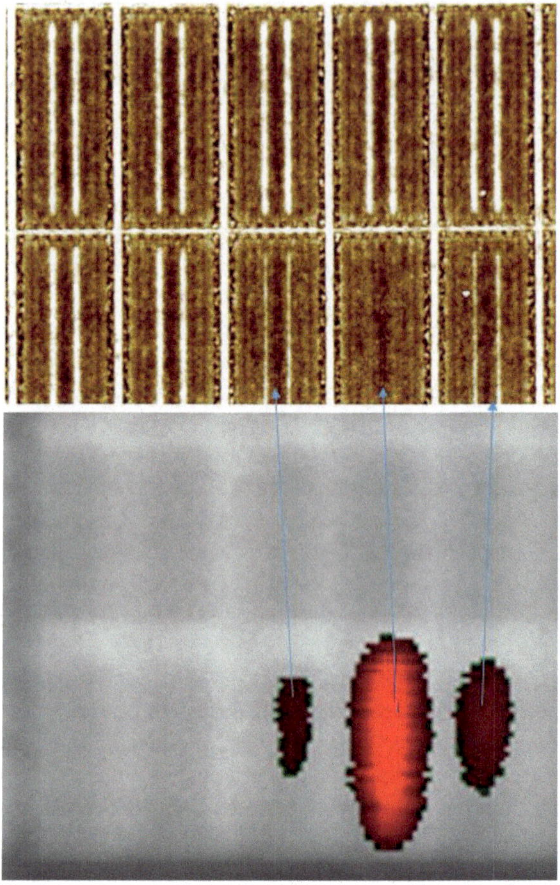

Fig. 4.21 AFM topographic image aligned with the OBIRCH hot spot analysis image

and missing p+ implant regions will be topographically lower than p+ regions in good diode areas. Therefore, the line profiles in Fig. 4.22 indicate that the shorter peaks and no peak areas manifest the partially blocked and shallow p-dopant or missing p-dopant regions.

Carrier profile analysis was also performed on the diodes in the OBIRCH hotspot area using SCM analysis. Figure 4.23 shows an SCM image and the corresponding line scan carrier profiles. Profile 1, shows the carrier profile along a line scanning through 5 good groups of diodes. It is seen that each group has 2 positive peaks ($\approx$10 mV) and 3 negative peaks ($\approx$25 mV), indicating 2 regions of p+ doping and 3 regions of n+ doping, respectively. This is consistent with the layout shown in Fig. 4.14. Profile 2 shows the carrier profile through the bad diode area. The 4 positive tall peaks ($\approx$80 mV) in the profile 2 are associated with the 4 bright lines in the SCM image, and indicate p-type doping with lower concentration at the group of diodes with small OBIRCH hotspot. In the group of diodes with a big OBIRCH hotspot, there is no positive peak in the carrier profile 2 in position located between 8 and 11

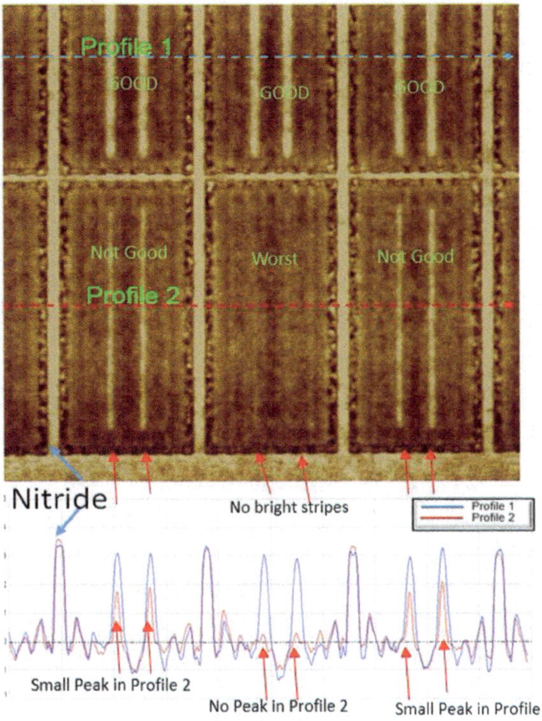

Fig. 4.22 AFM topographic image with corresponding scan line height profiles

on the X-axis. The absence of a positive peak at this location indicates a missing p+ dopant. This location corresponds to missing topographic peak in Fig. 4.22 (worst diode leakage).

4.2.2.1 Discussions and Conclusions

SCM analysis has shown that the pin leakage was due to blocked p+ implantation. But, what caused it? The OBIRCH hotspot showed that each hotspot was in oval shape and matched the rectangle area of a smallest group of diodes. The SCM image in Fig. 4.23 also showed that either the two bright lines at the center position between the red lines within a group of diodes are uniform and narrower than normal or both lines are totally missing.

It indicated that the p-type doping missing or with low concentration was always confined within a group of diodes and within the same group, two p-type doping missing or low doping was even. Thus, it was believed not to be random particle blocking implantation issue. The cause was likely an under-etch issue. Consider 3 situations of hard mask opening for p+ implantation as shown in Fig. 4.24 A: fully opening, B: partially opening and C: fully blocked, during the subsequent p+ implantation, the fully opening area will form normal p+ implantation, which reaches

Fig. 4.23 SCM analysis image with corresponding line scan carrier profiles

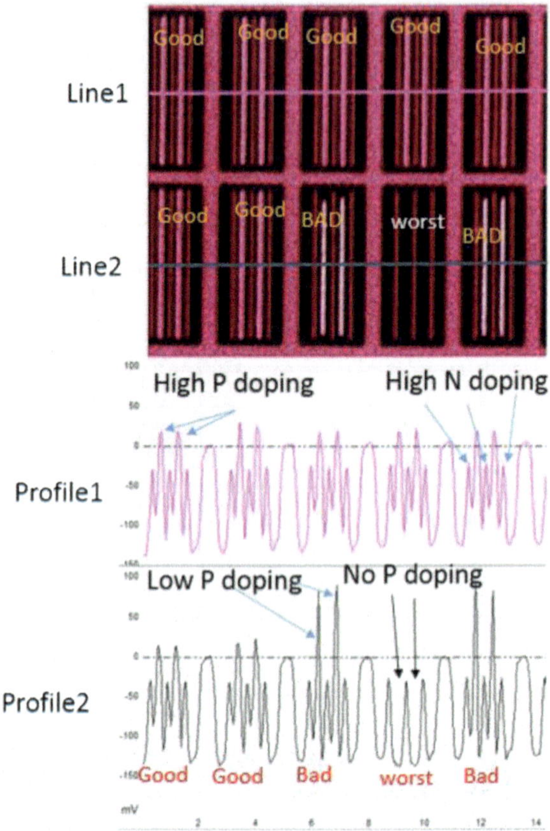

BOX; the fully blocked area will not have any implantation at all; while the partially opening area will have narrow and shallow p+ implantation, as shown in Fig. 4.25. Since the etch was symmetrically within the spacing between two dummy poly, the partially open area due to under-etch was always at the center between the two dummy poly. It resulted in the p-type doping with low concentration at the center, which was confirmed by the planar SCM image in Fig. 4.23. After p+ implantation was done, the physical evidence of p+ implantation hard mask under-etch was removed during hard mask removal. That was why top-down SEM inspection did not discover any abnormality. When the BOX was etched away by DHF during planar AFM sample preparation, due to differential etch dependent on dopant type and concentration [33, 34], the 3 situations of p+ doped regions were delineated at different height and width, with normal p+ doped region being higher and wider, as shown in Fig. 4.26. It was consistent with the observation in AFM topographic image (Fig. 4.22) and further confirmed with SCM analysis (Fig. 4.23).

Thus, in conclusion, a case study, demonstrating how to combine multiple techniques to identify underetch issue during p+ implantation hard mask opening for ESD

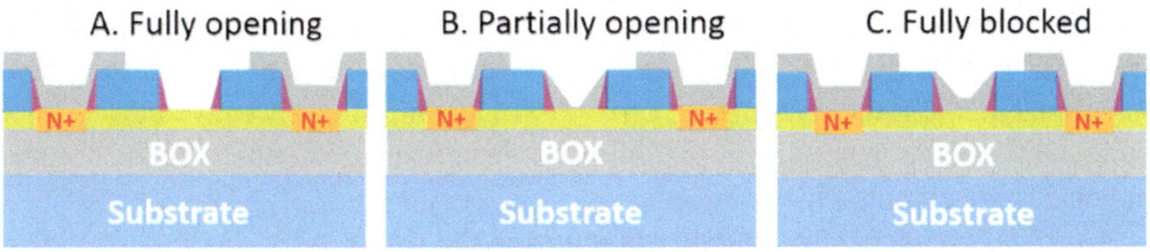

Fig. 4.24 The cross-section view of 3 situations of p+ implantation hard mask opening

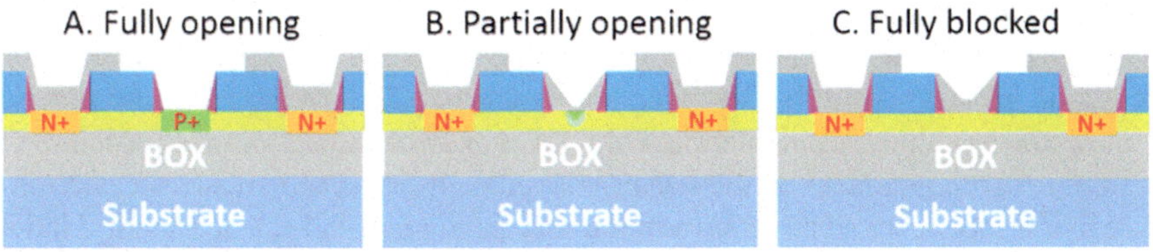

Fig. 4.25 The cross-section view of 3 situations of p+ implantation responding to the 3 situations of P+ implantation hard mark opening shown in Fig. 4.23

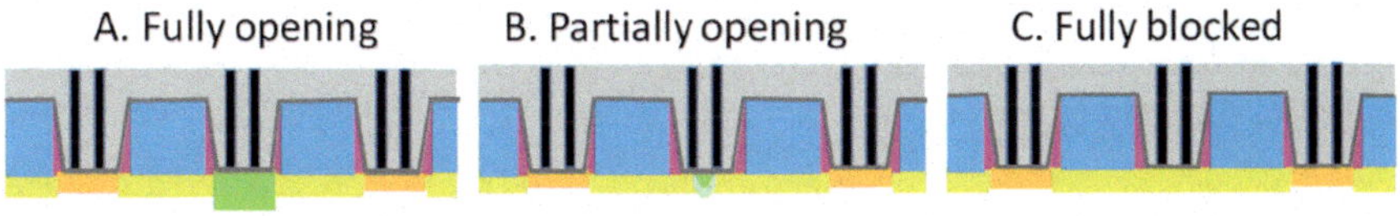

Fig. 4.26 The cross-section view of the 3 situations of p+ implantation after DHF etching away BOX

protection diode, responsible for the pin leakage issue of a device manufactured by advanced technology node.

4.2.3 Junction Profiling of Ge Photodetector Structures Using Scanning Capacitance Microscopy and Electron Holography

4.2.3.1 Background

Advanced high speed communication as well as increased data transfer rates and bandwidth devices are rapidly incorporating sophisticated optical components in the fabrication of silicon based integrated circuit, commonly called silicon photonics. Regular CMOS manufacturing and materials suitable for optical transmission waveguides and photon sensors are combined. A monolithic IC that integrates optical modulators and germanium photodetectors has been demonstrated on a 90 nm CMOS technology node [35, 36]. Recently, major microelectronic manufactures and research institutions have stepped up development efforts to optimize various aspects of silicon photonics to enable the integration of optical waveguides and efficient photodetectors into the challenging geometry of advanced technology nodes. Major challenges include material compatibility, co-design and co-simulation of photonic and electronic circuits [37].

A critical component of silicon photonics is the use of p-i-n diode photodetectors to convert optical energy into electrical signals. The material of choice for p-i-n diodes in this application is germanium because of its lower bandgap (0.67 eV) compared to silicon (1.1 eV). Figure 4.27a, b show illustrations of a photodiode and the energy band structure of a p-i-n photodiode, respectively. The p-i-n diode consists of a wide intrinsic layer ("i") between highly doped p and n regions. Under reverse bias conditions, the diode depletion width (W_d) extends throughout the intrinsic "i" region. A photon incident on the intrinsic layer is absorbed and creates an electron-hole pair that are excited and transferred to the conduction and valence bands. Under the electric field in the depletion region then drifts towards doped regions creating electric current. The advantage of a wide intrinsic width is large light absorption region and reduced junction capacitance. Photodetectors are optimized for high photon conversion efficiency and bandwidth, without affecting performance of regular CMOS devices. Diode photo conversion efficiency is characterized by its responsivity (R) that relates the output electrical current to the input optical power by the expression:

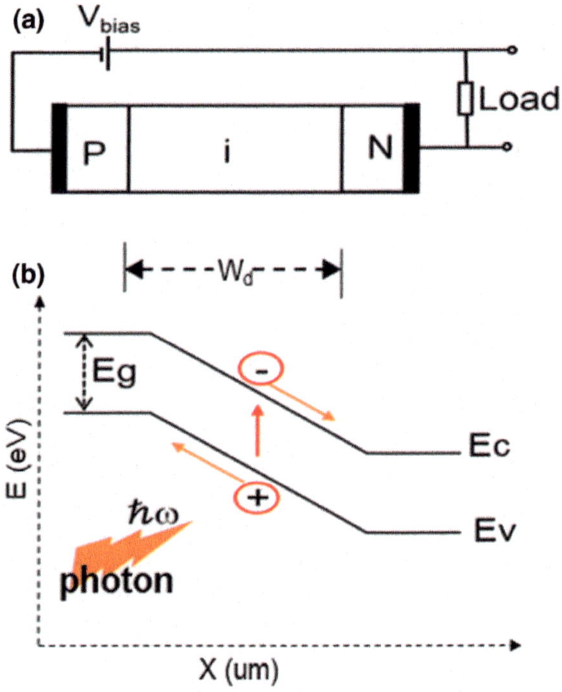

Fig. 4.27 P-i-N diode **a** doping and **b** reverse biased diode energy band diagram

$$Responsivity : R \propto I_p/P \quad (4.1)$$

where, I_p is the circuit photocurrent and P is incident optical power. The operation speed is characterized by the bandwidth and defined by:

$$Bandwidth : BW \propto 1/R_L C_J \quad (4.2)$$

where, R_L is the load resistance, and C_J is the diode junction capacitance.

In this section, Nxumalo et al. [38] describe a study of active dopant and junction profiles of germanium photodiode detectors used in a monolithic silicon photonics IC based on 90 nm SOI CMOS technology. Electrical measurements have revealed low responsivity of Ge diodes. To understand the root cause, we have used scanning capacitance microscopy (SCM) (Sect. 4.1) to map 2D carrier profiles and p/n junctions of Ge photodiode cross-section. We also measured the internal electric potential across the intrinsic region using electron holography [39, 40]. The combined results show that while internal electric potential suggest consistent "intrinsic" layer width between a good and poor responsivity diodes, carrier map reveal misalignment of p/n junction with respect to the waveguide between the input and output ends of a 15 μm long structure. Furthermore, p/n map reveal folded junction geometry, significantly different from design expectation. The junction geometry has increased junction area which may be responsible for low diode bandwidth. Evidence of boron segregation along the edges of the "intrinsic" Ge layer while phosphorus dominates the central portion is also revealed. To adequately characterize the intrinsic region of a p-i-n junction it is essential to combine SCM that to image majority carrier profiles in doped regions, with electron holography that is capable of imaging electric potential profiles within a depletion region of a p-i-n junction diode.

4.2.3.2 Germanium Photo Diode Process

Germanium photodiode analyzed in this paper were fabricated on SOI platform using a selective epitaxial growth and high temperature melt and regrowth (RMG) process [36]. Figure 4.28 shows an illustration of the RMG process flow. Following the formation of the front-end of the device CMOS process a window is open through silicon waveguide passivation; polycrystalline Ge is deposited; temperature raised to liquefy Ge layer; rapid cooling and selective epitaxial growth of Ge using waveguide Si as a seed; and finally sequential implantation of phosphorus and boron to form the p-i-n diode (Fig. 4.29).

The optical waveguide is aligned with the intrinsic layer and extends from the input end of the Ge photodetector to the window side where epitaxial growth begins. The designed p-i-n diode doping is illustrated by the cross-section shown in the insert.

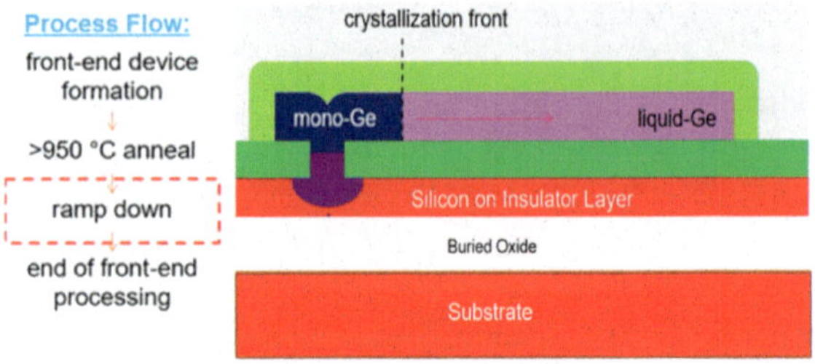

Fig. 4.28 Rapid melt regrowth (RMG) photodetector fabrication process flow

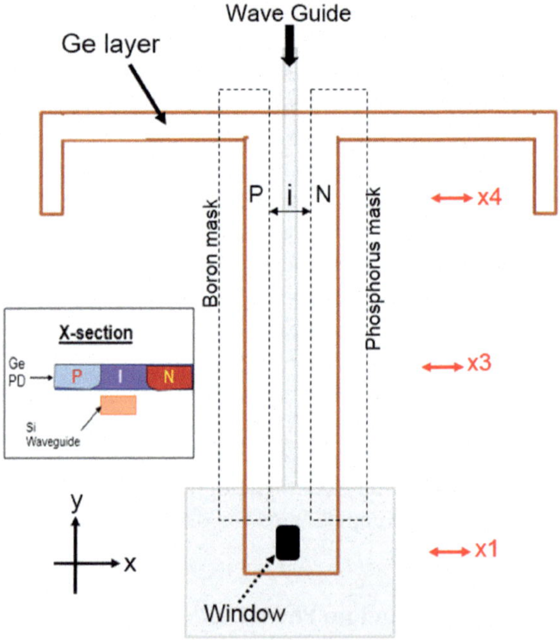

Fig. 4.29 Layout of Ge photodetector with insert illustrating designed doping at the x3/x4 cross-section

4.2.3.3 Photodiode Analysis and Conclusion

Optimization of Ge photodetectors involved performance characterization through electrical testing to measure the diode bandwidth and responsivity. Electrical results were followed by physical analyses including crystal defect imaging by TEM, silicon diffusion into germanium diode during RMG process by EDX, and active dopant distribution in the active p-i-n regions of the Ge photodiode using SCM and holography.

Low responsivity measurements on a product wafer initiated characterization of junction profiles in the Ge photodiode. High and low responsivity diodes were analyzed by SCM and holography to determine the root cause for low responsivity. Figure 4.30 shows a TEM cross-section of a Ge diode taken through X4 section shown in Fig. 4.29. The TEM micrograph shows a contrast in the Ge layer between the region

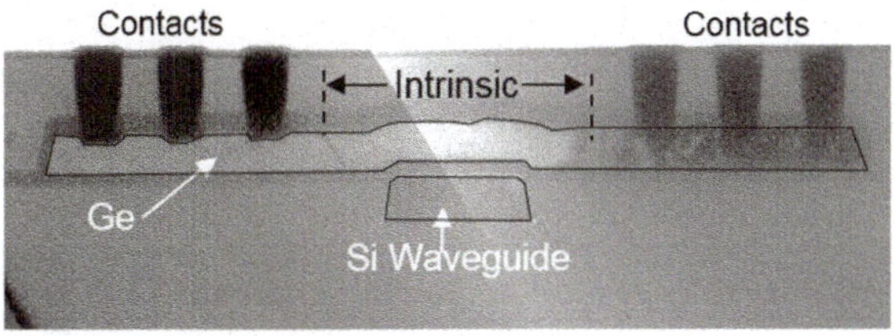

Fig. 4.30 TEM cross-section of a Ge photodiode

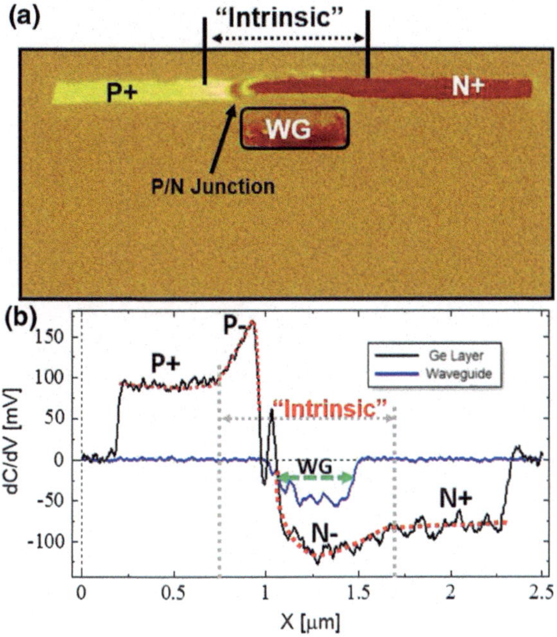

Fig. 4.31 SCM image of a Ge diode X3 cross-section showing **a** 2D carrier map and **b** lateral 1D carrier profile

under the right side contacts and left side. The contact regions were implanted with phosphorus on the right and boron on the left. No implantation was performed in the middle intrinsic region. Crystal defects are observed on the phosphorus implanted right side, and none on the boron implanted left side.

Figure 4.31a shows the active dopant profile of a Ge diode taken at section X3 shown in Fig. 4.29. Line profiles taken laterally along the central portion of Ge layer and through the waveguide are shown in Fig. 4.31b. It shows p-type on the left side of the diode and n-type in the "intrinsic" and right end of the Ge layer as well as the waveguide. The p/n junction is positioned near the left edge of the waveguide. The signal is lower (n−) in the "intrinsic" compared to the phosphorus implanted (n+) right side. After imaging the X3 plane, the sample was polished further to the X4 plane and imaged again. High magnification images of the intrinsic region at X3 and X4 are presented in Fig. 4.32a, b, respectively. More details of the "folded" junction geometry are highlighted with dashed lines. Lateral overlap of p-type on along the edges of Ge layer whilst n-type remains within the central portion. Junction

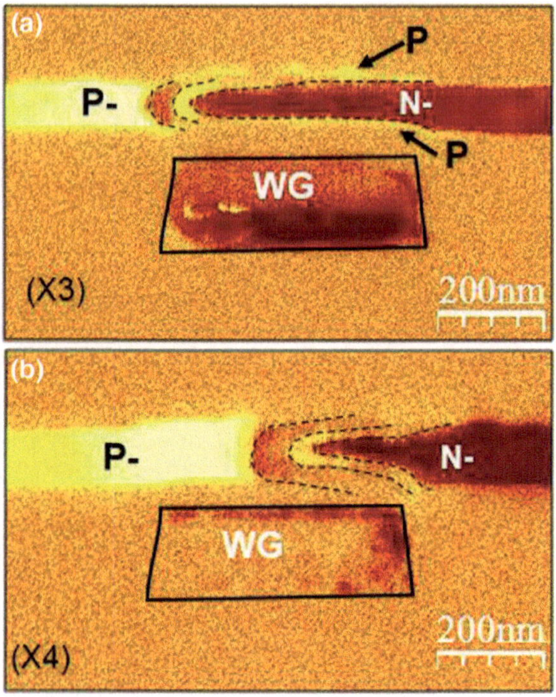

Fig. 4.32 SCM cross-section of the intrinsic) region of a low responsivity diode taken at **a** X3 and **b** X4 planes of Fig. 4.29

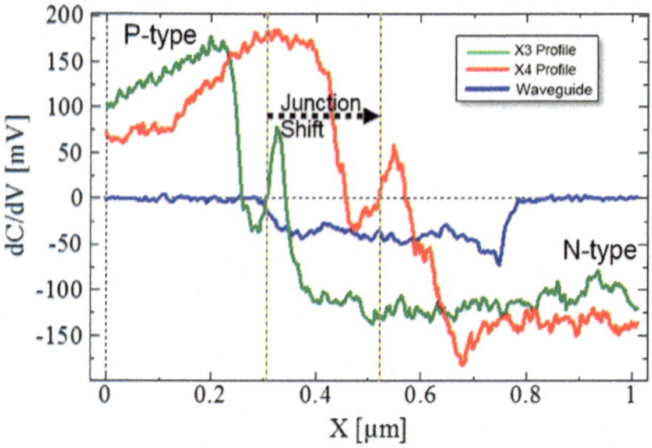

Fig. 4.33 Line carrier profiles showing p/n junction displacement between X3 and X4 planes in Ge photodetector

"folding" is limited to the region above the waveguide. The X4 plane shows more severe junction "folding" with the n-type clearly narrower compared to the X3 plane. Furthermore, the p/n junction position is centered over the waveguide in X4 whereas it is displaced to the left edge at the X3 plane. Line profiles through the two planes at the center of Ge in each of Fig. 4.32 are presented in Fig. 4.33.

The line profiles in Fig. 4.33 show that the p/n junction position is displaced by approximately 215 nm between X3 and X4 planes. Junction displacement is most likely the result of differential diffusion of phosphorus at the window side of the photodetector compared to the input side. In the RMG process, Ge recrystallization starts at the window where using silicon waveguide as a seed. Consequently, intermixing of Si with Ge occurs at this end of the diode and the Si mole fraction decreases towards the input side. Intermixing was confirmed by EDX mapping where the results showed a high silicon at the window and reduced below EDX detection limit at about 7um from the window. Furthermore, a longitudinal carrier profile analysis showed that the profile changes significantly at approximately the same location and the lowest detected silicon location.

Figure 4.34 shows electric potential maps acquired from two Ge diodes, one with poor responsivity and the other with good responsivity, using dual lens electron holography. Figure 4.34a, c are 2D map and 1D profiles, respectively, acquired from a diode with good responsivity, whereas Fig. 4.34b, c are the corresponding measurements from a diode with poor responsivity. The line profiles show a negative electric potential for p-type and positive potential for n-type. The central portion with approximately zero electric potential indicates that the fermi level is near the mid-gap of Ge energy band and therefore active dopant concentration is very low. The results show that the width of the low doped (labeled "intrinsic" here) regions for both good and bad are approximately the same (200 nm). The width is significantly lower that design dimensions ($\approx 1\,\mu m$).

Figure 4.35 is a comparison of electric potential with carrier profiles in the "intrinsic" region. The carrier line profiles are aligned with the corresponding electric potential profile, and the position of the waveguide is indicated. For the good

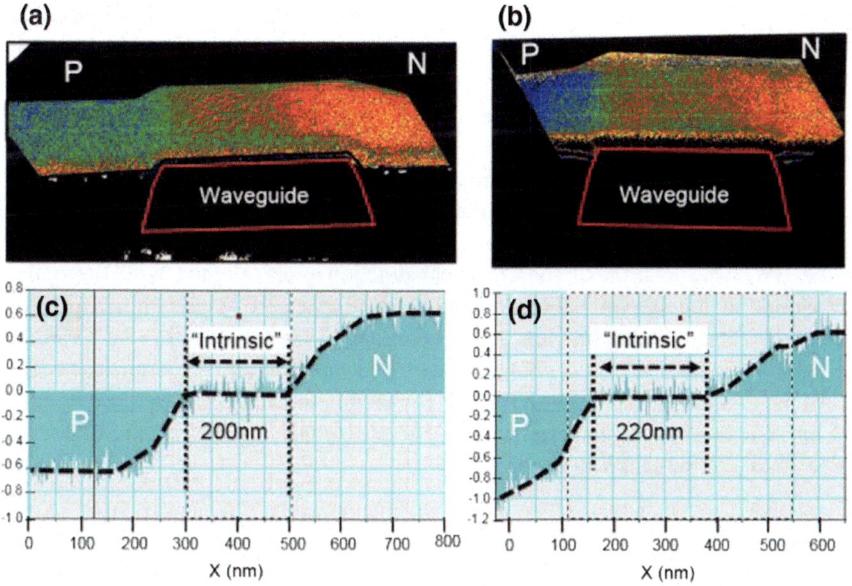

Fig. 4.34 Electron holography showing electric potential profiles from (**a** and **c**) a good and (**b** and **d**) bad responsivity diodes

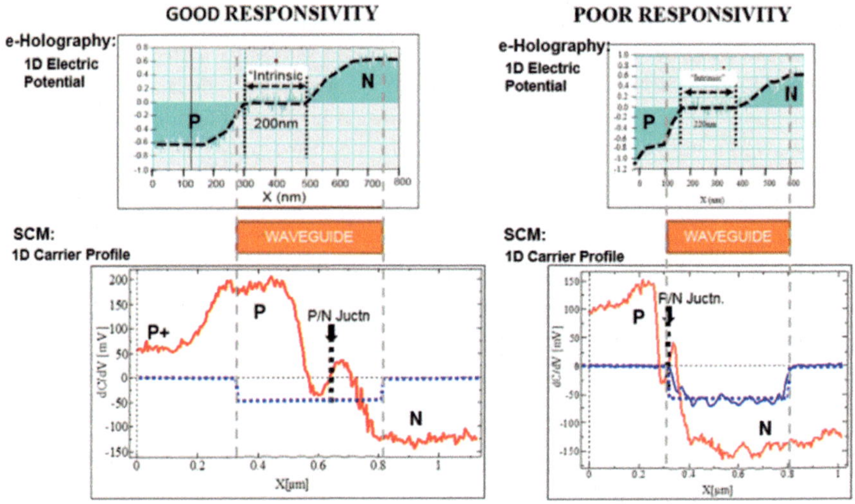

Fig. 4.35 Comparison of electric potential and carrier profiles in good versus poor responsivity diodes

responsivity diode, the intrinsic region indicated by electric potential corresponds to low p-type carrier concentration and the step on the right corresponds to electrical p/n junction, whilst the step on the left represents transition from low to high p-type carrier concentration. For the poor responsivity photodetector diode the intrinsic region corresponds to low n-type carrier concentration. In this case the left step in the electric potential profile corresponds to an electrical junction whilst the right step corresponds to change in concentration. Both SCM carrier profiles and electric potential results show that the designed "intrinsic" region has low active dopant concentration.

Thus, in conclusion scanning capacitance microscopy results demonstrate that a low responsivity Ge photodetectors manufactured with the RMG process has "folded p/n junction profile". This geometry increases the diode capacitance which reduces diode BW. The position of the electrical junction is misaligned with the optical waveguide at the silicon rich compared to the input end. A misaligned junction may reduce the quantum efficiency, and hence the diode responsivity. Both electron holography and SCM results show low carrier concentration in the intrinsic region of the p-i-n photodiode. Electron holography results show that the apparent width of the intrinsic layer is the same for good and poor responsivity diodes and SCM reveal more details of the junction geometry.

4.2.4 *Innovative Use of FA Techniques to Resolve Junction Scaling Issues at Advanced Technology Nodes*

4.2.4.1 Introduction

With new technology nodes (<14 nm), scaling introduces many new technical challenges, especially in junctions. Among these are the lateral straggle of counter-doping of adjacent n and p regions. Counter-doping of neighboring p/n regions by lateral straggle plays a more important role in device electrical performance as the size goes smaller. In finFET technology, the lateral straggle of dopants from Shallow Trench Isolation into the subfin region may play a role in SRAM leakage.

A typical process improvement methodology involves identifying an electrical problem, changing the process across multiple splits, and then choosing the option which best improves the electrical property. There are a number of possible limitations to this approach: (1) Too often there is a working assumption of complete process uniformity, without taking into consideration the possibility of local variations. Evidence that such an assumption has been relied upon is the making a random cuts of features, even if one does center/edge, etc. (2) Some process problems may not be completely visible by XTEM or XSEM sections. The implant distribution is one such example. (3) Process modeling by such means as TCAD not have adequate confirmation. (4) One may also presume to know the mechanism (thinness of layer changes electrical property in such a way) without having in-situ proof of the actual effect on the electrical property being studied. In this section, we demonstrate an integrated physical analysis that solves these problems. (5) Device engineering may rely on high currents to warn of a device problem when it is plausible that some cases may be due to physical defects (gate to tungsten stud shorts, etc.).

Johnson et al. [41] present a novel device engineering approach that combines three analysis methods to examine leakage in SRAM-like test structures at advanced technology nodes. This system exploits the physics of laser-based failure analysis, combined with carrier profile measurements by SCM, and TCAD simulation for verification of process conditions responsible for leakage. This system accounts for process non-uniformity, finds nonvisual implant differences, thoroughly proves which variations actually cause electrical problems, and eliminates spurious findings. They also provide two unique findings where different types of local implant distribution problems were shown to cause either reverse bias leakage or early diode turn-on at low voltage forward bias.

Scaling test structures, built like an SRAM up through tungsten stud, but wired to allow one to test specific components (V_{dd}, etc.) in parallel, were examined in process development. The I-V curves of various diode leakage analyses, depicted in Fig. 4.36, show three distinct camps in the current-voltage relationship. There were (1) massive ohmic shorts, (2) samples with considerable reverse bias leakage, and (3) samples with early diode turn on. A thorough approach was undertaken to examine all three camps. In contrast, a good diode would show a relatively flat curve in reverse bias.

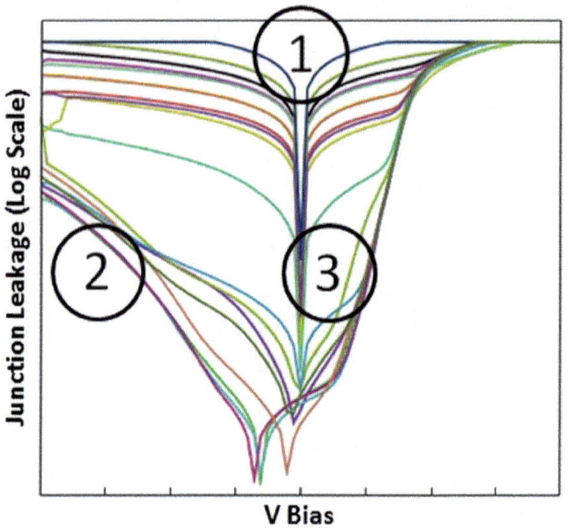

Fig. 4.36 Current-voltage relationships for a wide variety of chips examined in this study. Positive voltages here are forward bias

4.2.4.2 Massive Ohmic Short

A failure analysis regimen of OBIRCH scans, followed by passive voltage contrast on a sample delayered to tungsten stud, was able to identify gate to tungsten stud defects. There were two responses to this finding: (1) A lithography dose change fixed this problem of oversized gates. (2) A test structure design improvement (elimination of long gates) was undertaken to prevent this specific current path from being detected in the diode leakage test. Elimination of these samples enabled better concentration on the other two camps.

4.2.4.3 Diode Leakage

OBIRCH scans analyses in reverse bias voltage conditions were undertaken. The resolution of the image was sufficiently high to allow counting to specific rows in the test structure. Note, that there are large regions of the macro with apparently no leakage sites at all (compare the top and bottom areas of the scan as shown in Fig. 4.37. A consequence of this non-uniformity in leakage is that arbitrary center/edge cuts of the test structure would likely have completely missed the problem.

A cross-section was then made of this chip, carefully marking the rows of the structure with and without leakage sites (OBIRCH spots). These areas were then examined with SCM. Figure 4.38 shows typical results from a SCM analysis from a non-leaky area of the chip. In this image, dark brown regions are n-well (NW) and yellow is p-well (PW). One can see some variability to the strength of the wells, but hardly a definitive cause for action.

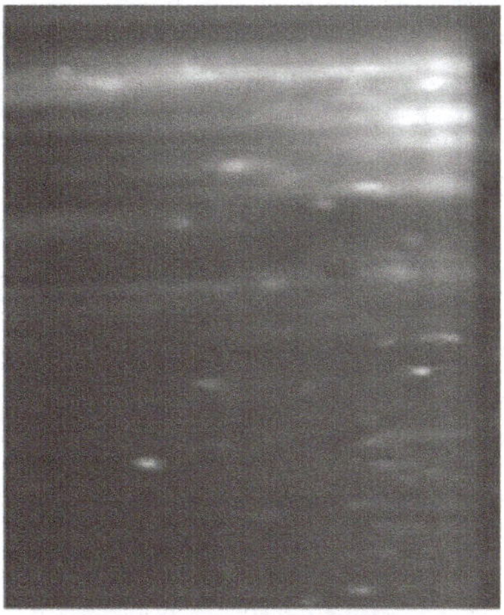

Fig. 4.37 Optical Beam Induced Resistance Change scan of failing macro when stressed under reverse bias conditions

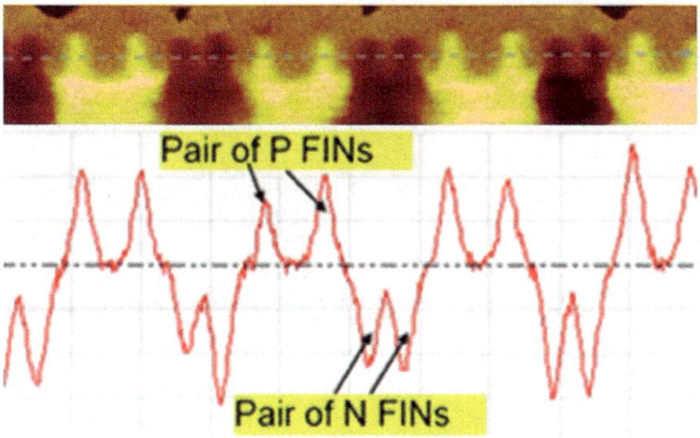

Fig. 4.38 Junction profiles in SRAM showing normal FIN doping occurring in pairs of alternating N and P types, captured from region with low OBIRCH activity

In contrast, Fig. 4.39 shows a SCM analysis of the area with very high OBIRCH activity. One can see that there is a continuous path (blue arrow) of PW in the PFET region from top to bottom. There should have been NW (brown material) underneath the p-fins. Thus there would locally be a source for p+ to PW leakage in this region.

This integrated approach provided not only pinpointed examples of where the leakage had occurred but also direct proof that the variability was associated with leakage. As a result, TCAD simulations were carried out with new process and the process correction was implemented and diode leakage corrected. Figure 4.40 shows a schematic with (a) old process showing diode leakage and (b) improved process that corrects leakage. The p/n junction profile in the old process correlates with SCM results.

Using this insight, new lots were run comparing the new and old processes. The red and blue curves below compare the leakage as a function of junction & well proximity for the two processes. A remarkable improvement is seen in leakage, as shown in Fig. 4.41.

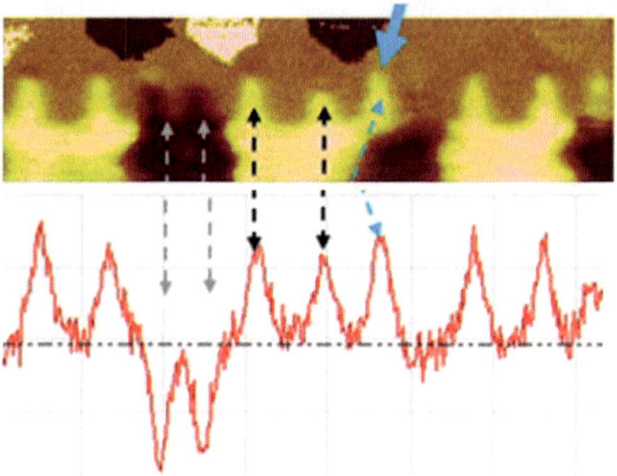

Fig. 4.39 Junction profile in SRAM shows Well CD variations and a Fin (blue arrows) counter-doped to p-type rather than intended n-type, captured from high OBIRCH activity site

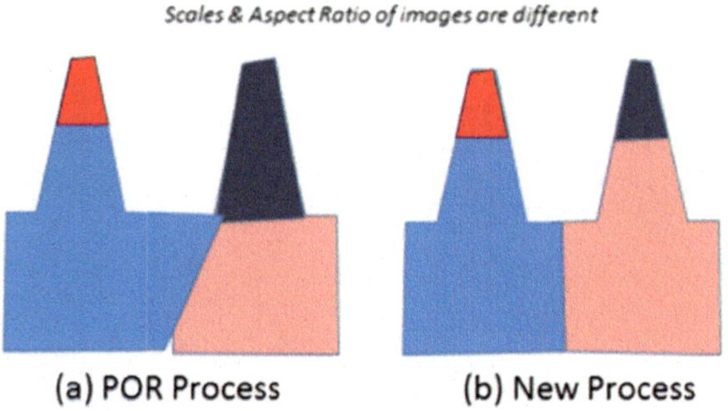

Fig. 4.40 Schematic illustrating process changes required to mitigate Diode Leakage

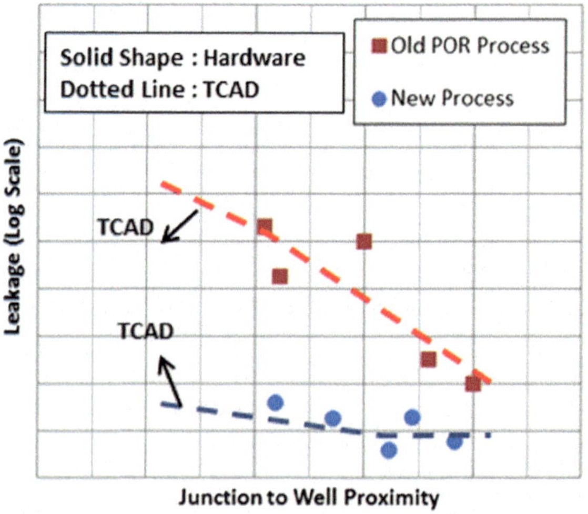

Fig. 4.41 Junction leakage as a function of junction and well proximity for the new and old process

4.2.4.4 Early Diode Turn-On in Forward Bias

A third, more subtle problem was also observed in the original data set, that of early leakage in diode turn-on. In the previous case of reverse bias leakage, we were relatively free to choose any voltage, since the current was so low. Here, we were interested in determining the leakage before diode turn-on. Typical OBIRCH scans are shown in Fig. 4.42. One can see that for −0.2, 0.0, and +0.2V, both the pattern of hotspots and the measured current are nearly identical. There are both current and hotspots at zero volts. Also note that the sign on the current is the same at ±0.2 V. This is due to the thermal promotion of carriers, or intrinsic conduction. The heating of a junction will result in a slight reverse bias current to do the thermal promotion of carrier. There may also be a very slight yield of creation of electron-hole pairs, which are then swept along the electric field of the junction, being detected by the OBIRCH system. This is an unuseful signal if we were need to find leakage in these voltage regimes.

If, however, we take the voltage slightly higher, black spots start to appear against the background of white OBIRCH signals, as shown in Fig. 4.43. On the left side, at 0.2 V of forward bias, there are white spots, which we have previously determined to be the reverse biased leakage from thermal promotion of carriers. The laser caused this. When we slowly ramp up in higher forward bias, here 0.5 V, we finally find a regime where a few black spots start to appear. This is due to the detection of current, and these spots constitute an "early" forward bias. These spots are the places to look for our forward bias leakage problem.

These forward-bias early spots were examined in SCM. On the left in the Fig. 4.44 below is a control or good area of fins. Again, n-type is brown and PW is yellow. On the right is a scan of the spot associated with forward bias leakage. One can see there are locally fins with deeper junctions than their neighbors, as seen in Fig. 4.44. This is believed to be the cause of the early turn-on. It is believed that this problem was also fixed by the process improvement suggested by TCAD modeling.

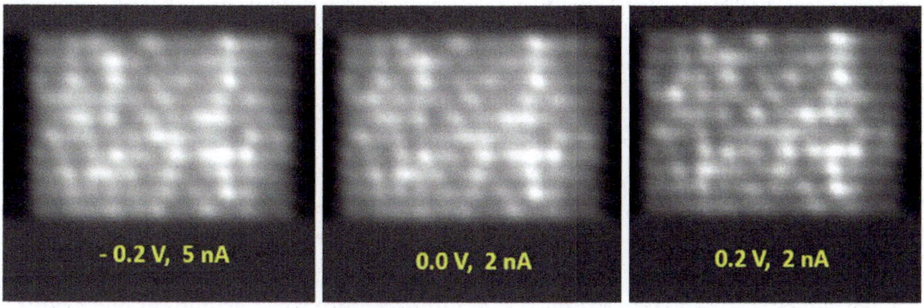

Fig. 4.42 OBIRCH scans of the diode macro at several voltages around 0 V

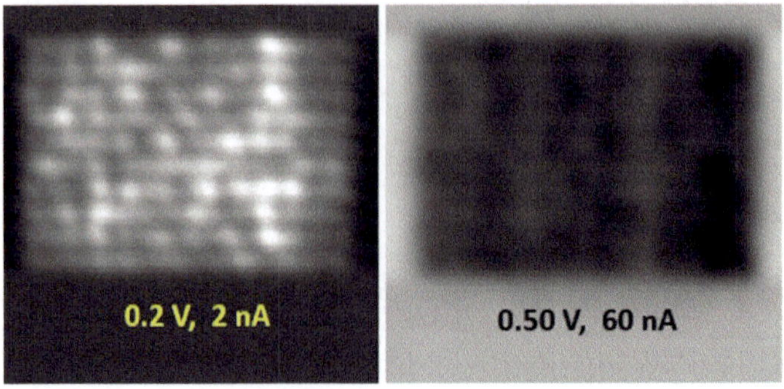

Fig. 4.43 Comparison of OBIRCH patterns for the junctions in 0.2 V versus 0.5 V, showing some "early" forward biased spots (black on right image)

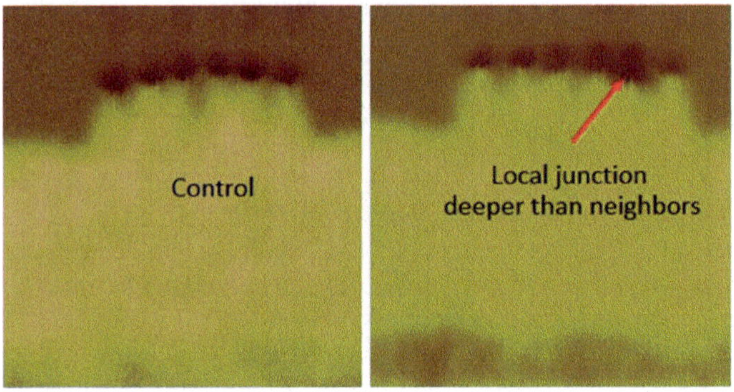

Fig. 4.44 SCM analyses comparing region from bulk of sample versus area shown to have higher forward bias current

Thus, in conclusion we have shown a novel process for the successful identification of implant distribution problems, associated with not only reverse bias leakage but also early diode turn on. The problems were isolated by an OBIRCH technique that needed to be aware of the physics of photon/device interactions, a technique that provides proof that mere anomalies were real contributors to an electrical problem. The extreme regionalization in the leakage profile provided by OBIRCH demonstrated the likely futility of a yield management system that relies on random, or even center/edge cuts. The leakage spots were thoroughly characterized for junction imaging using high-resolution SCM. A single fin with counter-doping which would provide direct p+ to PW leakage was identified. These data were then fed into a sophisticated TCAD model that not only duplicated the current profile but was able to run multiple process simulations until one likely to give substantial improvement in electrical performance was found. The improvement was verified on hardware.

4.3 Summary

With the capability of delineating junction and differentiating dopant types SCM as a technique has proved itself as an indispensable method for profiling carrier distribution of semiconductor devices to understand the root cause of fails. Used in conjunction with TCAD, SCM has been utilized to tweak process parameters for optimum device performances. In this chapter, SCM has been successfully demonstrated on planar silicon devices, silicon photonic devices and FinFETs.

Acknowledgements The authors acknowledge K. Barton, T. Su and C. Molella for their skillful sample preparation, Justin Clements and Danielle Clements for proof-reading and Globalfoundries for providing the support with this chapter.

References

1. J.R. Matey, J. Blanc, J. Appl. Phys. **57**(5), 1437 (1985). https://doi.org/10.1063/1.334506
2. C.D. Bugg, P.J. King, J. Phys. E: Sci. Instrum. **21**(2), 147 (1988)
3. Y. Martin, D.W. Abraham, H.K. Wickramasinghe, Appl. Phys. Lett. **52**(13), 1103 (1988). https://doi.org/10.1063/1.99224
4. C.C. Williams, W.P. Hough, S.A. Rishton, Appl. Phys. Lett. **55**(2), 203 (1989). https://doi.org/10.1063/1.102096
5. C.C. Williams, J. Slinkman, W.P. Hough, H.K. Wickramasinghe, Appl. Phys. Lett. **55**(16), 1662 (1989). https://doi.org/10.1063/1.102312
6. R.J. Hamers, D.G. Cahill, Appl. Phys. Lett. **57**(19), 2031 (1990). https://doi.org/10.1063/1.103997
7. J. Slinkman, C. Williams, D. Abraham, H. Wickramasinghe, in *Electron Devices Meeting, 1990. IEDM '90. Technical Digest., International* (San Francisco, CA, 1990), pp. 73–76. https://doi.org/10.1109/IEDM.1990.237223
8. C.C. Williams, J. Slinkman, W.P. Hough, H.K. Wickramasinghe, J. Vac. Sci. Technol. A Vac. Surf. Films **8**(2), 895 (1990). https://doi.org/10.1116/1.576936
9. D.W. Abraham, J. Vac. Sci. Technol. B Microelectron. Nanometer Struct. **9**(2), 703 (1991). https://doi.org/10.1116/1.585536
10. D. Boudreaux, F. Williams, A. Nozik, J. Appl. Phys. **51**(4), 2158 (1980)
11. C. Hu, S.C. Tam, F.C. Hsu, P.K. Ko, T.Y. Chan, K.W. Terrill, IEEE Trans. Electron Devices **32**(2), 375 (1985)
12. K. Cheng, J. Lee, Z. Chen, S. Shah, K. Hess, J.W. Lyding, Y.K. Kim, Y.W. Kim, K.P. Suh, Microelectron. Eng. **56**(3–4), 353 (2001)
13. P.K. Gopi, G. Li, G. Sonek, J. Dunkley, D. Hannaman, J. Patterson, S. Willard, Appl. Phys. Lett. **63**(9), 1237 (1993)
14. D. Quon, P.K. Gopi, G.J. Sonek, G. Li, IEEE Trans. Electron Devices **41**(10), 1824 (1994)
15. M. Panchore, J. Singh, S.P. Mohanty, IEEE Trans. Electron Devices **63**(12), 5068 (2016)
16. Y.Y. Wang, J. Nxumalo, D.P. Ioannou, A. Katnani, J. Brown, K. Bandy, M. Macdonald, R. Krishnasamy, J. Bruley, Appl. Phys. Lett. **112**(23), 233502 (2018). https://doi.org/10.1063/1.5009243
17. D. Cooper, C. Ailliot, J.P. Barnes, J.M. Hartmann, P. Salles, G. Benassayag, R.E. Dunin-Borkowski, Ultramicroscopy **110**(5), 383 (2010)
18. K.J. Chang, D. Chadi, Phys. Rev. Lett. **60**(14), 1422 (1988)
19. D. Belfennache, D. Madi, N. Brihi (2016)
20. K. Bergman, M. Stavola, S. Pearton, J. Lopata, Phys. Rev. B **37**(5), 2770 (1988)

21. N. Johnson, C. Herring, D. Chadi, Phys. Rev. Lett. **56**(7), 769 (1986)
22. N. Fukata, S. Sato, H. Morihiro, K. Murakami, K. Ishioka, M. Kitajima, S. Hishita, Dopant dependence on passivation and reactivation of carrier after hydrogenation (2007)
23. A. Chatterjee, A. Verma, B. Bhuva, E.D. Jansen, W.C. Lin, in *2001 IEEE International Reliability Physics Symposium, 2001. Proceedings. 39th Annual* (IEEE, 2001), pp. 200–205
24. M. De la Bardonnie, D. Jiang, S.E. Kerns, D.V. Kerns, P. Mialhe, J.P. Charles, A. Hoffman, IEEE Trans. Electron Devices **46**(6), 1234 (1999)
25. J. Lyding, K. Hess, I.C. Kizilyalli, Appl. Phys. Lett. **68**(18), 2526 (1996)
26. K. Hess, I.C. Kizilyalli, J.W. Lyding, IEEE Trans. Electron Devices **45**(2), 406 (1998)
27. K. Cooper, S. Schwartz, in *ISTFA 1992—International Symposium for Testing and Failure Analysis* (ASM International, 1992), pp. 219–219
28. N. Khurana, C.L. Chiang, in *24th International Reliability Physics Symposium* (IEEE, Anaheim, CA, USA, 1986), pp. 189–194. https://doi.org/10.1109/IRPS.1986.362132. http://ieeexplore.ieee.org/document/4208663/
29. K. Nikawa, in *Proceedings of International Symposium for Testing and Failure Analysis*, vol. 303 (ASM International, 1993)
30. E.J. Cole, P. Tangyunyong, D.L. Barton, in *1998 IEEE International Reliability Physics Symposium Proceedings, 1998. 36th Annual* (IEEE, 1998), pp. 129–136
31. R.A. Falk, in *ISTFA 2001—International Symposium for Testing and Failure Analysis* (ASM International, 2001), pp. 59–68
32. Z. Song, J. Nxumalo, M. Villalobos, S. Pendyala, in *ISTFA 2016—International Symposium for Testing and Failure Analysis* (ASM International, 2016)
33. L. Liu, F. Lin, M. Heinrich, A. Aberle, B. Hoex, ECS J. Solid State Sci. Technol. **2**(9), P380 (2013)
34. N. Valckx, D. Cuypers, R. Vos, H. Philipsen, J. Rip, G. Doumen, P.W. Mertens, M.M. Heyns, S. De Gendt, in *Solid State Phenomena*, vol. 187 (Trans Tech Publications, 2012), pp. 41–44
35. S. Assefa, S. Shank, W. Green, M. Khater, E. Kiewra, C. Reinholm, S. Kamlapurkar, A. Rylyakov, C. Schow, F. Horst, et al., in *2012 IEEE International Electron Devices Meeting (IEDM)* (IEEE, 2012), pp. 33–38
36. N. Feilchenfeld, F. Anderson, T. Barwicz, S. Chilstedt, Y. Ding, J. Ellis-Monaghan, D. Gill, C. Hedges, J. Hofrichter, F. Horst, et al., in *2015 IEEE International Electron Devices Meeting (IEDM)* (IEEE, 2015), pp. 25–27
37. W. Bogaerts, M. Fiers, P. Dumon, IEEE J. Sel. Top. Quantum Electron. **20**(4), 1 (2014)
38. J.N. Nxumalo, Y.Y. Wang, M. Iwatake, C. Molella, A. Katnani, J. Orcutt, J. Ayala, K. Nummy, in *2018 18th International Workshop on Junction Technology (IWJT)* (2018), pp. 1–4. https://doi.org/10.1109/IWJT.2018.8330282
39. H. Lichte, M. Lehmann, Rep. Prog. Phys. **71**(1), 016102 (2008)
40. U. Shaislamov, J.M. Yang, J.H. Yoo, H.S. Seo, K.J. Park, C.J. Choi, T.E. Hong, B. Yang, Microelectron. Reliab. **48**(10), 1734 (2008). https://doi.org/10.1016/j.microrel.2008.06.002
41. G. Johnson, J. Nxumalo, A. Arya, J. Johnson, Q. Gao, B. Greene, in *SEMI Advanced Semiconductor Manufacturing Conference (ASMC), 2018 29th Annual* (IEEE, 2018), pp. 370–373

Chapter 5
Oxidation and Thermal Scanning Probe Lithography for High-Resolution Nanopatterning and Nanodevices

Yu Kyoung Ryu and Armin Wolfgang Knoll

Abstract The strength of scanning probe lithography (SPL) lies in the operation at ambient conditions, sub-10 nm resolution capabilities, the in situ non-destructive inspection of the fabricated structures, the nanometric accuracy in positioning, the versatility in modifying any kind of materials, and the freedom in the patterning geometries. On the other hand, the tip size and lifetime-related issues hinder the achievable throughput, and a precise niche of application has yet to be determined for its implementation in technological applications. The complementarity of the high-resolution and precise positioning patterning by SPL and the high throughput and low-resolution patterning by other well-established lithographies (optical, electron beam, nanoimprint) can be achieved by the development of mix-and-match lithography strategies.

5.1 Introduction

Shortly after the invention of the scanning tunneling microscope (STM) by Gerd Binnig and Heinrich Rohrer in 1981, beyond the possibility of atomic resolution imaging, it was a straightforward matter to use a sharp probe to manipulate or change the composition of materials at the atomic and nano scales. Among early examples in this direction, we may mention the reversible pinning and unpinning of individual organic molecules on a graphite surface [1], the positioning of individual Xe atoms on a crystalline Ni surface [2], and the first local oxidation experiment on a hydrogen-passivated Si (111) surface [3].

The same path was followed with atomic force microscope (AFM), developed by Gerd Binnig, Calvin Quate, and Christoph Gerber in 1986. In fact, more or less at

Y. K. Ryu (✉)
Instituto de Ciencia de Materiales de Madrid,
Consejo Superior de Investigaciones Científicas (CSIC), Madrid, Spain
e-mail: yukyoung.ryu@csic.es

A. W. Knoll
IBM Research Laboratory Zurich, Zürich, Switzerland
e-mail: ark@zurich.ibm.com

U. Celano (ed.), *Electrical Atomic Force Microscopy for Nanoelectronics*,
NanoScience and Technology, https://doi.org/10.1007/978-3-030-15612-1_5

the same time, Majumdar et al. [4] were the first to point out an advantage of AFM over STM, namely, that it could control the force between the tip and the sample, holding it constant during lithography to avoid them entering into contact. Another point was the possibility of decoupling the imaging and patterning modes using an AFM. Day and Allee [5] were the first to report local oxidation on silicon. Excluding scientific problems whose specifications absolutely require the use of an STM, the atomic force microscope has become the tool of choice to perform scanning probe lithography due to its ability to operate with insulating samples and under ambient conditions. For this reason, only lithography performed with AFM will be described in this chapter.

Scanning probe lithography encompasses all the processes where a sharp tip modifies a small area of a material surface by means of electrical, thermal, mechanical, or chemical interactions. Some processes involve a combination of several of such interactions. The interaction chosen to modify a given surface will depend on both the properties of the material and the targeted application. In Levy's group, a 'write and erase nanowires' process with a conducting AFM tip has been developed [6]. The probe enters into contact with $LaAlO_3/SrTiO_3$ heterostructures. When it is biased positively (negatively) with respect to the surface, it creates a metallic (insulating) state. The process is reversible and has been used to fabricate several nanoelectronic devices [7–9]. In another recent study, a heated AFM tip reversibly modifies the magnetization on ferromagnetic/antiferromagnetic layers under the application of a magnetic field, without changing the material either chemically or structurally [10]. In a final example, since the insulator-to-metal phase transition of samarium sulphide (SmS), a heavy fermion compound material, is dependent on the pressure, an AFM tip has been used to fabricate arrays of periodic gratings by applying controlled force/pressure to create colour gradients by local strain modulation [11].

The present chapter will focus on two of the existing modalities: oxidation and thermal scanning probe lithography. Oxidation SPL relies on confined electrochemical reactions for direct local modifications at the nanoscale. Its patterning abilities extend to a wide range of materials including polymers, self-assembled monolayers, semiconductors, and metals. Up to now, this has been the modality of choice to fabricate high-resolution nanoelectronic devices such as quantum dots, quantum rings, single electron transistors, and nano-channel field effect transistors (FETs). On the other hand, among the scanning probe lithography modalities, thermal SPL has the most mature technological development, including a commercial tool (Nanofrazor, SwissLitho AG), and is the closest to CMOS processing integration. It is also this modality that presents the fastest nanopattern prototyping and has allowed a mix-and-match processing strategy with laser writing. The development of this hybrid lithography approach could further generalize the SPL technique for the fabrication of nanodevices. Some recent reviews of these modalities explain the details of the technique and will allow the reader to keep track of the latest results [12–15].

5.2 Oxidation Scanning Probe Lithography: Direct Chemical Modification at the Nanoscale

5.2.1 Mechanism and Growth Kinetics. Oxidation Parameters and Operation Modes

5.2.1.1 Reactions in an Electrochemical Nanocell. Role of the Water Meniscus and the Electric Field

Oxidation scanning probe lithography refers to the anodic oxidation process occurring at the surface of the material inside and in the direct vicinity of a water meniscus when the AFM tip is biased negatively (or the sample positively, depending on the instrumental setup). The tip, the sample, and the water meniscus form the analogue of an electrochemical cell at the nanoscale (Fig. 5.1).

In the case of silicon, the half (anodic) reaction occurring on the surface is expressed by

$$\mathrm{Si} + 2\mathrm{h}^{+} + 2(\mathrm{OH}^{-}) \rightarrow \mathrm{SiO_2} + 2\mathrm{H}^{+} + 2\mathrm{e}^{-} \quad (5.1)$$

and the half (cathodic) reaction occurring at the tip by

$$2\mathrm{H}^{+}(\mathrm{aq}) + 2\mathrm{e}^{-} \rightarrow \mathrm{H_2} \quad (5.2)$$

The anodic reaction of two other relevant semiconductor compounds oxidized by SPL, viz., GaAs [16] and MoS_2 [17], are, respectively,

$$2\mathrm{GaAs} + 12\mathrm{h}^{+} + 10\mathrm{OH}^{-} \rightarrow \mathrm{Ga_2O_3} + \mathrm{As_2O_3} + 4\mathrm{H_2O} + 2\mathrm{H}^{+} \quad (5.3)$$

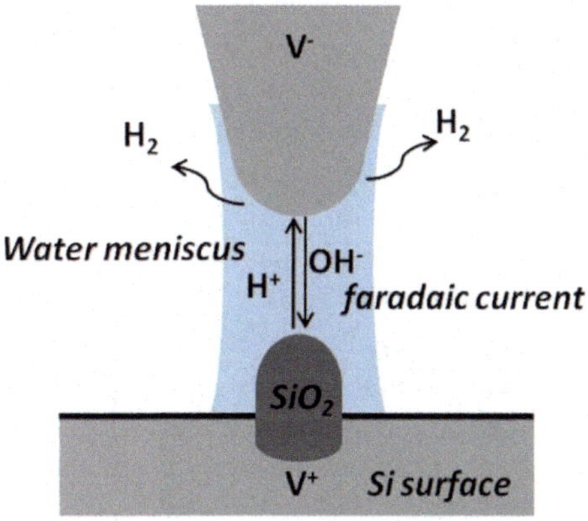

Fig. 5.1 Schematic view of the electrochemical process taking place between the AFM tip and the sample surface in oxidation-SPL

$$2MoS_2 + 9O_2 + 4H_2O \rightarrow 2MoO_3 + 4H_2SO_4 \tag{5.4}$$

In the case of the anodic oxidation of a metal, specifically niobium, the corresponding reaction is [18]

$$2Nb + 5H_2O \rightarrow Nb_2O_5 + 10H^+ + 10e^- \tag{5.5}$$

Generalizing to any elemental metallic surface, the anodic oxidation process can be expressed by

$$M + nH_2O \rightarrow MO_n + 2nH^+ + 2ne^- \tag{5.6}$$

Despite the chemical reactions described in (5.1–5.6), whose products are oxides with a specific stoichiometry, the exact composition of the resulting oxides are not always straightforward to determine. In addition, during the o-SPL process, several oxides can be formed simultaneously, depending on the characteristics of the sample or the patterning conditions. Therefore, different spectroscopic analyses have been performed: photoemission spectroscopy on n-type Si [19]; X-ray photoelectron emission microscopy on SiC_x [20], MoS_2 thin flakes [21], and graphene [22]; Raman spectroscopy on graphene [23] and MoS_2 [24]; and Auger electron spectroscopy on GaAs [25] and Nb [26]. Among these studies, it is worth noting a study of photoemission spectroscopy on epitaxial GaAs/AlAs/GaAs layers after performing o-SPL, where the formation of Al oxides together with the Ga oxides on the top GaAs layer was demonstrated [27], and a study of the desorption dynamics of oxides formed by o-SPL on GaAs under X-ray irradiation [28], in which it was found that the oxides desorb completely after 160 min.

The anodic oxide grows above and below the surface baseline. In Si (100), it was measured that 40% of the oxide thickness was grown below the surface [29] (Fig. 5.2a). This can be measured by etching the oxide with HF. However, a 1:1 ratio of above/below surface growth was observed in GaAs oxides [30] (Fig. 5.2b). In this case, the oxide was removed by a developer or HCl etching. This was used to calibrate the oxide thickness needed to reach a certain depth of the effective insulating barrier to avoid leakage currents in the final quantum device. Finally, in the case of WSe_2 [31] thin layers, the oxides formed by o-SPL were eliminated by deionized (DI) H_2O etching to fabricate nano-FET conduction channels (Fig. 5.2c).

Reduction processes (tip biased positively or sample biased negatively) have been much less exploited. Among the studies of this kind, we find one in which reduction SPL (negative sample bias) was performed to obtain Ag, Cr, Cu, and Pt nanostructures from various metal resists [32]. In another study, both reduction and oxidation SPL were used to apply a reversible writing-erasing process on organic semiconductor thin films without modifying the topography of the material, just switching from the conductive to the insulating state [33]. For the relevance of reduced graphene oxide (rGO) and graphene surfaces terminated with different chemical compositions and

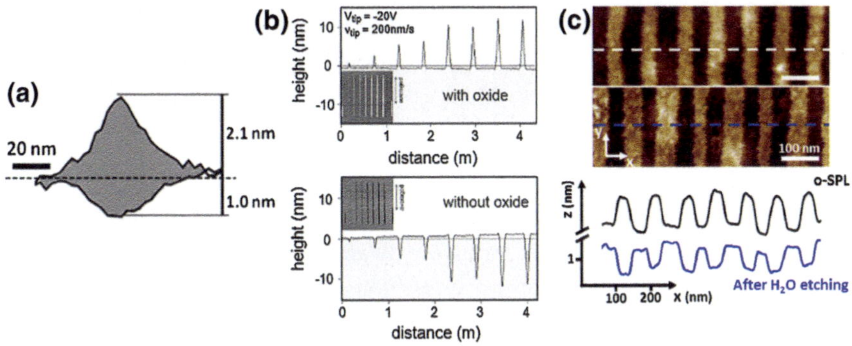

Fig. 5.2 Growth of o-SPL fabricated oxides above and below the material surface. The total thickness of an oxide is estimated as the sum of the height above the surface after patterning plus the depth of the groove left below the surface level after chemical wet etching. **a** AFM cross section of silicon oxide. The measured height of the oxide above the surface was 2.1 nm and the depth of the groove measured after HF etching was 1.0 nm [29]. **b** AFM cross-sections of GaAs oxide nanowires, shown in the insets. The top cross-section shows the measured height of the oxides above the surface, and the bottom cross-section the grooves left after etching with a developer [30]. **c** AFM topographic images of tungsten oxide nanowires fabricated by o-SPL on a WSe_2 thin layer (top) and the grooves left after deionized H_2O etching (middle). Their corresponding cross-sections, indicated by the dashed lines, are shown in the bottom image [31]

defect densities in a range of applications, the reduction of graphene oxide [34] and graphene [35, 36] by o-SPL have also been studied.

Sugimura and Nakagiri [37] were the first to note the role of adsorbed water in the electrochemical process triggered by o-SPL, ruling out a field-enhanced chemical mechanism. This role of the water meniscus is twofold: it supplies the oxidant species (principally OH^- ions) needed to modify the surface [38] and confines the reaction area laterally. This confinement represents one of the main parameters that defines the smallest pattern size and thereby determines the resolution of the technique.

The role of the electric field is also twofold: it dissociates the water molecules that form part of the meniscus to obtain the oxyanions and it is responsible for the migration of these anions towards the sample surface for oxidation to take place. This flux of the oxygen species and OH^- ions between the tip and the sample is associated with a Faradaic current that has been measured experimentally to be of the order of sub-pico amperes [39–41]. The high electric field formed during the o-SPL process is on the order of 10^9–10^{10} V/m. The shape of the electric field depends on the shape of the AFM tip and this will have an influence on the oxide geometry [42]. In addition, depending on the material, above certain values of relative humidity and applied bias, electrical breakdown occurs, and a conduction regime dominated by electrons will surpass the Faradaic ionic current [43]. Therefore, ring-like nano-explosions can occur, caused by shock wave propagation [44].

By tuning the chemical composition of the atmosphere in which oxidation takes place, nanomaterials different from oxides can be fabricated. In Garcia's group, several gases rich in organic solvent content have been used to create carbonaceous structures [45, 46]. Concretely, by performing o-SPL under an ethanol (CH_3CH_2OH) meniscus, SiC_x nanowires were produced [20] according to the reaction

$$Si + x/2CH_3 - CH_2OH \rightarrow SiC_x + xH_2 + x/2H_2O \tag{5.7}$$

Later, Lorenzoni et al. [47] used 1,3,5-trimethylbenzene as organic solvent to perform o-SPL. Playing with the polarity, two different reactions can be triggered at the silicon surface: silicon oxide formation by anodic oxidation (negatively biased tip) and carbonaceous nano-patterns by solvent decomposition (positively biased tip).

Finally, using hexadecane as organic solvent and applying o-SPL with the same polarity (positively biased sample), on a hydrophilic (OH^- terminated) and a hydrophobic (HMDS layer passivated) Si substrate, two different compounds were fabricated [48]. In the case of the hydrophilic Si surface, silicon oxide was formed by regular anodic oxidation. However, on the hydrophobic Si surface, a meniscus could not be formed, and due to the high electric field, solvent decomposition occurred and carbonaceous structures were deposited instead.

5.2.1.2 Growth Kinetics and Oxidation Parameters

In the o-SPL process, the experimental data of several groups have shown that the height of the oxide follows a power-of-time law [49–52]. Teuschler et al. [53] gave the empirical expression

$$h(t) \alpha\, bt^{\gamma} \tag{5.8}$$

where b is a constant which depends on the doping of the Si surface and the experimental conditions. The exponent γ ranges from 0.12 to 0.4.

In the case of non-contact AFM o-SPL, the dependence of the height on the oxidation time and voltage was calculated empirically to be [54]

$$h(t, V) = h_0(V) + h_1(V) \ln t \tag{5.9}$$

where $h_0 = -2.1 + 0.5V - 0.006V^2$ and $h_1 = 0.1 + 0.03V - 0.0005V^2$ for p-type Si (100).

The oxidation mechanism by SPL exhibits space-charge-limited growth behaviour [55–57]. For short oxidation times (in the range of μs–ms), there is fast initial vertical growth dominated by the conversion of the surface atoms and the anionic species supplied by the electric field. During this initial stage of anodic oxidation, unreacted OH^- ions generate a negative space charge [58, 59]. However, for longer oxidation times, the creation of positive charge defects at the oxide/surface interface involves the build-up of a space charge that decreases the effective electric field, hinders the diffusion of anionic oxidant species, and slows down the oxide growth.

Dagata et al. [52] proposed a model to relate the empirical power law found by Teuschler et al. (5.8) to the two growth regimes (fast and slow) mentioned above. According to this model, during the anodic oxidation process, two different reactions take place. There is a direct reaction at the initial stage of the process where the oxide grows fast due to the reaction of OH^- ions with the silicon surface:

$$Si + h^+ + 2OH^- \rightarrow SiO_2 + 2H^+ \quad (5.10)$$

In this stage, the γ that fits (5.10) is 0.4. However, for longer oxidation times, due to the creation of fixed charge traps at the Si/oxide interface, a slow growth rate with $\gamma \approx 0.2$ is reached and an indirect reaction dominates in this regime:

$$H^+ + OH^- \rightarrow H_2O \quad (5.11)$$

$$H - Si \equiv Si_3 + H_2O + h^+ \rightarrow {}^*Si \equiv Si_3 + H_3O^+ \quad (5.12)$$

$$H_3O^+ + OH^- \rightarrow [H_3O^+ - OH] \quad (5.13)$$

$$Si + h^+[H_3O^+ - OH] \rightarrow SiO_2 + 2H^+ \quad (5.14)$$

This model fits fairly well with almost all the existing experimental data from different groups, and not only for silicon.

In order to reduce or neutralize the build-up of the space charge, and in this way enhance the vertical growth rate of the oxide, several groups have proposed the use of techniques using ac voltage modulation [60–62] or polarity reversing cycles [63].

5.2.1.3 Amplitude Modulation Oxidation Scanning Probe Lithography

Contact AFM is the simplest modality for performing o-SPL. In this case, the water meniscus is formed spontaneously by capillary forces when the tip enters into contact with the surface. The patterning is achieved by displacing the tip with speeds in the range of 0.1–50 μm/s.

For o-SPL carried out in amplitude modulation AFM (AM-AFM) in the non-contact regime, a critical electric field must be applied to form the water meniscus. An experimental value of about 1.3 V/nm was measured as the ratio of the threshold voltage to the tip-sample distance by Garcia et al. [64]. A model was proposed to predict the value of the threshold voltage needed to form the meniscus as a function of the relative humidity and the tip-sample distance [65]. The lateral and vertical size of the water meniscus formed by the applied electric field can be estimated experimentally by measuring the snap-off separation [66].

In Garcia's group, one cycle of the oxidation process comprises (a) application of a pulse, (b) the formation of a water meniscus and the subsequent oxidation during the pulse, and finally (c) the breaking of the meniscus when the pulse is switched off.

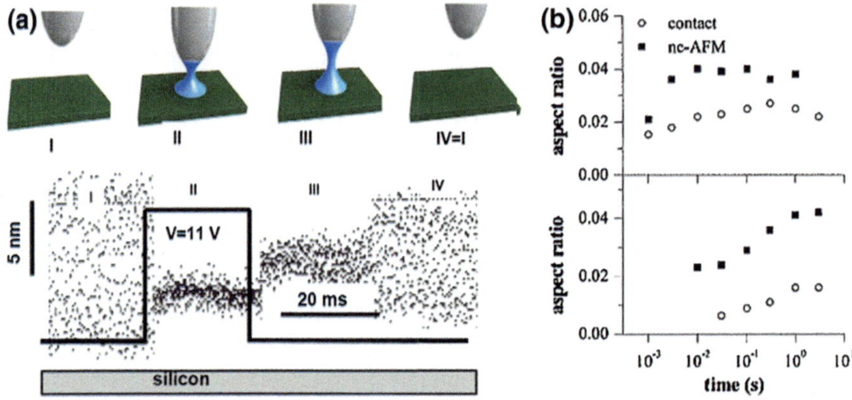

Fig. 5.3 **a** Schematic view (top) and corresponding experimental data (bottom) of the different steps occurring in an oxidation cycle. In I, the tip oscillates few nanometers above the surface. In II, under the application of a voltage pulse, the water meniscus is formed. Then, the sample is oxidized. The tip is pulled closer to the substrate as a result of electrostatic and capillary forces. In III, when the pulse is switched off, the tip is held by the capillary forces until the meniscus breaks. Finally, in IV, the tip is retracted by the z-piezo to position I [14]. **b** Comparison of the aspect ratio (height to width) as a function of the oxidation time between silicon oxides fabricated by contact (open circles) and non-contact (filled squares) o-SPL under $V_{ox} = 20$ V (top graph) and $V_{ox} = 14$ V (bottom graph). The experiments were performed under a RH of 36% [67]

This cycle can be monitored in real time by an oscilloscope during the lithography [64, 65]. An example is shown in Fig. 5.3a with a schematic of the process.

For similar oxidation parameters, the nanopatterns created by non-contact o-SPL have higher aspect ratios than the ones fabricated by contact o-SPL [67] (Fig. 5.3b). This is due to several factors. First, the mechanical work that the oxide has to do against the cantilever in order to grow reduces the achievable vertical growth in contact mode. Second, the o-SPL in non-contact mode adds the tip-sample distance/oscillation amplitude as an oxidation parameter, and this allows the experimenter to control the largest tip-surface separation that maintains a stable water meniscus. A larger tip-surface separation gives a smaller meniscus diameter and hence a narrower oxide. In addition, using the contact mode to perform o-SPL implies a faster degradation of the tip and therefore a limitation in the resolution and reproducibility achievable by this technique. The use of amplitude modulation non-contact o-SPL reduces this issue and gives higher resolution in the imaging mode to inspect the patterns after fabrication.

The common probes used for contact o-SPL are Si_3N_4 or Si cantilevers, with force constants and resonant frequencies in the range of 0.1–1 N/m and 10–50 kHz, respectively. The main probes used in amplitude modulation non-contact o-SPL are commercial n^+-type doped Si cantilevers with force constants and resonant frequencies of around 40 N/m and 350 kHz, respectively.

5.2.1.4 Patterning Parameters

The main o-SPL parameters are the **applied voltage** (V_{ox}), the **oxidation time** (t_{ox}), and the **relative humidity** (RH). Commonly used values are in the range 10–30 V, μs–ms, and 30–60%, respectively. The best combination of the three parameters is short time and high amplitude voltage pulses under the lowest relative humidity. Such combinations generally produce the highest and narrowest oxides, therefore, the patterns with the highest aspect ratio.

Other factors that influence the oxidation process are:

- The roughness of the surface has to be as low as possible, normally less than 2 nm peak to peak. The density of nanoparticles or organic residues on the surface must be minimized as well. Therefore, protocols are sometimes required to clean the surface before the o-SPL process. For example, on Si or silicon-on-insulator (SOI) substrates, variants of the RCA standard process reduce the amount of inorganic nanoparticles and make the surface more hydrophilic. If the o-SPL is performed after other lithographic steps, the resist residues can be removed by oxygen plasma or UV light exposure in an ozone atmosphere.
- The doping level and conductivity of the substrate affect the oxidation growth rate. It has been shown that, in p-doped silicon substrates, there is a faster vertical growth rate because the holes have a predilection to react with anionic species favouring anodic oxidation [53, 68]. Morimoto et al. [69] investigated the influence of the doping level on the oxide density using XTEM images. The nature of the substrate material, i.e., metal, semiconductor, organic, etc., is clearly relevant to the threshold voltage required to oxidize it. The influence of the sample conductivity on the width and shape of the oxides was studied by Cambel and Soltys [70].
- The hydrophilicity/hydrophobicity of the substrate surface influences the lateral confinement of the water bridge for a given relative humidity [71].
- Fang [72] has studied the influence of the crystalline orientation of the silicon on the oxidation rate.
- When the oxidation of a given surface requires a tip with higher conductivity, the commercial Si tip can be coated with a metallic layer, commonly Pt or Au. However, this means a loss in the lateral resolution of the oxides. Recently, Yamamoto et al. [73] reported the use of Pt-coated tips for local catalytic oxidation of Si immersed in water into water-soluble silicates.

5.2.2 Materials Modified by Oxidation Scanning Probe Lithography

5.2.2.1 o-SPL on Semiconductors and Metals

Despite the superior quality of III–V compound semiconductors for high mobility, high power, and high frequency applications, and the rise of 2D material-based electronics, **silicon** is still the main element of choice in micro- and nanoelectronics due to its abundance and cost, the ease with which high quality crystalline wafers can be obtained, its mechanical stability, the well-developed doping processes, and its oxide. For these reasons, it is also the most widely investigated material in o-SPL. There are several examples of early work in which Si field-effect transistors were fabricated by o-SPL [74–76]. The common approach uses the top Si thin layer of a SOI as a starting substrate. The Si FET fabrication steps vary depending on the research group, but mainly involve the definition of oxide hard masks by o-SPL and a subsequent wet or dry etching transfer process [77–79].

The steps of the fabrication process employed in Garcia's group are described below. First, two metal electrodes are defined by photolithography on the SOI Si top layer to be used as localization markers (Fig. 5.4a). Then, an oxide nanowire is fabricated between the markers by o-SPL (Fig. 5.4b). This oxide line constitutes the hard mask for the pattern transfer step. The oxide line cannot reach the metal electrodes, since the high electric field formed during o-SPL degrades them. Therefore, a second photolithography step is performed to contact the nanowire with the source and drain electrodes (Fig. 5.4c). Finally, the pattern is transferred to the Si top layer after processing by reactive ion etching (RIE) and a back-gate Si nanowire (SiNW) FET is produced (Fig. 5.4d). After rapid thermal annealing treatment, the o-SPL fabricated SiNW FET shows similar mobility and subthreshold swing values to those fabricated by electron beam lithography. Both have been fabricated with similar feature sizes and the same device configuration (Fig. 5.4e) [79]. The optimization of the RIE process allows the use of o-SPL oxide masks as thin as 0.3–1 nm to transfer to 2–12 nm thick SiNWs [29] with a SiO_2:Si width ratio close to 1 and an etching selectivity of about 11 (Fig. 5.4f). A TEM cross-section of one of the SiNWs fabricated using this process is shown in Fig. 5.4g [80]. SiNW FETs fabricated by o-SPL have been used as label-free biosensors to monitor the DNA repair process by the RecA protein [81]. Another group recently used similar o-SPL Ni-silicide/SiNWs as gas sensors [82]. Finally, o-SPL has been used to oxidize patterns at selective locations on a silicon photonic crystal. After the removal of the created oxides by HF etching, nanocavities with ultrahigh-Q (10^6) were demonstrated [83].

Various quantum devices based on **III–V compound semiconductors** have been produced by o-SPL [84–87]. The o-SPL step is performed to fabricate narrow dielectric barriers on high quality heterostructures to define the different point contacts and in-plane gates. Ensslin's group developed the multiple layer o-SPL [88] for the fabrication of complex quantum devices (Fig. 5.5a–c). In this approach, a first o-SPL step is performed to define in-plane gates and conduction leads on the GaAs-AlGaAs substrate (Fig. 5.5a). Then, a thin Ti film 6–7 nm thick is deposited on the structure and a second o-SPL is performed to define additional in-plane or top gates (Fig. 5.5b), which allow the individual tuning of each quantum point contact (QPC) or dot (QD) in a multicomponent device, such as the one shown in Fig. 5.5c.

Si_3N_4 is a relevant material in micro- and nanoelectronics as a dielectric and a hard transfer mask for its high etching and growth selectivity with Si. Gwo and coworkers have shown that Si_3N_4 could be converted locally to SiO_x by o-SPL [89, 90]. This circumstance was used [91] to fabricate SiO_2/Si_3N_4 bilayer thin masks

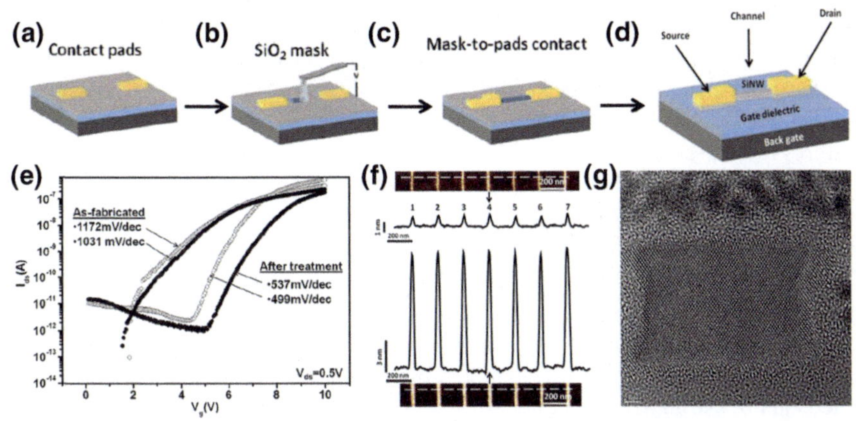

Fig. 5.4 SiNW FET fabrication process by o-SPL and its characterization. **a–d** Fabrication steps. **a** Metal contacts are fabricated by photolithography on the Si top layer of a SOI. **b** An oxide nanowire hard mask is defined by o-SPL between the electrodes, without reaching them with the AFM tip. **c** A second photolithography is performed to contact the hard mask with the contact electrodes. **d** After pattern transfer by RIE processing, the finalized back-gate SiNW FET is produced [79]. **e** Transfer curves of SiNW FETs fabricated by o-SPL (open circles) and electron beam lithography (filled circles) before and after a rapid thermal annealing treatment to improve the performance of the device. The two transistors were fabricated with similar feature sizes and the same device configuration to enable a comparison. It can be shown from the similar subthreshold swing values that the electrical characteristics of the SiNWs depend on their geometry but are independent on the lithographic method. **f** AFM topographic images and their corresponding cross-sections of a roughly 1 nm thick SiO_2 hard mask fabricated by o-SPL (top) and the SiNWs produced after RIE transfer onto a 12 nm thick Si layer of an ultra-thin SOI substrate [29]. **g** TEM cross-section of one of the SiNWs fabricated by the process depicted in (**f**) [80]

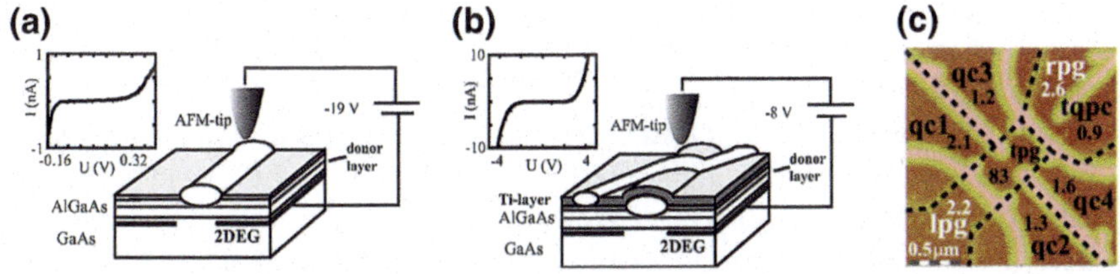

Fig. 5.5 Multiple layer o-SPL process [88]. **a** In the first step, the two-dimensional electron gas (2DEG) area located below the oxide lines fabricated on the top GaAs thin layer by o-SPL is depleted. Therefore, these oxide lines serve as narrow dielectric barriers to define the in-plane gates and conduction leads of multiple QPCs and QDs. **b** Then, a Ti thin film is deposited on top of the structure depicted in (**a**). Oxidation SPL is performed on the Ti layer to define additional top gates that will allow the individual control of each QPC and QD. **c** AFM topographic image of a double QD fabricated by the multiple layer o-SPL process

for the selective growth of Si nanodots and nanowires. Later, Pérez-Murano and coworkers [92] patterned arrays of lines on silicon nitride thin films (<10 nm) by o-SPL to use them as hard masks for transfer in silicon by wet etching. The potential application was a cheap and fast method for fabricating stamps with nanometric features to be used in nanoimprint lithography.

The strategy of using o-SPL to pattern nanostructures on the top layer of a transfer stack was explored by Rolandi et al. [93]. They developed a 4 nm thick **Mo**/35 nm thick PMMA [poly (methyl methacrylate)] bilayer transfer stack to fabricate 8 nm thick and 35 nm wide Ti electrodes after electron beam evaporation and lift-off. Later, Pellegrino et al. [94] used a very similar stack of 8 nm thick **Mo**/50 nm PMMA to obtain 100 nm resolution $(Fe,Mn)_3O_4$ nanostructures and measured the room-temperature ferromagnetism to explore spintronics applications. The same group opted later to use an **Mo** layer only 20 nm thick on top of an Al_2O_3 (0001) crystalline substrate, then after removal of the oxides formed by o-SPL by deionized water, deposition of 150 nm wide $Fe_{2.5}Mn_{0.5}O_4$ lines by pulsed laser deposition (PLD) and removal of the unmodified Mo by H_2O_2 etching, obtaining a structure for potential ferromagnetic random access memory applications [95].

SiC is a wide band gap semiconductor relevant for high-power and high-temperature microelectronics applications. Given the possibility of growing silicon oxide directly on SiC and thereby integrating this material in CMOS processing, the local oxidation of silicon carbide by o-SPL has been studied by various groups [96–98]. In Garcia's group, SiC_x nanowires were obtained by performing o-SPL on a Si (100) substrate in an ethanol (CH_3CH_2OH) atmosphere [20].

Niobium and its compounds are important materials in superconducting and electronic devices. Therefore, the process and kinetics of Nb oxidation by o-SPL has been studied [18], and various Nb/NbO_x or NbN/$NbNO_x$ based nanodevices have been fabricated: single electron transistors at room temperature [99], superconducting quantum interference devices (SQUIDs) [100, 101], and superconducting single-photon detectors [102].

One of the first o-SPL fabricated electronic nanodevices was based on **titanium** [103]. Moreover, some of the highest resolution patterns by o-SPL have been fabricated on this material. Using a SWNT grown by CVD onto the apex of a conventional Si tip to oxidize Ti thin films, line widths of 5 nm have been achieved [104]. This fabricated small oxide was used as the tunnel junction of a single-electron transistor. Later, using commercial conductive tips (NT-MDT, Russia) and by optimizing the oxidation parameters, 8 nm wide TiO_x lines were achieved by two different groups [105, 106].

The strategy for creating oxides by scanning probe lithography in selective areas of a substrate to place or grow nanostructures with determined periodicities has been explored in different studies. By patterning oxides by o-SPL and, after etching, creating holes in a silicon substrate, single PbS colloidal quantum dots were trapped inside these holes [107]. In another study, arrays of **Ni** oxide nanodots were fabricated on a 10 nm thick Ni film. Then, the unoxidized nickel was removed after etching in a nitric acid solution. The nickel oxide nanodots in the substrate were subsequently used as catalytic seeds for the growth of vertical carbon nanotubes with diameters of 30–80 nm by inductively coupled plasma chemical vapour deposition [108]. In a third group, an array of **GaAs** oxide nanodots with 2-μm pitch was fabricated by o-SPL. Then, these oxides were removed by HF etching. The nanoholes were used as selective sites for sub-100 nm diameter InAs quantum dot nucleation by molecular beam epitaxy [109, 110]. The optical properties of these InAs quantum dots were studied, showing single-photon emission capabilities [111].

The level of adhesion offered by a surface is important in nanoscale research for topics such as cell and biomolecule adhesion to a substrate or microelectromechanical systems (MEMS) engineering. With this in mind, oxide nanopillars with different geometries, viz., rectangular, triangular, circular, and ring-shaped, were fabricated by o-SPL on a **GaAs** surface and their adhesive force was measured by AFM force spectroscopy [112]. In another investigation, an array of o-SPL fabricated $\mathbf{GaO_x}$ nanostructures with a density of 10^7 cm^{-2}, patterned on top of a stack of InGaN/GaN multiple quantum wells (MQWs), showed enhanced photoluminescence intensity and a blue shift in the emission peak by acting as a graded-refractive index layer [113].

To conclude this section, the oxidation parameters and kinetics of o-SPL patterning have been studied in several other materials: **InN** and **GaN** [114], **Al** [115], **Co** and **Ni** [116], **Ni** [117, 118], **Hf** [119], **Zr** and **ZrN** [120], **Ge** [121], perovskite oxides such as $\mathbf{SrTiO_{3\text{-}\delta}}$ [122], and $\mathbf{La_{0.67}Ba_{0.33}MnO_{3\text{-}\delta}}$ [123].

5.2.2.2 o-SPL on 2D Materials

In the last few years, 2D materials have been extensively explored for electronic, optoelectronic, spintronic, and superconducting applications. These materials present an extremely high surface-to-volume ratio, due to their atomic monolayer or few-layer thickness. As a consequence, they possess a high chemical sensitivity that makes them appealing as sensors. Their properties can be tailored by the number of layers and by defect, edge, strain, and doping engineering. The standard lithographic processes available to define patterns on these materials generally involve the use of resist coatings and subsequent cleaning protocols with solvents or plasma treatments, which can sometimes introduce undesired modifications in device performance. In this case, the o-SPL technique provides a way to pattern the material directly into nanostructures with arbitrary geometry and nanoribbons with a chosen periodicity.

The strategy for generating narrow dielectric barriers to fabricate quantum nanodevices as mentioned in the previous section has been applied several times on graphene [124–126]. The quantum dots defined by the o-SPL insulating barriers on monolayer graphene (Fig. 5.6a) display the characteristic Coulomb blockade peak conductance (Fig. 5.6b) [127]. Later, to assess the quality of the o-SPL patterns on graphene for future high performance nanodevices, a combination of high resolution TEM (HRTEM) and spectroscopic analysis was performed [128]. One of the TEM cross-sections of an o-SPL nanowire is shown in Fig. 5.6c.

In the case of transition metal dichalcogenide (TMDC) thin layers, the oxides formed by o-SPL are soluble in deionized water. On the other hand, the oxides also

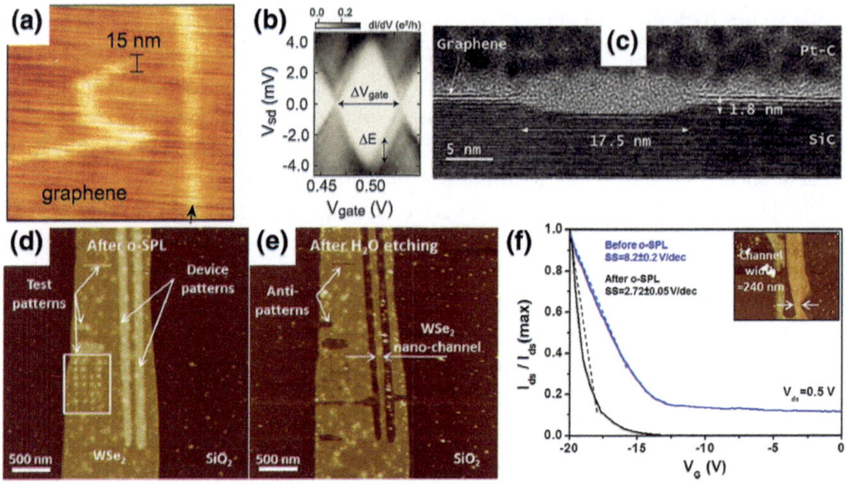

Fig. 5.6 2D material-based patterns and devices fabricated by o-SPL. **a** AFM topographic image of a graphene monolayer quantum dot with a diameter of about 70 nm. The brightest lines on the image correspond to the oxide lines defined by o-SPL, acting as insulating barriers [127]. **b** The electrical characterization of the device shown in (**a**) exhibits Coulomb blockade diamond behaviour, as expected from QDs [127]. **c** HRTEM cross-section of a graphene oxide nanowire fabricated by o-SPL, 17.5 nm wide. From the image, it can be seen that the o-SPL process has oxidized the graphene layer and also around 1–2 nm of the SiC substrate [128]. **d** AFM topographic image of oxide dots and lines defined by o-SPL on a few-layer thin WSe_2 flake [31]. **e** AFM topographic image of the flake in (**d**) after etching in DI H_2O for 30 s. The oxides have been removed, showing that they have penetrated the whole thickness of the flake. In this way, an electrically isolated 80 nm wide nano-channel has been obtained from the initial flake-wide channel of around 1 μm in its narrowest part [31]. **f** Transfer curves of a few-layer WSe_2 FET after photolithography (blue), with a channel width of around 1.3 μm and after o-SPL and wet etching (black), with a channel width of 240 nm [31]

grow below the surface of the material, as mentioned previously. Combining these two characteristics, field effect transistors with nano-channels have been fabricated by o-SPL on MoS_2 [129] and WSe_2 [31, 130]. The starting point of the process is the transfer of a TMDC thin layer flake on a SiO_2/Si substrate by mechanical exfoliation and subsequent FET fabrication by photolithography. In the next step, o-SPL is used to fabricate oxide lines that grow the whole thickness (4–12 nm) of the flake below the surface (Fig. 5.6d). Then, wet etching in deionized water for 30 s is performed to remove the oxides, ensuring the electrical isolation of the nano-channel FET from the rest of the flake (Fig. 5.6e). The transfer curves of a transistor before (1.3 μm wide channel) and after (240 nm wide channel) o-SPL are shown in Fig. 5.6f. The nano-channel FET shows an improved subthreshold swing.

Other recent work on TMDC thin layers patterned by o-SPL are have used NbS_2 [131] and TaS_2 [132].

5.2.2.3 o-SPL on Self-assembled Monolayers

Another major application of o-SPL is the selective modification of organic monolayers into chemically active radicals for template growth and fabrication of molecular architectures.

Sagiv's group developed the 'constructive nanolithography' approach [133–135]. In this process, a high quality OTS (n-octadecyltrichlorosilane, CH_3-$(CH_2)_{17}$-$SiCl_3$) monolayer deposited on a Si substrate is first modified in selective sites by o-SPL. The chemical reactions that take place at the tip (5.15) and at the surface (5.16) are, respectively:

$$2H_2O + 2e^- \rightarrow H_2 + 2OH^- \tag{5.15}$$

$$R{-}CH_3 + 2OH^- + 2h^+ \rightarrow RCOOH + 2H_2 \tag{5.16}$$

The oxidation process transforms the $-CH_3$ groups into carboxylic acid –COOH groups, which are chemically active. Then, successive functional monolayers can be deposited only in the OTS modified sites by o-SPL. Finally, metallic nanoparticles such as gold [136] or silver [137] can be attached selectively in the top functional monolayer. The additive hierarchical self-assembly nature of this process justifies the name 'constructive nanolithography'. This process was exploited later by the same group to fabricate an ionic transport device [138].

Shubert's group has studied in detail the two regimes of patterning present in the OTS monolayer/Si substrate [139], depending on the oxidation parameters. For small and short voltage pulses, which corresponds to the <8 V and ms regime, only the OTS monolayer is oxidized. For higher amplitude and longer time pulses, the monolayer is completely degraded, and the Si substrate is oxidized. The same group has applied constructive nanolithography to fabricate ring-like structures with a rim of organic molecules and a core of silicon oxide [140], nano-gap structures to trap individual metal nanoparticles [141], carbon nanotubes crossed in a pre-selected location [142], and Au NP cluster arrays for surface-enhanced Raman spectroscopy measurements [143].

Constructive nanolithography and variations on it have been employed in other groups to selectively deposit magnetic particles [144–146], for biomolecular patterning [147, 148], and to define pH-responsive polymer brushes [149].

Sometimes, o-SPL modification of different substrates can lead to the fabrication of direct guiding templates, without the need to deposit further organic monolayers. The electrostatic interaction between silicon oxide patterns and different molecules [150, 151] and the direct assembly of block copolymers [152] have been demonstrated.

5.3 Thermal Scanning Probe Lithography: Fast Turnaround Nanofabrication in Ambient Conditions Combining Thermal Probes and Focused Lasers

Thermal scanning probe lithography (t-SPL) is based on scanning probe cantilevers comprising integrated heaters, which have been developed at IBM for more than two decades. Soon after the invention of the AFM, heated tips were used to form indents on a PMMA substrate [153] for high density data storage applications. This idea was further developed in the so called "Millipede" project by the team led by Vettiger and Binnig [154] and produced a prototype [155] that demonstrated ultrahigh areal densities for data storage combined with massively parallel scanning probe devices. The technical know-how for writing high-resolution patterns on polymer surfaces using a heatable tip then provided the foundation for nanoscale lithography applications.

Owing to the two integrated heaters for tip heating and thermal topography sensing, the silicon cantilever used in t-SPL involves a specific design and fabrication process [156, 157], illustrated in Fig. 5.7a. The structure consists of highly doped silicon legs and integrated lower doped silicon heaters with a doping concentration of 10^{17}–10^{18} cm^{-3}. Three voltage inputs at the three legs selectively power the resistors. The central leg comprises a large area platform in order to provide additional capacitive force for actuation of the cantilever (force constant of 0.7 N/m). During patterning, the tip raster scans the substrate at a defined distance from the surface. For each programmed pixel, two different voltages are applied for a few microseconds to write the patterns: V_h to resistively heat the tip and V_f applied to the substrate to pull the tip into contact with the surface (Fig. 5.7b). The heater sustains temperatures of over 1000 °C for a time scale of days.

A TEM image of a pristine tip is shown in Fig. 5.7c. The most common resist on which the nanopatterns are generated by t-SPL is polyphthalaldehyde (PPA), a self-amplified depolymerisation polymer with a decomposition temperature of T_{unzip} = 120–150 °C (Fig. 5.7d). When the heated tip forms a mechanical contact with the polymer at sufficiently high temperature, PPA decomposes into volatile monomers [158], leaving behind a void of defined size and controllable depth. Using this fast decomposition process, fast turnaround scanning probe lithography has been developed to provide a complete patterning system featuring 4×10^4 $\mu m^2/s$ throughput at 15 nm resolution [159]. The speed is limited by the mechanical setup (Fig. 5.7e), which was optimized to achieve scan speeds of 20 mm/s and patterning rates of 2 μs/pixel to write 8×10^5 pixels in 12 s. Moreover, the technique has demonstrated nanometric depth level control, sub-10 nm lateral resolution capability, and sub-5 nm overlay accuracy [160–162]. An example that condenses some of the aforementioned capabilities is the Guinness world record for the smallest magazine cover, with a size of 11×14 μm^2, produced in 10 min.

Owing to the conical shape of the tip, highest resolution patterning is achieved at low writing depths of the patterns. This leads to a minimization of the resist thickness for pattern transfer, as seen similarly for other lithography methods [163].

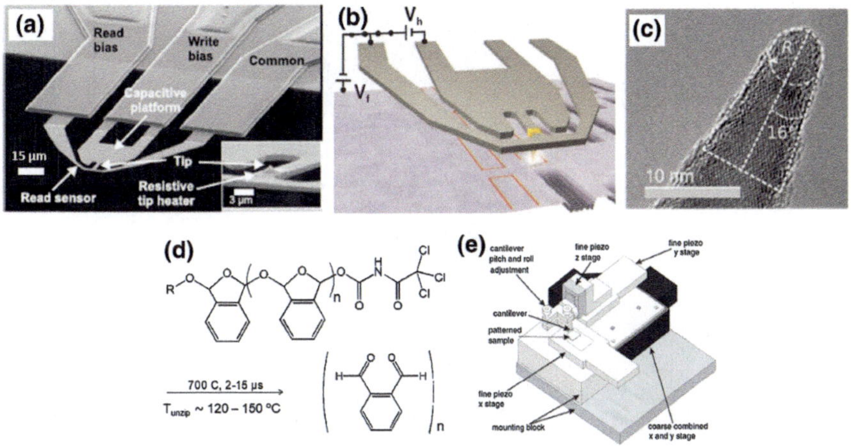

Fig. 5.7 **a** SEM image of a typical silicon cantilever specifically designed to perform t-SPL. The different components are indicated and identified by arrows [159]. **b** Schematic view of the t-SPL writing process. V_h is the voltage applied to the resistive heater to control the writing temperature and V_f is the voltage applied between the substrate and the cantilever to pull the tip into contact with the PPA layer [156]. **c** TEM image of a pristine tip with an apex radius of around 3 nm and an estimated half-angle of 16° [162]. **d** Molecular unit of polyphthalaldehyde (PPA). The time and temperature needed to decompose it into a monomer are indicated next to the arrow. Modified from [158]. **e** Tool setup for fast turnaround imaging and patterning by t-SPL [159]

To address the challenge of an ultrathin imaging layer, a specific high-resolution transfer stack was developed, together with high-aspect-ratio dry-etch processing [164, 165]. The transfer stack consists of three layers including, from top to bottom, a 9 nm thick PPA layer, a 2–3 nm thick thermally evaporated SiO_2 layer, and a 50 nm thick transfer polymer layer. A schematic view of the stack is shown in Fig. 5.8a. Using the same stack and processing it is possible either to fabricate tight half-pitch silicon nanowires (using HM8006) or to perform high-resolution metal lift-off for contacting nanostructures (using PMMA), just by replacing the polymeric transfer layer. Exploiting the high-resolution imaging capabilities of t-SPL to precisely detect the location of buried nanostructures beneath resist layers [161], metal contacts were fabricated on a 27 nm diameter InAs nanowire with 3 nm overlay accuracy [164].

Recently, small modifications have been introduced to further optimize the transfer stack for high-resolution patterning [162]. First, the minimum thickness of the PPA imaging layer that reliably provides the highest resolution after pattern transfer to silicon was found to be in the range 6–7 nm. In addition, a 2 nm thick PMMA layer was placed between the imaging layer and the 2.5 nm thick SiO_2 hard mask layer. The PMMA layer acts as a soft landing layer to minimize tip wear and provides additional thermal isolation. Moreover, the optimal geometry of the written lines leading to a successful pattern transfer was determined by screening the t-SPL patterning force and temperature. Under these conditions, 11 nm half-pitch dense lines were transferred into polymer (HM8006) and 14 nm half-pitch SiNWs with a

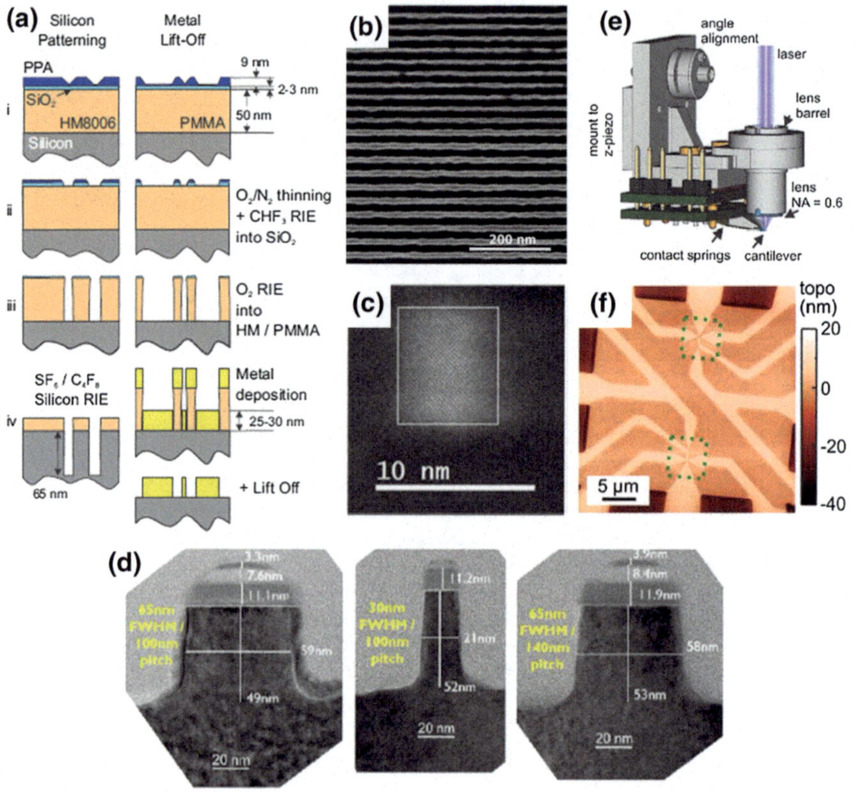

Fig. 5.8 High-resolution silicon patterning. **a** Scheme of the pattern transfer process using the three-layer stack configuration for silicon nanopatterning (left column) and metal lift-off (right column) [164]. **b** SEM image of an array of 14 nm (top) and 16 nm (bottom) half-pitch SiNWs [162]. **c** STEM cross-section of one of the 14 nm half-pitch SiNWs. The image shows that the nanowire has sub-10 nm feature sizes [162]. **d** TEM cross-sections of 50 nm deep SiNWs produced by t-SPL processing, using a novel transfer stack and conversion of the PPA imaging layer into an etch resistant Al_2O_3-C hard mask [163]. **e** Scheme of the mix-and-match laser writing + t-SPL tool setup. The laser beam is focused through a lens that is mounted on the holder of the AFM cantilever [166]. **f** AFM topographic image of a set of two single-electron transistors defined by the mix-and-match processing in a single step on the PPA layer. The high-resolution devices contained inside the green boxes were written by t-SPL. The remaining area was defined afterwards by laser writing in less than 30 s [166]

line-edge roughness (3σ) of 2.6 ± 0.4 nm (Fig. 5.8b) were fabricated. The STEM cross-section of one of these nanowires shows that they have sub-10 nm feature sizes (Fig. 5.8c).

In spite of the progress mentioned above, t-SPL resolution is still limited by the minimal PPA thickness of 6–7 nm, which is required due to the low etch resistance of the PPA imaging layer. For this reason, a strategy has been developed to transform the PPA layer into a highly etch-resistant mask by sequential infiltration synthesis [163].

The method simplifies the transfer stack and removes the need for the SiO_2 layer, which is considered to be a major factor for tip wear. A homogeneous sequential infiltration of PPA by tri-methyl-aluminium and water was successfully performed at room temperature. As a result, SiNWs with widths in the range of 10–60 nm and a depth up to 50 nm were fabricated (Fig. 5.8d), demonstrating >10× amplification and low surface roughness.

Recently, the high-resolution patterning and nanometer-accuracy overlay capabilities of t-SPL have been paired with high-throughput low-resolution laser writing for hybrid lithography of the same PPA resist layer in the same tool [166]. In this approach, an aspherical lens was mounted on the holder of the AFM tip to focus the laser beam on the sample (Fig. 5.8e). The setup provides a fixed offset between the laser focus and tip writing positions, this being determined by t-SPL imaging of laser written structures. Using the mix-and-match setup, device components can be fabricated by laser writing and by t-SPL in a single patterning step and aligned accurately relative to existing features on the substrate. To demonstrate this, a set of two room temperature single-electron transistors were fabricated within an area of $30 \times 30\ \mu m^2$ in 4.5 min (Fig. 5.2f).

Beyond high-resolution patterning, t-SPL also provides greyscale patterning with single-nanometer accuracy of the patterning depth in thicker PPA layers [14, 158]. The achievable depth accuracy and lateral resolution is high due to the absence of proximity or redeposition effects, characteristic for charged beam techniques such as electron beam lithography and focused ion beam [167]. By applying 3D t-SPL patterning, photonic molecules with Gaussian profiles [167] and nanofluidic ratchet-shaped Brownian motors [168, 169] have been fabricated.

Finally, mirroring the o-SPL applications, t-SPL has been employed for the fabrication of a wide range of nanodevices and nanopatterns on different types of materials. Using the above-mentioned high-resolution transfer stack processing [164], an array of 384 MoS_2 monolayer nanoribbons with 18 nm half-pitch and 350 nm length along the zigzag direction (Figs. 5.9a–c) have been fabricated at IBM. Other recent work includes nanochannel pentacene field-effect transistors [170], graphene-based nanodevices [171], guided assembly of gold nanorods [172] and block copolymers [173], and supramolecular polymer-based color-switchable QR codes [174] (Fig. 5.9e).

5.4 Conclusion. Strengths and Limitations of SPL

In this chapter we have illustrated the versatility of scanning probe lithography in terms of modified materials (polymers, self-assembled monolayers, Si and III–V compound semiconductors, 2D materials) and fabricated nanodevices (electronic, optoelectronic, spintronic, magnetic, superconducting). The main advantages of the technique are the sub-10 nm resolution patterning of arbitrary geometries in a maskless single-step, the ability of in situ non-destructive inspection after fabrication with the same tool, operation in ambient conditions, and the possibility of fabricating a wide range of nanostructures and nanodevices using the same tool with minimal

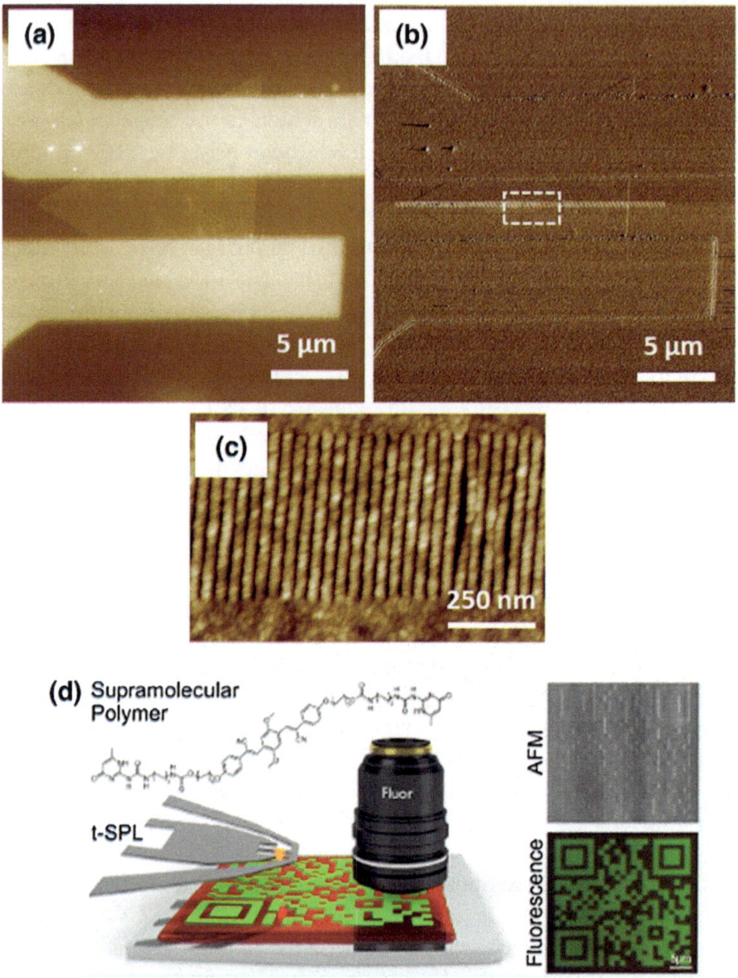

Fig. 5.9 **a** AFM topographic image of an array of 384 MoS_2 monolayer nanoribbons with 18 nm half-pitch and 350 nm length along the zigzag direction. This work was performed in collaboration with the LANES group, at EPFL. **b** AFM phase image of the device shown in (**a**). In this image the periodic nanoribbons patterned by t-SPL are clearly resolved, spanning the entire MoS_2 monolayer flake. **c** High-resolution image of the MoS_2 nanoribbons contained in the white dashed box marked in (**c**). **e** Fabrication of QR codes from a fluorescent supramolecular polymer film by t-SPL. The AFM topographic image (top right) shows very little modification of the surface by t-SPL, but the fluorescence microscope image (bottom right) clearly resolves the successful transformation of the polymer structure into a fluorescent material [174]

setup modifications. Oxidation SPL provides a way to perform direct writing on materials that are sensitive to resist residues, such as 2D material thin layers, or on organic monolayers to change their free radicals and hence their functionality, in order to build molecular architectures. Thermal SPL is the modality that has reached the fastest prototyping and the highest throughput by mix-and-match processing with laser writing. It has developed a high-resolution transfer stack that can be used either for sub-10 nm Si nanopatterning or lift-off processing for contact fabrication. It has also developed a greyscale lithography scheme with nanometer accuracy.

Despite the above-mentioned capabilities of the technique, there are two main drawbacks that prevent SPL from being used for technological applications. First, for sub-10 nm resolution, all the SPL modalities suffer from the relatively short endurance of the tips. This is due to blunting or the picking up of residues during the lithographic process, which change the AFM tip apex radius. A further enhancement in o-SPL resolution was explored by attaching a carbon nanotube (CNT) tip to the apex of conventional silicon tips [175, 176]. Some of the strategies that have been explored to reduce the wear of the probe are the use of carbon nanofibers attached to conventional tips [177] or the use of monolithic conductive polycrystalline diamond probes [178]. However, all these approaches share low viability due to the difficulty in fabrication, the high cost, and the insufficient endurance improvement of such probes.

The second drawback of the technique is its small throughput compared with other established lithographies. The serial nature of the technique slows down the patterning process and the size of the AFM probe imposes a limit in the maximum patterned area achievable. To overcome this issue, some strategies explored so far have been high-speed AFM implementation [159, 179] and parallel fabrication approaches using either arrays of probes [154, 180, 181] or stamps [182–184].

A third strategy to improve the throughput of the technique has come closest to implementing it for technological applications: mix-and-match lithography. With this idea, the fabrication of high-resolution nanometric patterns by SPL can be paired with the fabrication of contact electrodes with micron-squared areas using low-resolution high-throughput lithography. So far, this has been done with electron beam [185] and optical [166] lithographies.

References

1. J.S. Foster, J.E. Frommer, P.C. Arnett, Molecular manipulation using a tunnelling microscope. Nature **331**, 324–326 (1988)
2. D.M. Eigler, E.K. Schweizer, Positioning single atoms with a scanning tunnelling microscope. Nature **344**, 524–526 (1990)
3. J.A. Dagata, J. Schneir, H.H. Harary, C.J. Evans, M.T. Postek, J. Bennet, Modification of hydrogen-passivated silicon by a scanning tunnelling microscope operating in air. Appl. Phys. Lett. **56**, 2001–2003 (1990)
4. A. Majumdar, P.I. Oden, J.P. Carrejo, L.A. Nagahara, J.J. Graham, J. Alexander, Nanometer-scale lithography using the atomic force microscope. Appl. Phys. Lett. **61**, 2293–2295 (1992)

5. H.C. Day, D.R. Allee, Selective area oxidation of silicon with a scanning force microscope. Appl. Phys. Lett. **62**, 2691–2693 (1993)
6. C. Cen, S. Thiel, G. Hammerl, C.W. Schneider, K.E. Andersen, C.S. Hellberg, J. Mannhart, J. Levy, Nanoscale control of an interfacial metal–insulator transition at room temperature. Nat. Mater. **7**, 298–302 (2008)
7. P. Irvin, Y. Ma, D.F. Bogorin, C. Cen, C.W. Bark, C.M. Folkman, C.-B. Eom, J. Levy, Rewritable nanoscale oxide photodetector. Nat. Photonics **4**, 849–852 (2010)
8. P. Irvin, J.P. Veazey, G. Cheng, S. Lu, C.-W. Bark, S. Ryu, C.-B. Eom, J. Levy, Anomalous high mobility in LaAlO3/SrTiO3 nanowires. Nano Lett. **13**, 364–368 (2013)
9. M. Tomczyk, G. Cheng, H. Lee, S. Lu, A. Annadi, J.P. Veazey, M. Huang, P. Irvin, S. Ryu, C.-B. Eom, J. Levy, Micrometer-scale ballistic transport of electron pairs in $LaAlO_3/SrTiO_3$ nanowires. Phys. Rev. Lett. **117**, 096801-1–096801-6 (2016)
10. E. Albisetti, D. Petti, M. Pancaldi, M. Madami, S. Tacchi, J. Curtis, W.P. King, A. Papp, G. Csaba, W. Porod, P. Vavassori, E. Riedo, R. Bertacco, Nanopatterning reconfigurable magnetic landscapes via thermally assisted scanning probe lithography. Nat. Nanotechnol. **11**, 545–552 (2016)
11. S.N.G. Corder, X. Chen, S. Zhang, F. Hu, J. Zhang, Y. Luan, J.A. Logan, T. Ciavatti, H.A. Bechtel, M.C. Martin, M. Aronson, H.S. Suzuki, S. Kimura, T. Lizuka, Z. Fei, K. Imura, N.K. Sato, T.H. Tao, M. Liu, Near-field spectroscopic investigation of dual-band heavy fermion metamaterials. Nat. Commun. **8**, 2262-1–2262-7 (2017)
12. W.P. King, B. Bhatia, J.R. Felts, H.J. Kim, B. Kwon, B. Lee, S. Somnath, M. Rosenberger, Heated atomic force microscope cantilevers and their applications. Annu. Rev. Heat Transfer **16**, 287–326 (2013)
13. N. Kurra, R.G. Reifenberger, G.U. Kulkarni, Nanocarbon-scanning probe microscopy synergy: fundamental aspects to nanoscale devices. ACS Appl. Mater. Interfaces **6**, 6147–6163 (2014)
14. R. Garcia, A.W. Knoll, E. Riedo, Advanced scanning probe lithography. Nat. Nanotechnol. **9**, 577–587 (2014)
15. Y.K. Ryu, R. Garcia, Advanced oxidation scanning probe lithography. Nanotechnology **28**, 142003-1–142003-17 (2017)
16. S.-R. Jian, T.-H. Fang, D.-S. Chuu, Mechanisms of p-GaAs (100) surface by atomic force microscope nano-oxidation. J. Phys. D Appl. Phys. **38**, 2424–2432 (2005)
17. S. Ross, A. Sussman, Surface oxidation of molybdenum disulphide. J. Phys. Chem. **59**, 889–892 (1955)
18. A. Heidelberg, C. Rozenkranz, J.W. Schultze, Schäpers, G. Staikov, Localized electrochemical oxidation of thin Nb films in microscopic and nanoscopic dimensions. Surf. Sci. **597**, 173–180 (2005)
19. M. Lazzarino, S. Heun, B. Ressel, K.C. Prince, P. Pingue, C. Ascoli, Atomic force microscope anodic oxidation studied by spectroscopic microscopy. Appl. Phys. Lett. **81**, 2842–2844 (2002)
20. M. Tello, R. Garcia, J.A. Martín-Gago, N.F. Martínez, M.S. Martín-González, L. Aballe, A. Baranov, L. Gregoratti, Bottom-up fabrication of carbon-rich silicon carbide nanowires by manipulation of nanometer-sized ethanol menisci. Adv. Mater. **17**, 1480–1483 (2005)
21. M. Donarelli, F. Perrozzi, F. Bisti, F. Paparella, V. Feyer, A. Ponzoni, M. Gonchigsuren, L. Ottaviano, Few layered MoS_2 lithography with an AFM tip: description of the technique and nanospectroscopy investigations. Nanoscale **7**, 11453 (2015)
22. Y.-Z. Hong, H.-C. Tsai, Y.-H. Wang, J. Aumanen, P. Myllyperkiö, A. Johansson, Y.-C. Kuo, L.-Y. Chang, C.-H. Chen, M. Pettersson, W.-Y. Woon, Reduction-oxidation dynamics of oxidized graphene: functional group composition dependent path to reduction. Carbon **129**, 396–402 (2018)
23. F. Colangelo, V. Piazza, C. Coletti, S. Roddaro, F. Beltram, P. Pingue, Local anodic oxidation on hydrogen-intercalated graphene layers: oxide composition analysis and role of the silicon carbide substrate. Nanotechnology **28**, 105709-1–105709-6 (2017)

24. T.F.D. Fernandes, A.C. Gadelha, A.P.M. Barboza, R.M. Paniago, L.C. Campos, P.S.S. Guimaraes, P.-L. Assis, B.R.A. Neves, Robust nanofabrication of monolayer MoS_2 islands with strong photoluminescence enhancement via local anodic oxidation. 2D Mater. **5**, 025018-1–025018-6 (2018)
25. M. Lazzarino, M. Padovani, G. Mori, L. Sorba, M. Fanetti, M. Sancrotti, Chemical composition of GaAs oxides grown by local anodic oxidation: a spatially resolved Auger study. Chem. Phys. Lett. **402**, 155–159 (2005)
26. J. Shirakashi, M. Ishii, K. Matsumoto, N. Miura, M. Konagai, Surface modification of niobium (Nb) by atomic force microscope (AFM) nano-oxidation process. Jpn. J. Appl. Phys. **35**, L1524–L1527 (1996)
27. G. Mori, M. Lazzarino, D. Ercolani, G. Biasiol, L. Sorba, S. Heun, A. Locatelli, Evidence of material mixing during local anodic oxidation nanolithography. J. Appl. Phys. **98**, 114303-1–114303-6 (2005)
28. G. Mori, M. Lazzarino, D. Ercolani, L. Sorba, S. Heun, A. Locatelli, Desorption dynamics of oxide nanostructures fabricated by local anodic oxidation nanolithography. J. Appl. Phys. **97**, 114324-1–114324-8 (2005)
29. Y.K. Ryu, P.A. Postigo, F. Garcia, R. Garcia, Fabrication of sub-12 nm thick silicon nanowires by processing scanning probe lithography masks. Appl. Phys. Lett. **104**, 223112-1–223112-4 (2014)
30. A. Fuhrer, S. Dorn, S. Lüscher, T. Heinzel, K. Ensslin, W. Wegscheider, M. Bichler, Electronic properties of nanostructures defined in Ga(Al)As heterostructures by local oxidation. Superlattices Microst. **31**, 19–42 (2002)
31. A.I. Dago, Y.K. Ryu, R. Garcia, Sub-20 nm patterning of thin layer WSe_2 by scanning probe lithography. Appl. Phys. Lett. **109**, 163103-1–163103-4 (2016)
32. G. Kwon, H. Chu, J. Yoo, H. Kim, C. Han, C. Chung, J. Lee, H. Lee, Fabrication of uniform and high resolution copper nanowire using intermediate self-assembled monolayers through direct AFM lithography. Nanotechnology **23**, 185307-1–185307-6 (2012)
33. B.P. Brown, L. Picco, M.J. Miles, C.F.J. Faul, Conductive-AFM patterning of organic semiconductors. Small **11**, 5054–5058 (2015)
34. M. Lorenzoni, A. Giugni, E.D. Fabrizio, F. Pérez-Murano, A. Mescola, B. Torre, Nanoscale reduction of graphene oxide thin films and its characterization. Nanotechnology **26**, 285301-1 to 285301-10 (2015)
35. I.-S. Byun, D. Yoon, J.-S. Choi, I. Hwang, D.H. Lee, M.J. Lee, T. Kawai, Y.-W. Son, Q. Jia, H. Cheong, B.H. Park, Nanoscale lithography on monolayer graphene using hydrogenation and oxidation. ACS Nano **5**, 6417–6424 (2011)
36. Y.-Z. Hong, W.-H. Chiang, H.-C. Tsai, M.-C. Chuang, Y.-C. Kuo, L.-Y. Chang, C.-H. Chen, J.-D. White, W.-Y. Woon, Local oxidation and reduction of graphene. Nanotechnology **28**, 395704-1–395704-9 (2017)
37. H. Sugimura, N. Nakagiri, Chemical approach to nanofabrication: modifications of silicon surfaces patterned by scanning probe anodization. Jpn. J. Appl. Phys. **34**, 3406–3411 (1995)
38. J.A. Dagata, T. Inoue, J. Itoh, H. Yokoyama, Understanding scanned probe oxidation of silicon. Appl. Phys. Lett. **73**, 271–273 (1998)
39. H. Kuramochi, F. Pérez-Murano, J.A. Dagata, H. Yokoyama, Faradaic current detection during anodic oxidation of the H-passivated p-Si (001) surface with controlled relative humidity. Nanotechnology **15**, 297–302 (2004)
40. H. Kuramochi, K. Ando, T. Tokizaki, H. Yokoyama, *In situ* detection of faradaic current in probe oxidation using a dynamic force microscope. Appl. Phys. Lett. **84**, 4005–4007 (2004)
41. Y. Takemura, Y. Shimada, G. Watanabe, T. Yamada, J. Shirakashi, Measurement of faradaic current during AFM local oxidation of magnetic metal thin films. J. Phys: Conf. Ser. **61**, 1147–1151 (2007)
42. S. Djurkovic, C.B. Clemons, D. Golovaty, G.W. Young, Effects of the electric field shape on nano-scale oxidation. Surf. Sci. **601**, 5340–5358 (2007)
43. S.F. Lyuksyutov, P.B. Paramonov, I. Dolog, R.M. Ralich, Peculiarities of an anomalous electronic current during atomic force microscopy assisted nanolithography on n-type silicon. Nanotechnology **14**, 716–721 (2003)

44. X.N. Xie, H.J. Chung, C.H. Sow, K. Adamiak, A.T.S. Wee, Electrical discharge in a nanometer-sized air/water gap observed by atomic force microscopy. J. Am. Chem. Soc. **127**, 15562–15567 (2005)
45. M. Tello, R. Garcia, Giant growth rate in nano-oxidation of p-silicon surfaces by using ethyl alcohol liquid bridges. Appl. Phys. Lett. **83**, 2339–2341 (2003)
46. R.V. Martinez, R. Garcia, Nanolithography based on the formation and manipulation of nanometer-size organic liquid menisci. Nano Lett. **5**, 1161–1164 (2005)
47. M. Lorenzoni, A. Giugni, B. Torre, Oxidative and carbonaceous patterning of Si surface in an organic media by scanning probe lithography. Nanoscale Res. Lett. **8**, 75-1–75-9 (2013b)
48. I. Suez, M. Rolandi, S.A. Backer, A. Scholl, A. Doran, D. Okawa, A. Zettl, M.J. Fréchet, High-field scanning probe lithography in hexadecane: transitioning from field induced oxidation to solvent decomposition through surface modification. Adv. Mater. **19**, 3570–3573 (2007)
49. D. Stiévenard, P.A. Fontaine, E. Dubois, Nanooxidation using a scanning probe microscope: an analytical model based on field-induced oxidation. Appl. Phys. Lett. **70**, 327–329 (1997)
50. P. Avouris, T. Hertel, R. Martel, Atomic force microscope tip-induced local oxidation of silicon: kinetics, mechanism and nanofabrication. Appl. Phys. Lett. **71**, 285–287 (1997)
51. M. Calleja, J. Anguita, R. Garcia, K. Birkelund, F. Pérez-Murano, J. Dagata, Nanometre-scale oxidation of silicon surfaces by dynamic force microscopy: reproducibility, kinetics and nanofabrication. Nanotechnology **10**, 34–38 (1999)
52. J.A. Dagata, F. Pérez-Murano, G. Abadal, K. Morimoto, T. Inoue, J. Itoh, H. Yokoyama, Predictive model for scanned probe oxidation kinetics. Appl. Phys. Lett. **76**, 2710–2712 (2000)
53. T. Teuschler, K. Mahr, S. Miyazaki, M. Hundhausen, L. Ley, Nanometer-scale field-induced oxidation of Si(111): H by a conducting-probe scanning force microscope: doping dependence and kinetics. Appl. Phys. Lett. **67**, 3144–3146 (1995)
54. M. Calleja, R. Garcia, Nano-oxidation of silicon surfaces by noncontact atomic force microscopy: size dependence on voltage and pulse duration. Appl. Phys. Lett. **76**, 3427–3429 (2000)
55. E. Dubois, J.L. Bubendorff, Kinetics of scanned probe oxidation: space-charge limited growth. J. Appl. Phys. **87**, 8148–8154 (2000)
56. J.A. Dagata, T. Inoue, J. Itoh, K. Matsumoto, H. Yokoyama, Role of space charge in scanned probe oxidation. J. Appl. Phys. **84**, 6891–6900 (1998)
57. J.A. Dagata, F. Pérez-Murano, C. Martin, H. Kuramochi, J. Yokoyama, Current, charge, and capacitance during scanning probe oxidation of silicon. I. Maximum charge density and lateral diffusion. J. Appl. Phys. **96**, 2386–2392 (2004)
58. M. Chiesa, R. Garcia, Nanoscale space charge generation in local oxidation nanolithography. Appl. Phys. Lett. **96**, 263112-1–263112-3 (2010)
59. T. Baumgärtel, C.V. Borczyskowski, H. Graaf, Detection and stability of nanoscale space charges in local oxidation nanolithography. Nanotechnology **23**, 095707-1–095707-7 (2012)
60. F. Pérez-Murano, K. Birkelund, K. Morimoto, J.A. Dagata, Voltage modulation scanned probe oxidation. Appl. Phys. Lett. **75**, 199–201 (1999)
61. D. Graf, M. Frommenwiler, P. Studerus, T. Ihn, K. Ensslin, D.C. Driscoll, A.C. Gossard, Local oxidation of Ga(Al)As heterostructures with modulated tip-sample voltages. J. Appl. Phys. **99**, 053707-1–053707-7 (2006)
62. C.H. Park, S. Bae, H. Lee, Nano-oxidation of Si using ac modulation in atomic force microscope lithography. Colloids Surf. A **284–285**, 552–555 (2006)
63. X.N. Xie, H.J. Chung, C.H. Sow, A.T.S. Wee, Enhanced probe nano-oxidation by charge pump effect in swept tip voltage cycles. Appl. Phys. Lett. **91**, 243101-1–243101-3 (2007)
64. R. Garcia, M. Calleja, H. Rohrer, Patterning of silicon surfaces with noncontact atomic force microscopy: field-induced formation of nanometer-size water bridges. J. Appl. Phys. **86**, 1898–1903 (1999)
65. S. Gómez-Moñivas, J.J. Sáenz, M. Calleja, R. Garcia, Field-induced formation of nanometer-sized water bridges. Phys. Rev. Lett. **91**, 056101-1–056101-4 (2003)

66. M. Calleja, M. Tello, R. Garcia, Size determination of field-induced water menisci in noncontact atomic force microscopy. J. Appl. Phys. **92**, 5539–5542 (2002)
67. M. Tello, R. Garcia, Nano-oxidation of silicon surfaces: comparison of noncontact and contact atomic-force microscopy methods. Appl. Phys. Lett. **79**, 424–426 (2001)
68. A.E. Gordon, R.T. Fayfield, D.D. Litfin, T.K. Higman, Mechanisms of surface anodization produced by scanning probe microscopes. J. Vac. Sci. Technol. **B**, 2805–2808 (1995)
69. K. Morimoto, F. Pérez-Murano, J.A. Dagata, Density variations in scanned probe oxidation. Appl. Surf. Sci. **158**, 205–216 (2000)
70. V. Cambel, J. Soltys, The influence of sample conductivity on local anodic oxidation by the tip of atomic force microscope. J. Appl. Phys. **102**, 074315-1–074315-7 (2007)
71. P. Avouris, R. Martel, T. Hertel, R. Sandstrom, AFM-tip-induced and current-induced local oxidation of silicon and metals. Appl. Phys. A **66**, S659–S667 (1998)
72. T.-H. Fang, Mechanisms of nanooxidation of Si (100) from atomic force microscopy. Microelectron. J. **35**, 701–707 (2004)
73. K. Yamamoto, K. Sato, J. Sasano, M. Nagai, T. Shibata, Localized etching of silicon in water using a catalytically active platinum-coated atomic force microscopy probe. Precis. Eng. **50**, 344–353 (2017)
74. P.M. Campbell, E.S. Snow, P.J. McMarr, Fabrication of nanometer-scale side-gated silicon field effect transistors with an atomic force microscope. Appl. Phys. Lett. **66**, 1388–1390 (1995)
75. S.C. Minne, H.T. Soh, Ph Flueckiger, C.F. Quate, Fabrication of 0.1 μm metal oxide semiconductor field-effect transistors with the atomic force microscope. Appl. Phys. Lett. **66**, 703–705 (1995)
76. B. Legrand, D. Deresmes, D. Stiévenard, Silicon nanowires with sub 10 nm lateral dimensions: from atomic force microscope lithography based fabrication to electrical measurements. J. Vac. Sci. Technol. **B**, 862–870 (2002)
77. J. Martinez, R.V. Martínez, R. Garcia, Silicon nanowire transistors with a channel width of 4 nm fabricated by atomic force microscope nanolithography. Nano Lett. **8**, 3636–3639 (2008)
78. A. Dehzangi, A.M. Abdullah, F. Larki, S.D. Hutagalung, E.B. Saion, M.N. Hamidon, J. Hassan, Y. Gharayebi, Electrical property comparison and charge transmission in p-type double gate and single gate junctionless accumulation transistor fabricated by AFM nanolithography. Nanoscale Res. Lett. **7**, 381-1–381-9 (2012)
79. Y.K. Ryu, M. Chiesa, R. Garcia, Electrical characteristics of silicon nanowire transistors fabricated by scanning probe and electron beam lithographies. Nanotechnology **24**, 315205-1–315205-7 (2013)
80. The TEM cross-section was produced in the "Laboratorio de Microscopias Avanzadas" at the "Instituto de Nanociencia de Aragon (LMA-INA) - Universidad de Zaragoza" (Spain). The authors acknowledge the LMA-INA for offering access to their instruments and expertise
81. M. Chiesa, P.P. Cardenas, F. Otón, J. Martinez, M. Mas-Torrent, F. Garcia, J.C. Alonso, C. Rovira, R. Garcia, Detection of the early stage of recombinational DNA repair by silicon nanowire transistors. Nano Lett. **12**, 1275–1281 (2012)
82. H.-F. Hsu, C.-A. Chen, S.-W. Liu, C.-K. Tang, Fabrication and gas-sensing properties of Ni-silicide/Si nanowires. Nanoscale Res. Lett. **12**(182), 1–8 (2017)
83. A. Yokoo, T. Tanabe, E. Kuramochi, M. Notomi, Ultrahigh-Q nanocavities written with a nanoprobe. Nano Lett. **11**, 3634–3642 (2011)
84. A. Fuhrer, S. Lüscher, T. Ihn, T. Heinzel, K. Ensslin, W. Wegscheider, M. Bichler, Energy spectra of quantum rings. Nature **413**, 822–825 (2001)
85. M.C. Rogge, E. Räsänen, R.J. Haug, Interaction-induced spin polarization in quantum dots. Phys. Rev. Lett. **105**, 046802-1–046802-4 (2010)
86. Y. Komijani, M. Csontos, T. Ihn, K. Ensslin, Y. Meir, D. Reuter, A.D. Wieck, Origins of conductance anomalies in a p-type GaAs quantum point contact. Phys. Rev. B **87**, 245406-1–245406-9 (2013)
87. N. Ubbelohde, C. Fricke, F. Hohls, R.J. Haug, Spin-dependent shot noise enhancement in a quantum dot. Phys. Rev. B **88**, 041304-1–041304-4 (2013)

88. M. Sigrist, A. Fuhrer, T. Ihn, K. Ensslin, D.C. Driscoll, A.C. Gossard, Multiple layer local oxidation for fabricating semiconductor nanostructures. Appl. Phys. Lett. **85**, 3558–3560 (2004)
89. F.S.-S. Chien, J.-W. Chang, S.-W. Lin, Y.-C. Chou, T.T. Chen, S. Gwo, T.-S. Chao, W.-F. Hsieh, Nanometer-scale conversion of Si_3N_4 to SiO_x. Appl. Phys. Lett. **76**, 360–362 (2000)
90. F.S.-S. Chien, Y.C. Chou, T.T. Chen, W.-F. Hsieh, T.-S. Chao, S. Gwo, Nano-oxidation of silicon nitride films with an atomic force microscope: chemical mapping, kinetics, and applications. J. Appl. Phys. **89**, 2465–2472 (2001)
91. S. Gwo, Scanning probe oxidation of Si3N4 masks for nanoscale lithography, micromachining, and selective epitaxial growth on silicon. J. Phys. Chem. Solids **62**, 1673–1687 (2001)
92. I. Fernandez-Cuesta, X. Borrisé, F. Pérez-Murano, Atomic force microscopy local oxidation of silicon nitride thin films for mask fabrication. Nanotechnology **16**, 2731–2737 (2005)
93. M. Rolandi, C.F. Quate, H. Dai, A new scanning probe lithography scheme with a novel metal resist. Adv. Mater. **14**, 191–194 (2002)
94. L. Pellegrino, Y. Yanagisawa, M. Ishikawa, T. Matsumoto, H. Tanaka, T. Kawai, $(Fe, Mn)_3O_4$ nanochannels fabricated by AFM local-oxidation nanolithography using Mo/Poly (methyl methacrylate) nanomasks. Adv. Mater. **18**, 3099–3104 (2006)
95. N. Suzuki, H. Tanaka, T. Kawai, Epitaxial transition metal oxide nanostructures fabricated by a combination of AFM lithography and molybdenum lift-off. Adv. Mater. **20**, 909–913 (2008)
96. X.N. Xie, H.J. Chung, H. Xu, X. Xu, C.H. Sow, A.T.S. Wee, Probe-induced native oxide decomposition and localized oxidation on 6H-SiC (0001) surface: an atomic force microscopy investigation. J. Am. Chem. Soc. **126**, 7665–7675 (2004)
97. Y.-D. Jo, S.-H. Seo, W. Bahng, S.-C. Kim, N.-K. Kim, S.-S. Kim, S.-M. Koo, Improved local oxidation of silicon carbide using atomic force microscopy. Appl. Phys. Lett. **96**, 082105-1–082105-3 (2010)
98. M. Lorenzoni, B. Torre, Scanning probe oxidation of SiC, fabrication possibilities and kinetics considerations. Appl. Phys. Lett. **103**, 163109-1–163109-6 (2013a)
99. J. Shirakashi, K. Matsumoto, N. Matsumoto, M. Konagai, Single-electron charging effects in Nb/Nb oxide-based single-electron transistors at room temperature. Appl. Phys. Lett. **72**, 1893–1895 (1998)
100. V. Bouchiat, M. Faucher, C. Thirion, W. Wernsdorfer, T. Fournier, B. Pannetier, Josephson junctions and superconducting quantum interference devices made by local oxidation of niobium ultrathin films. Appl. Phys. Lett. **79**, 123–125 (2001)
101. M. Faucher, P.O. Jubert, O. Fruchart, W. Wernsdorfer, V. Bouchiat, Optimizing the flux coupling between a nanoSQUID and a magnetic particle using atomic force microscope nanolithography. Supercond. Sci. Technol. **22**, 064010-1–064010-5 (2009)
102. C. Delacour, B. Pannetier, J.C. Villegier, V. Bouchiat, Quantum and thermal phase slips in superconducting niobium nitride (NbN) ultrathin crystalline nanowire: application to single photon detection. Nano Lett. **12**, 3501–3506 (2012)
103. E.S. Snow, P.M. Campbell, AFM fabrication of sub-10-nanometer metal-oxide devices with in situ control of electrical properties. Science **270**, 1639–1641 (1995)
104. Y. Gotoh, K. Matsumoto, T. Maeda, E.B. Cooper, S.R. Manalis, H. Fang, S.C. Minne, T. Hunt, H. Dai, J. Harris, C.F. Quate, Experimental and theoretical results of room-temperature single-electron transistor formed by the atomic force microscope nano-oxidation process. J. Vac. Sci. Technol., A **18**, 1321–1325 (2000)
105. S. Lemeshko, S. Gavrilov, V. Shevyakov, V. Roschin, R. Solomatenko, Investigation of tip-induced ultrathin Ti film oxidation kinetics. Nanotechnology **12**, 273–276 (2001)
106. Z. Shen, S. Hou, H. Sun, X. Zhao, Z. Xue, Local oxidation of titanium thin films using an atomic force microscope under static and pulsed voltages. J. Phys. D Appl. Phys. **37**, 1357–1361 (2004)
107. K. Mukai, A. Hirota, S. Nakashima, Position control of PbS quantum dot using nanohole on silicon substrate processed by scanning probe lithography. Jpn. J. Appl. Phys. **54**, 04DJ02 (2015)

108. H.-N. Lin, Y.-H. Chang, J.-H. Yen, J.-H. Hsu, I.-C. Leu, M.-H. Hon, Selective growth of vertically aligned carbon nanotubes on nickel oxide nanostructures created by atomic force microscope nano-oxidation. Chem. Phys. Lett. **399**, 422–425 (2004)
109. J. Martín-Sánchez, Y. González, L. González, M. Tello, R. Garcia, D. Granados, J.M. García, F. Briones, Ordered InAs quantum dots on pre-patterned GaAs (001) by local oxidation nanolithography. J. Cryst. Growth **284**, 313–318 (2005)
110. J. Herranz, L. González, L. Wewior, B. Alén, D. Fuster, Y. González, Study of growth parameters for single InAs QD formation on GaAs (001) patterned substrates by local oxidation lithography. Cryst. Growth Des. **15**, 666–672 (2015)
111. J. Martín-Sánchez, G. Muñoz-Matutano, J. Herranz, J. Canet-Ferrer, B. Alén, Y. González, P. Alonso-González, D. Fuster, L. González, J. Martínez-Pastor, F. Briones, Single photon emission from site-controlled InAs quantum dots grown on GaAs (001) patterned substrates. ACS Nano **3**, 1513–1517 (2009)
112. Y. Mo, Z. Lu, A. Chau, F. Huang, Preparation and mechanics of nanotextures on adapting a low adhesive surface using local oxidation nanolithography. Appl. Mater. Interfaces **5**, 4356–4360 (2013)
113. C.-Y. Huang, Y.-C. Yao, Y.-J. Lee, T.-Y. Lin, W.-J. Kao, J.S. Hwang, Y.-J. Yang, J.-L. Shen, Local nanotip arrays sculptured by atomic force microscopy to enhance the light-output efficiency of GaN-based light-emitting diode structures. Nanotechnology **25**, 195401-1–195401-7 (2014)
114. J.S. Hwang, Z.S. Hu, Z.Y. You, T.Y. Lin, C.L. Hsiao, L.W. Tu, Local oxidation of InN and GaN using an atomic force microscope. Nanotechnology **17**, 859–863 (2006)
115. Z.J. Davis, G. Abadal, O. Hansen, X. Borisé, N. Barniol, F. Pérez-Murano, AFM lithography of aluminium for fabrication of nanomechanical systems. Ultramicroscopy **97**, 467–472 (2003)
116. K. Watanabe, Y. Takemura, Y. Shimazu, J. Shirakashi, Magnetic nanostructures fabricated by the atomic force microscopy nano-lithography technique. Nanotechnology **15**, S566–S569 (2004)
117. J. Shirakashi, Y. Takemura, Ferromagnetic ultra-small tunnel junction devices fabricated by scanning probe microscope (SPM) local oxidation. IEEE T. Magn. **40**, 2640–2642 (2004)
118. N. Lee, W. Jo, C. Liu, C. Mény, Size dependent bipolar resistance switching of NiO nanodots for low-power and multi-state operation. Nanotechnology **25**, 415302-1–415302-7 (2014)
119. S. Buyukkose, S. Okur, G. Aygun, Local oxidation nanolithography on Hf thin films using atomic force microscopy (AFM). J. Phys. D: Appl. Phys. **42**, 105302-1–105302-7 (2009)
120. N. Farkas, G. Zhang, E.A. Evans, R.D. Ramsier, J.A. Dagata, Nanoscale oxidation of zirconium surfaces: kinetics and mechanisms. J. Vac. Sci. Technol., A **21**, 1188–1193 (2003)
121. A.B. Oliveira, G. Medeiros-Ribeiro, A. Azevedo, Submicron fabrication by local anodic oxidation of germanium thin films. Nanotechnology **20**, 345301-1–345301-7 (2009)
122. L. Pellegrino, E. Bellingeri, A.S. Siri, D. Marré, Current-controlled lithography on conducting $SrTiO_3$-δ thin films by atomic force microscopy. Appl. Phys. Lett. **87**, 064102-1–064102-3 (2005)
123. G.J. Yong, W.E. Vanderlinde, E.K. Tanyi, D.M. Schaefer, C. Stumpf, R.M. Kolagani, Possible mechanisms in atomic force microscope-induced nano-oxidation lithography in epitaxial $La_{0.67}Ba_{0.33}MnO_{3-\delta}$ thin films. J. Vac. Sci. Technol. B **34**, 021601-1–021601-9 (2016)
124. L. Weng, L. Zhang, Y.P. Chen, L.P. Rokhinson, Atomic force microscope local oxidation nanolithography of graphene. Appl. Phys. Lett. **93**, 093107-1–093107-3 (2008)
125. S. Neubeck, L.A. Ponomarenko, F. Freitag, A.J.M. Giesbers, U. Zeitler, S.V. Morozov, P. Blake, A.K. Geim, K.S. Novoselov, From one electron to one hole: quasiparticle counting in graphene quantum dots determined by electrochemical and plasma etching. Small **6**, 1469–1473 (2010)
126. D.H. Lee, C.K. Kim, J.-H. Lee, H.-J. Chung, B.H. Park, Fabricating in-plane transistor and memory using atomic force microscope lithography towards graphene system on chip. Carbon **96**, 223–228 (2016)
127. R.K. Puddy, C.J. Chua, M.R. Buitelaar, Transport spectroscopy of a graphene quantum dot fabricated by atomic force microscope nanolithography. Appl. Phys. Lett. **103**, 183117-1–183117-4 (2013)

128. A.I. Dago, S. Sangiao, R. Fernández-Pacheco, J.M. De Teresa, R. Garcia, Chemical and structural analysis of sub-20 nm graphene patterns generated by scanning probe lithography. Carbon **129**, 281–285 (2018)
129. F.M. Espinosa, Y.K. Ryu, K. Marinov, D. Dumcenco, A. Kis, R. Garcia, Direct fabrication of thin layer MoS_2 field-effect nanoscale transistors by oxidation scanning probe lithography. Appl. Phys. Lett. **106**, 103503-1–103503-4 (2015)
130. A.I. Dago, Y.K. Ryu, F.J. Palomares, R. Garcia, Direct patterning of p-type-doped few-layer WSe_2 nanoelectronic devices by oxidation scanning probe lithography. ACS Appl. Mater. Interfaces **10**(46), 40054–40061 (2018)
131. H. Bark, S. Kwon, C. Lee, Bias-assisted atomic force microscope nanolithography on NbS_2 thin films grown by chemical vapour deposition. J. Phys. D: Appl. Phys. **49**, 484001-1–484001-6 (2016)
132. E. Pinilla-Cienfuegos, S. Mañas-Valero, E. Navarro-Moratalla, S. Tatay, A. Forment-Aliaga, E. Coronado, Local oxidation nanolithography on metallic transition metal dichalcogenide surfaces. Appl. Sci. **6**, 250-1-250-12 (2016)
133. R. Maoz, S.R. Cohen, J. Sagiv, Nanoelectrochemical patterning of monolayer surfaces: toward spatially defined self-assembly of nanostructures. Adv. Mater. **11**, 55–61 (1999)
134. R. Maoz, E. Frydman, S.R. Cohen, J. Sagiv, 'Constructive nanolithography': inert monolayers as patternable templates for in situ nanofabrication of metal-semiconductor-organic surface structures—a generic approach. Adv. Mater. **12**, 725–731 (2000)
135. S.R. Cohen, R. Maoz, J. Sagiv, *Constructive Nanolithography. Scanning Probe Microscopy: Characterization, Nanofabrication and Device Application of Functional Materials*. NATO Science Series II: Mathematics, Physics and Chemistry, vol. 186 (2005)
136. S. Liu, R. Maoz, J. Sagiv, Planned nanostructures of colloidal gold via self-assembly on hierarchically assembled organic bilayer template patterns with in-situ generated terminal amino functionality. Nano Lett. **4**, 845–851 (2004)
137. J. Berson, A. Zeira, R. Maoz, J. Sagiv, Parallel- and serial-contact electrochemical metallization of monolayer nanopatterns: a versatile synthetic tool en route to bottom-up assembly of electric nanocircuits. Beilstein J. Nanotechnol. **3**, 134–143 (2012)
138. J. Berson, D. Burshtain, A. Zeira, A. Yoffe, R. Maoz, J. Sagiv, Single-layer ionic conduction on carboxyl-terminated silane monolayers patterned by constructive lithography. Nat. Mater. **14**, 613–621 (2015)
139. D. Wouters, R. Willems, S. Hoeppener, C.F.J. Flipse, U.S. Schubert, Oxidation conditions for octadecyl trichlorosilane monolayers on silicon: a detailed atomic force microscopy study of the effects of pulse height and duration on the oxidation of the monolayer and the underlying Si substrate. Adv. Funct. Mater. **15**, 938–944 (2005)
140. T. Druzhinina, S. Hoeppener, N. Herzer, U.S. Schubert, Fabrication of ring structures by anodization lithography on self-assembled OTS monolayers. J. Mater. Chem. **21**, 8532–8536 (2011)
141. T.S. Druzhinina, S. Hoeppener, U.S. Schubert, New design concepts for the fabrication of nanometric gap structures: electrochemical oxidation of OTS mono- and bilayer structures. Small **8**, 852–857 (2012)
142. T.S. Druzhinina, C. Höppener, S. Hoeppener, U.S. Schubert, Hierarchical, guided self-assembly of preselected carbon nanotubes for the controlled fabrication of CNT structures by electrooxidative nanolithography. Langmuir **29**, 7515–7520 (2013)
143. H. Liu, A.M. Schwenke, F. Kretschmer, S. Hoeppener, U.S. Schubert, Gold nanoparticle cluster arrays for high-performance SERS substrates fabricated by electro-oxidative lithography. ChemNanoMat **2**, 781–785 (2016)
144. R.V. Martínez, F. García, R. Garcia, E. Coronado, A. Forment-Aliaga, F.M. Romero, S. Tatay, Nanoscale deposition of single-molecule magnets onto SiO_2 patterns. Adv. Mater. **19**, 291–295 (2007)
145. R.V. Martínez, J. Martínez, M. Chiesa, R. Garcia, E. Coronado, E. Pinilla-Cienfuegos, S. Tatay, Large-scale nanopatterning of single proteins used as carriers of magnetic nanoparticles. Adv. Mater. **22**, 588–591 (2010)

146. E. Coronado, A. Forment-Aliaga, E. Pinilla-Cienfuegos, S. Tatay, L. Catala, J.A. Plaza, Nanopatterning of anionic nanoparticles based on magnetic prussian-blue analogues. Adv. Funct. Mater. **22**, 3625–3633 (2012)
147. T. Yoshinobu, J. Suzuki, H. Kurooka, W.C. Moon, H. Iwasaki, AFM fabrication of oxide patterns and immobilization of biomolecules on Si surface. Electrochim. Acta **48**, 3131–3135 (2003)
148. G. Quin, J. Gu, K. Liu, Z. Xiao, C.M. Yam, C. Cai, Conductive AFM patterning on oligo(ethylene glycol)-terminated alkyl monolayers on silicon substrates: proposed mechanism and fabrication of avidin patterns. Langmuir **27**, 6987–6994 (2011)
149. E.M. Benetti, H.J. Chung, G.J. Vancso, PH responsive polymeric brush nanostructures: preparation and characterization by scanning probe oxidation and surface initiated polymerization. Macromol. Rapid Commun. **30**, 411–417 (2009)
150. R. Garcia, M. Tello, Size and shape controlled growth of molecular nanostructures on silicon oxide templates. Nano Lett. **4**, 1115–1119 (2004)
151. T. Baumgärtel, S. Rehm, F. Würthner, C.V. Borczyskowski, H. Graaf, Functional bisimide dyes bound via electrostatic interactions to oxide nanostructures generated by AFM lithography. Appl. Surf. Sci. **318**, 51–58 (2014)
152. M. Fernández-Regúlez, L. Evangelio, M. Lorenzoni, J. Fraxedas, F. Pérez-Murano, Sub-10 nm resistless nanolithography for directed self-assembly of block copolymers. ACS Appl. Mater. Interfaces **6**, 21596–21602 (2014)
153. H.J. Mamin, D. Rugar, Thermomechanical writing with an atomic force microscope tip. Appl. Phys. Lett. **61**, 1003-1–1003-3 (1992)
154. P. Vettiger, G. Cross, M. Despont, U. Drechsler, U. Dürig, B. Gotsmann, W. Häberle, M.A. Lantz, H.E. Rothuizen, R. Stutz, G.K. Binnig, The "Millipede"—nanotechnology entering data storage. IEEE Trans. Nanotechnol. **1**, 39–55 (2002)
155. A. Pantazi, A. Sebastian, T.A. Antonakopoulos, P. Baechtold, A.R. Bonaccio, J. Bonan, G. Cherubini, M. Despont, R.A. DiPietro, U. Drechsler, U. Duerig, B. Gotsmann, W. Haeberle, C. Hagleitner, J.L. Hedrick, D. Jubin, A. Knoll, M.A. Lantz, J. Pentarakis, H. Pozidis, R.C. Pratt, H. Rothuizen, R. Stutz, M. Varsamou, D. Wiesmann, E. Eleftheriou, Probe-based ultrahigh-density storage technology. IBM J. Res. Develop. **52**, 493–511 (2008)
156. D. Pires, J.L. Hedrick, A. de Silva, J. Frommer, B. Gotsmann, H. Wolf, M. Despont, U. Duerig, A.W. Knoll, Nanoscale three-dimensional patterning of molecular resists by scanning probes. Science **328**, 732–735 (2010)
157. M. Spieser, C. Rawlings, E. Lörtscher, U. Duerig, A.W. Knoll, Comprehensive modelling of Joule heated cantilever probes. J. Appl. Phys. **121**, 174503 (2017)
158. A.W. Knoll, D. Pires, O. Coulembier, P. Dubois, J.L. Hedrick, J. Frommer, U. Duerig, Probe-based 3-D nanolithography using self-amplified depolymerisation polymers. Adv. Mater. **22**, 3361–3365 (2010)
159. P.C. Paul, A.W. Knoll, F. Holzner, M. Despont, U. Duerig, Rapid turnaround scanning probe nanolithography. Nanotechnology **22**, 275306-1–275306-9 (2011)
160. C. Rawlings, U. Duerig, J. Hedrick, C. Coady, A.W. Knoll, Nanometer accurate markerless pattern overlay using thermal scanning probe lithography. IEEE Trans. Nanotechnol. **13**, 1204–1212 (2014)
161. C. Rawlings, H. Wolf, J.L. Hedrick, D.J. Coady, U. Duerig, A.W. Knoll, Accurate location and manipulation of nanoscaled objects buried under spin-coated films. ACS Nano **9**, 6188–6195 (2015)
162. Y.K. Ryu Cho, C.D. Rawlings, H. Wolf, M. Spieser, S. Bisig, S. Reidt, M. Sousa, S.R. Khanal, T.D.B. Jacobs, A.W. Knoll, Sub-10 nanometer feature size in silicon using thermal scanning probe lithography. ACS Nano **11**, 11890–11897 (2017)
163. J.-F. de Marneffe, B.T. Chan, M. Spieser, G. Vereecke, S. Naumov, D. Vanhaeren, H. Wolf, A.W. Knoll, Conversion of a patterned organic resist into a high performance inorganic hard mask for high resolution pattern transfer. ACS Nano **12**, 11152–11160 (2018)
164. H. Wolf, C. Rawlings, P. Mensch, J.L. Hedrick, D.J. Coady, U. Duerig, A.W. Knoll, Sub-20 nm silicon patterning and metal lift-off using thermal scanning probe lithography. J. Vac. Sci. Technol., B **33**, 02B102 (2015)

165. L.L. Cheong, P. Paul, F. Holzner, M. Despont, D.J. Coady, J.L. Hedrick, R. Allen, A.W. Knoll, U. Duerig, Thermal probe maskless lithography for 27.5 nm half-pitch Si technology. Nano Lett. **13**, 4485–4491 (2013)
166. C. Rawlings, Y.K. Ryu, M. Rüegg, N. Lassaline, C. Schwemmer, U. Duerig, A.W. Knoll, Z. Durrani, C. Wang, D. Liu, M.E. Jones, Fast turnaround fabrication of silicon point-contact quantum-dot transistors using combined thermal scanning probe lithography and laser writing. Nanotechnology **29**, 505302 (2018)
167. C.D. Rawlings, M. Zientek, M. Spieser, D. Urbonas, T. Stöferle, R.F. Mahrt, Y. Lisunova, J. Brugger, U. Duerig, A.W. Knoll, Control of the interaction strength of photonic molecules by nanometer precise 3D fabrication. Sci. Rep. **7**, 16502-1–16502-9 (2017)
168. M.J. Skaug, C. Schwemmer, S. Fringes, C.D. Rawlings, A.W. Knoll, Nanofluidic rocking Brownian motors. Science **359**, 1505–1508 (2018)
169. C. Schwemmer, S. Fringes, U. Duerig, Y.K. Ryu, A.W. Knoll, Experimental observation of current reversal in a rocking Brownian motor. Phys. Rev. Lett. **121**, 104102 (2018)
170. J.E. Shaw, P.N. Stavrinou, T.D. Anthopoulos, On-demand patterning of nanostructured pentacene transistors by scanning thermal lithography. Adv. Mater. **25**, 552–558 (2013)
171. Z. Wei, D. Wang, S. Kim, S.-Y. Kim, Y. Hu, M.K. Yakes, A.R. Laracuente, Z. Dai, S.R. Marder, C. Berger, W.P. King, W.A. de Heer, P.E. Sheehan, E. Riedo, Nanoscale tunable reduction of graphene oxide for graphene electronics. Science **328**, 1373–1376 (2010)
172. F. Holzner, C. Kuemin, P. Paul, J.L. Hedrick, H. Wolf, N.D. Spencer, U. Duerig, A.W. Knoll, Directed placement of gold nanorods using a removable template for guided assembly. Nano Lett. **11**, 3957–3962 (2011)
173. S. Gottlieb, M. Lorenzoni, L. Evangelio, M. Fernández-Regúlez, Y.K. Ryu, C. Rawlings, M. Spieser, A.W. Knoll, F. Perez-Murano, Thermal scanning probe lithography for the directed self-assembly of block copolymers. Nanotechnology **28**, 175301-1–175301-9 (2017)
174. S.T. Zimmermann, D.W.R. Balkenende, A. Lavrenova, C. Weder, J. Brugger, Nanopatterning of a stimuli-responsive fluorescent supramolecular polymer by thermal scanning probe lithography. ACS Appl. Mater. Interfaces. **9**(47), 41454–41461 (2017)
175. E.B. Cooper, S.R. Manalis, H. Fang, H. Dai, K. Matsumoto, S.C. Minne, T. Hunt, C.F. Quate, Terabit-per-square-inch data storage with the atomic force microscope. Appl. Phys. Lett. **75**, 3566–3568 (1999)
176. T.-H. Fang, K.-T. Wu, Local oxidation characteristics on titanium nitride film by electrochemical nanolithography with carbon nanotube tip. Electrochem. Commun. **8**, 173–178 (2006)
177. G. Rius, M. Lorenzoni, S. Matsui, M. Tanemura, F. Perez-Murano, Boosting the local anodic oxidation of silicon through carbon nanofiber atomic force microscopy probes. Beilstein J. Nanotechnol. **6**, 215–222 (2015)
178. A.J. Ulrich, A.D. Radadia, Conductive polycrystalline diamond probes for local anodic oxidation lithography. Nanotechnology **26**, 465201-1–465201-13 (2015)
179. J.A. Vicary, M.J. Miles, Real-time nanofabrication with high-speed atomic force microscopy. Nanotechnology **20**, 095302-1–095302-5 (2009)
180. S.C. Minne, J.D. Adams, G. Yaralioglu, S.R. Manalis, A. Atalar, C.F. Quate, Centimeter scale atomic force microscope imaging and lithography. Appl. Phys. Lett. **73**, 1742–1744 (1998)
181. F. Huo, Z. Zheng, G. Zheng, L.R. Giam, H. Zhang, C.A. Mirkin, Polymer pen lithography. Science **321**, 1658–1660 (2008)
182. J. Martinez, N.S. Losilla, F. Biscarini, G. Schmidt, T. Borzenko, L.W. Molenkamp, R. Garcia, Development of a parallel local oxidation nanolithography instrument. Rev. Sci. Instrum. **77**, 086106-1–086106-3 (2006)
183. A. Zeira, D. Chowdhury, R. Maoz, J. Sagiv, Contact electrochemical replication of hydrophilic-hydrophobic monolayer patterns. ACS Nano **2**, 2554–2568 (2008)
184. H. Sato, S.E. Vasko, M. Rolandi, Wafer scale direct-write of Ge and Si nanostructures with conducting stamps and a modified mask aligner. Nano Research **6**, 263–268 (2013)
185. M. Kaestner, M. Hofer, I.W. Rangelow, Nanolithography by scanning probes on calixarene molecular glass resist using mix-and-match lithography. J. Micro/Nanolith. MEMS MOEMS **12**, 031111-1–031111-13 (2013)

Chapter 6
Characterizing Ferroelectricity with an Atomic Force Microscopy: An All-Around Technique

Simon Martin, Brice Gautier, Nicolas Baboux, Alexei Gruverman, Adrian Carretero-Genevrier, Martí Gich and Andres Gomez

Abstract Atomic Force Microscopy (AFM) arises as an all-in-one characterization technique, capable of measuring several physical quantities by slight equipment's modifications. In particular, for piezo and ferroelectricity properties, the AFM overcame the limitations of macroscopic techniques. This chapter covers all the aspects of piezo and ferroelectricity measurements performed with an AFM. The chapter is divided in three main parts, one for each available technique: Piezoresponse Force Microscopy (PFM), Nano-PUND method and Direct Piezoelectric Force Microscopy (DPFM). While PFM method is based in the converse piezoelectric effect, nanoPUND measures polarization charges and DPFM measures the direct piezoelectric effect. The working principle and characteristics for each AFM mode is fully exploited and explained from entry level to more advanced users. The chapter also focuses in useful guidelines and practical hands-on explanation for maximizing the image quality and data acquisition. Finally, a set of different application based in the use of piezo and ferroelectric materials is depicted, in which the AFM characterization took an important role as the primary characterization technique.

S. Martin · B. Gautier · N. Baboux
Institut des Nanotechnologies de Lyon, INSA de Lyon, Université de Lyon, UMR CNRS 5270, 7, Avenue, Capelle, 69621 Villeurbanne, France

A. Gruverman
Department of Physics and Astronomy, University of Nebraska-Lincoln, 081 Jorgensen Hall, 855 N 16th Street, Lincoln, NE 68588-0299, USA

A. Carretero-Genevrier
Institut d'Electronique et des Systemes (IES), CNRS, Universite Montpellier, 2 860 Rue de Saint Priest, 34095 Montpellier, France

M. Gich · A. Gomez (✉)
Institut de Ciència de Materials de Barcelona (ICMAB-CSIC), Campus UAB, 08193 Bellaterra, Spain
e-mail: agomez@icmab.es

U. Celano (ed.), *Electrical Atomic Force Microscopy for Nanoelectronics*, NanoScience and Technology, https://doi.org/10.1007/978-3-030-15612-1_6

6.1 Introduction

While piezoelectric and ferroelectric materials started as laboratory rareness materials, during the 20th century their applications expanded from laboratory to industry and common electronic components. More recent industrial applications include their use as electromechanical systems, energy harvester for sensors in the Internet of Things (IoT), memory devices and as gate dielectric materials, among others. This broad range of applications relies on the same key: the understanding and control at the nanoscale of the piezoelectric effect.

In this sense, Atomic Force Microscopes have provided a platform from which both piezoelectricity and ferroelectricity can be studied. In this chapter we will review and analyze the state-of-the-art techniques from the family AFM advanced modes. The chapter covers from the most classical approach, exploiting the converse piezoelectric effect, to novel techniques relying into the use of the direct piezoelectric effect. More importantly, we describe not only methods to map ferroelectricity and piezoelectricity, but also an AFM-based technique that can be used to acquire the polarization charge density of the material, which is of special interest to capacitors and electronic components. A set of novel and promising applications closes the chapter, to give ideas into what are the uses and implications of piezoelectricity in the near future.

6.2 Piezoresponse Force Microscopy as a Domain Imaging Technique

Piezoresponse Force Microscopy (PFM) is an advanced Atomic Force Microscopy (AFM) mode capable of imaging ferroelectric domains through the converse piezoelectric effect, measuring the material displacement as a function of the applied bias. In order to exploit the technique, an introduction into the imaging mechanism is provided, followed by a hands-on step-by-step guide.

6.2.1 Principles of Imaging Ferroelectric Domains

Piezoelectricity includes two different and equivalent definitions [1]. For one side, piezoelectricity is the amount of charge density that is generated when a mechanical strain is applied to the material. This approach was given in the 19th century by Curie brothers [2] while in a later time, "converse" effect was discovered. The Converse Piezoelectric Effect (CPE) is defined as the mechanical strain as a function of the applied electric field [3]. The direct and converse piezoelectric effects are expressed by the following equations:

$$P_i = d_{ijk} \cdot X_{jk}, for\ direct\ effect \tag{6.1}$$

$$S_i = d_{ijk} \cdot E_{jk}, for\ converse\ effect \tag{6.2}$$

where P is the polarization, X the stress, S the strain, E the electric fields and d the "piezoelectric coefficient". In fact, since these equations are related both to vectors (polarization and electric field) and second rank tensors (stress and strain), d is a rank three tensor. Since stress and strain tensors are symmetric, d consists of 18 independent coefficients represented by a 3×6 matrix d_{ij} with $i = 1$–$3, j = 1$–6. If we simplify the equation for the out-of-plane perpendicular case, the equations take the form:

$$\sigma = d_{33} \cdot F, for\ direct\ effect \tag{6.3}$$

$$A = d_{33}V, for\ converse\ effect \tag{6.4}$$

In which σ is the surface charge density, F is the force per unit area applied by the tip to the sample in the direction perpendicular to its surface, V is the applied bias and A is the displacement. In both cases, the constant d_{33}, is called the piezoelectric constant, which has either units of pC/N or pm/V (1 pm/V = 1 pC/m) [3, 4]. The converse and direct piezoelectric effects offer two alternatives to study piezoelectric and ferroelectric materials. For each possibility, there exists one AFM-based technique, namely Piezoresponse Force Microscopy (PFM) which is based on the converse effect and Direct Piezoelectric Force Microscopy (DPFM), based on the direct effect.

6.2.2 *The Converse Piezoelectric Effect as Imaging Technique*

For a given Material-Under-Test (MUT), an expansion and contraction as a function of the applied bias is induced through the converse piezoelectric effect. For instance, for a given $d_{33} = 100$ pm/V and an applied bias of +1 V, the material expands 100 pm, if the polarization is parallel to the applied electric field, see Fig. 6.1, in which Pup denotes an upwards polarization state.

However, if the opposite bias is applied, −1 V, the material contracts by 100 pm. As a consequence, piezoelectricity can be studied by an electrical stimulus applied to the material, while the displacement is simultaneously measured with a suitable method. Typical piezoelectric d_{33} constant can range from 1 to 2000 pm/V, which represents a small displacement to be measured; hence a high precision instrument is required. An AFM is capable of measuring such tiny displacements with a suitable tip. The tip coated with a metallic layer or made of full-metal, touches the surface of the MUT. At the same time, a bias is applied to the tip, and hence, the material shrinks

Fig. 6.1 The CPE explained: without bias applied the material rests in its initial position, **a** For a given positive bias applied, the material expands, **b** while for a negative bias, **c** MUT contracts

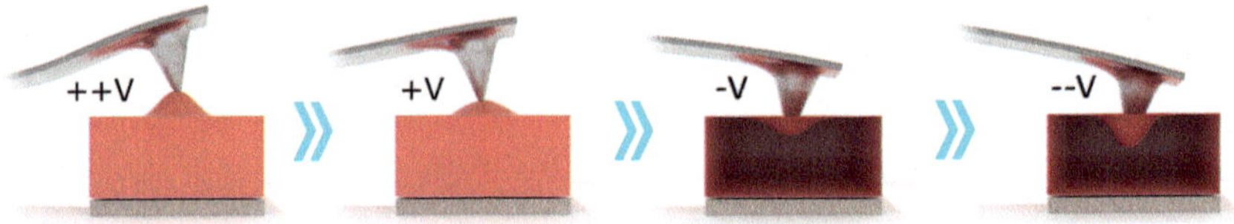

Fig. 6.2 3D scheme of the movement of the tip while performing PFM. If the applied bias to the tip is positive, the material will expand for an upward domain configuration. However, once the voltage is the opposite, the material will contract. The tip vibration is represented as an amplitude and phase signal, which gives information of the material properties

or expands as a function of the applied bias, see Fig. 6.2. In order to improve the signal-to-noise ratio, a Lock-In Amplifier (LIA) is employed in which its reference is directly connected to the tip. A sinusoidal reference bias, "V_{AC}", generates the following displacement, "A_{TIP}":

$$A_{TIP} = V\, d_{33} \sin(\omega t + \varphi) \tag{6.5}$$

In which $V_{AC} = V \sin(\omega t)$ is the voltage applied to the tip, with amplitude V, at frequency ω.

By measuring the deflection signal of the cantilever, it is possible to record the cantilever vibration by acquiring the amplitude and the phase signals. The amplitude signal, A, is related to the amount of vibration induced at the probe. The amplitude is an absolute value; it does not map expansion or contraction of MUT. The phase signal, φ, is related to the phase difference between two signals, the applied AC bias and the vibration of the probe. If the phase is 0°, the cantilever vibration signal is in phase with the applied stimulus, meaning the material will expand for a positive bias applied. However, a phase value of 180° represents that when the AC bias is positive the cantilever will move downwards, hence the material shrinks. Thus, the phase signal gives direct information about the ferroelectricity of the sample while the amplitude alone, is used to map only the piezoelectric property. Both the phase and the amplitude are used to study MUTs and typical studies performed in PFM may include the acquisition of topography, amplitude and phase image. In contrast to the vertical PFM (VPFM), there is also the possibility of mapping the lateral movement of the cantilever, which is called lateral PFM (LPFM), capable of measuring the

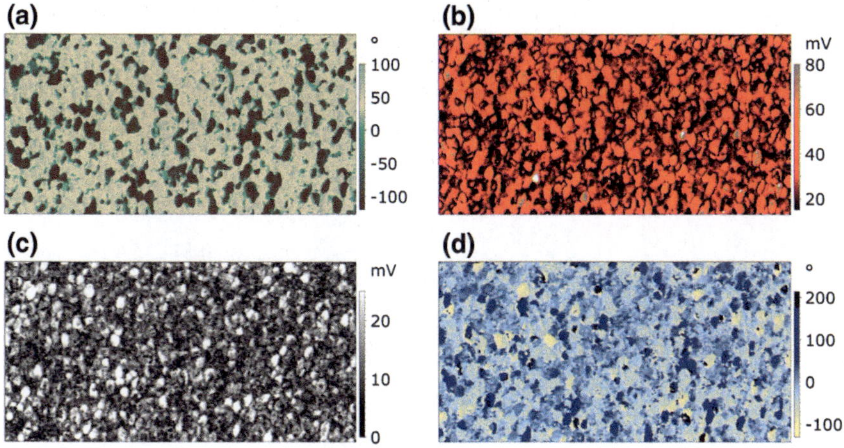

Fig. 6.3 **a** VPFM phase image, **b** VPFM amplitude image, **c** LPFM amplitude and **d** LPFM phase. The image depicts the natural domains of a 40 nm thick Bismuth Ferrite sample grown over a STO/LSMO bottom electrode. https://doi.org/10.1039/C8NR05737K

ferroelectric domains with the polarization lying in the same plane as the sample surface, see Fig. 6.3.

Another important application of PFM is the possibility of creating and manipulating ferroelectric domains, both with electrical stimulus and mechanical strain [5–8]. A common experiment comprises the recording of a specific domain pattern with a suitable dc bias applied to the AFM tip, see Fig. 6.4. In this experiment, an outer square is recorded with a given bias, in this case +7 V dc. After the recording event, a concentric, smaller square is recorded with an opposite bias, −7 V dc. A third image in PFM is carried out centered depicting the recorded domain structure. This same experiment can be synthesized into a graph; an example is shown in Fig. 6.4c. In this graph, an ac bias is applied to the tip, while, at the same time, an additional dc bias is applied. Through this approach, a spectroscopy graph is collected-it is shown in Fig. 6.4c. From this experiment a butterfly ferroelectric loop can be obtained, which may match with a ferroelectric like response [5, 9, 10].

6.2.3 Piezoresponse as a Quantitative Technique

Following (6.5), PFM can be interpreted as a quantitative technique [11, 12]. However, scientific community has proved that this simple equation may not be sufficient to describe the imaging mechanism [10, 12, 13]. Available bibliography has shown that in a few cases, the method has provided quantitative measurements following the aforementioned equation [14–17]. Despite such successful experiments, other authors have proved that a piezoelectric and ferroelectric like response can be mea-

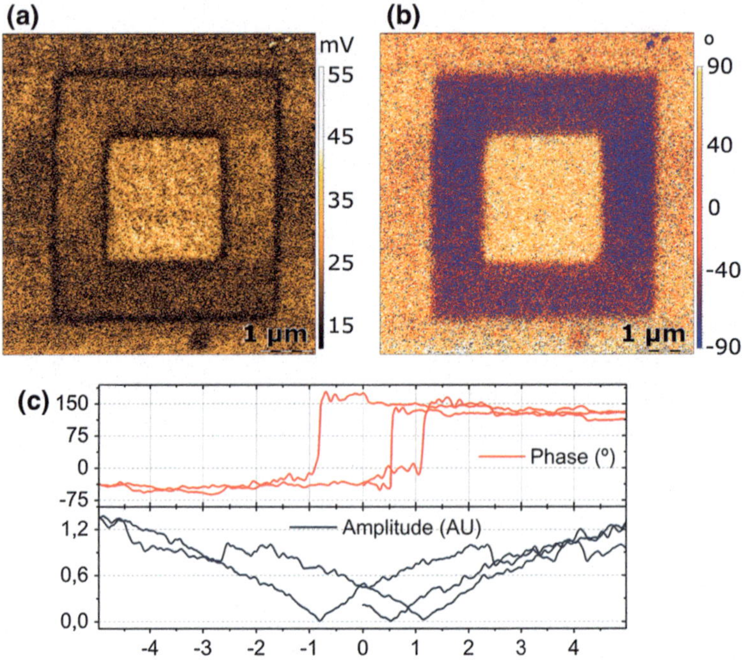

Fig. 6.4 **a** PFM amplitude image; **b** PFM phase image and **c** spectroscopy experiments. Data acquired for a thin film of 40 nm $BaTiO_{3-\delta}$ on silicon. A full 0–180º phase image is acquired while the domain wall is clearly depicted in the amplitude image

sured in non-piezoelectric materials [9, 10, 18]. Thus, at this point, "quantitativeness" cannot be granted for all kind of materials. This controversy in the scientific community has brought the attention of research to develop new solutions to overcome this situation which include: special sensor systems [19], longer tips [20], specifically designed probes with shielding [21], use of stiff cantilevers [22], data treating methods [23], among others [24]. Even though these innovations are extremely promising, there are many circumstances that may affect the measurements: ionic motion, different conduction states or trapped charges. The vast majority of this phenomena is related to parasitic electrostatic effects that could induce an electromechanical behavior to the PFM tip which can be interpreted as piezoelectricity. There is a fascinating horizon in investigating the influence of such effects in the electromechanical behavior and the possible solutions [13].

The reliability of PFM has been brought recently into the attention of the most prominent scientists in the field. Most of the materials, independently of piezoelectricity, may generate a tip vibration in PFM mode [13]. Hence, PFM cannot be used a technique to report piezoelectricity of an unknown material. A similar approach happens with ferroelectricity, as a ferroelectric like response can be read in PFM for a non-ferroelectric material [10]. As a conclusion, PFM cannot be used either to discern if a material is ferroelectric or piezoelectric, uniquely. Even with such limitation, PFM is a uniquely and powerful domain imaging technique to report domain structures at the nanoscale.

6.2.4 *Practical Aspects for Doing PFM*

This section is devoted to users that want a kick-off for their PFM measurements; in this section we summarize all the corks and tricks of an experienced PFM user.

1.4.a. **A testing platform**

A known ferroelectric sample is mandatory for the users carrying out PFM measurements. The most common testing sample, is called "Periodically Poled Lithium Niobate" (PPLN) that can be purchased in major optical retailers or AFM companies. This sample is used firstly to introduce you in PFM and acquiring your very first images but also, it is employed as a troubleshooting testing platform. An alternative sample may comprise a piezoelectric buzzer, commonly presented in electronic stores, polished and heated to create a natural domain structure. Both lithium niobate and the piezoelectric buzzer, which normally is composed of lead zirconate titanate material, are known to be piezoelectric and ferroelectric materials.

1.4.b. **The probe selection**

The probe is one of the most important parts for any AFM method. From all the available technologies, a conductive tip has to be employed. The most common one is the Pt/Ir coated AFM probe, however the coating tends to degrade easily. Another common approach is to use full metal tips, on which wearing does not affect its conductivity and a longer tip shank length reduces electrostatic interaction. The probe is commercialized mounted in different cantilevers, with different spring constants. The recommendation is to employ relatively stiff cantilevers, for instance, in Pt/Ir technology a common cantilever has a $k = 3$ N/m while, for a solid platinum tip, a common k constant of 18 N/m is found. A stiff cantilever is recommended to provide a good tip-sample mechanical contact and reduce electrostatic effects. Since typical ferroelectric materials are hard ceramics, there is no sample plastic deformation upon scanning. Other metallic probes may include doped single crystal diamond tips, titanium coated tips and doped diamond coated tips.

1.4.c. **The equipment**

The vast majority of AFM microscopes are equiped with components to perform PFM by feeding the probe with a suitable AC signal and routing the deflection signal into a Lock-In Amplifier (LIA) input. However, a common modification for PFM users is to bypass the commercial electronics and feed the AC voltage signal directly to the tip with an external coaxial cable. This improves the capacitive coupling between the signal cables and the AC generation that can influence the measured signals. An external lock-in amplifier can be used to overpass the equipment built-in LIA, if necessary. Other common modifications include improving the trans-impedance

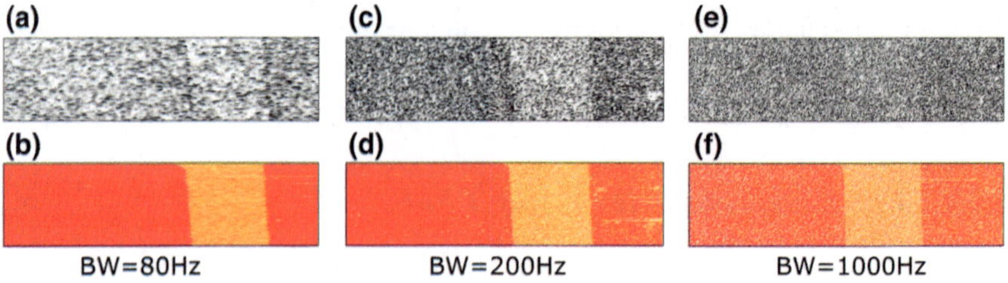

Fig. 6.5 Testing images performed in PPLN with different bandwidths, **a**, **b** PFM Amplitude and phase with BW = 80 Hz; **c**, **d** PFM Amplitude and phase with BW = 200 Hz and **e**, **f** PFM Amplitude and phase with BW = 1000 Hz

amplifier from the photodetector, upgrading the laser spot with a superluminescent diode or changing the spot size projected into the cantilever.

1.4.d. **The equipment's parameter**

The laser spot has to be localized into the exact top of the cantilever, as near as possible to the tip location. An incorrect laser position, for instance, in the middle of the cantilever, will result in possible cantilever buckling effect. The reflected spot has to be as circular as possible, and should cover all the photodetector. The photodetector has to be adjusted, so that the image is collected as near as possible to the 0.00 V deflection value, to maximize the signal acquired. The most important imaging parameters to be set are the force exerted by the probe and the LIA parameters. Typically, forces range between 0.1 and 2 μN. Using a high force normally results in a highly stable image due to the good mechanical contact between the tip and the sample. However, for a coated tip, it may result in tip wearing and loose of image, hence, it is preferably to switch to a full metallic tip. The LIA's parameters have to be set to maximize the input signal, through raising the internal gain, and reducing the LIA's bandwidth (or, equally, increasing its time constant). However, extremely high gain may substantially increase the noise of the images. The best gain value should be found by the user in each equipment by scanning the PPLN test sample with different LIA's parameters. The LIA's bandwidth (or time constant) is the next parameter to look at. A small bandwidth gives the best filtering option, and hence is the recommended starting point. However, a too slow LIA will not be sufficient to feed the necessary throughput data for the image. For instance, if you perform an image of 100 × 100 pixels, scanned at 1 line per second, you will need 200 points per second to completely feed the image with information. This means that the pixel time constant is 1/200 s. If the time constant of the amplifier is higher than this value, the LIA will be too slow to perform an image, and hence, horizontal lines may appear in the image, see Fig. 6.5a. If the BW is increased, data throughput is increased, but also, a noisier image is obtained. In this case, the best results can be acquired with a time constant that matches the pixel time constant.

1.4.e. **The environment**

To perform experiments in PFM, it is important to control the environmental conditions inside the AFM box. Typically, experiments are carried out in controlled atmospheric conditions with low humidity environment. If the AFM does not have a climate chamber, it is possible to fill the complete AFM box with compressed air, which reduces the ambient humidity to values of less than 6%. Another common approach is to fill the chamber with nitrogen gas to reduce the humidity further to 3% and getting rid of oxygen. However, security measurements have to be taken.

1.4.f. **Procedure**

A step-by-step guide into acquiring PFM images is provided in the following page:

a. Place the AFM tip into the tip holder. It has to be placed as centered as possible, which is very important for this mode. Laser spot is adjusted so that it is as close as possible to the top of the tip location-which is not the same as the very end of the cantilever. Laser photodetector is adjusted so that its value is as close as possible to 0.00 V, to maximize the sensitivity. A coaxial cable is connected directly to the tip-or a suitable switch to route the AC signal to the tip is configured. Sample is mounted into a conductive plate, which is connected to the DC generator of an AFM (typically, "sample bias" channel) with suitable silver paint. With the sample ready to scan, the box is closed and compressed air/nitrogen is filled inside. The imaging software has to be configured to acquire LIA's "Amplitude" and "Phase" channels.
b. Once humidity is below 10%, the tip can be approached to the sample. Once engaged, the LIA's generator can be feed into the tip, a typical starting amplitude is 2 VAC and frequency set to 20 kHz. The LIA's is configured with an aggressive time constant, for instance 10 ms (100 Hz BW) for images at 1 lines per second. The gain parameter is maximized so that the signal is not clipped. With these parameters, an image can be acquired. If natural domains can be seen, you can use this frame to adjust the time constant and gain parameters of the LIA. A common trick at this point is to reduce the scan speed, which enables the use of even higher LIA's time constant, hence a better filtering is achieved.
c. If at this point no stable signal can be obtained in the frames, you can increase the AC voltage applied to the tip up to the maximum value. In case the image is not stable, check the topography channel to see if topography image is obtained. In case the piezoelectric signal is not measured, you can change the applied AC frequency to the contact resonance frequency-which increases the signal-to-noise ratio. To do so, a frequency sweep over a range of 20–200 kHz is typically employed to look for the resonance peak. The resonance peak should be found between the first and second resonance of the cantilever, measured in air. Once the peak is found, the frequency can be changed to match such resonance frequency. After all these steps are carried out and there is still no signal, we can conclude that either natural domains are not present or they are too small to be imaged. If you can see the domains, you can set the Dual-Resonance Amplitude Tracking feature in your AFM, whenever such option is available.

d. The next step is to prove if you can create a domain configuration into the MUT. To do so, you can start by performing a spectroscopy experiment-similar to one that is found in Fig. 6.4c. With the same starting parameters described in **b** a curve for "Amplitude and "Phase" channels versus applied DC bias can be run. The typical sweep goes from 0 to +6VDC, then from +6VDC to −6VDC to back + 6VDC. The sweep rate can be set to ranges from 5–20 V/s with data throughputs of 80–500 points per second. If in the spectroscopy window you cannot obtain a similar graph like shown in Fig. 6.4c, you can switch to the resonance frequency as for the case **c**, and repeat the curves. If you are able to reproduce a butterfly-like ferroelectric loops you should be able to record a suitable domain configuration.
e. Recording the domain configuration is a multistep process. A typical domain recording is found in Fig. 6.3 consisting of 2 consequent squares. For one side, a bigger image, typical of 6 × 6 μm is performed with 0 VAC and applying a DC bias to the tip. From spectroscopy data obtained at **d** you can, approximately, calculate the necessary voltage to record the domains. Too much voltage, and the layer may undergo breakdown conditions, leading to the appearance of undesired topographic features. The voltage has to be high enough to record the domain, but not too high to break the sample. After the big square is recorded, a smaller one is created concentric to the bigger one. Typically sizes are 4 × 4 μm or 2 × 2 μm. These image sizes are enough for most common AFM to avoid drifting occurring in smaller images. After the two consequent squares are recorded, it is time to read the domain structure. To do so, a bigger than initial image has to be employed, for instance, 10 × 10 μm, to read the domain configuration following **b**.

6.3 The Nano-PUND Technique

Nano-PUND is a technique introduced in 2017 [25] which has been developed to be used as a complement to PFM analysis or as an alternative to it. PFM has become the most widely used technique to assess ferroelectricity at the nanoscale but suffers from numerous possible artifacts. Indeed, electrostatic [26] and electrochemical [27] contributions are responsible for parasitic and misleading PFM signals in e.g. leaky or oxygen vacancies-rich materials [18, 28–30]. It has become obvious that PFM images and loops could be obtained on non-ferroelectric materials, highlighting the need for a complementary technique which could assess ferroelectricity in a more reliable way in such cases. PUND technique, already implemented on several hundreds of microns large electrodes, has thus been adapted at the nanoscale to detect and measure remnant polarization with an atomic force microscope.

6.3.1 The Principle of PUND Measurement

PUND technique has been introduced for the first time by Scott et al. [31] in 1988. This technique is the best way to characterize ferroelectric materials because it deals with the remnant part of the signal i.e. the purely ferroelectric part. The Polarization-Voltage (P-V) or Polarization-Electric field (P-E) hysteresis loop which characterizes ferroelectric materials is obtained by measuring a transient current in response of specific excitation voltage pulses: two positives (P and U) and two negatives (N and D) pulses (see Fig. 6.6a). Figure 6.6b shows the typical current response of PUND voltage sequence measured on a PZT layer.

When a ferroelectric material is connected to a closed electrical circuit and the ferroelectric polarization is switched, a displacement current I_f appears. As ferroelectric materials are also dielectric (or because there are capacitive elements inherent to the measurement circuit), any variation of applied voltage at the sample terminals generates also a dielectric displacement current I_ε. In the case where the capacitance C of the structure remains constant, I_ε can be written:

$$I_\varepsilon = C\frac{\partial V}{\partial t} \tag{6.6}$$

Eventually, some leakage currents I_{leak} can be also present especially when samples have a very small thickness: tunneling or defect-induced currents are possible in this case. The leakage current is a location-dependent value, however all the cycles are performed in the exact same spot. As long as there is no drifting of the AFM tip, i.e. as long as the loops are performed fast, there is no reason to substract the leakage current as a background, following a similar procedure as in standard PUND method. As a result, the total current I_{total} obtained with a ferroelectric sample can be written:

$$I_{total} = I_f + I_\varepsilon + I_{leak} \tag{6.7}$$

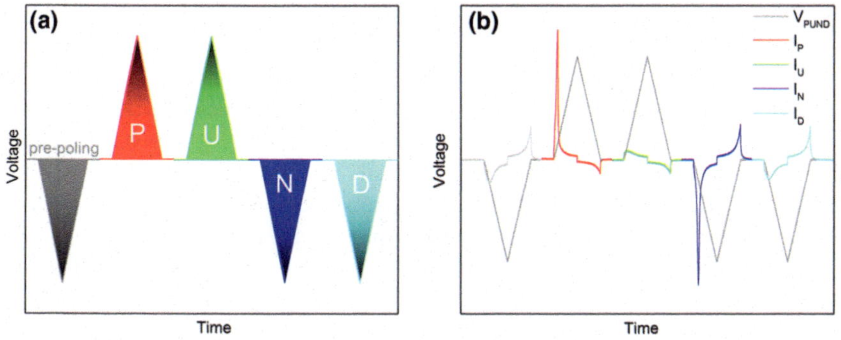

Fig. 6.6 PUND excitation signal (**a**), current response measured on PZT ferroelectric material (**b**)

During the measurement triangular voltage pulses are applied to the sample and the current measurement is operated in several steps:

- At the very beginning of the procedure, a negative voltage pulse is applied in order to switch all the ferroelectric dipoles.
- Then a first positive triangular voltage pulse, called "P" pulse is applied (called "switching pulse"), during which all the components of the current are detected: $I_P = I_f + I_\varepsilon + I_{\text{leak}}$.
- A second positive triangular voltage pulse, called "U" pulse is then applied (called "non-switching pulse"), during which only $I_U = I_\varepsilon + I_{\text{leak}}$ is recorded, since all ferroelectric dipoles have been switched during the "P" pulse and won't switch again.
- The same sequence is repeated with negative voltages: a "N" pulse (switching pulse) measures I_N composed of all contributions of the current. Eventually, a "D" pulse (non-switching pulse) leads to the measurement of only I_ε and I_{leak}.

If the dielectric and leakage current are the same during "P" and "U" pulses, respectively "N" and "D", then the ferroelectric switching current can be found from:

$$I_f^+ = I_P - I_U (I_{P-U}) \tag{6.8}$$

for positive applied voltages and:

$$I_f^- = I_N - I_D (I_{N-D}) \tag{6.9}$$

for negative applied voltages.

The time integration of currents allows calculating the corresponding charge Q_r^+ and Q_r^-. Then, the remnant charge Q_r is given by:

$$Q_r = \frac{Q_r^+ + Q_r^-}{2} \tag{6.10}$$

6.3.2 *Nano-PUND: PUND Method Implemented in an AFM*

Implementing the PUND technique in an AFM requires adjustments because of signal to noise ratio issues. For macroscopic measurements, $I_f > I_\varepsilon$ (Fig. 6.6b) whereas for nanoscale measurements $I_\varepsilon \gg I_f$. It can be easily understood: the total capacitance of the AFM system ($\approx 0.5 \times 10^{-12}$ F [32], including the tip/sample, lever/sample and chip/sample capacitances) is several orders of magnitude higher than the capacitance formed by the apex of the tip and the sample ($\approx 10^{-17}$ F). Therefore, the capacitive displacement current due to the setup geometry, which is proportional to the first capacitance, is dominating the ferroelectric switching current (which is proportional to the second capacitance) at the nanoscale. Tiedke et al. [33] have already suggested a technique to measure ferroelectric current switching on PZT sub-micron capacitors

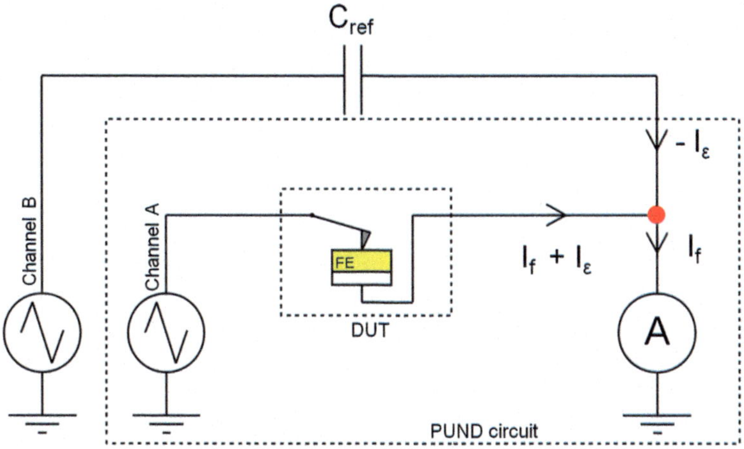

Fig. 6.7 Experimental setup for nano-PUND

with an AFM. However, this measurement requires a lot of nanofabrication operations and the compensation of the stray capacitance is not complete. Then, it leads to noisy measurements on the smallest capacitors. Prume et al. [34] have introduced a parasitic capacitance modeling of the AFM contact between the tip and the sample surface. This technique is not the best way to compensate stray capacitance because it does not solve experimentally the problem of the signal to noise ratio (the correction has to be done after the measurement). Schmitz et al. [35] presented an *in situ* compensation method, which is not completely described, but which allows to observe a polarization switching current on sub-micron capacitors with a pretty good signal to noise ratio.

Here, we suggest to significantly lower the capacitive current due to the parasitic capacitance of the setup. To do so, a compensation procedure is implemented to reduce the influence of capacitive current. An opposite PUND sequence (called NDPU) is applied on a reference capacitance C_{ref} in order to generate the exact opposite of the current due to stray capacitive current. This current is subsequently added to the signal obtained from the PUND measurement on the sample of interest. It must be noted that the synchronization of PUND (Channel A) and NDPU (Channel B) sequences is crucial for the whole process. Figure 6.7 illustrates the experimental setup.

6.3.3 *Examples and Applications of Nano-PUND Measurements*

Experimental measurements of nano-PUND on a 200 nm thick $PbZr_{0.52}Ti_{0.48}O_3$ sample are shown in Fig. 6.8. In this case, the leakage current is negligible. Figure 6.8a shows a nano-PUND measurement without any correction of the parasitic capacitive

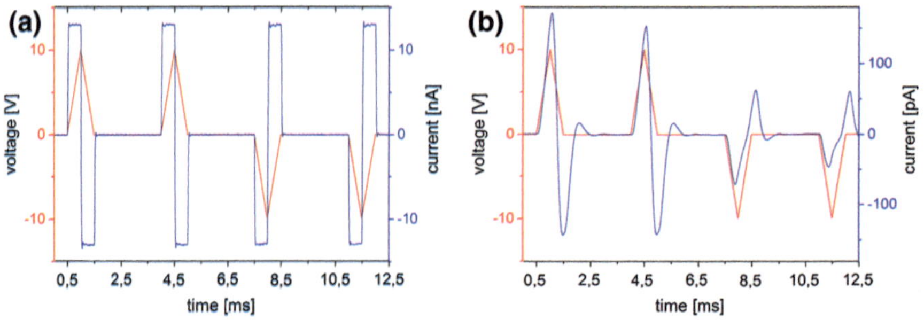

Fig. 6.8 Nano-PUND measurement without compensation of stray capacitive current on PZT (**a**), nano-PUND measurement with compensation on PZT (**b**)

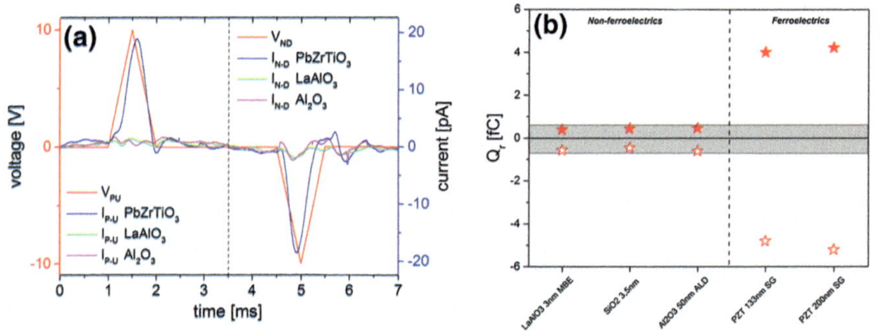

Fig. 6.9 Ferroelectric switching current measurement on several materials (**a**), average remnant charge on several materials (**b**)

current (corresponding to Channel A on Fig. 6.7). There is no difference between switching pulses (P and N) and non-switching pulses (U and D). The ferroelectric switching current is covered by the stray capacitive current which is several orders of magnitudes higher. Following (6.6), the total stray capacitance value can be calculated from this measurement. In this example, the parasitic capacitance is 0.65 pF. This step is important in order to choose the suitable capacitance C_{ref} and the amplitude of the voltage to apply on Channel B. Figure 6.8b displays a nano-PUND measurement after the compensation of the parasitic capacitance (applied through Channel B on Fig. 6.7). Consequently, the capacitive current has been lowered by a factor 100 compared to the first measurement. Now, a difference between I_P and I_U (resp. I_N and I_D) is clearly observed, proving that the ferroelectric switching polarization of the material can been evidenced (Fig. 6.8b). The ferroelectric switching current is plotted after the subtraction step in Fig. 6.9a (blue line).

The remnant charge can be obtained by time integration of I_f (I_{P-U} in the positive part and I_{N-D} in the negative part) and is estimated at $Q_r = 4.2$ fC. A similar value of remnant charge have been measured on two PZT layers of different thicknesses deposited with the same growth method, which shows the reproducibility of the technique. We have compared Q_r to the remnant polarization (P_r) measured at macroscopic scale on large Metal-Ferroelectric-Metal (MFM) structures. If we assume that P_r is scale-independent then the radius of the tip-sample contact can be deduced and estimated at 79 nm in this case. This pretty large radius is in accordance with the kind of AFM tip used for this experiment which have an average radius of 50 nm (diamond coated tips). Moreover, the experiments have been performed in ambient environment, therefore the presence of a water meniscus between the AFM tip and the sample surface can also account for the large calculated radius [36].

Additional measurements have been done on several non-ferroelectric materials ($LaAlO_3$, Al_2O_3, SiO_2) in order to determine the background noise of the technique (Fig. 6.9). The remnant part of the current (I_{P-U} and I_{N-D}) is displayed on Fig. 6.9a and shows no current peak for these non-ferroelectric materials. The measurements confirm the reliability of nano-PUND technique considering the fact that PFM images and loops have been obtained on the same $LaAlO_3$ sample [18]. Remnant charges are calculated by time integration of these current and results are plotted in Fig. 6.9b. From this graph, the extraction of the background level is possible and is estimated at 0.5 fC (gray stripe) which is therefore the average value obtained from paraelectric materials with nano-PUND. Although PZT remnant charge is almost ten times higher than this value, this background noise could become a limiting factor for materials with far lower remnant polarization.

6.3.4 Future Developments of Nano-PUND Technique

The decrease of the background noise is an important issue that must be solved if ferroelectric switching currents have to be detected on ferroelectric materials with a weak polarization. A high performance transimpedance amplifier with the lowest possible noise level must be used. Any solution which could decrease the stray capacitance of the tip-lever-chip/sample structure would help increasing the signal to noise ratio.

Another evolution of this technique would be to compensate not only displacement but also leakage current during measurement. Indeed, any voltage wave can be applied on Channel B, allowing not only to compensate dielectric displacement current but also any kind of parasitic contribution, including leakage current. This makes nano-PUND a very versatile method but imposes that the leakage current remains exactly the same for both P and U (respectively N and D) voltage pulses.

6.4 Direct Piezoelectric Force Microscopy as a Quantitative Tool

In part one of this chapter, we analyzed how the CPE is used as an imaging mechanism to map ferroelectricity in MUTs. In this part we will introduce and analyze how the direct piezoelectric effect provides a different and complementary view to the PFM information.

6.4.1 Principles of DPFM

The direct piezoelectric effect is used in DPFM as imaging mechanism by applying a mechanical stress and measuring the generated charge [3, 37]. The piezo-generated charge measuring protocol is not trivial, and it is not until 2017 in which the feasibility and implementation of DPFM was demonstrated [37]. The main limitation for this implementation arises from a technologically point of view as an appropriate operational amplifier (OA) was not available. As an example, if a suitable piezoelectric sample, with d_{33} value of 50 pC/N, is strained with 100 μN force, the generated charge is 50×10^{-4} pC, which indeed, it's an extremely tiny amount of charge for the existing sensing electronics. The direct piezoelectric effect is schematized in Fig. 6.10, for a given ferroelectric material with an upwards out-of-plane polarization, a pulling force generated a negative charge (see Fig. 6.10a). Reversing the applied force, inverses the generated charge sign (see Fig. 6.10b) [3, 38].

A correct selection of appropriate electronics to measure such charge is crucial. Typical OAs have a leakage current in the order of pA range, hence, the generated charge is likely to be absorbed by the leakage current of the OA [39–41]. In this configuration, the OA is configured like a transimpedance amplifier, in which the circuit acts as a current-to-voltage converter, with a current gain matching the feedback resistor of the circuit, R2, see Fig. 6.11 [42].

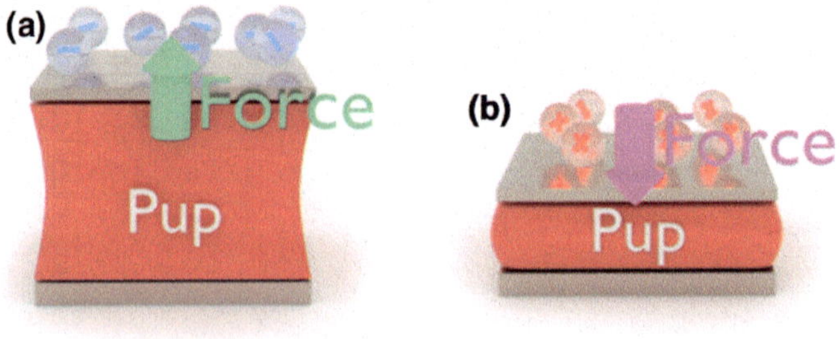

Fig. 6.10 Explanation of the direct piezoelectric effect phenomena. If a force pulls a material and expands it, a negative charge is built up on top of its surface as in (**a**). However, if the force is applied in the opposite direction, the generated charge sign inverses as in (**b**)

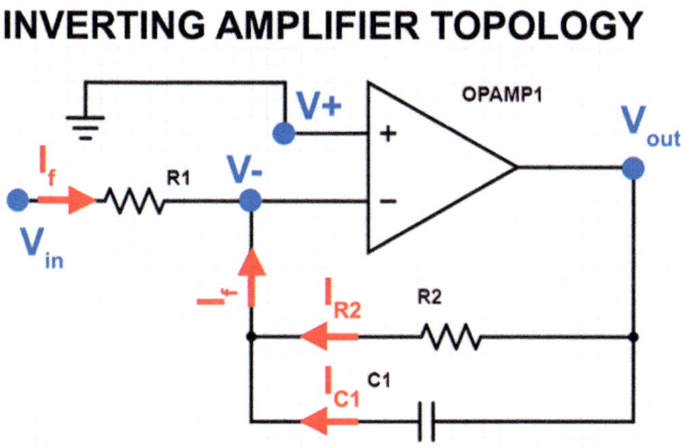

Fig. 6.11 Operational Amplifier (OA) configured as a transimpedance amplifier topology. In this configuration, the BandWidth (BW) of the system is limited by the low pass filter formed from the parallel configuration of R2 and C1

At this point, in order to maximize the signal-to-noise ratio, a higher feedback resistance is desirable in this case the feedback resistor value is 1 TΩ. Due to inherent parasitic capacitance present in the feedback path, the bandwidth (BW) of the system is determined by the RC low pass filter of the feedback, for the case of 1 TΩ and a parasitic capacitance of 0.1 pC, the BW can be approximated to 1.59 Hz, with the formula $1/(2\pi RC)$ [40]. For each value of R and C it is crucial to apply the RC filter formula to calculate the BW, to ensure that the amplifier is working in the transimpedance amplifier regime. The transimpedance amplifier output can be connected to a subsequent voltage amplifier stage to increase the total gain of the system however, it is preferable to use cascade amplifiers rather than a high gain voltage-to-voltage amplifier [43]. The very first image acquired in this mode is shown in Fig. 6.12 which depicts the domain configuration of the PPLN-which is the primary testing platform for PFM [37].

Figure 6.12c, d correspond to the PFM phase and PFM amplitude image obtained simultaneously with this mode. Such images, helps us to spatially locate the ferroelectric domains for Fig. 6.12e, f. For DPFM images, frames 6.12e, f, the current is only recorded when the AFM tip crosses from one domain to the other, however, when the tip scans in one single domain the current drops by one order of magnitude. The images can be interpreted through the understanding of the direct piezoelectric effect. As denoted by (6.3), a charge density is generated when a certain strain is applied to the material. In our case, we have two different domain structures, so it makes sense to divide (6.3) into two terms, a term for the pressure applied to the up polarization state and a term for the pressure applied to the down polarization state:

$$\frac{d\sigma}{dt} = d_{33}\left(\frac{dF^{UP}}{dt} - \frac{dF^{DW}}{dt}\right) \tag{6.11}$$

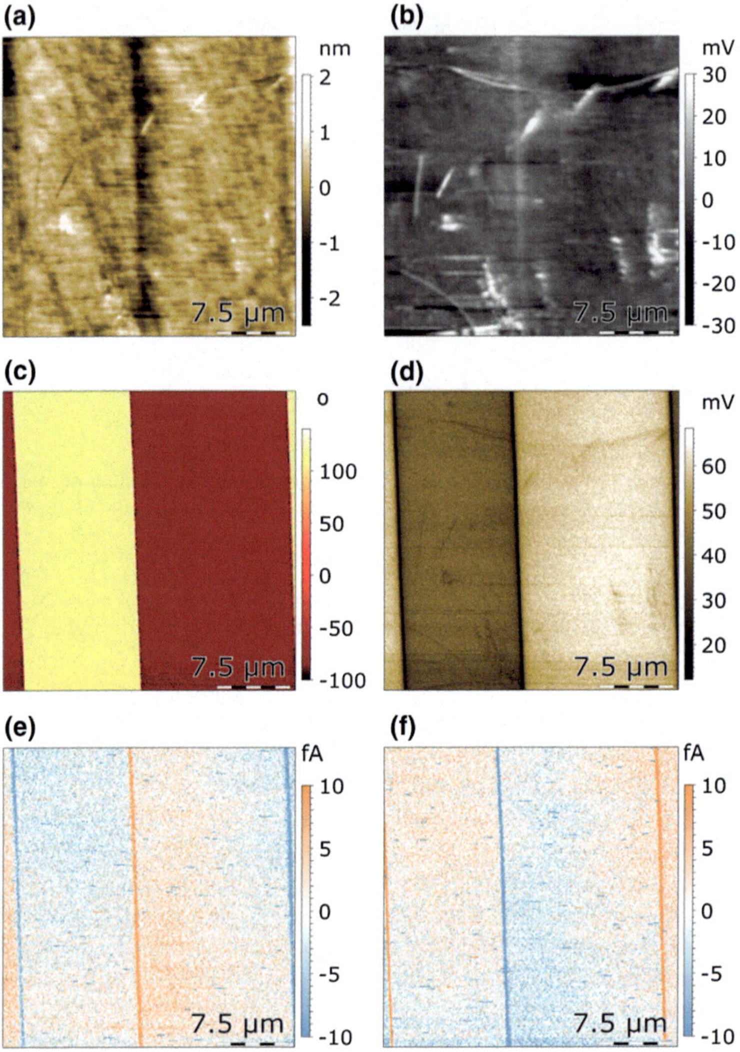

Fig. 6.12 DPFM images acquired for PPLN, Topography (**a**), and friction (**b**), are typical frame images acquired in contact mode. PFM Phase (**c**) and PFM amplitude (**d**), acquired with the settings depicted in point 1 of this chapter. DPFM-Si (**e**) and DPFM-So (**f**), from which the tip travels from left to right, and vice-versa, respectively, for the case of PPLN. All the six images are acquired simultaneously, consisting of a 256 × 256 pixels image

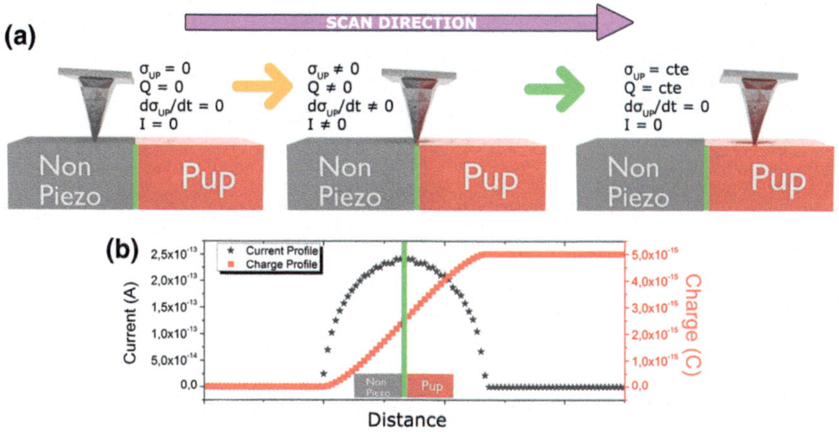

Fig. 6.13 3D model of non-piezoelectric and piezoelectric material scanned with DPFM, **a** While the tip scans the non-piezoelectric material, there is no charge and current generated, left. However, as the tip loads the piezoelectric material, generation of charge occurs, and hence, a current flows through the system, middle. Once the tip completely enters in the piezoelectric material, as the force is constant, the charge is maintained constant with time, and hence the generated current is zero. The process is described with graph showing the Charge and Current in **b** while the tip crosses from left to right

In which $d\sigma/dt$ is the current distribution, and $\frac{dF^{UP}}{dt}$, $\frac{dF^{DW}}{dt}$ are the pressure momentum for up and down domain areas. Hence, in order to record a current, time dependency of the terms d_{33} and F should occur. As the mode is employed in a constant-force contact mode, the force is maintained constant through-out the image, and, as consequence, dF/dt is always 0 [44]. However, when the tip crosses a domain wall, the generated charge changes its sing giving a generated current signal. In order to provide an intuitive view of this approach, a 3D model of the tip crossing from a non-piezoelectric material, to a ferroelectric material is given in Fig. 6.13.

Once the tip scans the non-piezoelectric material and the force is kept constant, there is no charge built up. Once the tip goes from left to right, the tip unloads the non-piezoelectric sample, and loads the piezoelectric part of the sample. Hence, as the tip is increasing the pressure into the piezoelectric material, there is a current flowing, which may depend upon the tip speed and tip load parameters. In this case, the term $\frac{dF^{UP}}{dt}$ of (6.11) is different from 0. Once the tip enters completely into the piezoelectric material, the generated charge is kept constant with time and, as a consequence, the generated current is zero. While the tip reverses its direction, the exact same reasoning can be employed to explain an inversion of the recorded current. This process is graphically depicted in Fig. 6.13b, in which the vertical green line depicts the middle of the tip apex crossing from one domain to the other. The charge at the side of the non-piezoelectric sample is 0, see red line of 6.13b, while at the side of the piezoelectric domain, the charge is positive. The current, in grey line is found by numerically differentiating the generated charge profile, red line. If we now substitute the non-piezoelectric material part, with a domain with reverse polarization, we find that a combination of two processes occurs. Hence, when the tip goes from left to right, the downwards polarization state is unloaded, and the upwards polarization is loaded. Hence, the (6.11), can be simplified by assuming $\frac{dF^{UP}}{dt} = -\frac{dF^{DW}}{dt}$:

$$\frac{d\sigma}{dt} = 2d_{33}\left(\frac{dF^{UP}}{dt}\right) \tag{6.12}$$

Substituting the tip speed, v, in the above equation,

$$\frac{d\sigma}{dt} = 2d_{33}v\left(\frac{dF^{UP}}{dx}\right) \tag{6.13}$$

From which we can interpret that the recorded current is directly proportional to the scan speed, v, the piezoelectric constant, d_{33} and to the applied force per unit area, $\frac{dF}{dt}$. For a tip crossing from antiparallel out-of-plane domain structures, such equation can be integrated to:

$$Q = 2d_{33}F \tag{6.14}$$

As expected, the generated charge is independent of the scan speed while the current is proportional to the applied load and the scan speed.

6.4.2 Quantitative Data in DPFM

The main application of DPFM is to acquire quantitative data from nanoscale piezo-generated charge images. At this point, the direct approach provides an accurate way of quantitatively measuring the piezoelectric constant of a MUT. In order to acquire the d_{33} value, [66] is used. Hence, both the collected charge and the applied force need to be calculated. In order to precisely calculate the applied force, the cantilever's spring constant is obtained through a direct measuring approach or a thermal tune approach [45, 46]. Through a standard Force-versus-Distance curve, the deflection sensitivity can be calculated and, with the spring constant, the applied force is obtained [47]. In order to measure the charge, the recorded current profile is time integrated, see Fig. 6.14.

From the current integral it is found that the generated charge is 5.9 and −6.4 fC for each DPFM-Si and DPFM-So images respectively. By applying the (6.13) the d_{33} value for the material can be obtained, with, for this case, it give the value of 14 ± 3 pC/N, which is inside the range obtained in literature [48]. The profile integration approach is a direct way of measuring the charge generated, however, it might not be convenient for the case of complex domain structures [49]. For such cases, it is possible to simply obtain an image data histogram from the current collected frames.

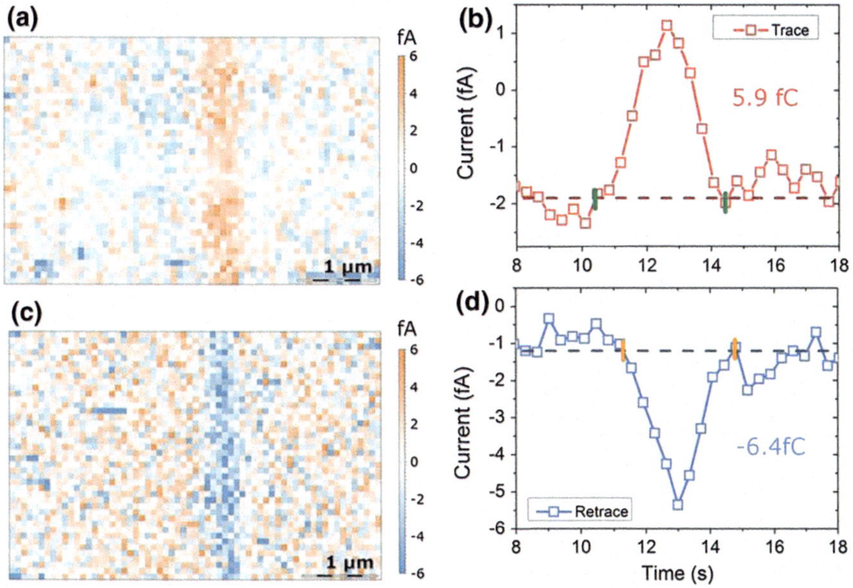

Fig. 6.14 Quantitative data analysis acquired from DPFM images. DPFM-Si image (**a**); average profile acquired from all the lines from DPFM-Si frame (**b**); DPFM-So image (**c**); average profile acquired from all the lines from DPFM-So frame (**d**)

For instance, such approach is applied in Fig. 6.15 for the case of a PZT with natural domains. In such complex systems, it is difficult to obtain a well-defined line as for the case of PPLN in which the tip crosses antiparallel out-of-plane domain structure.

The histogram is plotted for the different force values, named "F1" for 264 μN, "F2" for 384 μN and "F3" for 146 μN. It is seen that as the load increases, the corresponding current value increases accordingly. More importantly, both SI and SO images represent a similar amount of current for each of the applied loads, despite the highly localized measurements inherent to DPFM.

6.4.3 Practical How-to Guide for Imaging with DPFM

In this part of the chapter, you will learn how to get your fist DPFM image. In order to implement this mode into an AFM, you need the following elements: a low leakage amplifier ADA4530-1 from Analog Devices, a test sample, coaxial cables and the correct tip, currently "RMN-25PT200H".

Amplifier: The amplifier has to be configured as a transimpedance amplifier topology [39]. A feedback resistor of 10–100 GΩ is recommended as a starting point, while no feedback capacitor is placed in the feedback path. A suitable voltage regulated source with ±8 V can be used to power the amplifier. An additional Voltage-

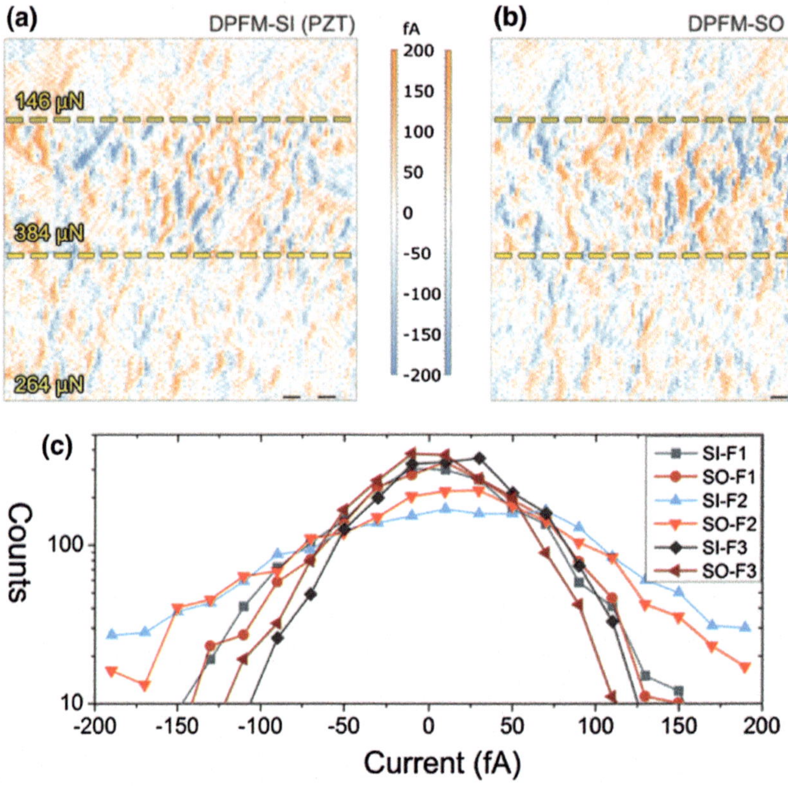

Fig. 6.15 DPFM images on PZT ceramic with natural domains structure. DPFM-Si image (**a**) and DPFM-So image (**b**). From these images, a histogram for different applied forces is calculated (**c**), where each count is referred to a single image pixel

to-Voltage amplifier can be used to increase the total gain of the system. A calibration step-by-step guide is given further in the text and more details can be found in the supplementary material of the following reference [37].

Test Sample: The same PPLN test sample, as for the PFM case, can be used for testing DPFM. However, this sample has a low d_{33} constant of ~10 pC/N so obtaining an additional sample with different d_{33} is mandatory. A low cost alternative is the use of PZT sample with natural domain structure as explained in point 1.5.a.

Coaxial cable: The AFM system used has to be modified so that a coaxial cable can be connected directly to the tip. This coaxial cable is a critical component; it has to be as short as possible, with coax cable reference RG174. One side of the cable is soldered to a high quality BNC connector, while the other side, coax core, can be directly connected to the tip or throughout another connector.

The tip: A specific tip made of full platinum and high spring constant is used. The references RMN-25PtIr200B-H and RMN-25Pt200B-H can be used. This specific tip is very important, because it allows you to apply loads of hundreds of μN without permanently deformation the cantilever. More importantly, the tip is made of full metal, preserving the conductivity of a degraded tip.

Procedure: Once all these components are available the analysis can start and DPFM images collected. The tip is placed in the tip holder, and the laser is configured as for the case of PFM. We recommend a feedback resistor of 10–100 GΩ for the case of the transimpedance amplifier. The photodetector is adjusted with a low level deflection signal-for instance, −3 V is a good starting point. With everything connected, the regulated voltage source can be connected to the low leakage current amplifier. The output of the transimpedance amplifier can be routed to an additional voltage-to-voltage amplifier before feeding the signal into a suitable AUX input channel of your AFM controller If using such additional amplifier, check that the signal read by the AFM is not saturated and correct to suitable REF input of the voltage amplifier. The PPLN sample is connected to the ground, typical "sample bias", through its bottom electrode. At this point, the tip can be approached towards the sample surface. For the first collected image, a set of recommended parameters are as follow: 128 × 128 pixels, 25 × 25 μm, speed of 0.1 ln/s and a deflection value of 0 V. The signal recorded is the AUX input channel to which the transimpedance amplifier output is connected, for both trace and retraces scan directions. This channel relates to the current that is circulating through the transimpedance amplifier, and hence, it is the current that you will measure. If topography channel looks like a PPLN surface, then you can increase the load to the sample. Setpoint deflection value up to +3 V can be typically used without losing the topography PID signal path. If after this procedure, you are not able to record an image like the one shown in Fig. 6.12e, f you can do the following:

1. Use the AFM motor to apply to the sample a much higher load, as compared to the force exerted by Z-piezo. The exact force value could reach the range of milliNewtons easily, making the tip blunt, and likely causing permanent damage to the cantilever. However, for testing purposes, you can do this in order to record a signal similar than in the case of Fig. 6.12e, f. In order to do this step, ensure that the piezo-scanner from the AFM is free to move, otherwise you may damage it. Use the AFM motor to manually approach the tip to the PPLN sample surface, until the cantilever is considerably bended. The laser spot should go out of the projection screen, but you could still enable the high voltage *X* and *Y* piezos and start a 25 × 25 μm scan. Topography, Friction and Deflection frames should be constant, but a suitable pattern similar to Fig. 6.12e, f would be recorded. An example of the frames recorded in this procedure is depicted in Fig. 6.16.
   This means that the setup is working correctly, but the leakage current is too high, and hence you have to find the leakage path and reduce it.
2. Increase the scan speed of the measurements to 1–2 ln/s. It is important, at this step, to note that the tip remaining time for each image pixel has to be larger than the coefficient of 1/BW to ensure that the amplifier still behaves as a transimpedance amplifier and not as a charge amplifier.
3. Place an external resistor in the tip holder, so that the external resistor completes the signal from the tip holder to ground. A suitable resistor, of 10–100 GΩ, can

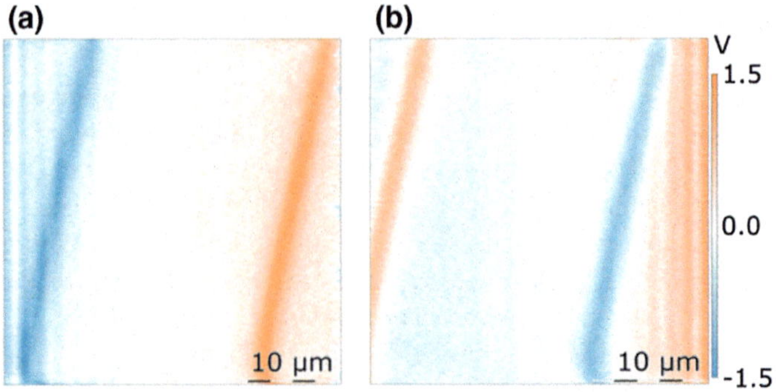

Fig. 6.16 Test images acquired for the case of PPLN with the procedure described. Image acquired while the tip is traveling from left to right (**a**), and from right to left (**b**). In this test the force applied is not controlled, however it is much easier to acquire a signal and hence, it can be used for troubleshooting purposes

be used as a test resistor. Route the DC bias to the tip holder-or sample bias- and record a curve depicting "AUX channel INPUT" versus applied DC bias. You should be able to see a linear straight curve. Otherwise the amplifier is not working correctly.

4. Start scanning the PZT sample, with a higher d_{33} value, the generated current should be of at least $10 \times$ the values of PPLN. Depending on its thickness, the domain structure should look like Fig. 6.15a, b.

6.5 Applications of Nanoscale Ferroelectric Characterization into Semiconductors

In this section, we will review how the proposed set of techniques has helped providing information in several other fields, giving special importance to the application of nanoscale characterization of solar cells, sensors, memories, transistors and wide-band gap semiconductors.

6.5.1 Solar Cells

Ferroelectric could boost the solar cell efficiency by favoring the separation of the photogeneration electron-hole pairs due to the internal polarization present in such materials. This property has been discussed for the case of the new technology of solar cells based on lead halide perovskite, and the role of ferroelectricity in such systems [50]. In this case, the specific domain patterning revealed with PFM technique is used to distinguish the ferroelectric property. Other authors have measured such domain

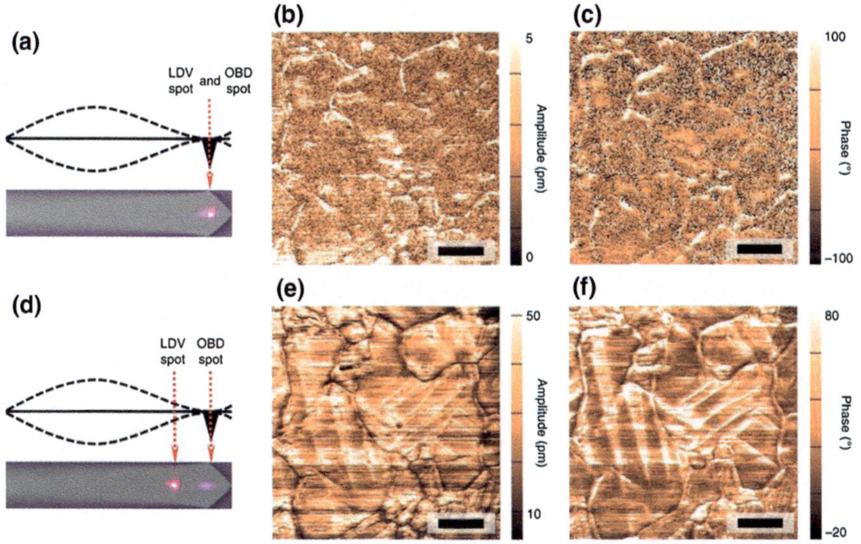

Fig. 6.17 **a**, **d**, The side-view illustrations of the expected mode shape of the cantilever show decoupling (**a**) and coupling (**d**) of the signal with cantilever dynamics; bottom insets show top-view images of the laser spot location on the cantilever. **b**, **e**, Amplitude images with the LDV spot in the locations a and d, respectively. **c**, **f**, Phase images with the LDV spot in the locations **a** and **d**, respectively. Scale bars, 1 μm. Reprinted from [51]

structure and correlated it to ferroelastic domains, rather than ferroelectric domains, see Fig. 6.17 [51].

By using a different sensor rather than the standard Optical Beam Deflection System, Yongtao Liu et al. were able to decouple artifacts coming from PFM measurements. Another important application of these techniques relies in the characterization of methods to increase the efficiency of the cell through the ferroelectric effect, for instance, by adding ferroelectric polymers [52]. Ferroelectric materials present absorption and electron-hole generation properties at wavelengths very close to visible light. One of the most promising fields of application of ferroelectrics is in the so-called "Abnormal Photovoltaic effect" in which the open circuit voltage of the cell is increased over the common material band gap. In this sense, the nanoscale patterning of ferroelectric surfaces with nanoscale domains is a key factor to enhance this abnormal effect. It has been shown by Akash Bhatnagar that Bismuth Ferrite, a common ferroelectric material, grown over a special substrate, generated a domain structure that can influence the efficiency of the cell [53]. Also, the efficient separation of electron-hole pairs can be combined with tunning the bandgap of multiferroic systems to increase the absorption as demonstrated by Nechache et al. [54].

6.5.2 Sensors

Ferroelectric materials can also provide solutions to other technological challenges including lower energy consumption devices and inertial sensors for distance, movement and acceleration detection. As an example, the Internet of Things (IoT) globally needs billions of sensors and WNS to remotely sensing and transferring information. Moreover, these billions of sensor nodes and bio-sensors will need to be powered. Using batteries for that purpose poses an unmanageable economic and environmental issue. Luckily, sensors for communications nodes or biomedical applications consume little power (~1 μW) and thus, powering them from environmental vibrations using the same ferroelectric sensor materials is a viable option. In general, ferroelectrics provide the most compact option, allowing micro- and nano-scaling energy harvesters for remote, autonomous and fit-and-forget powering of wireless network sensors [55]. In this direction, integrating high quality epitaxial ferroelectric films and nanostructures on silicon could boost the fabrication of a number of devices with the traditional Si-based complementary metal-oxide-semiconductor (CMOS) technology [56]. Moreover, advances in micro and nanofabrication technologies, open the possibility to implement a large scale integration of miniaturized piezoelectric materials into innovative electromechanical devices with nanosized moving parts (MEMS / NEMS) with prospective applications in electronics, biology and medicine [57–60]. For the first time, A. Gomez et al. established a hybrid chemical solution route to prepare nanostructured epitaxial lead free ferroelectric oxide materials on silicon wafers [5]. In particular, the authors designed a nano-structured multilayer to amplify the flexoelectric response of ferroelectric n-type semiconducting $BaTiO_{3-\delta}$ films. In this work, PFM technique demonstrated a simultaneous and independent mechanically-controlled switching of the ferroelectric polarization and the resistive state at the nanoscale. This nanoscale ferroelectric characterization is among the first examples that prove the integration on silicon of nonvolatile multilevel devices with mechanoelectric control and enhanced flexoelectric response, which could find applications in sensing and harvesting energy (Fig. 6.18).

6.5.3 Negative Capacitance

In a typical capacitor, an increased bias results in an increased charge accumulated at the capacitor electrodes. However, for the case of negative capacitance (NC), the opposite occurs, an increase in the applied bias voltage decreases the charge accumulated. NC has also been demonstrated experimentally by other authors, for instance, Asif Islam Khan et al. reported negative capacitance experimentally in a bilayer of ferroelectric $Pb(Zr_{0.2}Ti_{0.8})O_3$ and dielectric $SrTiO_3$ [61, 62]. The implications and uses of this effect include mainly the electronic industry, specifically to improve the dynamic capabilities of MOSFET technology. The effect of negative capacitance could enhance some of the specifications related to MOSFET technology [63]. From

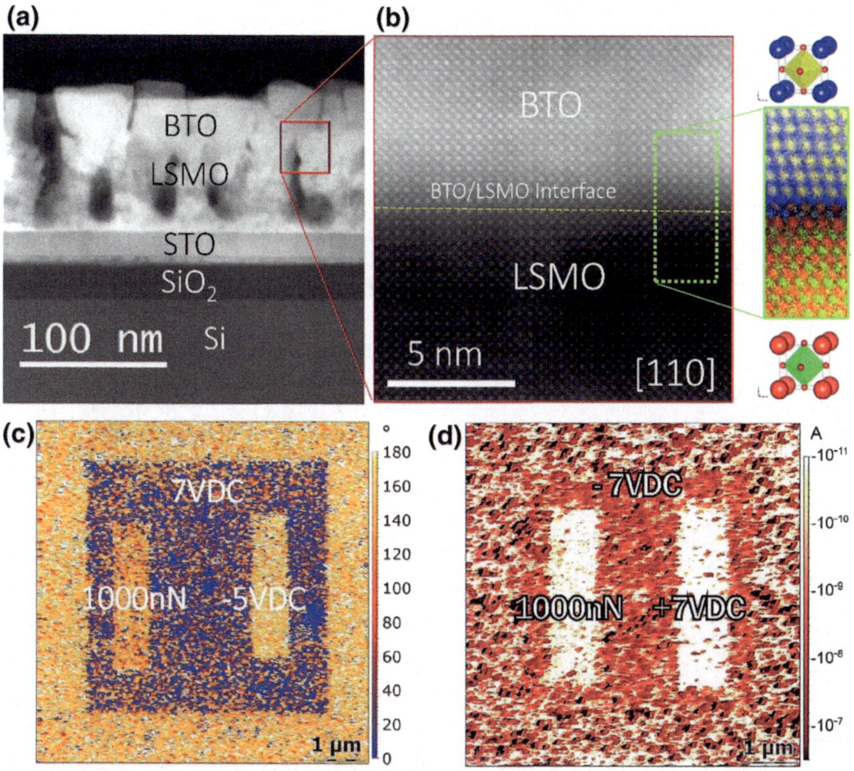

Fig. 6.18 **a** Low magnification Z-contrast image of a columnar epitaxial BTO/LSMO/STO/Si(001) multilayer viewed along the [110] Si axis. **b** High resolution Z-contrast image of the BTO/LSMO interface viewed along the Si [110]-crystallographic direction, and higher resolution Z-contrast image of the coherent interface between the BTO and LSMO layers. **c** Phase image of a 45 nm thick $BaTiO_3$ film after the generation of the downward-oriented ferroelectric domains by electrical poling and mechanical loading force. **e** Mapping of electrical current of the same film showing the piezo-driven resistance switching between two different resistance states by electrical poling and mechanical load. Reprinted from [5]

a theoretical point of view, with NC it would be possible to enhance the 60 mV/decade limit of typically employed gate oxides [63]. This possibility has been demonstrated recently by Mengwei Si et al., in a system using a ferroelectric hafnium zirconium oxide layer in the gate dielectric stack and molybdenum disulfide as semiconductor, see Fig. 6.19 [64].

By using MoS_2 NC-FET researchers report excellent on-off capabilities with a drain current of up to $5 \times 10\ \mu A\ \mu m^{-1}$. Another application of NC is using it as a step-up voltage converter to power up nanoscale devices [65]. In this application, the NC effect is used as a reliable and efficient voltage booster.

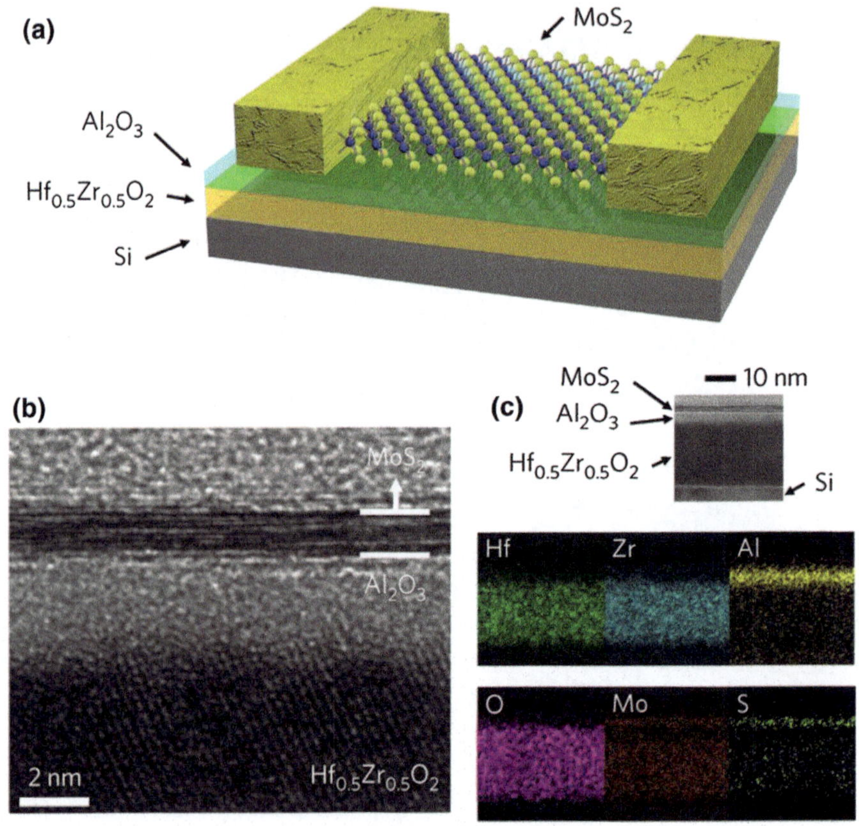

Fig. 6.19 **a** Schematic view of a MoS_2 NC-FET. The gate stack includes heavily doped Si as the gate electrode, 20 nm HZO as the ferroelectric capacitor, 2 nm Al_2O_3 as the capping layer and capacitance-matching layer. A 100 nm Ni layer was deposited using an electron-beam evaporator as the source–drain electrode. **b** Cross-sectional view of a representative sample showing the bilayer MoS_2 channel, amorphous Al_2O_3 and polycrystalline HZO gate dielectric. **c** Corresponding EDS elemental map showing the distribution of Hf, Zr, Al, O, Mo and S. Reprinted from [64]

References

1. C.R. Clayton, *Materials Science and Engineering: An Introduction* (1987)
2. J. Curie, P. Curie, Development by pressure of polar electricity in hemihedral crystals with inclined faces. Bull. Soc. Min. Fr. **3**, 90 (1880)
3. T.O. Ikeda, *Fundamentals of Piezoelectricity* (Oxford University Press. Oxford University Press, Oxford, 1990)
4. W.G. Cady, J. Valasek, *BOOK REVIEWS: Piezoelectricity: An Introduction to the Theory and Applications of Electro-mechanical Phenomena in Crystals* (1965)
5. A. Gómez, J.M. Vila-Fungueiriño, R. Moalla, G. Saint-Girons, J. Gázquez, M. Varela, R. Bachelet, M. Gich, F. Rivadulla, A. Carretero-Genevrier, Electric and mechanical switching of ferroelectric and resistive states in semiconducting $BaTiO_{3-\delta}$ Films on silicon. Small. **13** (2017). https://doi.org/10.1002/smll.201701614

6. H. Lu, C.W. Bark, D. Esque De Los Ojos, J. Alcala, C.B. Eom, G. Catalan, A. Gruverman, Mechanical writing of ferroelectric polarization. Science (80-.). **335**, 59–61 (2012). https://doi.org/10.1126/science.1218693
7. H.D. Espinosa, R.A. Bernal, M. Minary-Jolandan, A review of mechanical and electromechanical properties of piezoelectric nanowires. Adv. Mater. **24**, 4656–4675 (2012). https://doi.org/10.1002/adma.201104810
8. H. Lu, D.J. Kim, C.W. Bark, S. Ryu, C.B. Eom, E.Y. Tsymbal, A. Gruverman, Mechanically-induced resistive switching in ferroelectric tunnel junctions. Nano Lett. **12**, 6289–6292 (2012). https://doi.org/10.1021/nl303396n
9. P. Sharma, S. Ryu, Z. Viskadourakis, T.R. Paudel, H. Lee, C. Panagopoulos, E.Y. Tsymbal, C.B. Eom, A. Gruverman, Electromechanics of Ferroelectric-Like Behavior of $LaAlO_3$Thin Films. Adv. Funct. Mater. **25**, 6538–6544 (2015). https://doi.org/10.1002/adfm.201502483
10. R.K. Vasudevan, N. Balke, P. Maksymovych, S. Jesse, S.V. Kalinin, Ferroelectric or non-ferroelectric: why so many materials exhibit "ferroelectricity" on the nanoscale. Appl. Phys. Rev. **4** (2017). https://doi.org/10.1063/1.4979015
11. S. Xie, A. Gannepalli, Q.N. Chen, Y. Liu, Y. Zhou, R. Proksch, J. Li, High resolution quantitative piezoresponse force microscopy of $BiFeO_3$ nanofibers with dramatically enhanced sensitivity. Nanoscale **4**, 408–413 (2012). https://doi.org/10.1039/C1NR11099C
12. S.V. Kalinin, E. Karapetian, M. Kachanov, Nanoelectromechanics of piezoresponse force microscopy. Phys. Rev. B. **70**, 184101 (2004). https://doi.org/10.1103/PhysRevB.70.184101
13. D. Seol, B. Kim, Y. Kim, Non-piezoelectric effects in piezoresponse force microscopy. Curr. Appl. Phys. **17**, 661–674 (2017). https://doi.org/10.1016/j.cap.2016.12.012
14. E. Soergel, T. Jungk, A. Hoffmann, Quantitative analysis of ferroelectric domain imaging with piezoresponse force microscopy. Cond-Mat/0510005. (2005)
15. S. Xie, A. Gannepalli, Q.N. Chen, Y. Liu, Y. Zhou, R. Proksch, J. Li, High resolution quantitative piezoresponse force microscopy of $BiFeO_3$ nanofibers with dramatically enhanced sensitivity. Nanoscale **4**, 408 (2012). https://doi.org/10.1039/c1nr11099c
16. C. Harnagea, A. Pignolet, M. Alexe, D. Hesse, Piezoresponse scanning force microscopy: what quantitative information can we really get out of piezoresponse measurements on ferroelectric thin films. Integr. Ferroelectr. **44**, 113–124 (2002). https://doi.org/10.1080/10584580215148
17. J. Stephen, S. Jesse, J. Stephen, Quantitative mapping of switching behavior in piezoresponse force microscopy. Rev. Sci. Instrum. **77**, 73702 (2006)
18. A.S. Borowiak, N. Baboux, D. Albertini, B. Vilquin, G. Saint Girons, S. Pelloquin, B. Gautier, Electromechanical response of amorphous $LaAlO_3$ thin film probed by scanning probe microscopies. Appl. Phys. Lett. **105** (2014). https://doi.org/10.1063/1.4889853
19. A. Labuda, R. Proksch, Quantitative measurements of electromechanical response with a combined optical beam and interferometric atomic force microscope. Appl. Phys. Lett. **106** (2015). https://doi.org/10.1063/1.4922210
20. A. Gomez, T. Puig, X. Obradors, Diminish electrostatic in piezoresponse force microscopy through longer or ultra-stiff tips. Appl. Surf. Sci. **439**, 577–582 (2018). https://doi.org/10.1016/j.apsusc.2018.01.080
21. E.A. Eliseev, S.V. Kalinin, S. Jesse, S.L. Bravina, A.N. Morozovska, Electromechanical detection in scanning probe microscopy: tip models and materials contrast. J. Appl. Phys. **102** (2007). https://doi.org/10.1063/1.2749463
22. S.V. Kalinin, D.A. Bonnell, Imaging mechanism of piezoresponse force microscopy of ferroelectric surfaces. Phys. Rev. B—Condens. Matter Mater. Phys. **65**, 1–11 (2002). https://doi.org/10.1103/physrevb.65.125408
23. N. Balke, P. Maksymovych, S. Jesse, A. Herklotz, A. Tselev, C.B. Eom, I.I. Kravchenko, P. Yu, S.V. Kalinin, Differentiating ferroelectric and nonferroelectric electromechanical effects with scanning probe microscopy. ACS Nano **9**, 6484–6492 (2015). https://doi.org/10.1021/acsnano.5b02227
24. G.A. MacDonald, F.W. DelRio, J.P. Killgore, Higher-eigenmode piezoresponse force microscopy: a path towards increased sensitivity and the elimination of electrostatic artifacts. Nano Future **2**, 015005 (2018). https://doi.org/10.1088/2399-1984/aab2bc
25. S. Martin, N. Baboux, D. Albertini, B. Gautier, A new technique based on current measurement for nanoscale ferroelectricity assessment: nano-positive up negative down. Rev. Sci. Instrum. **88** (2017). https://doi.org/10.1063/1.4974953

26. S.V. Kalinin, D.A. Bonnell, Electric scanning probe imaging and modification of ferroelectric surfaces. (2004), pp. 1–43. https://doi.org/10.1007/978-3-662-08901-9_1
27. S.V. Kalinin, S. Jesse, A. Tselev, A.P. Baddorf, N. Balke, The role of electrochemical phenomena in scanning probe microscopy of ferroelectric thin films. ACS Nano **5**, 5683–5691 (2011). https://doi.org/10.1021/nn2013518
28. S. Martin, N. Baboux, D. Albertini, B. Gautier, Interpretation of multiscale characterization techniques to assess ferroelectricity: the case of GaFeO3. Ultramicroscopy **172**, 47–51 (2017). https://doi.org/10.1016/j.ultramic.2016.10.012
29. H. Miao, C. Tan, X. Zhou, X. Wei, F. Li, More ferroelectrics discovered by switching spectroscopy piezoresponse force microscopy? Epl. **108** (2014). https://doi.org/10.1209/0295-5075/108/27010
30. N. Balke, P. Maksymovych, S. Jesse, A. Herklotz, A. Tselev, C. Eom, I.I. Kravchenko, P. Yu, S.V. Kalinin, Differentiating ferroelectric and effects with scanning probe microscopy. ACS Nano **9**, 6484–6492 (2015). https://doi.org/10.1021/acsnano.5b02227
31. J.F. Scott, L. Kammerdiner, M. Parris, S. Traynor, V. Ottenbacher, A. Shawabkeh, W.F. Oliver, F. Scott, Switching kinetics of lead zirconate titanate submicron thin-film memories. **64**, 787 (2007). https://doi.org/10.1063/1.341925
32. I. Estevez, P. Chrétien, O. Schneegans, F. Houzé, Specific methodology for capacitance imaging by atomic force microscopy: a breakthrough towards an elimination of parasitic effects. Appl. Phys. Lett. **104** (2014). https://doi.org/10.1063/1.4866607
33. S. Tiedke, T. Schmitz, K. Prume, A. Roelofs, T. Schneller, U. Kall, R. Waser, C.S. Ganpule, V. Nagarajan, A. Stanishevsky, R. Ramesh, Direct hysteresis measurements of single nanosized ferroelectric capacitors contacted with an atomic force microscope. Appl. Phys. Lett. **79**, 3678–3680 (2001). https://doi.org/10.1063/1.1421638
34. K. Prume, A. Roelofs, T. Schmitz, B. Reichenberg, S. Tiedke, R. Waser, Compensation of the parasitic capacitance of a scanning force microscope cantilever used for measurements on ferroelectric capacitors of submicron size by means of finite element simulations. Japan. J. Appl. Phys., Part 1 Regul. Pap. Short Notes Rev. Pap. **41**, 7198–7201 (2002). https://doi.org/10.1143/jjap.41.7198
35. T. Schmitz, K. Prume, B. Reichenberg, A. Roelofs, R. Waser, S. Tiedke, In-situ compensation of the parasitic capacitance for nanoscale hysteresis measurements. J. Eur. Ceram. Soc. **24**, 1145–1147 (2004). https://doi.org/10.1016/S0955-2219(03)00583-1
36. A.V. Ievlev, A.N. Morozovska, V.Y. Shur, S.V. Kalinin, Humidity effects on tip-induced polarization switching in lithium niobate. Appl. Phys. Lett. **104** (2014). https://doi.org/10.1063/1.4867979
37. A. Gomez, M. Gich, A. Carretero-Genevrier, T. Puig, X. Obradors, Piezo-generated charge mapping revealed through direct piezoelectric force microscopy. Nat. Commun. **8** (2017). https://doi.org/10.1038/s41467-017-01361-2
38. K. Uchino, *Piezoelectric Actuators and Ultrasonic Motors* (Springer Science & Business Media, 1997)
39. P.R. Gray, R.G. Meyer, *Analysis and Design of Analog Integrated Circuits* (1984)
40. I. Hickman, *Analog Electronics* (1999)
41. J.R. Singer, Operational amplifiers, theory and practice. Proc. IEEE **64**, 1456 (1976). https://doi.org/10.1109/PROC.1976.10358
42. F. Tavernier, M. Steyaert, *High-Speed Optical Receivers with Integrated Photodiode in Nanoscale CMOS* (2011)
43. P.T. Krein, *Elements of Electronics* (1998)
44. P.E. West, *Atomic Force Microscopy* (2007)
45. J.L. Hutter, J. Bechhoefer, Calibration of atomic-force microscope tips. Rev. Sci. Instrum. **64**, 1868–1873 (1993). https://doi.org/10.1063/1.1143970
46. N.A. Burnham, X. Chen, C.S. Hodges, G.A. Matei, E.J. Thoreson, C.J. Roberts, M.C. Davies, S.J.B. Tendler, Comparison of calibration methods for atomic-force microscopy cantilevers. Nanotechnology **14**, 1–6 (2003). https://doi.org/10.1088/0957-4484/14/1/301
47. H.J. Butt, B. Cappella, M. Kappl, Force measurements with the atomic force microscope: Technique, interpretation and applications. Surf. Sci. Rep. **59**, 1–152 (2005). https://doi.org/10.1016/j.surfrep.2005.08.003

48. R.S. Weis, T.K. Gaylord, Lithium niobate: Summary of physical properties and crystal structure. Appl. Phys. A Solids Surf. **37**, 191–203 (1985). https://doi.org/10.1007/BF00614817
49. A. Gruverman, O. Auciello, H. Tokumoto, Scanning force microscopy for the study of domain structure in ferroelectric thin films. J. Vac. Sci. Technol., B **14**, 602–605 (1996)
50. H. Röhm, T. Leonhard, M.J. Hoffmann, A. Colsmann, Ferroelectric domains in methylammonium lead iodide perovskite thin-films. Energy Environ. Sci. **10**, 950–955 (2017). https://doi.org/10.1039/C7EE00420F
51. Y. Liu, L. Collins, R. Proksch, S. Kim, B.R. Watson, B. Doughty, T.R. Calhoun, M. Ahmadi, A.V. Ievlev, S. Jesse, S.T. Retterer, A. Belianinov, K. Xiao, J. Huang, B.G. Sumpter, S.V. Kalinin, B. Hu, O.S. Ovchinnikova, Chemical nature of ferroelastic twin domains in $CH_3NH_3PbI_3$ perovskite. Nat. Mater. 1 (2018). https://doi.org/10.1038/s41563-018-0152-z
52. Y. Yuan, T. Reece, P. Sharma, Efficiency enhancement in organic solar cells with ferroelectric polymers. Nat. Mater. **10**, 296–302 (2011). https://doi.org/10.1038/nmat2951
53. A. Bhatnagar, A. Roy Chaudhuri, Y. Heon Kim, D. Hesse, M. Alexe, Role of domain walls in the abnormal photovoltaic effect in $BiFeO_3$. Nat. Commun. **4** (2013). https://doi.org/10.1038/ncomms3835
54. R. Nechache, C. Harnagea, S. Li, L. Cardenas, W. Huang, J. Chakrabartty, F. Rosei, Bandgap tuning of multiferroic oxide solar cells. Nat. Photonics **9**, 61–67 (2014). https://doi.org/10.1038/nphoton.2014.255
55. Kompis, C.: Energy harvesting technologies to enable remote and wireless sensing. Security. 20, 2010 (2008)
56. R. Guo, Z. Wang, S. Zeng, K. Han, L. Huang, D.G. Schlom, T. Venkatesan, Ariando, J. Chen, Functional ferroelectric tunnel junctions on silicon. Sci. Rep. **5** (2015). https://doi.org/10.1038/srep12576
57. M.P. Warusawithana, C. Cen, C.R. Sleasman, J.C. Woicik, Y. Li, L.F. Kourkoutis, J.A. Klug, H. Li, P. Ryan, L.-P. Wang, M. Bedzyk, D.A. Muller, L.-Q. Chen, J. Levy, D.G. Schlom, A ferroelectric oxide made directly on silicon. Science (80-.). **324**, 367–370 (2009). https://doi.org/10.1126/science.1169678
58. R. Ramesh, MATERIALS SCIENCE: Orienting ferroelectric films. Science. **296**, 1975–1976 (2002). https://doi.org/10.1126/science.1072855
59. S.H. Baek, J. Park, D.M. Kim, V.A. Aksyuk, R.R. Das, S.D. Bu, D.A. Felker, J. Lettieri, V. Vaithyanathan, S.S.N. Bharadwaja, N. Bassiri-Gharb, Y.B. Chen, H.P. Sun, C.M. Folkman, H.W. Jang, D.J. Kreft, S.K. Streiffer, R. Ramesh, X.Q. Pan, S. Trolier-McKinstry, D.G. Schlom, M.S. Rzchowski, R.H. Blick, C.B. Eom, Giant piezoelectricity on Si for hyperactive MEMS. Science (80-.). **334**, 958–961 (2011). https://doi.org/10.1126/science.1207186
60. H.G. Craighead, Nanoelectromechanical systems. Science (80-.). **290**, 1532–1535 (2000). https://doi.org/10.1126/science.290.5496.1532
61. A. Islam Khan, D. Bhowmik, P. Yu, S. Joo Kim, X. Pan, R. Ramesh, S. Salahuddin, Experimental evidence of ferroelectric negative capacitance in nanoscale heterostructures. Appl. Phys. Lett. **99** (2011). https://doi.org/10.1063/1.3634072
62. A.I. Khan, K. Chatterjee, B. Wang, S. Drapcho, L. You, C. Serrao, S.R. Bakaul, R. Ramesh, S. Salahuddin, Negative capacitance in a ferroelectric capacitor. Nat. Mater. **14**, 182–186 (2015). https://doi.org/10.1038/nmat4148
63. A.I. Khan, C.W. Yeung, C. Hu, S. Salahuddin, Ferroelectric negative capacitance MOSFET: capacitance tuning & antiferroelectric operation. Tech. Dig. Int. Electron. Devices Meet. IEDM. (2011). https://doi.org/10.1109/iedm.2011.6131532
64. M. Si, C.-J. Su, C. Jiang, N.J. Conrad, H. Zhou, K.D. Maize, G. Qiu, C.-T. Wu, A. Shakouri, M.A. Alam, P.D. Ye, Steep-slope hysteresis-free negative capacitance MoS_2 transistors. Nat. Nanotechnol. **13**(1) (2017). https://doi.org/10.1038/s41565-017-0010-1
65. S. Salahuddin, S. Datta, Use of negative capacitance to provide voltage amplification for low power nanoscale devices. Nano Lett. **8**, 405–410 (2008). https://doi.org/10.1021/nl071804g
66. A. Gomez, M. Gich, A. Carretero-Genevrier, T. Puig, X. Obradors, Piezo-generated charge mapping revealed through direct piezoelectric force microscopy. Nat. commun. **8**(1), 1113 (2017)

Chapter 7
Electrical AFM for the Analysis of Resistive Switching

Stefano Brivio, Jacopo Frascaroli and Min Hwan Lee

Abstract Resistive switching (RS), the property of reversible changes in electrical resistance of a metal/insulator/metal cell upon electrical stimulation, has been widely studied in the last few decades for non-volatile memories and, more recently, for logic, alternative computation and sensor purposes. Atomic force microscopy (AFM) has been widely used to characterize switching behaviors and understand their underpinning mechanisms due to its unique capability and versatility for highly localized in situ and ex situ studies. The present chapter provides a brief introduction to the physics of RS and AFM schemes used to study RS, followed by an overview of recent research on RS performed by means of AFM. A particular emphasis is given to innovative AFM techniques and AFM-based studies of significant scientific contribution to the field of RS in the last few decades.

7.1 Introduction to Resistive Switching

Resistive switching (RS) usually refers to the phenomenon of electrical resistance transition in a metal/insulator/metal (MIM) structure caused by electrical stimuli. RS, observed in a plethora of different combinations of materials, can be ascribed to several physical processes, including trapping/de-trapping of charges, drift and diffusion of ionic species and phase transitions. In this section, the devices exploiting RS and their applications are introduced, and various physical mechanisms behind RS are presented.

S. Brivio (✉)
CNR-IMM, Unit of Agrate Brianza, via C. Olivetti 2, 20864 Agrate Brianza, Italy
e-mail: stefano.brivio@mdm.imm.cnr.it

J. Frascaroli
STMicroelectronics, via C. Olivetti 2, 20864 Agrate Brianza, Italy
e-mail: jacopo.frascaroli@st.com

M. H. Lee
Department of Mechanical Engineering, University of California—Merced,
5200 N. Lake Rd., Merced CA 95343, USA
e-mail: mlee49@ucmerced.edu

U. Celano (ed.), *Electrical Atomic Force Microscopy for Nanoelectronics*,
NanoScience and Technology, https://doi.org/10.1007/978-3-030-15612-1_7

7.1.1 Devices and Applications

Two terminal MIM structures can feature resistance variations upon application of voltage stimuli. An RS effect can be appreciated when two distinct resistance values result from a reading at a relatively low voltage (<1 V), before and after applying a relatively high voltage (from some Volts to tens of Volts).

An archetypal distinction of RS is made on the basis of the voltage polarity needed for a complete RS cycle. The bipolar RS requires both polarities of voltage biases, while the non-polar RS does not pose any constraint on the polarity for a complete switching cycle to be performed [1–4]. Typical current–voltage (*I*–*V*) characteristics of bipolar and non-polar switching operations are presented in Figs. 7.1a, b, respectively.

The operation bringing the device from high to low resistance (solid line in Fig. 7.1) is called SET; the operation causing a low-to-high resistance transition (dashed line in Fig. 7.1) is called RESET. High and low resistance states, HRS and LRS, respectively, can be maintained for several years with minimal degradation [4–6]. Switching operations can be performed more than billions of times before an irreversible breakage of the cell [4, 7, 8]. Furthermore, low voltage and low power operation [9, 10] and scalability below 10 nm [10, 11] have been successfully achieved by employing fab-friendly materials [1, 4]. All these features have made RS devices attractive for memory applications (resistive switching random access memories, RRAMs), mainly as a storage class memory, i.e. a low latency, long retention memory capable of filling the gap left open in the computing architecture between fast and volatile working memory and slow and non-volatile storage [12].

In addition, the recent classification of RRAMs as memristive devices has resulted in a broadening of their application spectrum. Recently, memristive devices have been proposed as key elements for alternative computing schemes, as logic in memory computing [13], stateful logic [14], parallel computing for neural network accelerators [15, 16], or bio-inspired neuromorphic computing [17–19].

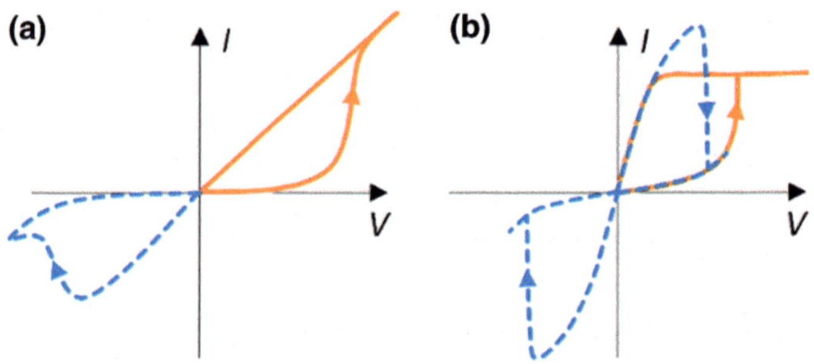

Fig. 7.1 Representative *I*–*V* characteristics of a bipolar switching device (**a**) and of a non-polar switching device (**b**). SET operations are indicated by a continuous line and RESET operations are indicated by a dashed line. In (**b**), RESET operation can be alternatively performed with positive or negative voltages

7.1.2 Physics of Resistive Switching

This short section aims at providing an introduction to the basic physical processes underlying RS phenomena that are necessary for the understanding of this chapter. Even though a clear comprehension and a unified theory of RS phenomena have not been established yet, detailed discussions of the physics of RS can be found in various review papers published in recent years [1–3, 20].

RS phenomena are usually classified into interface- and filamentary-type RS. In the interface-type RS, physical processes modifying the device resistance involve the entire surface of one or both metal-insulator interfaces of the MIM structure. As a consequence, the resistance of every undertaken state linearly scales with the device area. Conversely, there are devices in which the resistance is changed by the formation and (partial or complete) dissolution of conductive paths that disconnect the two electrodes. These conductive paths are generally known, and will be called throughout the chapter, as conductive filaments (CFs). In filamentary switching, resistance and switching properties are independent of the device area, at least in a first approximation. For the sake of completeness, it must be mentioned that a resistance change can occur over the entire volume of the insulating material in a MIM structure (e.g. phase transition-based RS). The most representative class of this type is phase change memories (PCMs), in which a chalcogenide material can switch between insulating amorphous and polycrystalline conductive phases upon short temperature shocks driven by Joule heating [20]. PCMs are usually considered a device class by themselves and their physical processes are well understood. Therefore, PCM will not be discussed in the present chapter.

For both interface- and filamentary-type devices, it is usually considered that migration of ionic species (and defects such as ionic vacancies) produces the resistance change. Indeed, a high voltage bias allows a temperature increase by Joule heating, which activates ionic mobility. The rearrangement of ionic species also modifies the electronic conduction.

In devices in which the electronic conduction is limited by the presence of an interfacial barrier against electrons at a metal/insulator interface (e.g. a Schottky barrier), a sufficiently high voltage with the suitable polarity can cause an accumulation of electrically active defects at an interface and lower the electronic barrier, thus the overall device resistance. An application of voltage with the opposite polarity is usually able to restore the previous condition of blocking interface and the high resistance state. This general picture is at the basis of interface-type switching devices, which display bipolar switching operation [1–3]. Usually, insulating materials featuring interface-type switching are oxides and, therefore, interface-type RRAMs belong to the general class of oxide RRAMs.

The second class of RS devices is identified as filamentary-type and their operation is based on migration of either anionic or cationic species. Let us present the former class that belongs to the oxide RRAM class. In every insulating material, defects, which introduce localized electronic levels within the forbidden energy gap, can be created or are already present after synthesis and fabrication. A close packing of defects produces a superimposition of such localized states and a formation of a band close to the Fermi energy that enables electronic conduction. Indeed, a CF is constituted by electrically active defects with a density that instates a delocalized band for electronic conduction [21, 22]. The initial creation of a CF in a fresh device is called electroforming (or simply forming), which is distinguished from subsequent switching operations that necessitate a much smaller voltage. The forming process can be considered a soft oxide breakdown. For filamentary devices, a technical trick is required to prevent a complete and irreversible device breakdown by limiting the current flow, which can be achieved by defining the so-called compliance current (i.e. maximum allowed current) in the measurement system or by using a load resistor or a series transistor [23]. The CFs can be partially dissolved through a RESET operation by an application of voltages of either polarity. If an application of voltage with the polarity used for electroforming is effective in restoring a high resistance state, the device is non-polar, and the process of filament dissolution is mostly thermally driven [24, 25]. This type of devices are also referred to as fuse/anti-fuse type [2, 3]. Conversely, in bipolar switching, the electric field during RESET causes electrically active defects to migrate in the direction opposite to the one induced by the forming operation and partially dissolves the CFs. The SET process brings the resistance back to a low value by reinstating the CFs. As the RESET process only partially dissolves the CFs, the voltage values needed for SET are lower than those needed for electroforming operations [6, 22, 26, 27].

Filamentary operation can also be based upon the motion of cationic species, typically positively charged Cu or Ag ions, which enter the insulator after an oxidation process from an active electrode of the same element. Afterwards, they diffuse through an electronically insulating and ionic conducting film (oxide or chalcogenide) and eventually deposit onto the opposite (inert) electrode after a reduction into metallic phase, instating a CF between active and inert electrodes [2, 3, 20]. These cation-based devices, named conductive bridge RAMs (CBRAMs), typically exhibit bipolar switching because cation migration is driven by the electric field, and thus dependent upon the applied voltage polarity.

It is noted that filamentary RRAMs, especially CBRAMs, when programmed at a very low current significantly limiting CF growth, are very likely to show the so-called threshold switching behavior, a *volatile* transition from a high to a low resistance state. That is, in a short time after the transition, the device switches back to the original high resistance state [2, 28, 29].

7.2 AFM Experimental Setup for Resistive Switching Characterization

7.2.1 Advantages of AFM

Although RS behavior has been actively studied over the past several decades, many details of the switching and conduction mechanisms still remain unclear [1, 2, 4]. This is partially because the active RS regions are tiny in both the lateral (e.g. the size of CFs) and vertical directions (e.g. effective thickness where interfacial switching occurs) and not exposed to accessible surfaces. These factors make the direct characterization of the RS region highly challenging. While RS is often dominated by nanoscale phenomena, most studies have usually relied on bulk-type (i.e., area/volume averaged) measurements through pad-type devices with areas much larger than the dimensions of interest (e.g. CF width). Transmission electron microscopy (TEM) can provide information on the localized static and evolutionary information on the geometry, phase, and chemistry of active switching areas in the most direct fashion [28, 30, 31]. However, TEM necessitates elaborate and destructive sample preparation processes. The setup also precludes a continuous characterization during repeated RS processes. Furthermore, the environment of characterization is also significantly different from operational conditions due to the requirement of ultrahigh vacuum condition [32].

Atomic force microscopy (AFM) may be the tool of choice for in situ characterization of RS behavior at a true nanoscale by integrating with conventional electrical measurement schemes (e.g. dc voltammetry, impedance spectroscopy, current interrupt). In addition, by leveraging the versatility and flexibility of AFM, a variety of characteristics including electrostatic [33, 34], ferroelectric [35] and mechanical/chemical properties [31] can be also quantified at the nanoscale, and the characterization can be achieved simultaneously under a controlled electrical, thermal and gas environments if needed. Furthermore, sample preparation is much simpler than other nanoscale analysis schemes [36]. Due to these significant merits, there have been numerous AFM-based studies of RS behavior.

7.2.2 Contact AFM Techniques

The mode of scanning probe-based electrical measurements can be categorized depending upon whether the tip is in direct contact with the sample surface or detached from it. In the so-called conductive AFM (C-AFM) scheme, local electrical current is measured while a controlled voltage bias (either ac or dc) is applied between the tip and the sample (see Fig. 7.2a). During the current measurement, the tip needs to make a consistent solid contact with the sample surface. Therefore, the conventional force feedback scheme is in action (i.e., contact-mode AFM) [37], which maintains a constant tip-sample force during the measurement. The C-AFM

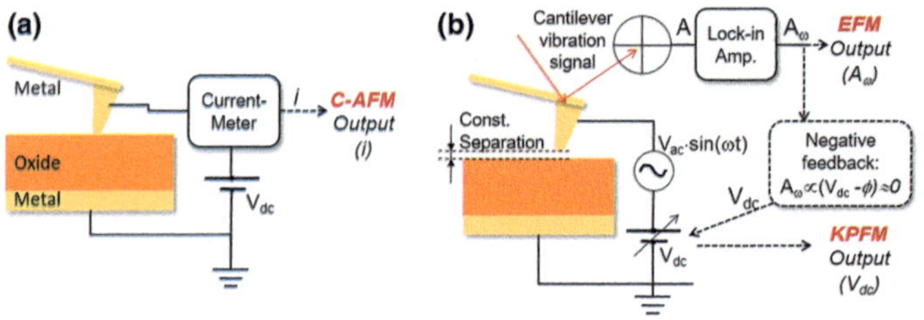

Fig. 7.2 Simplified schematic diagrams depicting the setups for **a** C-AFM, **b** EFM and KPFM

can be performed by obtaining *I*–*V* spectra at a local contact of choice or mapping a spatial variation of electrical current under a constant voltage bias while scanning a specific area of interest on a sample.

For RS characterization, the C-AFM can be used either to *incur* RS in a selected area of substrate surface (or at a tip-sample point contact) by applying a pre-designed amount of voltage bias or to *characterize* the RS states of the substrate at various locations by measuring electrical current under a small controlled voltage bias. In most cases, a RS MIM structure is implemented by placing a metal-coated tip on a bi-layered insulator/metal substrate in an AFM setup. A sequence of in situ RS incurrence followed by a localized C-AFM characterization is one of the most widely used approach for AFM-based RS studies. While in situ RS study allows a convenient and sequential RS behavior without a limit in the number of switching processes, the configuration of RS is significantly different from the usual device cell configuration, and thus the observation from this AFM-based configuration may not be directly applicable to the interpretation of RS behaviors in actual devices. The acute geometry of the tip enhances the electric field around the tip/substrate contact, which is likely to facilitate surface electrochemical reaction(s) at the contact [38]. To avoid complications in interpreting AFM-based data, there have been efforts to perform RS processes (forming, SET and/or RESET switching) in a usual device configuration before using C-AFM for a local characterization. For this, the top electrode used for the device-level switching should be removed by ion [39, 40], chemical etching [41, 42], or by mechanical treatments [43, 44], before a C-AFM-based characterization. Examples can be found in the *Filamentary Switching* section in the present chapter.

The tip-sample force in contact mode normally ranges from 1 nN to 1 μN [45]. If the force between the tip and sample surface is too weak, the current information during C-AFM would not be stable [46–48]. On the other hand, a strong contact force tends to result in an accelerated tip-coating wear, hence the interfacial nature and contact area are not conserved [47]. A pressing force as small as ~50 nN can cause a plastic deformation of metal-coated tips, which makes a quantitative evaluation of a physical quantities by C-AFM challenging [36]. A wise choice of pressing force is therefore required to achieve a stable contact while preserving the tip geometry. The tip-sample adhesion force is also known to largely affect the rate of tip degradation [47].

Tip degradation can be additionally caused by an electrical stress on the tip-sample interface. Even when a mild electrical bias is applied, an electrostatic discharge upon contact and/or an unexpected high charge flux (for example, caused by a sudden change in the device resistance) can damage the tip geometry significantly [36]. Given the tiny tip-sample contact area (a few to tens of square nanometers), even a low voltage bias would induce a current density and Joule heating significant enough to locally melt the tip apex. Considering dramatic variations in electrical resistance are expected in an RS system, it is necessary to have a self-limiting current scheme integrated in the measurement setup (e.g. inserting an electrical resistor in the loop) [36, 49]. Artefacts from morphological convolution between the tip and sample surface are another source of complication [50]. As some degree of surface roughness and/or surface contaminants cannot be avoided, it is generally appropriate to take a statistical approach by comparing the averaged resistances (or current level) obtained in an area with those from other areas.

7.2.3 Non-contact AFM Techniques

While the contact AFM techniques, i.e. C-AFM, measures the actual electrical currents flowing through a tip-substrate contact, non-contact AFM techniques senses electrostatic interactions between the tip and substrate at a tip-substrate separation of a few tens of nanometers. The local variation of electrostatic (capacitive) interaction is largely affected by the work function and charge distribution of the substrate [33]. The non-contact electrical measurement scheme can be divided into electrostatic force microscopy (EFM) [51] and Kelvin probe force microscopy (KPFM) [52] depending upon the output signal.

The EFM provides a map of local electrostatic force in a relative term while the KPFM renders a map of surface potentials of the substrate (see Fig. 7.2b). A combination of dc and ac bias with a fixed oscillatory frequency ($V = V_{dc} + V_{ac} \sin \omega t$) is provided between the tip and cantilever while the tip is positioned off the surface by a controlled distance. The applied voltage induces oscillatory capacitive tip-substrate interaction, which results in a cantilever vibration at the same frequency, ω. The vibratory amplitude should be proportional to the amplitude of capacitive force, F:

$$F = -\frac{1}{2}\frac{\partial C}{\partial z}(V - \phi)^2$$

where C is the tip-sample capacitance, z is vertical spatial vector from the sample surface, V is the tip-sample voltage bias, and ϕ is the surface potential. An application of combined dc and ac-voltage bias therefore incurs an oscillatory capacitive interaction:

$$
\begin{aligned}
F &= -\frac{1}{2}\frac{\partial C}{\partial z}\left\{(V_{dc}-\phi)^2+\frac{1}{2}V_{ac}^2\right\}-\frac{\partial C}{\partial z}V_{ac}(V_{dc}-\phi)\sin\omega t \\
&\quad+\frac{1}{4}\frac{\partial C}{\partial z}V_{ac}^2\cos 2\omega t \\
&= F_{dc}+F_{\omega}+F_{2\omega}
\end{aligned}
$$

The capacitive interaction then results in corresponding cantilever deflection/vibration with dc and oscillatory terms, which turns into a laser photo-detector signal, A, of the AFM system. The first harmonic signal, A_{ω}, which is expected to be proportional to F_{ω} and thus to $(V_{dc}-\phi)$ is extracted by the use of a lock-in amplifier. When the first harmonic signal (A_{ω}) itself is measured and displayed as a function of surface location by fixing V_{dc} at a constant value, we call the scheme as EFM. In the KPFM mode, on the other hand, local surface potential is directly quantified through a negative feedback scheme, in which the V_{dc} is actively tuned to nullify the first harmonic A_{ω} signal during the scan. When A_{ω} is nullified, V_{dc} coincides with ϕ, and can be collected to obtain a map of the surface potential of the sample (see Fig. 7.2b).

During the electrostatic measurement, the tip is supposed to be at a constant distance of a few or a few tens of nanometers away from the surface. The separation control is realized in the (1) tapping, (2) frequency-modulation or (3) lift modes. In the *tapping mode*, the AFM cantilever is mechanically vibrated close to the resonance frequency of the cantilever by applying an oscillatory electrical bias to a piezoelectric plate attached to the cantilever. The amplitude of the cantilever vibration is reduced as the tip comes close to the substrate and possibly taps on its surface. A negative feedback scheme finely controls the tip-surface distance, indirectly evaluated from the cantilever vibration amplitude. The frequency of ac voltage applied to the piezoelectric material should be away from the frequency used for EFM/KPFM so that the signal from electrostatic interaction can be differentiated from that used to control tip-substrate separation. In the *frequency modulation mode*, the used signal is the resonance frequency of cantilever vibration as opposed to the vibratory amplitude. This mode leverages the fact that an attractive/repulsive force causes a shift in resonance frequency of a structure. The *lift mode* does not rely on a feedback scheme unlike the former two modes. During the electrostatic measurement, the tip is lifted above the surface by a fixed amount (usually a few tens of nanometers) by tracing the surface morphology of the substrate, which was acquired in a preceding scan in contact, tapping or frequency-modulation mode.

7.3 Noteworthy AFM Scientific Results

In this section, we present recent noteworthy scientific findings in the field of RS enabled by versatile AFM techniques. The section is divided into the following subsections: *Interfacial Switching*, *Filamentary Switching* and *C-AFM as a Nano-probe for Critical Morphologies*. The second subsection is the heftiest because the merit of AFM in spatial resolution makes AFM more suitable for a direct observation of nanoscale filaments.

7.3.1 Interfacial Switching

Interface-type RS in an MIM cell is closely related to the valence state of the RS film in the vicinity of the interface with metal electrodes [1, 53]. The height and width of a Schottky-like barrier at the oxide/metal interface is modulated by the change of local valence state in the oxide film. Anions (oxygen ions or oxygen vacancies, $V_{\ddot{O}}$) are usually more mobile than cations in oxides, and thus they play a determining role in the electronic barrier modulation [54].

Figure 7.3a depicts the charge transfer at the interface and charge transport within the RS film caused by an electrical voltage bias application. A positive tip potential would result in an oxygen evolution reaction ($V_{\ddot{O}}$ formation), electron (e′) extraction and/or hole (h˙) injection at the tip/substrate interface. Simultaneously, the positive bias would repel positively charged $V_{\ddot{O}}$ and holes away from the contact. In view of local charge redistribution, especially in the vicinity of the tip/surface interface, the aforementioned two processes are competing against each other. For example, if the surface charge transfer process contributes more than the charge migration process to the space charge formation, positive charges will be accumulated at the interface by a positive tip bias. In the opposite case, the interface will be negatively charged.

In the aforementioned reasoning framework, a combination of C-AFM and KPFM was used to understand the mechanism of interface-type RS [55]. Figure 7.4 shows surface potential (by KPFM) and current maps (by C-AFM) of a Pt tip/TiO_2/Ti back electrode (BE) and a Pt tip/NiO/Pt BE cell. TiO_2 and NiO were chosen as

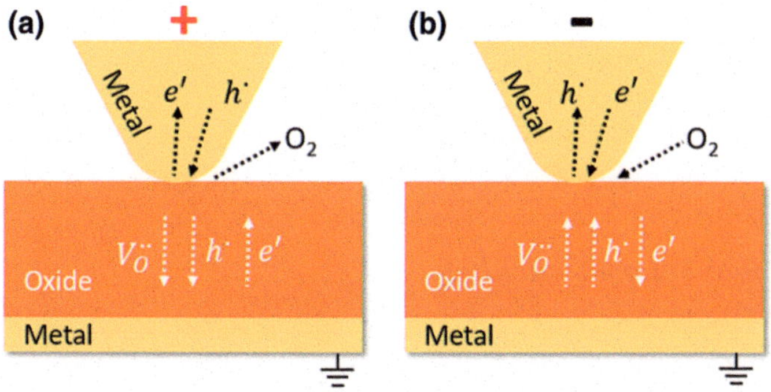

Fig. 7.3 Schematic drawings of oxygen exchange, ionic migration, and hole (electron) injection under an electrical bias of **a** positive and **b** negative potential in an AFM-based MIM configuration. Reformatted [55]. Copyright 2011 Springer

they are representative n-type and p-type conducting metal oxides widely studied for RS. Before KPFM and C-AFM measurements, a RS operation was performed by scanning a positively biased Pt-coated tip on a rectangular area of NiO/Pt surface followed by another scan of smaller rectangular area with a negative tip bias. From the KPFM maps (Fig. 7.4a, b), a positive (negative) tip bias application was found to increase (decrease) the surface potential, indicating that the surface charge exchange, rather than the charge migration, plays a dominant role in determining the charge state in the vicinity of the interface. This was the case regardless of the conducting character of the RS films (n-type or p-type). However, the KPFM images alone could not reveal whether the RS is caused by the movement of ions or electronic species because oxygen vacancy formation and hole injection, both of which can be caused by a positive tip bias, would produce the same KPFM result. A C-AFM observation can complement the KPFM analysis. If electronic charge injection/extraction were the determining factor for the RS, a C-AFM would have resulted in maps with the opposite contrast to those shown in Fig. 7.4c, d. For example, in the case of n-type cell (Pt tip on TiO_2/Ti), a positive tip bias would extract electrons from the RS film and thus reduces the concentration of the majority carrier (electron), which would shift the resistive state to an HRS. For a p-type cell, on the other hand, a positive tip bias increases the concentration of the majority carrier (hole) at the interface, changing the state to an LRS. However, as shown in Fig. 7.4c, d, a positive tip bias resulted in an HRS on the n-type cell and an LRS on the p-type cell, indicating that those exchanged at the interface are mainly ionic species (oxygen ions), not electronic species (electrons or holes).

The important role of oxygen exchange in RS was also revealed earlier by a combination of C-AFM and time-of-flight secondary ion mass spectroscopy (TOF-SIMS) [56]. In a Rh-tip/NiO/Pt BE system, either +7 or −7 V were applied during a tip scan over a specific area of the oxide surface, giving results well aligned with that of [55]; a positive (negative) tip bias incurred an HRS (LRS). In addition, the TOF-SIMS elemental analysis showed that a high concentration of ^{18}O ions are accumulated in the area where a negative tip bias was applied in ^{18}O containing environment. This proves that a significant oxygen incorporation was achieved by a negative tip bias application. Oxygen incorporation is considered as a prerequisite of RS because in oxygen deficient ambient (i.e. vacuum) RS cannot be detected [56, 57]. There have been other studies using similar approaches for TiO_2 [33], Cu-O/Si [58] and Nb:$SrTiO_3$ [59] based systems.

7.3.2 *Filamentary Switching*

C-AFM has been employed as a research tool of choice to investigate the filamentary nature of RS devices by virtue of its unrivalled combined topographic and electric resolution. The studies of conductive regions formed after electroforming date back to the experiments addressing the reliability of oxide barrier for electronic applications [60–62] and unleashed many studies aimed at the investigation of filamentary

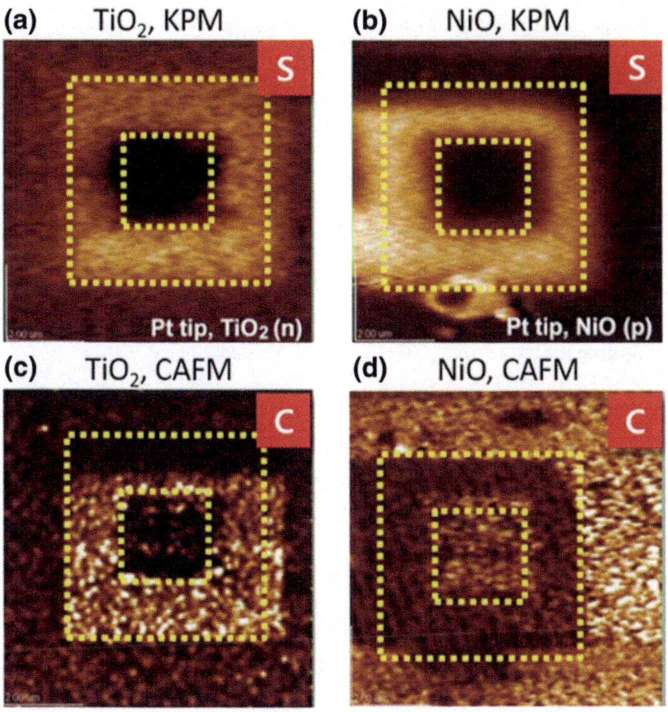

Fig. 7.4 **a**, **b** Surface potential and **c**, **d** current images measured through **a**, **c** Pt tip/TiO_2/Ti BE and **b**, **d** Pt tip/NiO/Pt BE. The measurements were performed in a high vacuum ($\sim 10^{-6}$ torr range). Reprinted with permission from [55]. Copyright 2011 Springer

operation of RS devices exhibiting both bipolar and non-polar operation [38, 39, 43, 63–66].

In the following, we present the investigations of filamentary RS that span throughout all the stages from the pristine conduction to CF creation and dissolution and to CF characterization.

7.3.2.1 Pristine Conduction Through Insulating Films

The investigation of an RS process starts from the pristine state of a functional dielectric material and its electroforming process, which is the initial step of a filamentary switching. C-AFM allows to identify nanometric inhomogeneities in the conductivity of pristine dielectric materials, which are believed to be the precursors of local CF development. A local high conductivity spot provides a path for a vertical current flow and local temperature increase by Joule heating, which promotes the migration of preexisting ionic defects, generation of new ones and the eventual formation of CF.

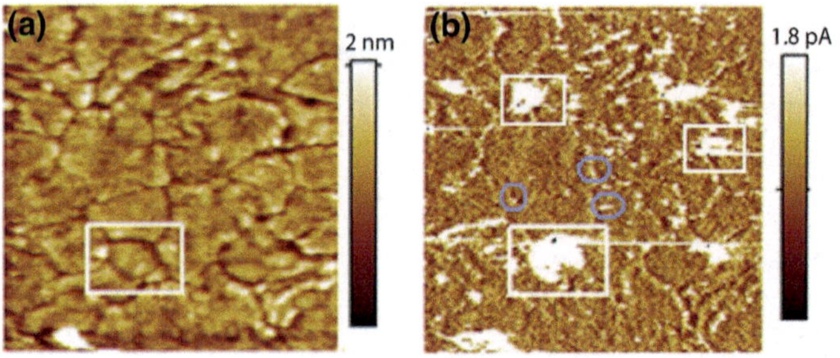

Fig. 7.5 Topography (**a**) and current map (**b**) of the surface a $HfO_2/SiO_2/Si$ structure. Reprinted with permission from [61]. Copyright 2011 American Vacuum Society

Prior studies revealed the critical role of grain boundaries of polycrystalline dielectrics in localized electrical breakdown [60, 61]. An incipient leakage conduction and thermal Joule heating tend to occur at the regions localized at grain boundaries, where the formation and migration of defects are promoted. The pristine conduction properties are usually investigated on a dielectric film grown on a conductive substrate and an AFM tip is used as the top mobile electrode. Figure 7.5 shows a surface topography exhibiting a polycrystalline nature in a HfO_2 film (Fig. 7.5a), and the corresponding electrical map (Fig. 7.5b) showing higher conductivity along the grain boundaries [61]. A remarkable special case of incipient conductive region that can give rise to filamentary RS is a dislocation in single crystalline insulating films. For instance, [67] demonstrates that dislocations give rise to CF formation in a $SrTiO_3$ film.

Subsequent studies extended the analysis of the pristine conduction properties to amorphous dielectric films. Against the intuitive expectation that amorphous films would exhibit a uniform conductivity, conductivity inhomogeneity has been observed, which acts as the precursor of CF formation [60, 68]. Despite the fact that defects in amorphous structures are hardly definable [69], localized variation of stoichiometry or unpaired bondings are considered to introduce energy levels within the forbidden gap, which would assist hopping conduction through the bulk of the dielectric film.

7.3.2.2 Filament Formation and Dissolution

The application of a large enough voltage in close proximity of the incipient conductive paths (being associated to grain boundaries or other generic defects in amorphous films) in a pristine material produces the formation of large and stable CFs. It has been demonstrated that a moderate voltage is sufficient to generate CFs close to incipient conductive paths, whereas a voltage as high as 20 V (maximum voltage that can be applied by a commercial C-AFM setup) would not produce any appreciable resistance change at locations away from them [70].

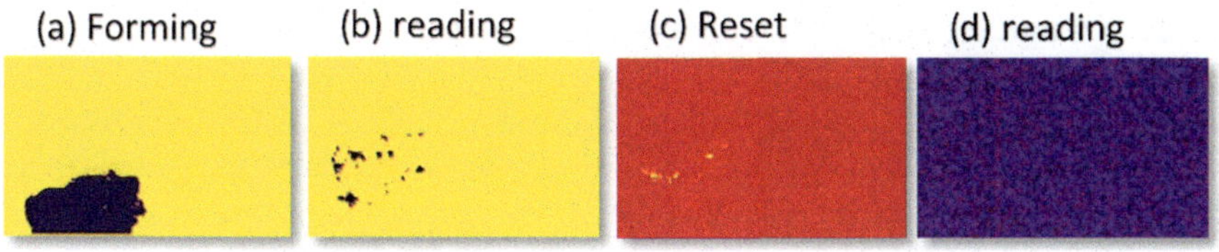

Fig. 7.6 a–d Current map acquired sequentially during a forming ($V_{substrate} = -3.5$ V), a reading, a RESET ($V_{substrate} = 3.5$ V) and another reading ($V_{substrate} = -0.5$ V) scans. Re-arranged from [68]

An electroforming step can either enable subsequent RS processes or lead to an irremediably destructive breakdown. The use of a voltage divider (a series resistance) or a prolonged low voltage application, instead of an abrupt high voltage application, can prevent irreversible breakdown [23, 49, 68]. Figure 7.6 shows a current map acquired during a slow scan at relatively low voltage that produces the formation of CFs (Fig. 7.6a). The subsequent reading scan at a voltage lower than that or forming (Fig. 7.6b) certifies the CF formation and allows to resolve the details of the CFs [68]. A current map acquired applying an opposite voltage polarity (Fig. 7.6c) resulted in a RESET operation, as verified in a subsequent reading scan (Fig. 7.6d). The presence of multiple current spots in the current map of Fig. 7.6b is attributed to the formation of ramifying branches originated from a common root, which is localized in proximity to electrically active defects at the bottom interface. According to the study, such filament geometry is the consequence of the formation of CFs driven by a mobile electrode that scans throughout the surface with an electrical voltage bias [68].

On the other hand, another AFM study reported an observation of multiple *independent* CFs formed during an electroforming process in a standard MIM button cell. The C-AFM measurement was performed after the top electrode was removed [64]. Different approaches have been followed to access the top metal/insulator interface underneath the top electrode: the use of a Hg drop as a removable electrode [64], the opening of a window through the top electrode by ion etching [39, 40], the removal of top electrode by chemical etching [41, 42], mechanical delamination procedures [43] and mechanical tip-induced material removal [71].

The combination of nanometric lateral resolution and high current sensitivity makes C-AFM suitable also for quantitative measurements at a scale hardly accessible by a conventional electrical setup. For instance, by contacting individual CFs in a HfO_2 film by a C-AFM tip, *I–V* characteristics have been measured at low current levels, allowing to reveal the relationship between RESET current and CF resistance [72]. The phenomenon of conductance quantization, which affects filamentary RRAMs with an additional source of telegraphic noise, was analyzed at the level of single CF again in a HfO_2 film [73]. C-AFM has been used also to study the retention of individual CFs [74], enabling a link between nano- and macro-scopic scale measurements.

7.3.2.3 Characterization of Conductive Filaments

In this section, novel techniques for detailed characterization of CFs are discussed. In all the aforementioned studies, CFs in vertically stacked metal/oxide structures are detected only if they extend up to the free surface where a C-AFM tip can establish an electric contact. In addition, the use of other SPM variants and/or of alternative cell geometries can provide further insight into the switching mechanism.

Recent studies based upon a combination of C-AFM and KPFM observations, in which the former is sensitive to flowing current and the latter responds to trapped charges, suggested that CFs are comprised of ionic species that render a fluent electric conduction between top and bottom electrodes [33, 74, 75]. In the experiment on an Al_2O_3/TiN structure reported in Fig. 7.7, different voltages are applied at different location as specified by the yellow spots in the topography map of Fig. 7.7a. The voltage application leads to CF formation only for voltages above ~10 V, i.e. 11–13 V spots in Fig. 7.7a. Interestingly, KPFM reveals the permanent presence of electrostatic charges only in correspondence of those points (Fig. 7.7b, c). The result attests the role of charge injection in the establishment of a CF [74].

The filamentary switching phenomena can be better understood by an investigation of CF geometry. The study of vertically stacked metal/insulator structures allows to probe only the very end of vertical CFs exposed on the surface, making the CF geometry inferred indirectly [68, 76]. In order to visualize a CF in their growth direction, two solutions have been attempted: the use of planar device structures and C-AFM tomography.

Lateral structures, in which two electrodes are patterned on each end of a RS material, are used as an alternative configuration for CF characterization (Fig. 7.8a). The structure enables a voltage application by two standard electric probes through sufficiently large contact pads while leaving room for an AFM tip to detect electrically induced morphology changes after electroforming (see Fig. 7.8b, c acquired before and after electroforming, respectively) [77]. On the other hand, planar geometries induce a non-optimal electric field layout, usually sufficient for driving the cationic motion of CBRAMs. For instance, Sun et al. used planar $Pt/SiO_2/Ag$ devices to monitor the filament evolution during the transition from threshold switching and memory switching through topographic and KPFM characterization [28]. They found that volatile threshold switching is associated with the presence of disconnected Ag

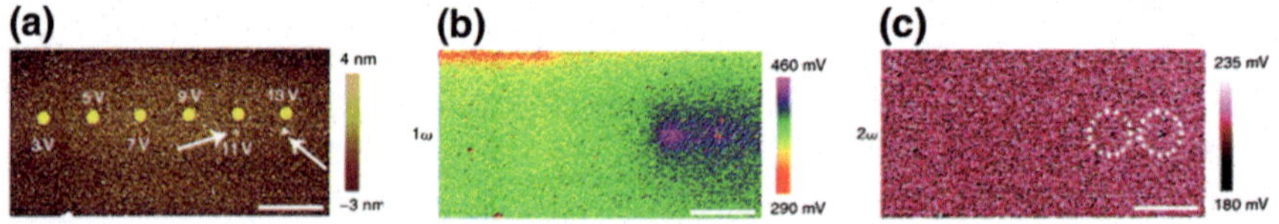

Fig. 7.7 Topography (**a**), 1ω component (**b**) and 2ω (**c**) of the KPFM signal. Circles in (**a**) indicate the location where the indicated voltage has been applied (C-AFM). Re-arranged from [74] in agreement with Creative Commons CC BY license

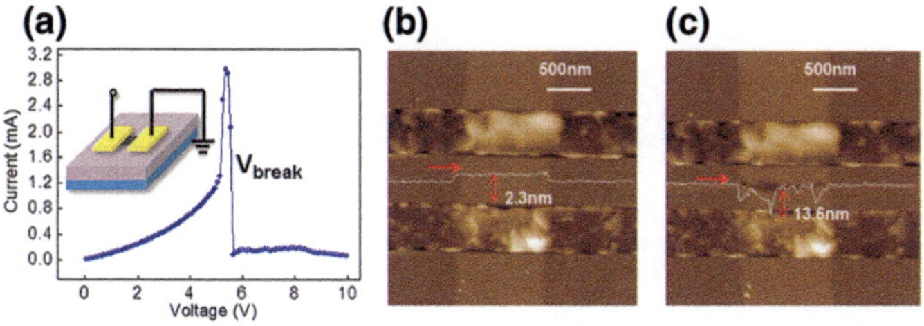

Fig. 7.8 **a** I–V characteristic of a forming process of a graphene-based planar whose structure is reported in the inset. AFM image of the whole device and height profiles of the graphene sheet before (**b**) and after (**c**) the electroforming. Reprinted with permission from [77]. Copyright 2012 American Chemical Society

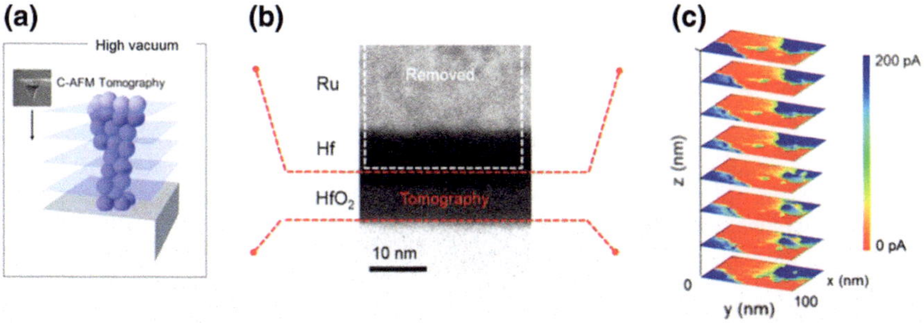

Fig. 7.9 **a** Sketch of a CF sectioned in various layers; **b** image of the oxide RRAM stack in cross section and **c** layer-by-layer conductivity maps acquired while progressively removing material from the device. Reprinted (adapted) with permission from [79]. Copyright 2015 American Chemical Society

islands in between the two electrodes. Conversely, memory switching occurs because of the formation of a stable continuous CF [28].

Recently, a different technique has been developed to access the complete CF length in vertical devices. Celano et al. [44, 78, 79], demonstrated that a conductive AFM tip can be used as a scalpel to etch a material progressively. They also demonstrated that it is possible to acquire localized electrical information during the etching process and thus build a tomography map of the CFs, i.e. layer-by-layer-conductivity maps, as shown in Fig. 7.9 [44, 79].

The technique has the potential to unveil the full geometry of CFs in 3D, which has significance for the interpretation of the physics related to the RS behavior. For instance, in [44, 79], the authors study the orientations of conical shaped CFs in order to attest the roles of the processes of ion migration and injection for the CF formation in both ionic and cationic filamentary RRAMs. Part of the team also proved that the presence of multiple CFs in a device leads to different RESET dynamics in CBRAMs [80]. The tomography technique has proved even more powerful than TEM-based investigations in that C-AFM can detect localized electrical changes resulting from slight chemical/compositional modifications of the insulator material

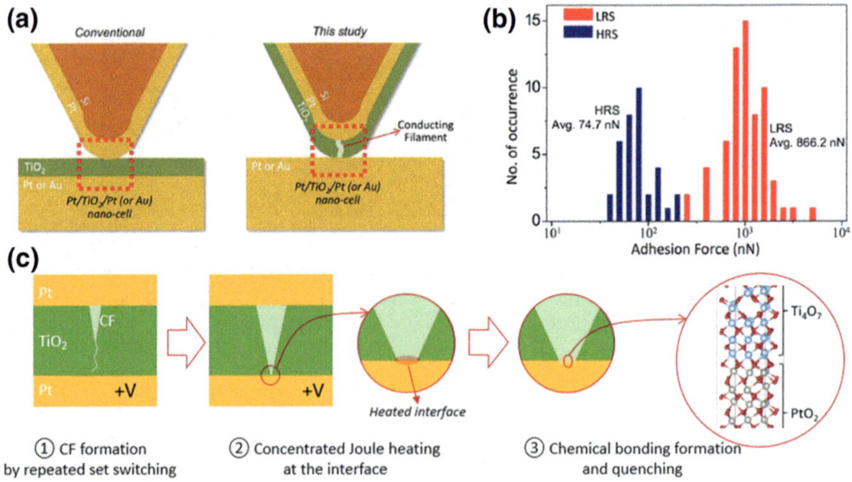

Fig. 7.10 **a** Conventional (left) and alternative (right) AFM-based configurations for RS study. **b** Distribution of measured adhesion forces after SET (with 20 μA compliance current) and RESET processes. **c** A schematic drawing to depict the conjectured SET process. Reprinted (adapted) [31]. Copyright 2017 Nature Publishing Group (an open access article)

that would produce negligible contrast in TEM maps. As a consequence, C-AFM-based tomography has been adopted for other device types [81] and also by other groups [82].

An insight on the chemical nature of CFs and their interfaces has been obtained by combining an innovative definition of device geometry together with an exploitation of the AFM versatility to acquire signals of diverse physical properties. In a recent study, Moon et al. demonstrated the critical impact of chemical bonding at the interface of insulator/metal on the RS behavior of an MIM cell by leveraging the AFM capability of quantifying nano-Newton level adhesion force in addition to electrical properties [31]. The unprecedented observation was made by employing an unconventional configuration, in which an AFM-based nanoscale MIM cell was implemented between a metal/insulator-coated tip and a metal-coated substrate, as opposed to the conventional coupling of metal coated tip and insulator/metal substrate (see Fig. 7.10a). Due to the uncontrollable drift (1–10 nm/min) of the tip location with respect to substrate, it was not possible to repeat RS processed through the same MIM cell in an AFM setup. Even in the new configuration, the location of metal/insulator-coated tip with respect to the substrate changes uncontrollably over time. However, at least the area within the insulator layer that would be affected by any previous RS process is still part of the nano-cell of characterization wherever the tip is located.

In the new AFM setup, a non-polar RS could be initiated by a repeated SET switching with a positive bias on the substrate at a compliance current of $\gtrsim$20 μA. After reaching a non-polar RS regime, the adhesion force between the RS film (TiO_2) and the metal substrate (Pt) was measured by utilizing the so-called force spectroscopy scheme [83], a widely used functionality of AFM. From the observation, it was found that the adhesive force at the interface in a unipolar LRS is much higher than those measured in a HRS cell by more than an order of magnitude on average (Fig. 7.10b). The interface was also found to return to a low adhesive state after a RESET switching regardless of the switching polarity. In addition, once a tip-substrate interface in an LRS cell is mechanically detached, the cell behaves like an HRS cell even after the tip is re-attached to the substrate; the cell can be electrically SET to an LRS again. The LRS comes with a high adhesive state again. These observations prove that the non-polar RS behavior is highly coupled with the interfacial bonding at the active interface. Based upon the amplitude of adhesive force and density function theory (DFT) calculations, the nature of the adhesion at the active interface of an LRS cell was proposed to be an oxygen-mediated chemical bonding, rather than a van der Waals interaction or electrostatic force [31].

The following RS procedure was proposed based upon the aforementioned observations. The repeated switching processes is conjectured to have formed a CF within the insulating layer (process ① of Fig. 7.10c) as observed by other groups. The CF formed in TiO_2 layer is known to be in a conical structure with Magnéli phase [30, 84]. For a n-type RS material, it is believed that the cone-shaped CFs are thicker at the interface where a negative voltage bias is applied and form an ohmic contact with the neighboring electrode [85]. The other interface where the CFs are narrower plays the determining role in RS and thus is referred to as the "active" interface. When the CF grows enough to 'touch' the bottom Pt, most of the electric field will be concentrated around the interface causing a significant Joule heating there (process ②). The thermal energy will facilitate the rearrangement of the interface in a thermodynamically stable configuration and the formation of a stable chemical bonding. Once it is achieved, a unipolar LRS is attained and thus the electric field that used to be concentrated at the interface will be spread throughout the whole CF and electrodes. This results in a quenching effect (i.e., an abrupt temperature drop) at the interface and forms a quasi-permanent chemical bonding to maintain the resultant RS state (process ③).

7.3.3 C-AFM as a Nano-probe for Critical Morphologies

Several patterning techniques are available for the fabrication of nanostructures which can be used to investigate RS properties in highly scaled morphologies. However, one of the greatest challenges is to provide an electric contact to those nanostructures with sufficient flexibility and lateral precision. The sharp conductive tip of C-AFM suits the task of accessing highly scaled and compact morphologies with a lateral precision of a few nanometers. The high lateral resolution of morphology maps allows locating the structures of interest, while the scanning probe can act as a mobile electrode to achieve a selective contact on the scanned surface. Contact mode is usually necessary to achieve an electrical contact between the tip and the sample. A notable exception is the methodology developed by Otsuka et al. [86], called point-contact current-imaging AFM. Tapping mode was adopted to locate the nanostructures with high accuracy and avoiding any destructive tip – sample interaction. Then, cantilever oscillation was interrupted and the tip was approached in the desired position to achieve the electric contact.

In many situations, C-AFM can be preferred over conventional contact pad patterning by lithographic techniques to perform electrical testing on nanostructures. The main advantage is undoubtedly the flexibility to access multiple positions on the sample surface in a non-predetermined and non-destructive way. In case of submicron features, the ease of contact formation by C-AFM significantly lowers process complexity, time and costs associated with advanced lithography, which makes it the tool of choice for explorative lab-level investigations. In addition, the electric contact is reversible since the contact location can be changed by simply moving the tip position. This aspect is significant for example in the in-plane switching configuration for the investigation of the switching phenomenon.

As an example, a lateral switching morphology can be applied to contact RS elements grown by bottom-up techniques in the form of nanowires and then dispersed on an insulating substrate. This permits to exploit well established bottom-up growth techniques to synthesize high quality materials. C-AFM allows to investigate multiple portions of the switching element, extract the resistivity of the formed structure, to inspect wire continuity and to locate the switching end. Oka et al. [87] fabricated single crystalline core/shell nanowires composed of a MgO core and a NiO shell. After dispersion of the nanowires on a SiO_2 substrate, a metal contact was created at one edge by Pt patterning. The other nanowire extremity was located by morphology maps operated in tapping mode, then the Ir/Pt coated tip was contacted at about 1 μm from the Pt electrode and I-V curves demonstrated a bipolar switching operation.

The application of C-AFM find its main motivation in highly compact morphologies, especially when the density of the nanostructures makes contact formation challenging when adopting advanced patterning methods. An example can be found in the investigation of individual nano-features. A large interest has raised in the last few years about switching properties of individual nanoparticles in order to prove the fundamental scalability limit of different switching materials, in particular when filamentary-type switching is concerned. Several examples can be found of RS in nanostructures down to a few nanometers fabricated by various methods, both based on chemical synthesis [88] or template-assisted methods [89, 90]. In a notable example, Uenuma et al. [89]. exploited a biological process involving a supramolecular protein as a reaction cage to form magnetite nanostructures with a diameter or 6 nm. C-AFM was used to address individual nanoparticles and to prove bipolar switching operations.

Fabrication by template-assisted methods attracted large interest because of the ability to fabricate RRAM devices in a controlled fashion. Self-assembled templates in particular offer the potential to form large ordered arrays of nanostructures with size and spacing controlled by the process parameters. Although self-assembled patterning techniques still pose significant challenges in terms of process integration and features placing and alignment, at the laboratory level C-AFM offers a unique combination of lateral resolution, flexibility and cost effectiveness to fully exploit the possibilities offered by self-assembled patterning techniques. For a short review about bottom-up template-assisted fabrication of RS devices, see [4].

A largely employed fabrication method to form arrays of RS elements uses anodized aluminum oxide (AAO) membranes. Regular pores with high aspect ratio are formed in the Aluminium material with hexagonally packed symmetry that can be filled with the desired material. Brivio et al. [49, 91], fabricated vertical arrays of hetero-structured unipolar Au/NiO/Au nanowires by electrodeposition through the AAO template followed by thermal oxidation (Fig. 7.11). Selective addressing of individual nanowires by the C-AFM tip demonstrated the unipolar behavior at very low switching current (Fig. 7.11c, d). Interestingly, by evolving the nanowires structure to Au/NiO_x/Ni/Au, the switching behaviour changed from unipolar to bipolar. A similar approach was developed by other authors to prove unipolar switching in Ni/NiO/Ni nanowires [92], bipolar switching operations in a Cu/CuO_x/Cu patterned structure due to Cu migration in the CuO_x film [93] and unipolar switching in Au/HfO_2/Pt nanostructured devices [94]. In this latter work, multiple devices in close proximity were tested to provide device distributions of high and low resistance states, as well as of the SET and RESET switching voltages.

Concerning the investigation of multiple nanostructure, C-AFM has the disadvantage of being relatively slow when compared to patterned electrodes that can be accessed externally in parallel. However, it offers the flexibility of inspecting multiple devices even at densities usually difficult to contact with top-down lithography. This particular property made possible the analysis of multiple devices in the initial state in a work by Frascaroli et al. [76]. C-AFM scans of the sample surface allowed to build the distribution of the conductivity level of different devices, which showed

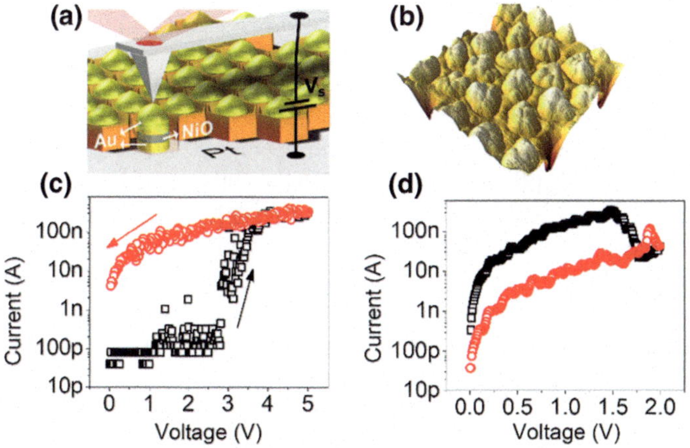

Fig. 7.11 **a** Pictorial view of a C-AFM tip contacting Au/NiO/Au nanowires fabricated in regular arrays by AAO template. **b** Topographic AFM image of the nanowires protruding out of the sample surface. **c**, **d** Examples of set and reset I-V switching curves obtained on a single nanowire. Adapted from [49]

a clear correlation with the density of electrically active defects in the HfO_2 film. In this work, a self-assembled template formed by block copolymers was applied to pattern the top metal electrodes of Pt/Ti/HfO_2/TiN memory elements in regular arrays. This self-assembly method offers higher flexibility and easier process integration when compared to other methodologies and allowed the fabrication of arrays of metal electrodes down to 12 nm [95]. However, it should be noted that the typical curvature radius of conductive metal-coated or diamond-coated C-AFM tips hinders the clear identification of features at this dimension.

Block copolymer self-assembly allows to reach a density of patterned features above 10^{10} cm^{-2} in a relatively straightforward way [96]. In [76] when probing bipolar switching on nanoscale electrodes, different devices showed a crosstalk behavior, which was assigned to connected conduction paths in the underlying continuous films.

A different patterning approach was adopted by Ye et al. [97], which makes use of self-assembled nanospheres of different diameters on the sample surface. By depositing the top electrode though the space left open by the nanospheres, electrode sizes from 50 to 300 nm were realized on a NiO film and tested by C-AFM contacting. The experimental results, supported by Monte Carlo simulations, unveiled a strong dependence at nanoscale of the switching probability on the device size and filament dimension.

In summary, C-AFM allows to perform investigations on nanoscale systems which would be hardly accessible by means of other techniques. The unique features and the ease of use of C-AFM instrument provided the first indication of the high scalability of filamentary based RRAM down to a few nanometers and unveiled peculiar phenomena that only arise in highly scaled devices or at high density.

7.4 Conclusions

The chapter presents an overview of recent AFM-based studies of RS in addition to a short introduction to the application and operating fundamentals of RS. AFM technique has led to significant advancement in the understanding of RS behaviour at the nanoscale, taking advantage of the combination of localized electric characterization (of both electrical conduction and charge distribution) and measurement of other physical properties (e.g. morphology and adhesion force, to name a few). Notable RS studies employing different sample geometries, operative modalities and various AFM schemes are reviewed, demonstrating the potentiality of AFM in probing RS phenomena and unveiling the related mechanisms.

7.5 Perspectives

Since RS attracted a renewed interest at the beginning of the twentieth century, AFM has played a crucial role in studying RS phenomena thanks to the operational simplicity, easy sample preparation, and versatility. Over the past decade or so, RS has dramatically expanded its applicability; while RS was initially considered predominantly for non-volatile memory applications, global semiconductor industry trend towards "diversification", in terms of product functionalities and computational paradigms, has widened the potential applications of RS [12]. The realization that RS is at the base of the memristor device operation opened new opportunities for RS devices ranging from logic-in-memory and neuromorphic computing [13–18], to sensing and signal detection [98]. New device paradigms have driven researchers to pay attention to different aspects of RS phenomenology, including its dynamics, stochastic operation and sensitivity to external perturbations of various types (e.g. mechanical, chemical and other). Furthermore, research on RS materials has recently expanded even to organic and biological substances [99].

Aligned with the trend, we envision that AFM will continue to play a crucial role in RS studies. First, the fundamental studies of non-linear dynamics or stochasticity of RS transitions, which are ascribed to nanoscale effects, will be pursued actively. Further, the physical principles underlying the sensitivity of RS to various external perturbations (optical, mechanical and thermal, to cite a few) in a controlled environment can be investigated with AFM analysis, in view of alternative applications for RS devices. On the other hand, the conduction and RS behavior of organic or biological materials, a very recent research field, requires the investigation of novel conduction mechanisms and switching processes [99]. For these reasons, AFM is likely to continue to be one of the main experimental techniques for innovative investigation of RS.

References

1. A. Sawa, Resistive switching in transition metal oxides. Mater. Today **11**, 28–36 (2008)
2. R. Waser, R. Dittmann, G. Staikov, K. Szot, Redox-based resistive switching memories—nanoionic mechanisms, prospects, and challenges. Adv. Mater. **21**, 2632–2663 (2009)
3. D.S. Jeong et al., Emerging memories: resistive switching mechanisms and current status. Rep. Prog. Phys. **75**, 076502 (2012)
4. H.-Y. Chen et al., Resistive random access memory (RRAM) technology: From material, device, selector, 3D integration to bottom-up fabrication. J. Electroceramics **39**, 21–38 (2017)
5. J. Frascaroli, F.G. Volpe, S. Brivio, S. Spiga, Effect of Al doping on the retention behavior of HfO_2 resistive switching memories. Microelectron. Eng. **147**, 104–107 (2015)
6. S. Brivio, J. Frascaroli, S. Spiga, Role of Al doping in the filament disruption in HfO_2 resistance switches. Nanotechnology **28**, 395202 (2017)
7. M. Azzaz et al., Improvement of performances HfO_2-based RRAM from elementary cell to 16 kb demonstrator by introduction of thin layer of Al_2O_3. Solid-State Electron. **125**, 182–188 (2016)

8. M.-J. Lee et al., A fast, high-endurance and scalable non-volatile memory device made from asymmetric Ta_2O_{5-x}/TaO_{2-x} bilayer structures. Nat. Mater. **10**, 625–630 (2011)
9. F. Nardi et al., Sub-10 μA reset in NiO-based resistive switching memory (RRAM) cells, in *2010 IEEE International Memory Workshop*, (2010), pp. 1–4. https://doi.org/10.1109/imw.2010.5488317
10. B. Govoreanu et al., $10 \times 10\,nm^2$ Hf/HfO_x crossbar resistive RAM with excellent performance, reliability and low-energy operation, in *2011 International Electron Devices Meeting* (2011), pp. 31.6.1–31.6.4. https://doi.org/10.1109/iedm.2011.6131652
11. K.-S. Li et al., Utilizing Sub-5 nm sidewall electrode technology for atomic-scale resistive memory fabrication, in *2014 Symposium on VLSI Technology (VLSI-Technology): Digest of Technical Papers* 1–2 (2014). https://doi.org/10.1109/vlsit.2014.6894402
12. International Technology Roadmap for Semiconductors, 2013 [online]. Available: http://www.itrs2.net/
13. D.S. Jeong, K.M. Kim, S. Kim, B.J. Choi, C.S. Hwang, Memristors for energy-efficient new computing paradigms. Adv. Electron. Mater. **2**, 1600090 (2016)
14. J. Borghetti et al., 'Memristive' switches enable 'stateful' logic operations via material implication. Nature **464**, 873–876 (2010)
15. S. Agarwal et al., Energy scaling advantages of resistive memory crossbar based computation and its application to sparse coding. Front. Neurosci. **9** (2016)
16. G.W. Burr et al., Neuromorphic computing using non-volatile memory. Adv. Phys. X **2**, 89–124 (2017)
17. E. Covi et al., Analog memristive synapse in spiking networks implementing unsupervised learning. Front. Neurosci. **10**, 482 (2016)
18. M. Prezioso et al., Training and operation of an integrated neuromorphic network based on metal-oxide memristors. Nature **521**, 61–64 (2015)
19. S. Brivio et al., Extended memory lifetime in spiking neural networks employing memristive synapses with nonlinear conductance dynamics. Nanotechnology **30**, 015102 (2019)
20. S. Menzel, U. Böttger, M. Wimmer, M. Salinga, Physics of the switching kinetics in resistive memories. Adv. Funct. Mater. **25**, 6306–6325 (2015)
21. B. Magyari-Köpe, M. Tendulkar, S.-G. Park, H.D. Lee, Y. Nishi, Resistive switching mechanisms in random access memory devices incorporating transition metal oxides: TiO_2, NiO and $Pr_{0.7}Ca_{0.3}MnO_3$. Nanotechnology **22**, 254029 (2011)
22. A. Padovani, L. Larcher, O. Pirrotta, L. Vandelli, G. Bersuker, Microscopic modeling of HfO_x RRAM operations: from forming to switching. IEEE Trans. Electron Devices **62**, 1998–2006 (2015)
23. K. Kinoshita et al., Reduction in the reset current in a resistive random access memory consisting of NiO_x brought about by reducing a parasitic capacitance. Appl. Phys. Lett. **93**, 033506 (2008)
24. J. Kim et al., Switching power universality in unipolar resistive switching memories. Sci. Rep. **6**, 23930 (2016)
25. U. Russo, D. Ielmini, C. Cagli, A.L. Lacaita, Self-accelerated thermal dissolution model for reset programming in unipolar resistive-switching memory (RRAM) devices. IEEE Trans. Electron Devices **56**, 193–200 (2009)
26. S. Brivio, S. Spiga, Stochastic circuit breaker network model for bipolar resistance switching memories. J. Comput. Electron. **16**, 1154–1166 (2017)
27. B. Gao et al., Oxide-based RRAM switching mechanism: a new ion-transport-recombination model, in *Electron Devices Meeting, 2008. IEDM 2008. IEEE International* (2008), pp. 1–4. https://doi.org/10.1109/iedm.2008.4796751
28. H. Sun et al., Direct observation of conversion between threshold switching and memory switching induced by conductive filament morphology. Adv. Funct. Mater. **24**, 5679–5686 (2014)
29. U. Celano et al., Tuning the switching behavior of conductive-bridge resistive memory by the modulation of the cation-supplier alloys. Microelectron. Eng. **167**, 47–51 (2017)
30. D.-H.H. Kwon et al., Atomic structure of conducting nanofilaments in TiO_2 resistive switching memory. Nat. Nanotechnol. **5**, 148–153 (2010)

31. H. Moon et al., Interfacial chemical bonding-mediated ionic resistive switching. Sci. Rep. **7**, 1264 (2017)
32. M. Kudo et al., Real-time resistive switching of Cu/MoO_x ReRAM observed in transmission electron microscope, in *2014 Silicon Nanoelectronics Workshop (SNW)*(2014), pp. 1–2. https://doi.org/10.1109/snw.2014.7348556
33. Y. Du et al., The resistive switching in TiO_2 films studied by conductive atomic force microscopy and Kelvin probe force microscopy. AIP Adv. **3** (2013)
34. E. Strelcov et al., Space- and time-resolved mapping of ionic dynamic and electroresistive phenomena in lateral devices. ACS Nano **7**, 6806–6815 (2013)
35. S. Hong et al., Large resistive switching in ferroelectric $BiFeO_3$ nano-island based switchable diodes. Adv. Mater. **25**, 2339–2343 (2013)
36. V. Zade, H.-S. Kang, M.H. Lee, Effect of mechanical and electrical stimuli in conductive atomic force microscopy with noble metal-coated tips. J. Appl. Phys. **123**, 015301 (2018)
37. F.J. Giessibl, Advances in atomic force microscopy. Rev. Mod. Phys. **75**, 949–983 (2003)
38. M.H. Lee, C.S. Hwang, Resistive switching memory: observations with scanning probe microscopy. Nanoscale **3**, 490–502 (2011)
39. B.J. Choi et al., Resistive switching mechanism of TiO_2 thin films grown by atomic-layer deposition. J. Appl. Phys. **98**, 033715 (2005)
40. B. Singh, B.R. Mehta, D. Varandani, A.V. Savu, J. Brugger, CAFM investigations of filamentary conduction in Cu_2O ReRAM devices fabricated using stencil lithography technique. Nanotechnology **23**, 495707 (2012)
41. H. Woo, S.K. Vishwanath, S. Jeon, Excellent resistive switching performance in Cu-Se-based atomic switch using lanthanide metal nanolayer at Cu-Se/Al_2O_3 interface. ACS Appl. Mater. Interfaces (2018). https://doi.org/10.1021/acsami.7b18055
42. S. Claramunt et al., Non-homogeneous conduction of conductive filaments in Ni/HfO_2/Si resistive switching structures observed with CAFM. Microelectron. Eng. **147**, 335–338 (2015)
43. R. Münstermann et al., Morphological and electrical changes in TiO_2 memristive devices induced by electroforming and switching. Phys. Status Solidi RRL Rapid Res. Lett. **4**, 16–18 (2010)
44. U. Celano et al., Three-dimensional observation of the conductive filament in nanoscaled resistive memory devices. Nano Lett. **14**, 2401–2406 (2014)
45. M. Lanza, U. Celano, F. Miao, Nanoscale characterization of resistive switching using advanced conductive atomic force microscopy based setups. J. Electroceramics **39**, 94–108 (2017)
46. K.M. Lang et al., Conducting atomic force microscopy for nanoscale tunnel barrier characterization. Rev. Sci. Instrum. **75**, 2726–2731 (2004)
47. M.A. Lantz, S.J. O'Shea, M.E. Welland, Characterization of tips for conducting atomic force microscopy in ultrahigh vacuum. Rev. Sci. Instrum. **69**, 1757–1764 (1998)
48. R. O'Hayre, M. Lee, F.B. Prinz, Ionic and electronic impedance imaging using atomic force microscopy. J. Appl. Phys. **95**, 8382–8392 (2004)
49. S. Brivio et al., Low-power resistive switching in Au/NiO/Au nanowire arrays. Appl. Phys. Lett. **101**, 223510 (2012)
50. Lee, M., Lee, W. & Prinz, F. B. Geometric artefact suppressed surface potential measurements. Nanotechnology **17** (2006)
51. J.E. Stern, B.D. Terris, H.J. Mamin, D. Rugar, Deposition and imaging of localized charge on insulator surfaces using a force microscope. Appl. Phys. Lett. **53**, 2717–2719 (1988)
52. M. Nonnenmacher, M.P. Oboyle, H.K. Wickramasinghe, Kelvin probe force microscopy. Appl. Phys. Lett. **58**, 2921–2923 (1991)
53. S. Bagdzevicius, K. Maas, M. Boudard, M. Burriel, Interface-type resistive switching in perovskite materials. J. Electroceramics **39**, 157–184 (2017)
54. R. Waser, M. Aono, Nanoionics-based resistive switching memories. Nat. Mater. **6**, 833–840 (2007)
55. M.H. Lee et al., Surface redox induced bipolar switching of transition metal oxide films examined by scanning probe microscopy. Appl. Phys. A **102**, 827–834 (2011)

56. C. Yoshida, K. Kinoshita, T. Yamasaki, Y. Sugiyama, Direct observation of oxygen movement during resistance switching in NiO/Pt film. Appl. Phys. Lett. **93**, 042106 (2008)
57. M.H. Lee et al., Scanning probe based observation of bipolar resistive switching NiO films. Appl. Phys. Lett. **97**, 062909 (2010)
58. M. Kumar, T. Som, Structural defect-dependent resistive switching in Cu-O/Si studied by Kelvin probe force microscopy and conductive atomic force microscopy. Nanotechnology **26**, 345702 (2015)
59. Z. Fan et al., Resistive switching induced by charge trapping/detrapping: a unified mechanism for colossal electroresistance in certain Nb:$SrTiO_3$-based heterojunctions. J. Mater. Chem. C **5**, 7317–7327 (2017)
60. S. Kremmer et al., Nanoscale morphological and electrical homogeneity of HfO_2 and ZrO_2 thin films studied by conducting atomic-force microscopy. J. Appl. Phys. **97**, 074315 (2005)
61. V. Iglesias et al., Dielectric breakdown in polycrystalline hafnium oxide gate dielectrics investigated by conductive atomic force microscopy. J. Vac. Sci. Technol. B Microelectron. Nanometer Struct. **29**, 01AB02 (2011)
62. S.J. O'Shea, R.M. Atta, M.P. Murrell, M.E. Welland, Conducting atomic force microscopy study of silicon dioxide breakdown. J. Vac. Sci. Technol. B Microelectron. Nanometer Struct. Process. Meas. Phenom. **13**, 1945–1952 (1995)
63. A. Pradel, N. Frolet, M. Ramonda, A. Piarristeguy, M. Ribes, Bipolar resistance switching in chalcogenide materials. Phys. Status Solidi A **208**, 2303–2308 (2011)
64. J.Y. Son, Y.-H. Shin, Direct observation of conducting filaments on resistive switching of NiO thin films. Appl. Phys. Lett. **92**, 222106 (2008)
65. D. Deleruyelle et al., Direct observation at nanoscale of resistance switching in NiO layers by conductive-atomic force microscopy. Appl. Phys. Express **4**, 051101 (2011)
66. D. Deleruyelle et al., $Ge_2Sb_2Te_5$ layer used as solid electrolyte in conductive-bridge memory devices fabricated on flexible substrate. Solid-State Electron. **79**, 159–165 (2013)
67. K. Szot, W. Speier, G. Bihlmayer, R. Waser, Switching the electrical resistance of individual dislocations in single-crystalline $SrTiO_3$. Nat. Mater. **5**, 312–320 (2006)
68. S. Brivio, G. Tallarida, E. Cianci, S. Spiga, Formation and disruption of conductive filaments in a HfO_2/TiN structure. Nanotechnology **25**, 385705 (2014)
69. M. Schie, M.P. Müller, M. Salinga, R. Waser, R.A. De Souza, Ion migration in crystalline and amorphous HfO_X. J. Chem. Phys. **146**, 094508 (2017)
70. V. Iglesias et al., Degradation of polycrystalline HfO_2-based gate dielectrics under nanoscale electrical stress. Appl. Phys. Lett. **99**, 103510 (2011)
71. U. Celano et al., Filament observation in metal-oxide resistive switching devices. Appl. Phys. Lett. **102**, 121602 (2013)
72. F. Nardi et al., Switching of nanosized filaments in NiO by conductive atomic force microscopy. J. Appl. Phys. **112**, 064310 (2012)
73. A. Ranjan et al., Analysis of quantum conductance, read disturb and switching statistics in HfO_2 RRAM using conductive AFM. Microelectron. Reliab. **64**, 172–178 (2016)
74. Y. Yang et al., Probing nanoscale oxygen ion motion in memristive systems. Nat. Commun. **8**, 15173 (2017)
75. J.Y. Son, Y.-H. Shin, H. Kim, J.H. Cho, H. Jang, Kelvin probe force microscopy for conducting nanobits of NiO thin films. Nanotechnology **21**, 215704 (2010)
76. J. Frascaroli et al., Resistive switching in high-density nanodevices fabricated by block copolymer self-assembly. ACS Nano **9**, 2518–2529 (2015)
77. C. He et al., Multilevel resistive switching in planar graphene/SiO_2 nanogap structures. ACS Nano **6**, 4214–4221 (2012)
78. U. Celano et al., Switching mechanism and reverse engineering of low-power Cu-based resistive switching devices. Nanoscale **5**, 11187 (2013)
79. U. Celano et al., Imaging the three-dimensional conductive channel in filamentary-based oxide resistive switching memory. Nano Lett. **15**, 7970–7975 (2015)
80. U. Celano et al., Nanoscopic structural rearrangements of the Cu-filament in conductive-bridge memories. Nanoscale **8**, 13915–13923 (2016)

81. J. Luria et al., Charge transport in CdTe solar cells revealed by conductive tomographic atomic force microscopy. Nat. Energy **1**, 16150 (2016)
82. M. Buckwell, L. Montesi, S. Hudziak, A. Mehonic, A. Kenyon, Conductance tomography of conductive filaments in intrinsic silicon-rich silica RRAM. Nanoscale **7**, 18030–18035 (2015)
83. J.E. Hudson, H.D. Abruña, Electrochemically controlled adhesion in atomic force spectroscopy. J. Am. Chem. Soc. **118**, 6303–6304 (1996)
84. K.M. Kim, C.S. Hwang, The conical shape filament growth model in unipolar resistance switching of TiO_2 thin film. Appl. Phys. Lett. **94**, 122109 (2009)
85. J.J. Yang, I.H. Inoue, T. Mikolajick, C.S. Hwang, Metal oxide memories based on thermochemical and valence change mechanisms. MRS Bull. **37**, 131–137 (2012)
86. Y. Otsuka, Y. Naitoh, T. Matsumoto, T. Kawai, Point-contact current-imaging atomic force microscopy: Measurement of contact resistance between single-walled carbon nanotubes in a bundle. Appl. Phys. Lett. **82**, 1944–1946 (2003)
87. K. Oka, T. Yanagida, K. Nagashima, H. Tanaka, T. Kawai, Nonvolatile bipolar resistive memory switching in single crystalline NiO heterostructured nanowires. J. Am. Chem. Soc. **131**, 3434–3435 (2009)
88. D.O. Schmidt et al., Resistive Switching of Individual, Chemically Synthesized TiO_2 Nanoparticles. Small **11**, 6444–6456 (2015)
89. M. Uenuma et al., Memristive nanoparticles formed using a biotemplate. RSC Adv. **3**, 18044–18048 (2013)
90. X.Y. Sun et al., A simple implement of submicron meter RRAM array based on porous SiO_2 film with uniform and large pores. Vacuum **132**, 119–122 (2016)
91. S. Brivio et al., Bipolar resistive switching of Au/NiO_x/Ni/Au heterostructure nanowires. Appl. Phys. Lett. **103**, 153506 (2013)
92. J.-M. Song, J.-S. Lee, Self-assembled nanostructured resistive switching memory devices fabricated by templated bottom-up growth. Sci. Rep. **6**, 18967 (2016)
93. U.-B. Han, J.-S. Lee, Bottom-up synthesis of ordered metal/oxide/metal nanodots on substrates for nanoscale resistive switching memory. Sci. Rep. **6**, 25537 (2016)
94. S.-H. Lyu, J.-S. Lee, Highly scalable resistive switching memory cells using pore-size-controlled nanoporous alumina templates. J. Mater. Chem. **22**, 1852–1861 (2012)
95. J. Frascaroli, G. Seguini, S. Spiga, M. Perego, L. Boarino, Fabrication of periodic arrays of metallic nanoparticles by block copolymer templates on HfO_2 substrates. Nanotechnology **26**, 215301 (2015)
96. W.I. Park et al., Self-assembly-induced formation of high-density silicon oxide memristor nanostructures on graphene and metal electrodes. Nano Lett. **12**, 1235–1240 (2012)
97. J.Y. Ye et al., Nanoscale resistive switching and filamentary conduction in NiO thin films. Appl. Phys. Lett. **97**, 132108 (2010)
98. I. Tzouvadaki et al., Label-free ultrasensitive memristive aptasensor. Nano Lett. **16**, 4472–4476 (2016)
99. K. Nagashima et al., Cellulose nanofiber paper as an ultra flexible nonvolatile memory. Sci. Rep. **4**, 5532 (2014)

Chapter 8
Magnetic Force Microscopy for Magnetic Recording and Devices

Atsufumi Hirohata, Marjan Samiepour and Marco Corbetta

Abstract By coating a tip of a non-contacting scanning probe with a magnetic material, scanning probe microscopy can become sensitive to a stray field from the surface of magnetic materials and devices, magnetic force microscopy. The behaviour of such magnetic samples is well-known to be controlled by the formation and reversal of magnetic domains, each of which has a uniform magnetic moment separated by a region with moment rotation, a magnetic domain wall, to minimise total energy. The formation of the magnetic domains and walls is dependent upon size changes even at an atomic scale, which defines a critical length scale in much more strict manner than a semiconductor and metallic sample. It is therefore important to image magnetic domain structures of the magnetic samples precisely to reveal the corresponding performance. Magnetic force microscopy is one of the most convenient techniques for magnetic imaging with nanometric resolution as detailed in this chapter.

8.1 Introduction

In the development of new advanced magnetic materials and devices, it is critical to evaluate their properties not only in their bulk form but also in the form to be used, such as multilayered junctions, nanoparticles and nanocomposites. Recent progress in nanotechnology enables the production of atomically flat or abrupt interfaces in multi-layered junctions. The progress has been, for instance, allowed us to continue the increase in the number of transistors in a processor, as known as Moore's law. Similar increase has been achieved in the other electronic devices, such includes data storage and communication. However, we have not achieved uniform electron

A. Hirohata (✉) · M. Samiepour
Department of Electronic Engineering, University of York, Heslington, York YO10 5DD, UK
e-mail: atsufumi.hirohata@york.ac.uk

M. Samiepour
e-mail: marjan.samiepour@york.ac.uk

M. Corbetta
NanoScan AG, Überlandstrasse 129, Dübendorf 8600, Switzerland
e-mail: m.corbetta@nanoscan.ch

U. Celano (ed.), *Electrical Atomic Force Microscopy for Nanoelectronics*, NanoScience and Technology, https://doi.org/10.1007/978-3-030-15612-1_8

transport across the entire interfacial area in such junctions due to the existence of local defects, which introduce differences in resistance. The defects are hence known to cause local heating and reduction in transport efficiency. To date, structural junction uniformity has been predominantly assessed by cross-sectional transmission electron microscopy (TEM) in atomic resolution. On the other hand, the corresponding imaging of magnetic information with similar atomic resolution is still under development.

8.1.1 Magnetic Imaging

The magnetic imaging has been achieved via interactions between a probe and magnetic fields (or magnetisation) of a specimen [1]. In the Bitter method, which is one of the most commonly used magnetic powder patterns, colloidal magnetite particles are used as a probe to form domain patterns due to the stray field from the sample surface. This is still an important method because of the simplicity and high sensitivity (~sub-micron resolution within a ~sub-micron subsurface region), however, it makes sample surface dirty.

For the deeper imaging, magneto-optical (MO) imaging can be used as listed in Fig. 8.1. MO effects, MO Faraday effect and Kerr effect (MOKE) can be used to observe magnetic domain configurations. Samples need to be flat and transparent for the Faraday effect observation. The MOKE also requires a flat surface but can measure local hysteresis and dynamic behaviour. These techniques can offer magnetic information up to about 20 nm (on metallic materials) below the surface with a resolution down to approximately 200 nm.

In order to improve the resolution of magnetic imaging, electron-beam and X-ray techniques have been used as an imaging probe. Among them, one of the most popular technique is Lorentz transmission electron microscopy (Lorentz TEM). In out-of-focus (defocused) mode, electrons are refracted within a sample due to the Lorentz force, and the interference contrast corresponds to the magnetisation distributions of the sample.

Using a similar TEM setting, electron holography can also be realized. The phase distributions of electrons transmitted through a specimen are recorded as an off-axis hologram using a collimated electron beam. The hologram can be reconstructed by using an interferometer [2].

On the other hand, using scanning electron microscopy (SEM), an electron beam (<10 keV) introduced perpendicular to the conductive sample generates secondary electrons, whose intensity corresponds to the stray field from the sample. Here, backscattered electrons can also be used for magnetic imaging.

Scanning electron microscopy with polarisation analysis (SEMPA) has also been used for high-resolution magnetic imaging. Since the secondary electrons from a sample are spin-polarised, these electrons are collected and measured as a normalised difference signal in a Mott detector. This method is sensitive to the surface magnetisation vector, and can be *in situ* [3].

By replacing electron-beam with X-ray, X-ray topography can be achieved. Aligning a slit to fulfil the Bragg condition with a monochromatic X-ray, the crystalline imperfection due to magnetostriction can be detected, *e.g.*, X-ray magnetic circular dichroism (XMCD) and X-ray photo-emission electron microscopy (XPEEM),

8.1.2 Magnetic Force Microscopy

In comparison to the above techniques listed in Sect. 8.1.1, magnetic force microscope (MFM) is one of scanning probe microscopes (SPMs) and is sensitive to a magnetic stray field from the sample (within 100 nm below the surface) by coating a tip for an atomic force microscope with a ferromagnetic layer. The development of MFM has started in 1980s, followed by the introduction of a commercial system in the early 1990. MFM is an easy method with the resolution better than ~10 nm without requiring any preparation for imaging among conventional magnetic imaging methods. This is why MFM has been widely used for magnetic imaging in fundamental research and device development. The details of the working principles and image examples will be described in Sects. 8.2 and 8.3, respectively.

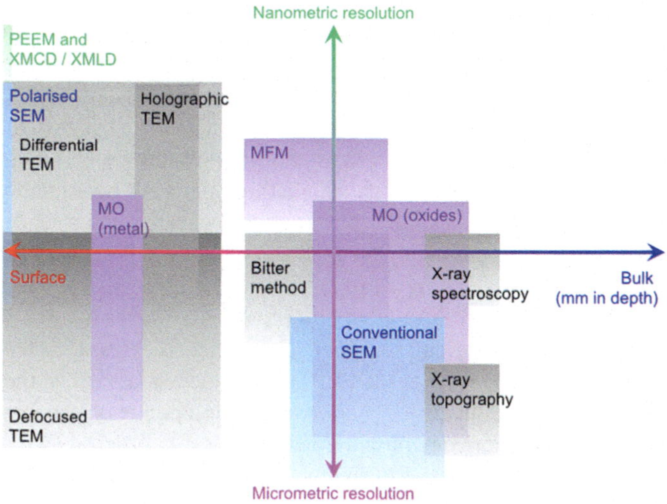

Fig. 8.1 Major techniques for the magnetic imaging against penetration depth of probes. After [1]

8.1.3 Other SPM-Based Magnetic Microscopy

8.1.3.1 Spin-Polarised Scanning Tunnelling Microscopy

Based on scanning tunnelling microscopy (STM) with a ferromagnetic tip, spin-polarised (SP-STM) has been developed. SP-STM was proposed in 1990 by Johnson and Clarke [4] and Wiesendanger et al. [5]. Spin-polarised secondary electrons can be generated using STM in the field-emission mode [6], which holds promise for nanoscale magnetic imaging. By comparing the signals of ferromagnetic (FM) and non-magnetic (NM) STM tips, atomic scale magnetic imaging can be achieved [5]. These reports were theoretically supported by Molotkov [7] and Laiho and Reittu [8] in 1993. This technique uses a direct-gap semiconductor tip. This is expected to be used to observe the surface magnetic configurations with an almost atomic resolution. Spin-polarised electron tunnelling from a Ni STM tip into a GaAs substrate was first demonstrated by Alvarado and Renaud [9]. The Ni tip is magnetised by an electromagnet and is used as a spin injector. It scans over the sample surface in its measurement state. Spin-polarised electron tunnelling through the vacuum is detected as circularly polarised EL signals in which the change is ~30% at room temperature (RT). This value corresponded to a minority electron spin polarisation of Ni(001) at the Fermi level. This suggests that minority spin electrons provide the dominant contribution to the tunnelling current.

After the first photoexcitation measurement by Prins et al. [10], modulated circularly polarised light has been used to excite spin-polarised electrons in a semiconductor (e.g., GaAs). Although optically excited electrons are scattered mainly at the semiconductor surface with back illumination [11], Sueoka et al. have demonstrated the possibility of detecting spin-polarised signals by scanning a Ni STM tip over a GaAs film with circularly polarised light shone through an AlGaAs membrane [12]. Suzuki et al. have performed a similar observation by scanning a *p*-GaAs STM tip over a Co film with back illumination through a mica/Au/Co film, and have obtained magnetic domain images [13]. GaAs tips are fabricated using photolithography and anisotropic etching to prevent limitations due to facets {105}. A three monolayer (ML) Co film exhibits perpendicular magnetisation and shows less than the magnetic circular dichroism (MCD) effect of 0.14%. This is much smaller than the observed polarisation response of about 10%. Polarisation modulation response images of the SP-STM show very good agreement with MFM images. In order to avoid the MCD effect and possible light scattering through the sample structures, Kodama et al. introduced photon helicity into a GaAs tip in the vicinity of the sample, which is equivalent of front illumination [14]. They detected a change of approximately 7% in Current-voltage (*I*-*V*) curves between right and left circular light irradiation of NiFe films.

8.1.3.2 Scanning Hall Probe Microscopy

A submicron Hall probe associated with precise STM positioning can be used to sense a stray field from a sample for scanning Hall probe microscopy (SHPM) [15]. This method is very sensitive to a magnetic field (~10e), however, it can be operated only at low temperature ($\leq$90 K). This technique can be used to detect a local perpendicular magnetic field from the surface of a sample [16]. The SHPM can also image magnetic induction under applied fields up to ~1 kOe and over a wide range of temperatures (a few 100 mK–300 K) which make this method suitable to study the effect of temperature on some phenomenon such as a domain wall structure in a ferromagnetic semiconductor and conventional vortex patterns on superconductors. In SHPM a nano-Hall probe with dimensions of ~50 nm $\times$ 20 nm has been fabricated by a combination of optical lithography and focused ion beam milling [17]. The probe is mounted on a piezoelectric tube (PZT) at a small angle with respect to the surface of a sample to minimise the distance between the probe and the sample surface during the measurement. By decreasing the distance between the Hall probe and the sample surface to 500–700 nm, the field resolution of the Hall can achieve 0.1Oe. The scanning area of a single image is between 14 $\times$ 14 (at 4 K) and 24 $\times$ 24 (at 30 K) μm^2 [18].

8.1.3.3 Scanning Superconducting Quantum Interference Device Microscopy

A superconducting quantum interference device (SQUID) can be used as a field detector. This SQUID microscopy is typically only valid at 4.2 K [19]. SQUID is a direct method to detect magnetic flux using a pick-up loop at low temperature (4 K). By measuring the magnetic flux in a mesoscopic ring, the current which is first derivative of the free energy of the ring dependent on the magnetic flux can be extracted as a fundamental thermodynamic property. SQUID is one of the most reliable instruments to measure a magnetic field. Traditionally the pick-up loop and SQUID sensor are used in flip-chip geometry. SQUID susceptometer has first been produced by Ketchen et al. [20]. The smallest current measured is reported to be 100 mA at 4.2 K. In 1993, a special SQUID technique has been developed where the sample and the SQUID sensor are patterned on the same chip [21]. The measured current of (4 $\pm$ 2) nA is comparable to the computed value of 5 nA in liquid Helium. In all these experiments, both the SQUID sensor and the sample are maintained at 4.2 K initially but only the sample is elevated up to room temperature. At room temperature, the sample has been separated from the SQUID loop with a thin window in a cryostat. Here, the spatial resolution plays an important role. In 2001, a German group has achieved a spatial resolution of 1 $pT/\sqrt{Hz}$ using a 50 μm thick sapphire window [22].

Furthermore, SQUID is a non-destructive technique which can also be used to detect the remanent magnetisation such as scratches, dents and mechanical stress in some stainless steels. A scanning SQUID susceptometer on a micrometric scale objects has been reported in 2008 [23]. In this design, the pick-up loop (4 μm in diameter) and a modulation coil are separated to reduce cross coupling between them, allowing better coupling between the pick-up loop and the sample. The ability to isolate the sensor from the samples and background subtraction maximises the device utility and sensitivity. The completed model of the response of SQUID with sub-micron pick-up loops to different sources of magnetic fields has been calculated [24]. In this method, the coupling with London's and Maxwell's equations for a full geometry of measured magnetic flux is solved. The result of this calculation is in good agreement with experimental imaging with a deep sub-micron sized pick-up loop.

8.1.3.4 Scanning Near-Field Optical Microscopy

By introducing light through a submicron aperture, the diffraction limit of light can be circumvented, which provides an image 10–50 times better than a conventional optical image, known as scanning near-field optical microscopy (SNOM). Combining the SNOM with MOKE optics, both imaging and writing magnetic domains can be achieved [25]. SNOM is used to overcome diffraction in classical optical microscopy which limits spatial resolution. In SNOM, the excitation laser focuses in an aperture with diameter smaller than the excitation wavelength. The sample locates at small distance (near-field) below the aperture. The optical resolution is in the order of 30–50 nm [26]. This method made good progress on nanotechnology research, surface structures and bio-materials. By the microfabrication of SNOM probes, the resolution of the images can be better by a factor of 5–10 as compared with the classical microscopy [27]. Even though SNOM can achieve high-resolution imaging, SNOM requires challenging processes for the fabrication of the aperture since its pioneering demonstration in 1989. Fisher and Pohl have used the scattering of surface plasmons in a metallic film instead of aperture to image a sample in the near-field [28]. The main difficulty of this method is a very weak signal due to large illumination background and the influence of an extended part of tips other than just far extreme of the tip. One of reliable procedures is to use a single gold nanoparticle on very sharp glass fibre, achieving the resolution of an optical measurement of 100 nm [29]. A new magnetic microscope was introduced by a combination of SNOM and magneto-optical contrast which is called time-resolved near-field scanning magneto-optical microscope [30]. This microscope is used to study ultrafast magnetisation dynamics on sub-nanoseconds. J. Rudge et al. have measured a gyrotropic resonance frequency of vortex core in a micron-sized CoFeB disc at ~240 MHz. Recently, time-resolved scanning Kerr microscopy in the near-filed became one of intense methods to investigate magnetisation dynamics [31].

8.2 Principles of Non-contact/Tapping Mode

8.2.1 Principles of Non-contact Atomic Force Microscopy

Scanning probe microscopy (SPM) provides opportunities to obtain various kinds of information, such as height, magnetic pole and friction, using interactions between a probe and sample surface. As detailed in Chap. 1, the first SPM is scanning tunnelling microscopy (STM) invented by Binnig et al. [32]. By scanning a very sharp conducting tip in the vicinity of a sample surface with a bias application between the tip and the sample, a tunnelling current I_t can be detected as $I_t \sim V\exp(-Cd)$, where V is the bias voltage, C is a constant of the sample material and d is the distance between the lowest atom of the tip and the highest atom on the sample surface. Accordingly, I_t can be used in a feedback loop to control the precise movement of a piezoelectric scanner, which can be directly converted as topographic information. The ability of STM to measure surface morphology is clearly outstanding (atomic resolution). However, the most severe limitation is that both the tip and the sample must be almost perfect conductors.

In order to avoid the limitation in STM, Binnig et al. have invented atomic force microscopy (AFM) [33]. AFM also uses a very sharp tip for scanning over a sample but with a long cantilever with a low spring constant (~1 N/m). Since the tip senses interatomic force (van der Waals force), the cantilever is bent, the magnitude of which is detected as an electrical signal from a 2-segment photodiode using laser light reflection.

When the tip approaches the sample surface, the tip first feels the attractive force and is bent towards the sample surface. With further approach the sample surface, the tip encounters the repulsive force from the sample. At a certain distance these two forces are balanced, at which distance the AFM measures the topographical profile. One way of measuring the profile is to maintain the distance between the tip and the sample, and detect the amplitude of the force via the magnitude of the cantilever distortion (constant height mode). Another way is to keep the force constant, and measure the movement of the tip (constant force mode).

In AFM observation, well-known problems which one should be aware of are the large repulsive force from a contamination layer on the sample surface, the influence of statistic electronics from either the sample or the probe, and the friction between the sample and the tip during the AFM scan. To avoid such issues, Martin et al. have developed non-contact AFM (NC-AFM) [34]. When a probe resonating with small amplitude ($\ll$1 nm) is scanned 5–10 nm above the sample surface, the resonating frequency ω_0 shifts to ω_0' due to the atomic force F from the sample. If we take the sample plane normal as z axis, ω_0' is given by

$$\omega_0' = \omega_0\sqrt{1 - \frac{1}{k_z}\left(\frac{\partial F}{\partial z}\right)},$$

where k_z is the spring constant of the cantilever. In the Q-curve of the cantilever, a frequency ω_D^+, which satisfies

$$\omega_D^+ = \omega_0\left(1 + \frac{1}{\sqrt{8Q}}\right).$$

ω_D^+ is typically used for non-contact AFM observation. This provides an amplitude shift of

$$\Delta A = \frac{2A_0 Q}{3\sqrt{3}k_z}\left(\frac{\partial F}{\partial z}\right).$$

When the probe is scanned as ΔA constant, topographical information can be obtained similarly to AFM. The resolution of NC-AFM typically depends upon the tip height from the sample, which is slightly less than that for contact AFM (5–10 nm [35]).

8.2.2 Principles of Magnetic Force Microscopy

With a magnetic tip magnetised in one direction, NC-AFM can become sensitive to magnetic force [36]. There are relationships between tip-sample distance z and force gradient $F' = \partial F/\partial z$ as shown in Fig. 8.2 [37]. In the distance range of $5 < z < 20$ nm, the tip mainly detects atomic force (van der Waals' force), while the tip is sensitive to magnetic force with $50 < z < 500$ nm. Magnetic force microscopy (MFM) senses magnetic force gradients from a sample due to the perpendicular component of the sample stray field. The cantilever is excited at its first longitudinal resonance (typically 50 kHz) and magnetic interaction with the surface is detected as shift of the resonance (typically of a few Hz, detected using phase-locked loop technology).

A magnetic probe, whose tip is magnetised perpendicular to sample surface, should be mounted on a tip holder. A schematic phase curve of the cantilever is shown in Fig. 8.3. The phase curve decreases with increasing frequency, and the 90° phase lag point (at the centre of the curve) corresponds to the peak frequency of the cantilever. The phase curve then measures the phase lag between the drive voltage and the cantilever response, and hence the vertical gradients in the magnetic force cause a shift ΔF_0 in the resonance frequency. This resonance shift ΔF_0 gives rise to a phase shift $\Delta\phi$, which provides an image of the magnetic force gradients (phase detection). Typically, phase detection is used to obtain MFM images because phase detection results are generally superior to those obtained with amplitude detection. Frequency modulation, which provides quantitative images of force gradients by directly detecting the drive frequency at 90° phase lag, can also be achieved. Further details about the tip-sample interactions can be found in [38]. The ΔF_0 also gives variations in oscillation amplitude ΔA, which can be used for imaging the magnetic force gradient (amplitude detection) (see Fig. 8.4 [39]).

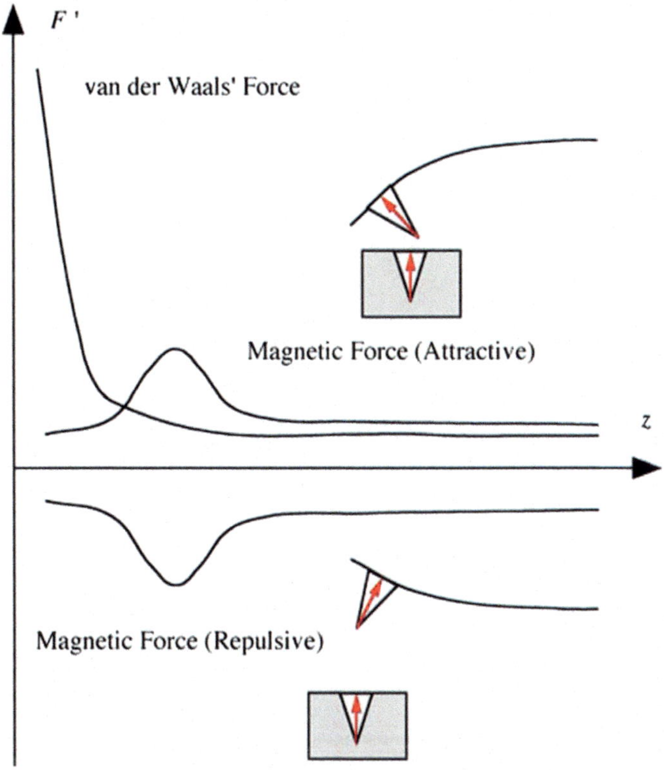

Fig. 8.2 Schematic energy diagram of force gradients for both van der Waals' and magnetic force

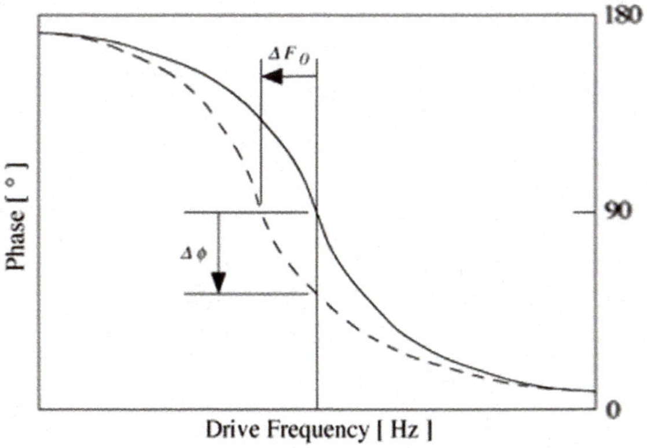

Fig. 8.3 Phase shift of a MFM cantilever at fixed drive frequency. After [39]

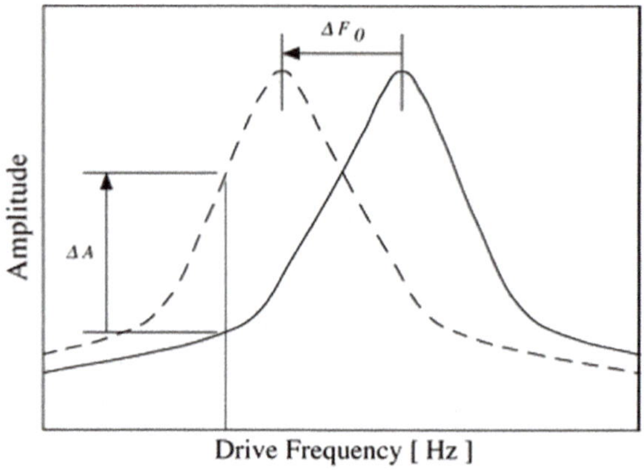

Fig. 8.4 Amplitude shift of a MFM cantilever at fixed drive frequency. After [39]

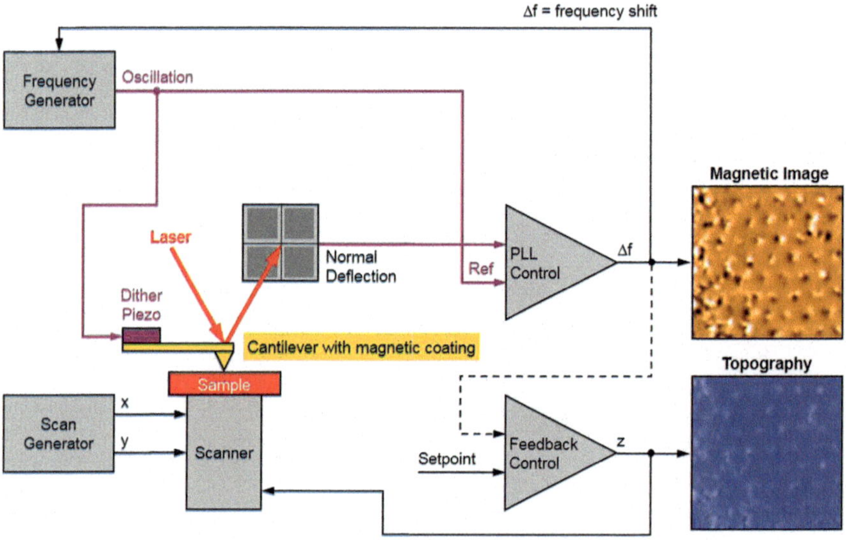

Fig. 8.5 Block diagram for MFM measurements

In MFM observation (see Fig. 8.5), a tapping cantilever first scans over the sample surface to obtain topographic information. Using lift mode as shown in Fig. 8.6, the tip is then raised just above the sample surface, where lift scan height h_{lift} is typically set as 100 nm unless samples have excessive surface height variations, and the surface topography is profiled while responding to magnetic influence on the same trace line.

In the lift mode, the total tip-sample distance h_{total} is the sum of the average tip-sample distance in the tapping mode h_{t} and the lift scan height h_{lift} as shown in Fig. 8.6. h_{t} is equal to the oscillation amplitude of the MFM cantilever, which is

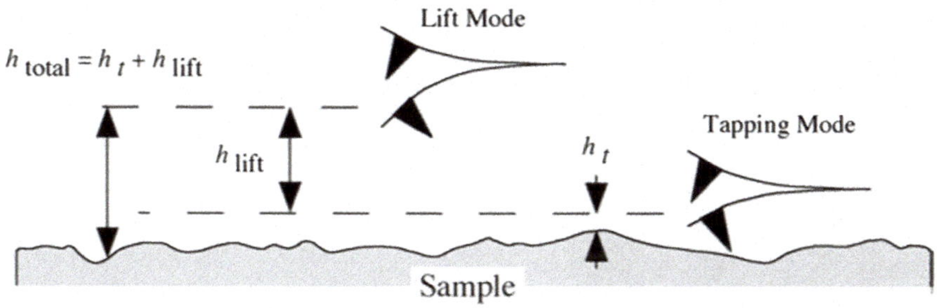

Fig. 8.6 Tip-sample distance and oscillation amplitude in both the tapping and the lift mode for MFM measurements

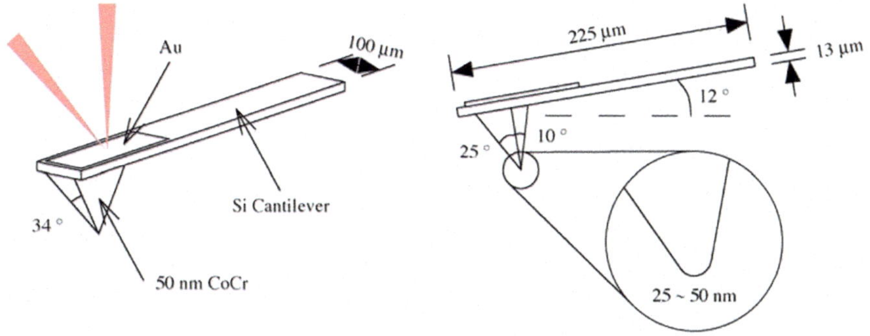

Fig. 8.7 Schematic views of typical magnetic force microscopy probe

determined by a setpoint. Typically, the setpoint of the measurements was chosen to be 1 V, which typically corresponds to h_t ~ 15–20 nm.

8.3 Magnetic Tips and Specifications

8.3.1 Magnetic Tips

Figure 8.7 shows schematics of a typical MFM probe. A standard etched Si probe with a CoCr coated tip is commonly used for MFM imaging. Specifications are listed in Table 8.1. The back of the cantilever has a reflective coating for better resolution. Here, the laser spot should be introduced around the centre of the cantilever to avoid optical interference.

In MFM, a Si-based cantilever is bent by attractive force between the different polarity, *i.e.*, between the N- and S-poles, and repulsive force between the same polarity, i.e., between the N- and N-poles or the S- and S-poles, occurring between the ferromagnetic tip located at the end of the cantilever and a stray field from

Table 8.1 List of commercial MFM tips

	Cantilever thickness (μm)	Cantilever width (μm)	Cantilever length (μm)	Force constant (N/m)	Resonance frequency (kHz)	Coating	Tip radius (nm)	Tip height (μm)	Coercivity (Oe)	Remanence magnetisation (emu/cm^3)	Effective magnetic moment (emu)	Imaging resolution (nm)
Bruker												
MESP	3 (2–3.5)	28 (23–33)	225 (200–250)	2.8 (1–5)	75 (50–100)	Co-Cr	35	10–15	~300		~10^{-13}	
MESP-V2	2.80 (2.05–3.55)	35 (33–37)	225 (215–235)	3.0 (1.5–6.0)	75 (50–100)	Co-Cr	35	10–15	~300		~10^{-13}	
Standard moment MFM probe with a conductive coating (MESP-CPMT)	2.75 (2–3.5)	28 (23–33)	225 (200–250)	2.8 (1–5)	75 (50–100)	Co-Cr	20	10–15	~400		~10^{-13}	
High moment MFM probe (MESP-HM)	2.75 (2–3.5)	28 (23–33)	225 (200–250)	2.8 (1–5)	75 (50–100)	Co-Cr	80	10–15	~400		3×10^{-13}	

(continued)

Table 8.1 (continued)

	Cantilever thickness (μm)	Cantilever width (μm)	Cantilever length (μm)	Force constant (N/m)	Resonance frequency (kHz)	Coating	Tip radius (nm)	Tip height (μm)	Coercivity (Oe)	Remanence magnetisation (emu/cm^3)	Effective magnetic moment (emu)	Imaging resolution (nm)
High moment MFM probe (MESP-HM-V2)	2.80 (2.05–3.55)	35 (33–37)	225 (215–235)	3.0 (1.5–6.0)	75 (50–100)	Co-Cr	80	10–15	~300		~10^{-13}	
High moment MFM probe with a conductive coating (MESP-HMW)	2.75 (2–3.5)	28 (23–33)	225 (200–250)	2.8 (1–5)	75 (50–100)	Co-Cr	80	10–15	~400		~10^{-13}	
Low coercivity MFM probe with a conductive coating (MESP-LC)	2.75 (2–3.5)	28 (23–33)	225 (200–250)	2.8 (1–5)	75 (50–100)		35	10–15	<10		<10^{-13}	

(continued)

Table 8.1 (continued)

	Cantilever thickness (μm)	Cantilever width (μm)	Cantilever length (μm)	Force constant (N/m)	Resonance frequency (kHz)	Coating	Tip radius (nm)	Tip height (μm)	Coercivity (Oe)	Remanence magnetisation (emu/cm^3)	Effective magnetic moment (emu)	Imaging resolution (nm)
Low moment MFM probe with a conductive coating (MESP-LMW)	2.75 (2–3.5)	28 (23–33)	225 (200–250)	2.8 (1–5)	75 (50–100)		25	10–15	<400		3×10^{-14}	
MESP-LM-V2	2.80 (2.05–3.55)	35 (33–37)	225 (215–235)	3.0 (1.5–6.0)	75 (50–100)		25	10–15	<400		3×10^{-14}	
Nanosensor												
PointProbe® Plus Magnetic Force Microscopy Probe (PPP-MFMR)	3	28	225	2.8	75	Hard magnet	<50		~300	~300	~10^{-13}	<50

(continued)

Table 8.1 (continued)

	Cantilever thickness (μm)	Cantilever width (μm)	Cantilever length (μm)	Force constant (N/m)	Resonance frequency (kHz)	Coating	Tip radius (nm)	Tip height (μm)	Coercivity (Oe)	Remanence magnetisation (emu/cm^3)	Effective magnetic moment (emu)	Imaging resolution (nm)
PointProbe® Plus Low Moment Magnetic Force Microscopy Probe (PPP-LM-MFMR)	3	28	225	2.8	75	Hard magnet	<30		~250	~150	1/2 of PPP-MFMR	<35
PointProbe® Plus Low Coercivity Magnetic Force Microscopy Probe (PPP-LC-MFMR)	3	28	225	2.8	75	Soft magnet	<30		~0.75	~225	3/4 of PPP-MFMR	<35

(continued)

Table 8.1 (continued)

	Cantilever thickness (μm)	Cantilever width (μm)	Cantilever length (μm)	Force constant (N/m)	Resonance frequency (kHz)	Coating	Tip radius (nm)	Tip height (μm)	Coercivity (Oe)	Remanence magnetisation (emu/cm^3)	Effective magnetic moment (emu)	Imaging resolution (nm)
SuperSharp SiliconTM High Resolution Magnetic Force Microscopy Probe (SSS-MFMR)	3	28	225	2.8	75	Hard magnet	<15		~125	~80	1/4 of PPP-MFMR	<25
MikroMasch												
High Quality (HQ): Magnetic Noncontact (NSC) silicon probe 18	3 ± 0.5	27.5 ± 3	225 ± 5	2.8 (12–5.5)	75 (60–90)	Co(60 nm)/Cr (20 nm)	<60	12–18	300–400			

(continued)

Table 8.1 (continued)

	Cantilever thickness (μm)	Cantilever width (μm)	Cantilever length (μm)	Force constant (N/m)	Resonance frequency (kHz)	Coating	Tip radius (nm)	Tip height (μm)	Coercivity (Oe)	Remanence magnetisation (emu/cm^3)	Effective magnetic moment (emu)	Imaging resolution (nm)
High Quality (HQ): Magnetic Noncontact (NSc) silicon probe 36	3 ± 0.5	32.5 ± 3	110 ± 5 90 ± 5 130 ± 5	1.0 (0.1–4.6) 2.0 (0.2–9.0) 0.6 (0.06–2.7)	90 (30–16) 130 (45–240) 65 (25–115)	Co(60 nm)/Cr (20 nm)			300–400			
NT-MDT												
High resolution long lifetime magnetic probe (MFM01)	2.5 ± 0.5	32 ± 5	225 ± 10	3 (1–5)	70 (47–90)	Co-Cr (30–40 nm)						20–30
Low Moment High Resolution Magnetic tip (MFM_LM)	2.5 ± 0.5	32 ± 5	225 ± 10	3 (1–5)	70 (47–90)	Co-Cr (15–20 nm)			~20			

(continued)

Table 8.1 (continued)

	Cantilever thickness (μm)	Cantilever width (μm)	Cantilever length (μm)	Force constant (N/m)	Resonance frequency (kHz)	Coating	Tip radius (nm)	Tip height (μm)	Coercivity (Oe)	Remanence magnetisation (emu/cm^3)	Effective magnetic moment (emu)	Imaging resolution (nm)
High Coercivity Magnetic Probe (MFM_HC)	2.5 ± 0.5	32 ± 5	225 ± 10	3 (1–5)	70 (47–90)	Co-Cr (~ 60 nm)			500–800 in vertical and >2 k in horizontal directions			
NanoWorld												
Magnetic force microscopy - reflex coating (MFMR)	3 (2.5–3.5)	28 (22.5–32.5)	225 (200–230)	2.8 (1.2–5.5)	75 (60–90)	Co alloy						
Soft magnetic coating - magnetic force microscopy - reflex coating (S-MFMR)	3 (2.5–3.5)	28 (22.5–32.5)	225 (200–230)	2.8 (1.2–5.5)	75 (60–90)	Soft magnet			~0.75	~225		

(continued)

Table 8.1 (continued)

	Cantilever thickness (μm)	Cantilever width (μm)	Cantilever length (μm)	Force constant (N/m)	Resonance frequency (kHz)	Coating	Tip radius (nm)	Tip height (μm)	Coercivity (Oe)	Remanence magnetisation (emu/cm^3)	Effective magnetic moment (emu)	Imaging resolution (nm)
Team Nanotec												
High resolution magnetic probe tip (ML 1)				3.0 or 0.7	75 or 45	~20 nm thick Co-alloy hard magnet	15	>9				
High resolution magnetic probe tip (ML 3)				3.0 or 0.7	75 or 45	~37 nm thick Co-alloy hard magnet	20	>9				

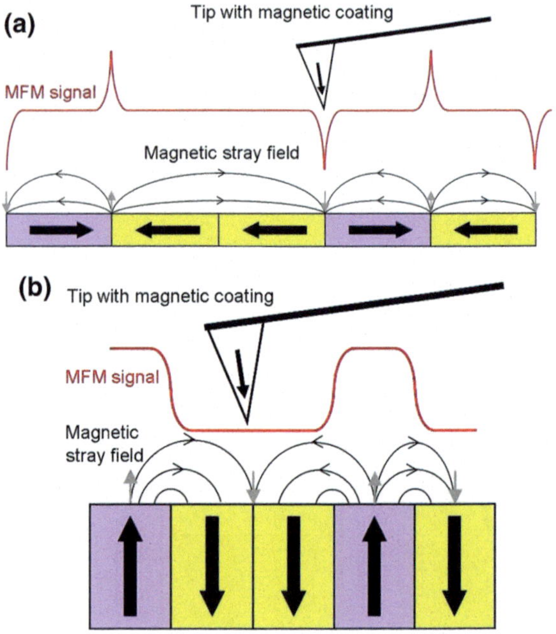

Fig. 8.8 Principle of magnetic force microscopy over a sample surface with **a** in-plane and **b** perpendicular magnetisation. The magnetically-coated tip detects the vertical component of the stray field (grey arrows) emanating from the surface. Hence, the MFM signal (in red) exhibits peaks at the domain boundaries

a specimen. The tip has the height of ~10 μm with the curvature of 25–50 nm, which is coated by 10–100 nm thick ferromagnetic layers, *e.g.*, CoCrPt and CoTa, as detailed in Sect. 8.2. By detecting the displacement of the cantilever by the attractive and repulsive force via a photodetector, magnetic imaging can be achieved. Here, the force F between the MFM tip with the magnetisation m and the stray field from the specimen H can be defined as $F = \nabla(m \cdot H)$. By directly detecting the force, static MFM can be achieved, which is relevant for a large stray field (>6 kOe [40]) from a specimen, e.g., bulk permanent magnets. However, a stray field from a thin-film device (<2 Oe [40]), such as hard disk media and magnetic memory, is too small to be detected. Hence, dynamic MFM has been developed by vibrating the cantilever perpendicular to the specimen plane to detect changes in the amplitude, phase and frequency of the vibration as described in Sect. 8.2.2. Here, the gradient of a magnetic force ∇F_z from an in-plane magnetised specimen is detected as shown in Fig. 8.8a. As shown in Fig. 8.8a, the MFM signals are reversed depending on the attractive and repulsive force. For the case of perpendicularly magnetised specimen, the representative schematics are shown in Fig. 8.8b.

In MFM images, brighter contrast means positive frequency shift which in turn depicts repulsive interaction or a magnetic stray field of the surface anti-parallel to the magnetisation of the tip. Darker contrast translates as negative frequency shift, or attractive interaction and a magnetic stray field of the surface parallel to the tip magnetisation.

8.3.2 Improvement of Specifications

Magnetic imaging is performed by detecting displacement of Si-based cantilever with a MFM tip coated with a ferromagnetic layer [41–46]. In order to increase the imaging resolution, the distance between the MFM tip and the specimen needs to be closer but it increases the contributions from the specimen topology. Hence, for the high-resolution imaging, a sharper tip [43–47] and improvement of the field gradient [45, 46] are necessary. As listed in Table 8.1, commercial MFM tips have their resolution of 20–40 nm, which requires further improvement for high-density recording media. In addition, for highly anisotropic recording media and permanent magnets, the magnetisation of the MFM tip is affected by the stray fields from these specimens [48]. It is therefore essential to achieve both high resolution and high coercivity for imaging [49, 50], requiring a highly anisotropic material as a ferromagnetic coating layer on the MFM tip.

By introducing an alternative magnetic field to a ferromagnetic nanoparticle attached on a MFM tip, alternating magnetic force microscopy (A-MFM) has been developed with higher resolution. The alternative field has a frequency different from the resonant frequency of the cantilever, so that alternating gradient field can be generated by the nanoparticle, which can be detected from the highly coercive MFM tip independently. This method is sensitive to a stray field near the specimen surface with the resolution better than 10 nm [51].

8.4 Applications for Magnetic Recording

Since the 1970s, commercial mass data storage has been available with continuously diminishing costs and increasing storage capacity. The old eight-inch floppy disk displayed a storage capacity of 250 kB, which today might seem like an incredibly low capacity, however large-scale production was a technological prowess back then.

From the first floppies to the currently available terabyte hard disk drives (HDDs), the principle of data storage continues to be the same: the active component is a magnetic layer that stores information in the form of 0 s and 1 s by being magnetised locally in one direction or its opposite. The layer is placed on a spinning disk that can be written and read by a mobile head.

One remarkable progress in magnetic media recording was accomplished with the discovery of the Giant Magnetoresistance effect (GMR) by Albert Fert and Peter Grünberg, which got him the Nobel Prize in 2007. This effect facilitated moving from in-plane magnetic media recording to perpendicular media recording (PMR), thus expanding the storage capacities by over three orders of magnitude. Technologists today are discovering new tricks to improve density and pack more terabytes in hard disks measuring just a couple of inches wide.

Fig. 8.9 Photograph of a floppy disk

Fig. 8.10 75 × 200 μm^2 MFM image of a floppy disk. Magnetisation is in-plane, contrast arises from the magnetic stray field at the domain boundaries. The orange square illustrates the dimensions of Fig. 8.12 (25 × 25 μm^2 MFM image of a zip drive)

8.4.1 3.5-in. Floppy Disk Introduced in 1987

In the 1970s, magnetic storage made its appearance and a decade later became a widespread consumer good (Fig. 8.9). The bits of an 8 in. floppy disk were macroscopic so that they can hardly be imaged with an MFM. Since 1987, the next-generation floppies, the 3.5-in. disks, already had bits that had shrunk to a microscopic 2 × 100 μm^2, as shown in the MFM image as shown in Fig. 8.10.

Here, magnetisation in this case is oriented in the plane of the layer and hence the MFM detects the vertical magnetic stray field originating at domain boundaries, as shown in Fig. 8.8a.

The floppy disk shown here is a "High Density" disk with 1.44 MBytes of storage, or 11.52 Mbits (1 Byte = 8 bits) dispatched on a 3.5 in. platter. Taking into consideration the one-inch rotor, that makes an area of 8.84 in.^{2} of active layer. However, not the entire area is available for data storage: for example the rims of the disk cannot be written and formatting also occupies space. Counting the bits in the 75 × 200 μm^2 MFM image (approximately 37 bits), a density of 1.6 Mbit/in.^{2} can be established. Thus, 7.2 in.^{2} or 81% of the whole surface is requested to store 11.52 Mbits of infor-

Fig. 8.11 Photograph of a zip disk

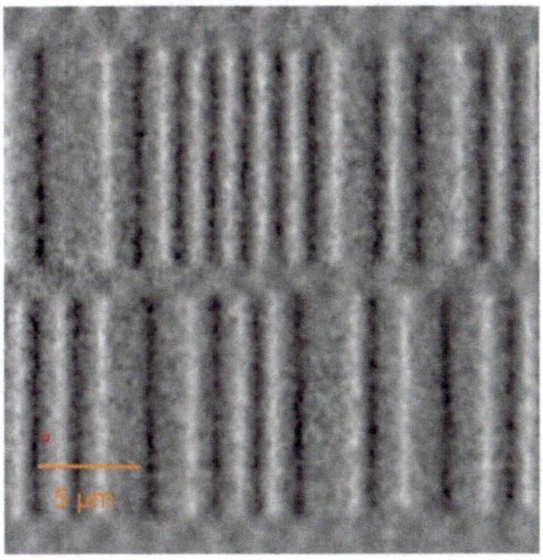

Fig. 8.12 $25 \times 25\ \mu m^2$ MFM image of a zip drive. Magnetisation is again in-plane. The red square illustrates the scale of Figs. 8.14, 8.15, 8.16 and 8.17 ($500 \times 500\ nm^2$ MFM image of HDDs)

mation. This leads to the areal density of 1.6 Mbit/in.^{2}, to put the number to scale with the following Hard Disk Drives that is a density of 0.0016 Gbit/in^2.

8.4.2 *Zip Drive Introduced in 1994*

After the arrival of the floppy disk, the zip drive was introduced in 1994. The size of the medium was still a 3.5 in. platter, and magnetic information was still stored in-plane, but due to its decreased bit size, the storage capacity could be expanded to 100 MB or 800 Mbits (Fig. 8.11).

In Fig. 8.12 showing a MFM image of $25 \times 25\ \mu m^2$, the bit dimension is assessed to be $1 \times 10\ \mu m^2$, which is 20 times smaller than in the floppy disk.

On the MFM data, 54 bits are counted, amounting to a density of 56 Mbit/in.^{2}, which makes 403 Mbits on the available 7.2 in.^{2} (assuming the same active area as on a floppy disk) or approximately 800 Mbits when considering both faces of the

Fig. 8.13 Photograph of HDD

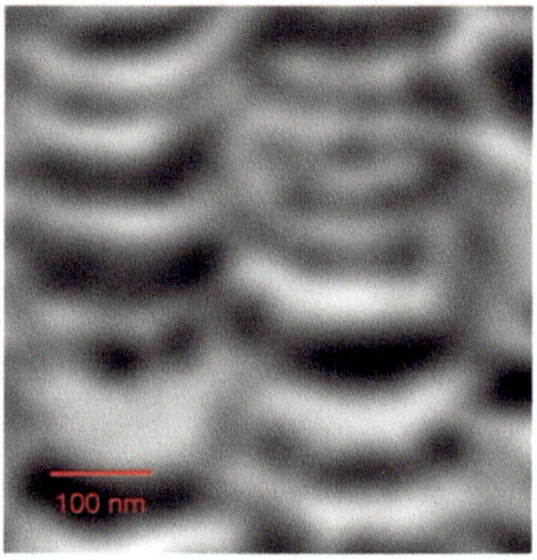

Fig. 8.14 500 × 500 nm^2 MFM image of a Fujitsu HDD from 2007. Magnetisation is perpendicular to the surface

zip. This figure is in seamless agreement with the Manufacturer's specifications. This leads to the areal density of 56 Mbit/in.^{2}, that is just 0.0056 Gbit/in.^{2}.

8.4.3 Fujitsu HDD Introduced in 2007

After the discovery of the GMR effect, the bits could be written vertically in the medium and much more information could be packed into the same area. From the early 2000s onwards, Perpendicular Magnetic Recording (PMR) became the norm and Hard Disk Drives were launched (Fig. 8.13).

The following 500 × 500 nm^2 MFM image was captured on a Fujitsu HDD from 2007 comprising of a single 2.5 in. platter and stipulating a total storage capacity of 40 GB or 320 Gbits. The bit dimension derived from the image is approximately 25 × 200 nm^2 which makes it 2000 times smaller than in a zip drive. Logically, the density also makes a massive leap: on average 37.4 bits are counted on this MFM data, on an area of only 0.25 μm^2: that comprises a density of 100 Gbit/in.^{2}.

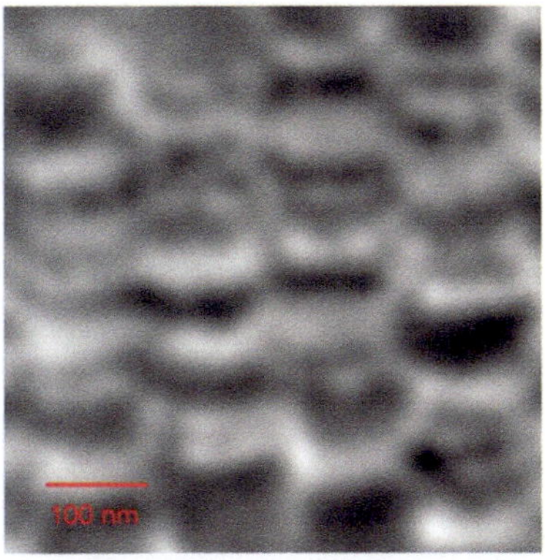

Fig. 8.15 $500 \times 500\ \text{nm}^2$ MFM image of a Seagate HDD from 2009

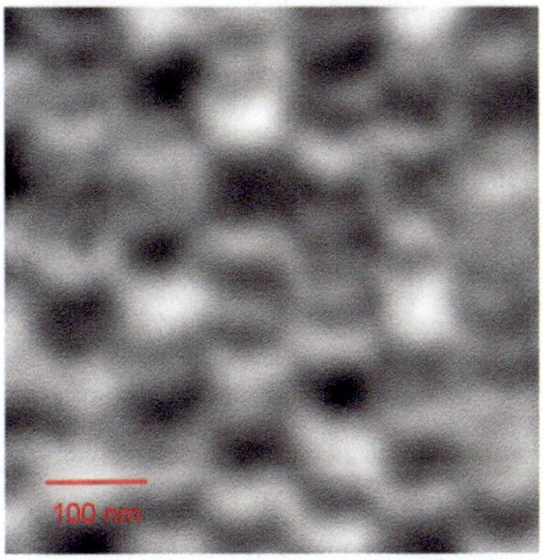

Fig. 8.16 $500\times500\ \text{nm}^2$ MFM image of a Western Digital HDD from 2012

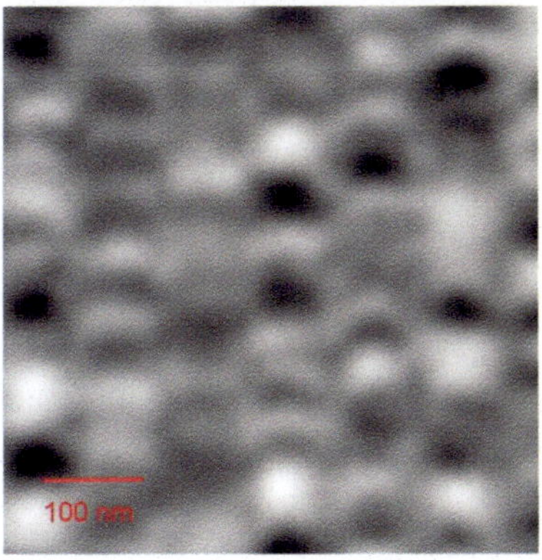

Fig. 8.17 $500 \times 500\ \text{nm}^2$ MFM image of a Seagate HDD from 2016

8.4.4 Seagate HDD Introduced in 2009

Since then, the race for ever-shrinking bits was propelled. In the spring of 2009, Seagate commercialized a Hard Disk Drive totaling 500 GB (4000 Gbits or 4 Terabits) split on two 2.5 in. platters.

The 500×500 nm^2 MFM image above represents 83.2 bits on average, the bit sizes amounts to ca. 20×125 nm^2. This signifies a gain of over a factor of two in bit size over the Fujitsu HDD. Logically, the bit density increases to 215 Gbit/in.^{2}.

8.4.5 Western Digital HDD Introduced in 2012

In 2012, Western Digital commercialized a HDD that had 1 TB or 8 Tbits of storage capacity divided over two 2.5 in. platters. The 500×500 nm^2 MFM image below displays 190 bits on average, the bit sizes totals to ca. 17×80 nm^2, which is becoming hard for an MFM to image. The bit area represents again a gain of over a factor of two in bit size over the earlier Seagate HDD, the bit density increasing to 490 Gbit/in.^{2}.

8.4.6 Seagate HDD Introduced in 2016

Since 2015, the battle for bit size reduction became harder as the physical limitations to the stability of the bits were increasingly reached. HDD Manufacturers have examined two tricks as workarounds in the existing Perpendicular Magnetic Recording technology. Both have the objective of reducing the width of the bits:

- Shingled Magnetic Recording (SMR): similar to roof tiles, a track is written overlapping the earlier neighbouring track.
- Two-Dimensional Magnetic Recording (TDMR): so as to prevent inter-track interference during readout of very narrow tracks, an array of read-heads is used to read adjacent tracks and correlate data. This technology, combined to SMR would gain another 5–10% of storage capacity.

The HDD from Seagate illustrated below is from the BarraCuda series (up to 5 TB of storage on three 2.5 in. platters) and according to the Manufacturer employs the SMR technology. However, from the MFM image, only a weak increase of storage density can be noticed. The 500×500 nm^2 MFM image below displays 208 bits on average, the bit size amounts to ca. 16×77 nm^2: compared to the earlier Western Digital HDD there is only a minor improvement of the bit size of a factor of 1.1. The bit density is 540 Gbit/in.^{2}.

Both TDMR and SMR technologies are thought to be stop-gap measures where read heads can be made smaller but write heads not. Moreover, SMR highlights the issue of re-writing a track, which has an incidence on the following adjacent tracks.

8.4.7 Outlook

Beating the 1 Tbit/in^2 limit is the next challenge HDD Manufacturers are currently facing. Bear in mind: a storage density of 1 Tbit/in^2, that is a bit size of just 12.7 × 50 nm^2.

The active layer of present day HDDs is made up of magnetic grains (usually CoPtCr alloys) clustered together, with dimensions of just few nanometers, typically 8 nm. The actual bit size is so small that less than 20 grains make up each bit.

Intuitively, to expand capacity it would be enough to reduce additional bit size and hence have fewer grains composing a bit. But this would weaken the Signal-to-Noise ratio and would make readout more complex.

Alternatively, one could think of decreasing the grain size. However physics stalls this: a smaller grain cannot hold its magnetisation anymore as thermal activation at room temperature will serve to flip its spin.

However, there are technological solutions, which are currently under investigation. One possible technology enabler would be to employ other alloys with higher anisotropy and hence greater thermal stability. The downside to it is that writing would become a lot more laborious and would require to local energy introduction, such as heat and microwave, onto the active layer to provisionally lower the magnetisation barrier. This so-called heat-assisted magnetic recording (HAMR) [52, 53] and microwave-assisted magnetic recording (MAMR) [54] would hence not only need new materials but also new and intricate read/write heads.

In HAMR, a laser beam is employed to heat the media up locally and reduce the thermal stability of the data bit to be written. A scanning near-field optical microscope is implemented into a writing head and a successful demonstration has been reported [55]. In 2012, TDK demonstrated a new HAMR head with the areal density of 1.5 Tbit/in.^{2} and bit-error rate of 10^{-2}. Seagate also demonstrated a new HAMR drive in 2012. HAMR contains many properties, such as grain size, characterisation and optimisation of a laser unit, media thermal stack and magnetic distributions. The optimisation of these properties can improve the density of HAMR and can reduce the corresponding energy consumption. In 2015, Seagate demonstrated a HDD using the HAMR technology by laser pulse instead of continuous wave [56]. Fruchart et al. have suggested the utilisation of thermal gradients generated in HAMR can determine the polarisation of tunnelling electrons across the media grain boundaries [57]. Matlak and Komvopoulos have reported a role of pulsed substrate bias voltage and a C^+ ion incident angle as well as the thickness of 1–4 nm thick amorphous C films deposited by filtered cathodic vacuum arc deposition method (FCVA) [58]. They show FCVA can produce extremely thin and uniform protective *a*-C films with high sp^3 contents ideal for HAMR heads.

Microwave-assisted magnetic recording (MAMR) was proposed by Zhu et al. [54]. The MAMR utilizes microwaves to reduce the switching field by an order of magnitude [59]. Western Digital announced that they will introduce a MAMR system within a few years. Instead of a conventional one-dimensional transverse microwave field, an ac field with a circular trace can increase efficiently of MAMR [60]. By

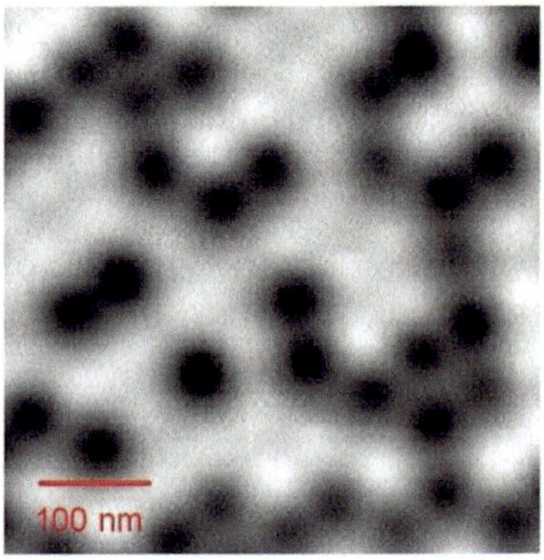

Fig. 8.18 500 × 500 nm^2 MFM image of a BPM. Sample courtesy of Seagate

utilising a circular ac field, a damping constant increases by an order of magnitude as compared with a linear ac field which achieves the areal density of 1.9 Tbit/in.^{2}. Following the utilisation of a circular ac field in MAMR a new micromagnetic modelling on MAMR shows the areal recording density can exceed 3 Tbit/in.^{2} with medium signal-to-noise and thermal stability at $K_u V = 60 k_B T_{300K}$. This model uses a perpendicular spin oscillator (PSTO) to generate localised a circular ac field [61]. Muraoka et al. have showed a stable spin torque oscillator (STO) requires a large and well-shaped gap field [62].

Another solution that can be considered is to pattern the active layer so as to physically define the shape of the storage unit: these are called bit-patterned media (BPM) [63] and are also under investigation as shown in Fig. 8.18. By patterning nano-scale island-shaped bits typically consisting of 10–20 grains, the media can achieve 10 Tbit in.$^{-2}$. Production of the layers would then need additional steps, which is a huge disadvantage in keeping production costs low. According to a recent calculation model, the areal density can even achieve beyond 10 Tbit/in.^{2} with a grain size of 5 nm in diameter and 10 nm in height [64]. As Advanced Technology Consortium reported that by utilising BPM in HAMR, the areal density can lead to 10 Tbit/in.^{2} by 2025. Nano-imprint lithography is suggested to use for high-density data storage applications. In BPM, the metallopolymer precursor is used as photoresist, which can be imprinted on the substrate [65].

In Fig. 8.18, there are on average 92.5 dots over an area of 500 × 500 nm^2 which makes a density of just 240 Gbit/in^2. The pitch between each dot is 50 nm in this sample from 2009. To realise a density of 1 Tbit/in.^{2}, the pitch between each dot would have to be halved to 25 nm.

HAMR, MAMR, BPM, or a combination thereof: in the days to come it will be known which path technologists have chosen to take to solve the challenge of data storage capacity. The Advanced Storage Technology Consortium (ASTC) roadmap, issued by the International Disk Drive Equipment and Materials Association (IDEMA) provides a hint of what is anticipated in the next 10–15 years (Fig. 8.19).

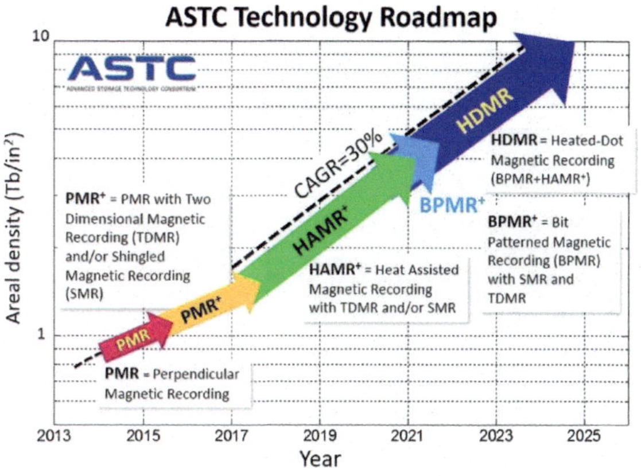

Fig. 8.19 ASTC Technology Roadmap gives an insight of the evolution of storage density in the coming 5–10 years in relation with the development of new technologies [66]

8.5 Applications for Magnetic Memories and Devices

8.5.1 Magnetic Random Access Memory and Spin Random Access Memory

After the commercial success of the HDD heads and media as described in Sect. 8.4, one of the major ongoing device studies is to realise a magnetic random access memory (MRAM) and/or spin random access memory (SpinRAM) [67]. MRAM can achieve the following features [68]: (i) non-volatility similar to HDD without using a mechanical head, (ii) read and write times of the order of nano-seconds, (iii) high density and (iv) low-power consumption. Hence, MRAM/SpinRAM is a good candidate for a universal memory. MRAM/SpinRAM can also avoid the delay and unnecessary power consumption due to circuit leads, and can provide a high-density circuit by the converting passive current leads in conventional electronic devices into memory cells.

For the next-generation MRAM/SpinRAM, the following issues need to be solved: the reduction of the switching field or critical current density and a reduction of the fabrication rule. Based on the scalability of MRAM/Spin RAM bits, spin-transfer torque (STT) must be used as discussed above. For the reduction of power consumption, the thermal stability of the free layer has to be compromised. A synthetic ferromagnet, such as CoFeB/Ru/CoFeB, has been implemented as the free layer by AIST [69], resulting in a five times increase in thermal stability with only an 80% increase in the critical current density. This configuration can achieve a 10 Gbit SpinRAM device. The remaining issue to realize a high-density SpinRAM is the reduction of the fabrication rule, which may be achieved by replacing the additional transistor with a possible new device design.

For the design improvement, MFM imaging can play an important role to remove any edge domains and magnetic defects to minimise the corresponding magnetisation

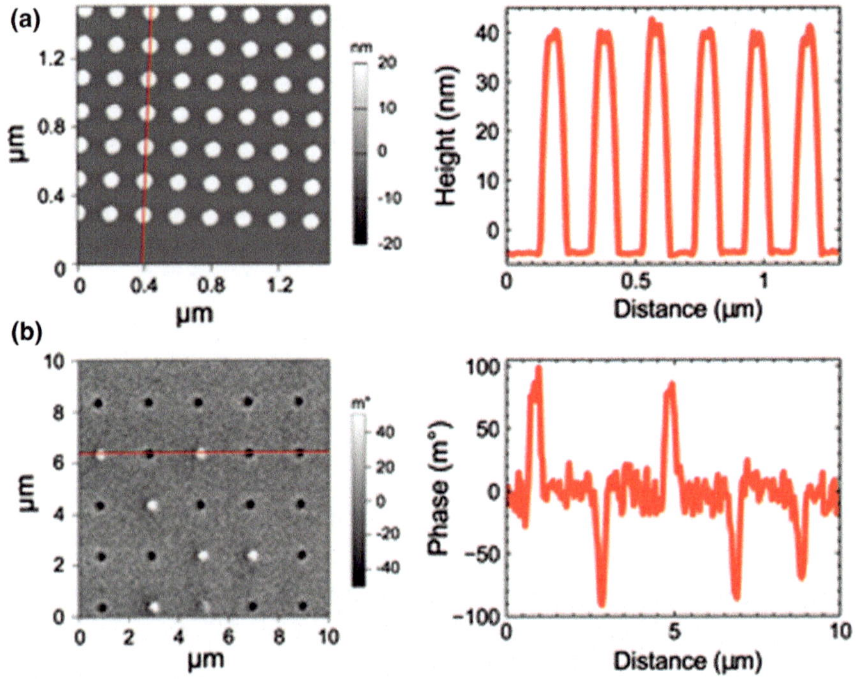

Fig. 8.20 **a** MFM topography image of patterned Mo/CoFeB/MgO/CoFeB/Mo p-MTJ pillars of 125 nm in diameter and **b** phase images from the magnetic structure of the same sample of p-MTJ pillars of 500 nm in diameter, representing the pillars after out-of-plane ac-demagnetisation [70]

damping. MFM was used for imaging perpendicular magnetic tunnel junction (p-MTJ) pillars after ac-demagnetisation in a perpendicular field [70]. Figure 8.20a, b show AFM topography and corresponding MFM images of an array of pillars made from a Mo/CoFeB/MgO/CoFeB/Mo film. The topography shows a uniform set of pillars with height of about 40–50 nm, corresponding to the thickness of the Mo/CoFeB/MgO/CoFeB/Mo/Ru. As shown in Fig. 8.20b, the ac-demagnetisation generates up and down magnetisation of the ferromagnetic layers in p-MTJ. The images for 500 nm pillars sizes show single domain structure even at this large size.

8.5.2 Racetrack Memory

In order to minimize the total magnetic energy of a ferromagnetic structure, a domain wall (DW) with a width of the order of 100 nm can be formed between magnetic domains. The competition among the energy terms, magnetostatic, exchange and anisotropy, produces magnetic domains, mainly due to the magnetostatic energy in a mesoscopic ferromagnet. Within a DW, the spins gradually rotate from the direction of the domain to the direction in the neighbouring domain. If the ferromagnet is thick, the spin rotation within the wall is out-of-plane, forming a Bloch wall. For thin ferromagnet films, in-plane rotation, known as a Néel wall, occurs. For an intermediate ferromagnetic film thickness, a combination of these two walls, a cross-tie wall, forms [71].

Following Slonczewski and Berger [72, 73], STT has also been applied to displace a DW by flowing an electrical current (see Fig. 8.21). A ferromagnetic micro-wire has been prepared with a DW, giving a current density between 10^{11} and 10^{12} A/m^2 for wall motion [74, 75]. The velocity of the wall motion has also been estimated to be 2–6 m/s [76]. A similar experiment has also been performed in a dilute magnetic semiconducting wire [77].

8.5.3 Magnetic Skyrmion Logics

A similar concept to Sect. 8.5.2 was demonstrated by employing a magnetic skyrmion [78] and magnon [79] instead of a DW, which are expected to warrant reproducible motion by a lower current density as compared with the studies described as above. A magnetic skyrmion is a topologically protected structure by forming a rotating structure of spins. Depending on the rotating structure, the skyrmions can be categorised into three types, Néel, Bloch and anti-skyrmions. They have these chirality and the polarity of the centre spin, which can be used as information carrier. Due to the diameter of the skyrmion (typically ~10 nm), the corresponding critical current density can be reduced by five orders of magnitude.

The displacement of a magnetic skyrmion can also be imaged using MFM [80]. In Fig. 8.22, the velocity of the skyrmion generated on the Pt (5 nm)/FM/Au (3 nm)/FM/Pt (5 nm) multilayer, where FM = Ni (0.4 nm)/Co (0.7 nm)/Ni (0.4 nm), was measured by applying a 3-ns-long current pulses with a density of $j = 3.9 \times 10^{11}$ A/m^2. Under a magnetic field application of −60 Oe, the velocity was measured to be 60 m/s.

These results unambiguously confirm the importance of MFM imaging for the development of magnetic and spintronic devices. Due to the easy use of MFM, it can be utilised in broader fields in research and development.

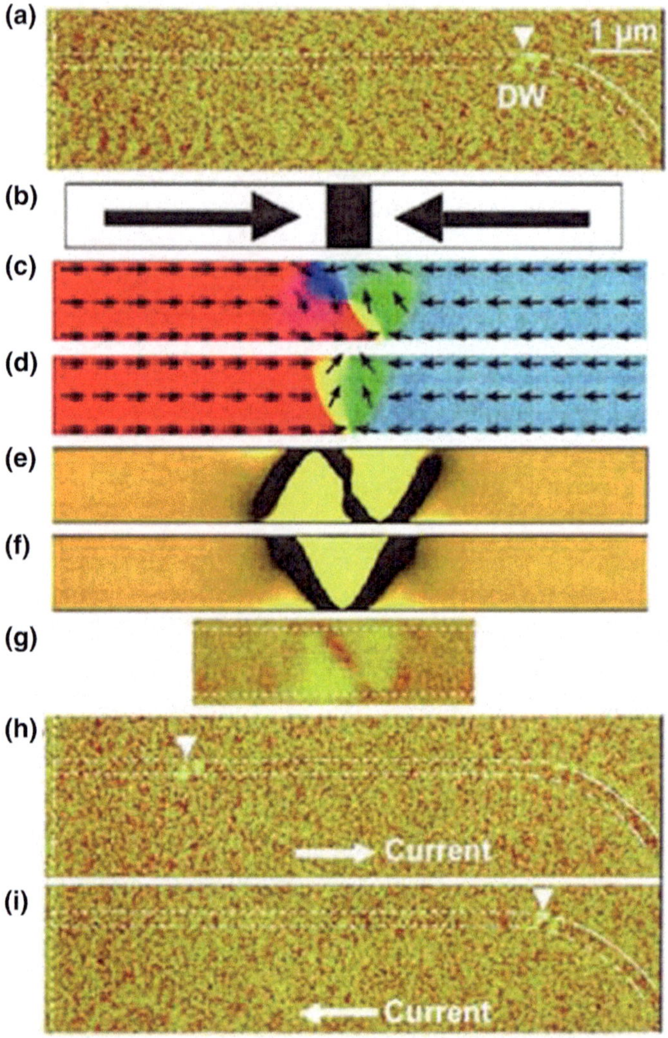

Fig. 8.21 **a** MFM image after the introduction of a DW. DW is imaged as a bright contrast, which corresponds to the stray field from positive magnetic charge. **b** Schematic illustration of a magnetic domain structure inferred from the MFM image. DW has a head-to-head structure. **c** Result of micromagnetics simulation (vortex DW). **d** Result of micromagnetics simulation (transverse DW). **e** MFM image calculated from the magnetic structure shown in (**c**). **f** MFM image calculated from the magnetic structure shown in (**d**). **g** Magnified MFM image of a DW. **h** MFM image after an application of a pulsed current from left to right. The current density and pulse duration are 1.2×10^{12} A/m^2 and 5 μs, respectively. DW is displaced from right to left by the pulsed current. (i) MFM image after an application of a pulsed current from right to left. The current density and pulse duration are 1.2×10^{12} A/m^2 and 5 μs, respectively. DW is displaced from left to right by the pulsed current [77]

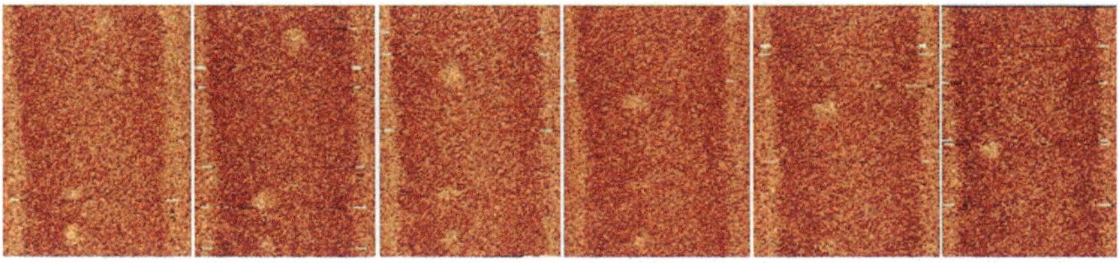

Fig. 8.22 Series of images showing magnetic skyrmions shift along the track between 3 ns, $j = 3.9 \times 10^{11}$ A/m^2 electric pulses. Scale bar is 500 nm

References

1. A. Hubert, R. Schäfer, *Magnetic Domains* (Springer, Berlin, 2000)
2. A. Tonomura, T. Matsuda, J. Endo, T. Arii, K. Mihama, Phys. Rev. Lett. **44**, 1430 (1980)
3. K. Koike, K. Hayakawa, Appl. Phys. Lett. **45**, 585 (1984)
4. M. Johnson, J. Clarke, J. Appl. Phys. **67**, 6141 (1990)
5. R. Wiesendanger, H.J. Güntherodt, G. Güntherodt, R.J. Gambino, R. Ruf, Phys. Rev. Lett. **65**, 247 (1990)
6. R. Allenspach, A. Bischof, Appl. Phys. Lett. **54**, 587 (1989)
7. S.N. Molotkov, Surf. Sci. **287–288**, 1098 (1993)
8. R. Laiho, H.J. Reittu, Surf. Sci. **289**, 363 (1993)
9. S.F. Alvarado, P. Renaud, Phys. Rev. Lett. **68**, 1387 (1992)
10. M.W.J. Prins, H. van Kempen, H. van Leuken, R.A. de Groot, W. van Roy, J. de Boeck, J. Phys.: Condens. Matter **7**, 9447 (1995)
11. S.M. Sze, *Physics of Semiconductor Devices*, 2nd edn. (Wiley, New York, 1981)
12. K. Sueoka, K. Mukasa, K. Hayakawa, Jpn. J. Appl. Phys. **32**, 2989 (1993)
13. Y. Suzuki, W. Nabhan, R. Shinohara, K. Yamaguchi, K. Mukasa, J. Magn. Magn. Mater. **198–199**, 540 (1999)
14. H. Kodama, T. Uzumaki, M. Oshiki, K. Sueoka, K. Mukasa, J. Appl. Phys. **83**, 6831 (1999)
15. A.M. Chang, H.D. Hallen, L. Harriot, H.F. Hess, H.L. Kao, R.E. Miller, R. Wolfe, J. van der Ziel, Appl. Phys. Lett. **61**, 1974 (1992)
16. T. Shono, T. Hasegwa, T. Fukumura, H. Ohno, Appl. Phys. Lett. **77**, 1363 (2000)
17. A. Sandhu, K. Kurosawa, M. Dede, A. Oral, Jpn. J. Appl. Phys. **43**, 2 (2004)
18. J. Gutierrez, B. Raes, A.V. Silhanek, L.J. Li, N.D. Zhigadlo, J. Karpinski, J. Tempere, V.V. Moshchalkov, Phys. Rev. B **85**, 094511 (2012)
19. J.R. Kirtley, M.B. Ketchen, K.G. Stawiasz, J.Z. Sun, W.J. Gallagher, S.H. Blanton, S.J. Wind, Appl. Phys. Lett. **66**, 1138 (1995)
20. M.B. Ketchen, T. Kopley, H. Ling, Appl. Phys. Lett. **44**, 1008 (1984)
21. D. Mailly, C. Chapelier, A. Benoit, Phys. Rev. Lett. **70**, 2020 (1993)
22. F. Grhhl, M. Mück, M. Kreutzbruck, J. Dechert, Rev. Sci. Instrum. **72**, 2090 (2001)
23. M.E. Huber, N.C. Koshnick, H. Bluhm, L.J. Archuleta, T. Azua, P.G. Björnsson, B.W. Gardner, S.T. Halloran, E.A. Lucero, K.A. Moler, Rev. Sci. Instrum. **79**, 053704 (2008)
24. J.R. Kirtley, L. Paulius, A.J. Rosenberg, J.C. Palmstrom, D. Schiessl, C.L. Jermain, J. Gibbons, C.M. Holland, Y. Fung, M.E. Huber, M.B. Ketchen, D.C. Ralph, G.W. Gibson Jr., K.A. Moler, Supersond. Sci. Technol. **29**, 124001 (2016)
25. E. Betzig, J.K. Trautman, R. Wolfe, E.M. Gyorgy, P.L. Finn, M.H. Kryder, C.-H. Chang, Appl. Phys. Lett. **61**, 142 (1992)
26. B. Hecht, H. Bielefeldt, Y. Inouye, D.W. Pohl, L. Novotny, J. Appl. Phys. **81**, 2492 (1997)
27. B. Hecht, B. Sick, U.P. Wild, V. Decert, R. Zenobi, O.J.F. Martin, D. Pohl, J. Chem. Phys. **112**, 7761 (2000)
28. U.Ch. Fischer, D.W. Pohl, *Phys. Rev. Lett.* **62**, 4 (1989)

29. T. Kalkbrenner, M. Ramstein, J. Mlynek, V. Sandoghdar, J. Microscopy **202**, 72 (2000)
30. J. Rudge, H. Xu, J. Kolthammer, Y.K. Hong, B.C. Choi, Rev. Sci. Instrum. **86**, 023703 (2015)
31. P.S. Keatley, T.H.J. Loughran, E. Hendry, W.L. Barnes, R.J. Hicken, J.R. Childress, J.A. Katin, *Rev. Sci. Instrum.* **88**, 123708 (2017)
32. G. Binnig, H. Rohrer, C. Gerber, E. Weibell, Phys. Rev. Lett. **49**, 57 (1982)
33. G. Binnig, C.F. Quate, C. Gerber, Phys. Rev. Lett. **56**, 930 (1986)
34. Y. Martin, C.C. Williams, H.K. Wickramasinghe, J. Appl. Phys. **61**, 4723 (1987)
35. J. Szymońska, F. Krok, Int. J. Bio. Macromol. **33**, 1 (2003)
36. Y. Martin, H.K. Wickramasinghe, Appl. Phys. Lett. **50**, 1445 (1987)
37. K. Sueoka, K. Okuda, N. Matsubara, F. Sai, J. Vac. Sci. Technol., B **9**, 1313 (1991)
38. D. Rugar, H.J. Mamin, P. Guethner, S.E. Lambert, J.E. Stern, I. McFadyen, T. Yogi, J. Appl. Phys. **68**, 1169 (1990)
39. Digital Instruments, *Support Note No. 229, Rev. B Magnetic Force Microscopy (MFM)* (Santa Barbara, USA, 1996)
40. H. Kaiju, H. Kasa, T. Komine, S. Mori, T. Misawa, T. Abe, J. Nishii, J. Appl. Phys. **117**, 17738 (2015)
41. A.J. den Boef, *Appl. Phys. Lett.* **56**, 2045 (1990); Z. Liu, Y. Dan, Q. Jinjun, Y. Wu, *J. Appl. Phys.* **91**, 8843 (2002)
42. G.N. Phillips, M. Siekman, L. Abelmann, J.C. Lodder, Appl. Phys. Lett. **81**, 865 (2002)
43. L. Gao, L.P. Yue, T. Yokota, R. Skomski, S.H. Liou, H. Saito, S. Ishio, IEEE Trans. Magn. **40**, 2194 (2004)
44. N. Amos, R. Ikkawai, R. Haddon, D. Litvinov, S. Khizroev, Appl. Phys. Lett. **93**, 3116 (2008)
45. T. Yamaoka, K. Watanabe, Y. Shirakawabe, K. Chinone, J. Magn. Soc. Jpn. **27**, 429 (2003)
46. K. Nagano, K. Tobari, K. Soneta, M. Ohtake, M. Futamoto, J. Magn. Soc. Jpn. **36**, 109 (2012)
47. Z. Liu, Y. Dan, Q. Jinjun, Y. Wu, *J. Appl. Phys.* **91**, 8843 (2002); G.N. Phillips, M. Siekman, L. Abelmann, J. C. Lodder, *Appl. Phys. Lett.* **81**, 865 (2002)
48. M.A. Al-Khafaji, W.M. Rainforth, M.R.J. Gibbs, J.E.L. Bishop, H.A. Davies, IEEE Trans. Magn. **32**, 4138 (1996)
49. T. Hasegawa, W. Pei, T. Wang, Y. Fu, T. Washiya, H. Saito, S. Ishio, Acta Mater. **56**, 1564 (2008)
50. T. Yamaoka, H. Tsujikawa, R. Hirose, A. Ito, H. Kawamura, T. Sakon, J. Magn. Soc. Jpn. **35**, 60 (2011)
51. H. Saito, H. Ikeya, G. Egawa, S. Ishio, S. Yoshimura, J. Appl. Phys. **105**, 07D524 (2009)
52. H. Katayama, S. Sawamura, Y. Ogimoto, J. Nakajima, K. Kojima, K. Ohta, J. Magn. Soc. Jpn. **23**, 233 (1999)
53. R.E. Rottmayer, S. Batra, D. Buechel, W.A. Challener, J. Hohlfeld, Y. Kubota, L. Li, B. Lu, C. Mihalcea, K. Mountfield, K. Pelhos, C. Peng, T. Rausch, M.A. Seigler, D. Weller, X.-M. Yang, IEEE Trans. Magn. **42**, 2417 (2006)
54. J.-G. Zhu, H. Zhu, Y. Tang, IEEE Trans. Magn. **44**, 125 (2008)
55. M.A. Seigler, W.A. Challener, E. Gage, N. Gokemeijer, G. Ju, B. Lu, K. Pelhos, C. Peng, R.E. Rottmayer, X. Yang, H. Zhou, T. Rausch, IEEE Trans. Magn. **44**, 119 (2008)
56. T. Raush, A.S. Chu, P. Lu, S. Puranam, D. Nagulapally, T. Lammers, J.W. Dykes, E.C. Gage, IEEE Trans. Magn. **51**, 4 (2015)
57. O. Fruchart, B. Dieny, M. Chshiev, B. Charles, N. Strelkov, A. Truong, A. Hallal, J. Wang, Y.K. Takahashi, T. Mizuno, K. Honi, IEEE Trans. Magn. **54**, 0800111 (2018)
58. J. Matlak, K. Komvopoulos, Sci. Rep. **8**, 9647 (2018)
59. T. Seki, K. Utsumiya, Y. Nozaki, H. Imamura, K. Takanashi, Nature Commun. **4**, 1726 (2013)
60. Y. Wang, Y. Tang, J. Zhu, J. Appl. Phys. **105**, 07B902 (2009)
61. J.-G. Zhu, Y. Wang, IEEE Trans. Magn. **46**, 751 (2010)
62. Y. Kanai, R. Itagaki, S.J. Greaves, H. Muraoka, IEEE Trans. Magn. **53**, 11 (2017)
63. B.D. Terris, T. Thomson, J. Phys. D Appl. Phys. **38**, R199 (2005)
64. C. Vogler, A. Abert, F. Bruckner, D. Suess, D. Praetorius, Appl. Phys. Lett. **108**, 102406 (2018)
65. Z. Meng, G. Li, H. Wong, S. Ng, S. Yiu, C. Ho, C. Leung, I. Manners, W. Wong, Nanoscale **9**, 731 (2017)

66. http://idema.org/?page_id=5868
67. I.L. Prejbeanu, S. Bandiera, J. Alvarez-Hérault, R.C. Sousa, B. Dieny, J.-P. Nozières, J. Phys. D Appl. Phys. **46**, 074002 (2013)
68. A. Hirohata, K. Takanashi, J. Phys. D Appl. Phys. **47**, 193001 (2014)
69. S. Yakata, H. Kubota, T. Seki, K. Yakushiji, A. Fukushima, S. Yuasa, K. Ando, IEEE Trans. Magn. **46**, 2232 (2010)
70. L. Tryputen, K.-H. Tu, S.K. Piotrowski, M. Bapna, S.A. Majetich, C. Sun, P.M. Voyles, H. Almasi, W. Wang, P. Vargas, J.S. Tresback, C.A. Ross, Nanotechnology **27**, 185302 (2016)
71. S. Chikazumi, *Physics of Ferromagnetism* (Oxford University Press, Oxford, 1997)
72. J. Slonczewski, J. Magn. Magn. Mater. **159**, L1 (1996)
73. L. Berger, Phys. Rev. B **54**, 9353 (1996)
74. J. Grollier, P. Boulenc, V. Cros, A. Hamzić, A. Vaurès, A. Fert, G. Faini, Appl. Phys. Lett. **83**, 509 (2003)
75. M. Tsoi, R.E. Fontana, S.S.P. Parkin, Appl. Phys. Lett. **83**, 2617 (2003)
76. A. Yamaguchi, T. Ono, S. Nasu, K. Miyake, K. Mibu, T. Shinjo, Phys. Rev. Lett. **92**, 077205 (2004)
77. M. Yamanouchi, D. Chiba, F. Matsukura, H. Ohno, Nature **428**, 539 (2004)
78. J.S. Gardner, J. Phys. Condens. Matter **23**, 160301 (2011)
79. U. Rössler, *Spin Waves: Magnons* (Springer, Berlin, 2009)
80. A. Hrabec, J. Sampaio, M. Belmeguenai, I. Gross, R. Weil, S.M. Chérif, A. Stashkevich, V. Jacques, A. Thiaville, S. Rohart, Nat. Commun. **8**, 15765 (2016)

Chapter 9
Space Charge at Nanoscale: Probing Injection and Dynamic Phenomena Under Dark/Light Configurations by Using KPFM and C-AFM

Christina Villeneuve-Faure, Kremena Makasheva, Laurent Boudou and Gilbert Teyssedre

Abstract Fine description of the electrical properties of solids at their surfaces is a very old problem, difficult to tackle because the surface of a solid itself represents a break in the periodic structure of crystallized materials, hence a defect, and most importantly because of the potential impact of surface oxidation, contamination, humidity, atmosphere, etc., on the material response (Galembeck et al. in Polymer 42:4845, 2001 [1]). For dielectrics, electrical charging of the surface leads to the build-up of a surface potential. The occurring mechanisms depend on the kind of charges being deposited, e.g. by triboelectrification, and are particularly difficult to anticipate (Lacks and Sankaran in J Phys D Appl Phys 44:453001, 2001 [2], Shinbrot et al. in Phys Rev E 96:032912, 2017 [3]). Aside these difficulties in defining the surface properties, nanosciences and nanomaterials have brought us new paradigms with the tremendous increase of the amount of interfaces between particles and host matrix, and with the variety in the material nature and interface linked to the different elaboration processes. In a way it may constitute a chance to better describe what interfaces on an electronic properties standpoint are, because materials are better controlled. Besides the nanostructuration of materials, the miniaturization of devices is a further challenge to face. When dealing with thin layers (thicknesses of less than 100 nm) the rules for bulk properties behavior are broken. Obviously, in both cases the experimental approach is more demanding, since the tools that are implemented for the study must have a spatial resolution compatible with the scale at which phenomena should be probed. In this Chapter we illustrate on a few examples the

C. Villeneuve-Faure (✉) · K. Makasheva · L. Boudou · G. Teyssedre
LAPLACE, Université de Toulouse, CNRS, UPS, INPT, Toulouse, France
e-mail: christina.villeneuve@laplace.univ-tlse.fr

K. Makasheva
e-mail: kremena.makasheva@laplace.univ-tlse.fr

L. Boudou
e-mail: laurent.boudou@laplace.univ-tlse.fr

G. Teyssedre
e-mail: gilbert.teyssedre@laplace.univ-tlse.fr

U. Celano (ed.), *Electrical Atomic Force Microscopy for Nanoelectronics*, NanoScience and Technology, https://doi.org/10.1007/978-3-030-15612-1_9

need for ever lower scale characterization of the electrical properties of dielectric materials.

9.1 Context

Fine description of the electrical properties of solids at their surfaces is a very old problem, difficult to tackle because the surface of a solid itself represents a break in the periodic structure of crystallized materials, hence a defect, and most importantly because of the potential impact of surface oxidation, contamination, humidity, atmosphere, etc., on the material response [1]. For dielectrics, electrical charging of the surface leads to the build-up of a surface potential. The occurring mechanisms depend on the kind of charges being deposited, e.g. by triboelectrification, and are particularly difficult to anticipate [2, 3]. Aside these difficulties in defining the surface properties, nanosciences and nanomaterials have brought us new paradigms with the tremendous increase of the amount of interfaces between particles and host matrix, and with the variety in the material nature and interface linked to the different elaboration processes. In a way it may constitute a chance to better describe what interfaces on an electronic properties standpoint are, because materials are better controlled. Besides the nanostructuration of materials, the miniaturization of devices is a further challenge to face. When dealing with thin layers (thicknesses of less than 100 nm) the rules for bulk properties behavior are broken. Obviously, in both cases the experimental approach is more demanding, since the tools that are implemented for the study must have a spatial resolution compatible with the scale at which phenomena should be probed. In this Chapter we illustrate on a few examples the need for ever lower scale characterization of the electrical properties of dielectric materials.

9.1.1 Miniaturization of Thin Dielectric Layers

The specifications from the semiconductor technology sector are particularly demanding for what concerns gate dielectrics for transistors, with equivalent oxide thickness (EOT) of less than 1 nm in applications such as microprocessors (MPUs), or with low leakage current combined with EOT of less than 2 nm in applications such as cell phones, where the static power consumption is the major limiting factor [4]. This is an example of the break of bulk properties where the leakage current is through tunneling across the layer. To overcome this issue for gate dielectrics, high-k dielectrics are searched for replacement of the relatively low permittivity value of SiO_2, with both fundamental (intrinsic material properties) and manufacturing limitations.

In the era of miniaturization, many benefits are being brought by Micro-ElectroMechanical Systems (MEMS), and now Nano-ElectroMechanical Systems (NEMS), to the downsizing to nanoscale, to exploit their capabilities in device size reduction, in response speed, energy conservation, autonomy, integration of complexity with combining sensors, actuators, energy harvesting, etc. There seems to exist a limitless possibility in MEMS-NEMS development. Because of their complexity, and of the necessity of motion of piece parts, the MEMS-NEMS are naturally exposed to ambient stresses such as vibration, humidity or pollution, which makes their reliability an issue [5]. The stresses endowed by materials can be relatively strong, an example being of electrostatic origin due to the charging of materials that can lead to malfunction of voltage-driven actuators. Though driving voltages are relatively small, the electric fields are high and the device geometry often imposes high diverging field such as the charge deposition is highly effective. Another important issue for the reliability of MEMS is tribological effect particularly appearing on the contacts [6]. For probing devices with such outcomes, appropriate techniques and methodologies are needed to provide information on charges and forces at the relevant scale [7–9].

Whereas the charge trapping represents an issue for MEMS devices, it is exploited in non-volatile memory devices to store the information. Here again the objectives are to miniaturize and to improve reliability. Indeed, in order to keep up with the demand for increased memory capacities, flash memory devices have been continuously scaled down. The main benefits of down-scaling the cell size and increasing integration are enabling lower manufacturing cost while keeping high performance. Charge trapping memory is regarded as one of the most promising flash memory technologies at further down-scaling strategies. High-k dielectrics are usually preferred as charge trapping layer, blocking layer, and tunneling layer [10]. SiO_2 is the first dielectric used as blocking layer in the flash memory. However, as for transistors, the large tunneling current through SiO_2 is not acceptable upon continually scaling down the dimensions of the flash memory.

9.1.2 Interfaces

In organic solar cells, the low diffusion length of excitons imposes that donor and acceptor materials segregate to form small size domains with high interface area to reach reasonable efficiency. To optimize the efficiency, intermixing of the donor and the acceptor moieties on the nanometer scale is essential. This insight led to the development of the so-called bulk-heterojunction concept [11]. Connected domains with a typical size of several tens of nanometers are formed in the film. At the same time a small amount of the acceptor material may be dissolved in the donor domains or vice versa. Obviously, control and optimization of such structures require adequate tools [12].

Parameters of outmost value, when dealing with interfaces, are the energy level estimations of one material relative to the other one as they determine the working conditions of the device. When going to thin films, these energy levels become modified and very new properties are being introduced with the use of 2D heterostructures made of few atom thick layers [13]. Besides the physical properties of materials and junctions, the behavior of materials under stress should also be evaluated. With the new tools brought by scanning probe techniques, Atomic Force Microscopy (AFM) in particular, it becomes now possible to evaluate in details the mechanisms at the origin of charges generation into dielectrics, and to assess and revise the models for charge injection currently under use [14, 15]. Moreover, it is important to characterize how such charges are dissipated in the bulk or at the surface.

Specific dielectric properties of nanocomposite materials, made of nanoparticles (not necessarily insulating) dispersed into insulating matrices have been recently reported [13, 16]. The claimed effect has been associated with the structuration of the matrix under the impact of the nanoparticles or to "charge trap" formation at the interfaces [17]. It has also been shown that differences imposed by the polymer dielectric interface processing can lead to substantial changes in the macroscopic response of the material [18]. However, future developments should contribute to assess the different hypotheses put behind changes in the charging behavior of bulk materials.

The above given examples are only a flavor of how broad can be the field of electronic properties and electrical charges profiling into materials. Certainly many other aspects could be addressed. Our purpose in the following is to present newly developed methods likely to provide information at pertinent scale to address these problems of electrostatics: local information is needed either because the investigated process is actually at the interface like for charge injection or energy levels estimation, or because the structures have been downsized and existing methodologies to probe the properties cannot be easily adapted. This is for example the case of charge density distribution measurements.

9.2 KPFM and C-AFM Measurement Under Dark and Light Configurations

9.2.1 Introduction to Surface Potential

In electrochemistry the surface potential in a solid is classically defined as the difference between the internal and the external electric potentials i.e. Galvani and Volta potentials, respectively [19]. Consequently, it depends on the material properties (crystallographic structure…) and the surface features (adsorbed molecules…) and for the different materials is defined as follow: For a metal (Fig. 9.1a), the surface potential therefore corresponds to the work function ϕ_m which is the amount of energy required to extract an electron from fundamental state in the material to the vacuum level [20]. For a semiconductor, the surface potential corresponds to the energy of Fermi level (Fig. 9.1b). It is influenced by the interface states, which induce band curvature, and by the level of doping [21]. Indeed, *n*-doping type decreases the Fermi level and consequently, the surface potential (9.1c). For an insulator, the definition of surface potential is scarcely reported in the literature. However, the surface potential reflects the charge distribution present in the insulation [22]. In the presence of electrons (holes) the surface potential is decreased (increased). So, as for the semiconductors, the surface potential should be considered as the energy difference between the vacuum level and the last energy level under the conduction band (CB).

9.2.2 *Surface Potential Measurement in AM-KPFM*

Kelvin Probe Force Microscopy (KPFM) permits to probe the surface potential difference between an AFM tip and the sample surface. This difference, named $\Delta\phi$, corresponds to Contact Potential Difference V_{CPD} which is expressed by:

$$\Delta\phi = V_{CPD} = \phi_m - V_S, \tag{9.1}$$

where ϕ_m is the work function of the conductive AFM probe and V_S is the sample surface potential. To enable surface potential measurements a lot of different KPFM modes were developed during the past decade. In this section, Amplitude Modulation KPFM (AM-KPFM) is presented to introduce the principles of surface potential measurement and their limitations.

9.2.2.1 Principles

The AM-KPFM is performed in lift mode, that is to say the measurement runs in two steps. During the first pass, surface topography is acquired in tapping mode (with mechanical oscillation of the AFM probe at its resonance frequency ω_0). During the second pass, the tip is moved away from the surface by a fixed height, h (called lift height) and the mechanical oscillation at ω_0 is stopped. Instead, a voltage V_{bias} is

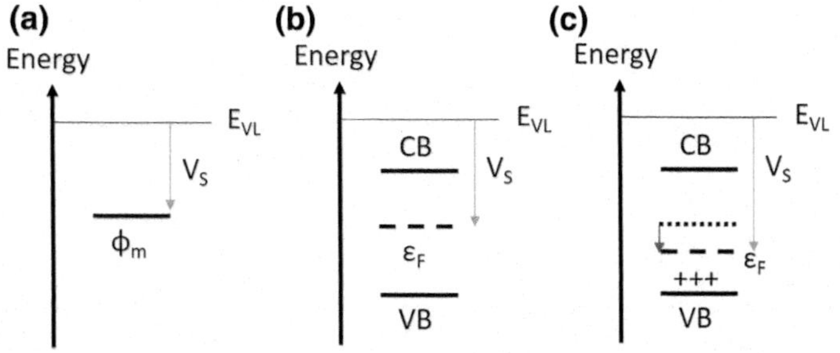

Fig. 9.1 Energy level of **a** metal and **b** semiconductor or dielectric material with the corresponding surface potential V_S. **c** In presence of doping (semiconductor) or charges (dielectric) the surface potential is modified. Positive charges induce surface potential increase. ε_F is the Fermi level. CB and VB are the conduction and valence bands, respectively. E_{VL} is the vacuum level

applied to the AFM probe constituted by a DC component V_{dc} and an AC component V_{ac} at the same pulsation ω_0 as the mechanical oscillations:

$$V_{bias} = V_{dc} + V_{ac}\sin(\omega_0 t) \tag{9.2}$$

During KPFM measurement, the AFM probe scans the material surface at lift height and the surface potential difference $\Delta\phi$ is superimposed to the applied bias. For V_{bias} applied on the tip, an electrostatic force F_e is induced on the AFM probe that depends on the sample tip distance z [23].

$$F_e = -\frac{1}{2}\frac{dC}{dz}(V_{dc} - \Delta\phi + V_{ac}\sin(\omega_0 t))^2, \tag{9.3}$$

with C being the capacitance formed by the AFM probe and the sample. The electrostatic force can be split into three components:

$$F_e = F_{DC} + F(\omega_0) + F(2\omega_0), \tag{9.4}$$

with

$$\text{the DC component}: F_{DC} = \frac{1}{2}\frac{dC}{dz}(\Delta\phi - V_{dc})^2 + \frac{V_{ac}^2}{2}, \tag{9.5}$$

$$\text{the } \omega_0\text{-component: } F(\omega_0) = -\frac{dC}{dz}(\Delta\phi - V_{dc})V_{ac}\sin(\omega_0 t), \tag{9.6}$$

$$\text{the } 2\omega_0\text{-component: } F(2\omega_0) = \frac{1}{4}\frac{dC}{dz}\cdot V_{ac}^2\cos(2\omega_0 t + 2\varphi) \tag{9.7}$$

In AM-KPFM mode, the V_{DC} bias is adjusted as to cancel the electrostatic force at ω_0. It is determined at each point and corresponds to the potential difference $\Delta\phi$ between the tip and the surface of the sample. Using this mode, it is therefore possible to probe simultaneously surface topography and potential difference between the tip and the surface.

9.2.2.2 AM-KPFM Performances and Limitations

According to the literature, the smallest surface potential value measured by AM-KPFM is of 5 mV [24]. It is more difficult to determine a common lateral resolution for all kinds of samples. Indeed, the lateral resolution depends on several parameters [24, 25] such as the AFM tip curvature radius, the lift height, the surface topography, the nature of material, etc. The lateral resolution is estimated between 10 and 100 nm [26, 27]. However, it is mainly limited by three parasitic contributions: (i) measurement environment, (ii) probe contribution due to parasitic capacitance and (iii) topography. These effects are detailed in the following:

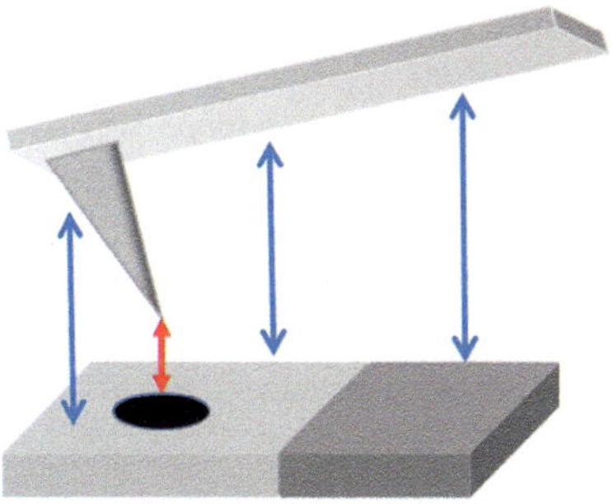

Fig. 9.2 Contribution of parasitic capacitance (blue arrow) compare to main contribution (red arrow) during KPFM measurement over a spot

- Environment
  KPFM measurements in air environment exhibit the worst resolution and the lowest potential sensitivity. This is mainly due to the water layer formation on the sample surface. Measurements performed under controlled environment (dry air or N_2 atmosphere) prevent from water layer formation and improve KPFM performances. However, measurements done under vacuum present the best spatial resolution and the highest sensitivity due to quality factor improvement and sample surface cleanness. So, this configuration is required for absolute surface potential measurements. Moreover, under vacuum, the surface topography is measured no more in tapping mode but in Non-Contact mode [28]. The later was extensively used to investigate thin dielectric films and managed to reach atomic resolution [29, 30].
- KPFM tip contribution
  Parasitic capacitance is the main issue during surface potential measurement by KPFM. Indeed, as shown on Fig. 9.2, surface features localized close to the measurement point (here a spot) contribute to the KPFM signal. Jacobs et al. [24] demonstrated that the parasitic capacitance reduces the measured potential and deteriorates the lateral resolution due to averaging effects. This phenomenon is amplified when the lift height is increased. To determine the real potential from the KPFM measurements, an approach based on deconvolution of the measured potential profile by a Gaussian type function was proposed [31]. This function, called Point Spreading Function (PSF) of the system represents the transfer function of the AFM probe. The Full-Width at Half maximum (FWHM) and the maximum of the PSE are calculated for given height of the lift and for curvature radius of the AFM probe [32] (Table 9.1).
- Topography influence
  The crosstalk between topography and surface potential mapping is mainly related to the dC/dz contribution in (9.6). For samples with flat surfaces, presenting roughness of only few nanometers, the contribution due to the crosstalk remains in the same range as the spatial resolution and the topography influence are quite small. However, topography features such as steps and grooves are often present. Zerweck et al. demonstrated that the real magnitude of a potential step is not reached for distances as large as 500 nm away from the step [27]. To weigh this influ-

ence, Sadewasser et al. have shown that step and groove topography has different contributions to the potential peak/step (Table 9.2) [33]. Moreover, this effect is strengthened when increasing lift height and tip oscillation amplitude during potential measurements [34]. Consequently, these two parameters should be maintained as small as possible to reach the best spatial resolution and to avoid artefacts. So, even if the AM-KPFM presents some drawbacks, it remains the most used KPFM mode.

9.2.3 Surface Potential Measurement in FM-KPFM

Another common mode for surface potential measurements is Frequency Modulation KPFM (FM-KPFM). In this mode, the surface potential can be measured in lift [35] or single pass [27, 36] configurations. The single pass mode permits to probe topography and surface potential simultaneously. In this mode, the frequency of AC excitation ω is lower than the mechanical resonance frequency ω_0. For single pass-measurement, this difference is exploited to discriminate topographical and surface potential contributions.

Contrary to AM-KPFM, the FM-KPFM close-loop tunes V_{DC} to null force gradient at ω, which is linked to the frequency shift Δf_ω by the following relation:

$$\Delta f_\omega \propto \frac{\mathrm{dF}(\omega)}{\mathrm{d}z} = -\frac{\mathrm{d}^2 C}{\mathrm{d}z^2}(\Delta\phi - V_{dc})V_{AC}\sin(\omega t) \tag{9.8}$$

The use of the force gradient allows better spatial resolution and measurement of the surface potential close to the real one. Indeed, this effect is confirmed by the corresponding PSF parameters that present lower FWHM and higher amplitude than the AM ones (Table 9.1). Moreover, the FM-KPFM mode has lower noise level compared to the AM-KPFM, thus leading to an increase of the sensitivity and allowing detection of smaller surface potential [35].

Relative merit of the AM and FM modes was studied in details by Meliz et al. [28] and Ziegler et al. [35]. In summary, the spatial resolution of V_{CPD} when measuring in FM-KPFM mode is higher than in AM-KPFM mode. In contrary, the energy resolution of V_{CPD} in measurements under FM-KPFM mode is lower than in AM-KPFM mode. However under AM-KPFM mode the V_{CPD} is measured from the resonance peak of the oscillating cantilever which greatly enhances the signal-to-noise ratio.

9.2.4 Surface Potential Measurement in PF-KPFM

Peak Force-KPFM (PF-KPFM) is a quite new mode introduced in 2010 by Bruker©. It combines the advantages of Peak Force Quantitative NanoMechanical (PF-QNM)

Table 9.1 PSF FWHM and maximum (peak) computed for AFM probe with curvature radius of 30 nm for different lift heights in AM and FM mode [32]

Lift height (nm)		2	5	10	20	50	100
AM-KPFM	PSF HWHM (nm)	9	20	33	48	82	165
	PSF peak	1.8×10^{-3}	3×10^{-4}	1×10^{-4}	4.5×10^{-5}	7×10^{-6}	1.5×10^{-6}
FM-KPFM	PSF HWHM (nm)	6	17	25	35	65	110
	PSF peak	8×10^{-2}	1×10^{-3}	7×10^{-4}	3×10^{-4}	1×10^{-4}	3×10^{-5}

Table 9.2 Influence of step and groove topography on potential peak/step profile

	Small topography step	Groove topography
Potential peak	Increase of the peak width Peak maximum remains constant	Maximum is slightly decreased
Potential step	Potential step shifts toward lower terrace	No general effect

mode and the high spatial resolution of FM-KPFM [37]. The PF-KPFM is performed in lift mode in two measurements steps:

(i) Topography is measured in PF-QNM which allows improvement of the lateral resolution, decrease of the interaction force while probing simultaneously surface mechanical properties [38].
(ii) Surface potential is measured in lift mode using FM-KPFM.

The main advantages of PF-KPFM are related to the AFM probe which is specially designed with high quality factor Q, low spring constant k and without coating (very small tip radius). As a consequence the surface potential profile can be probed by PF-KPFM with high spatial resolution and with close to the theoretical V_{CPD} values [37]. However, until now this mode is dedicated to surface potential mapping and does not appear suitable to investigate charge distribution in dielectric films. Indeed, a strong limitation of the method seems to be the repetitive contact of the AFM tip that can modify the charge distribution during contact step in PF-QNM topography measurement.

9.2.5 Photoconductive and Photo-KPFM Modes

Essentially three different current measurement methods using a conductive AFM tip have been developed depending on the available current range:

(i) Scanning Spreading Resistance Microscopy (SSRM) which uses logarithmic amplifier to probe current in the range from 100 nA to 100 μA. This is an indirect current measurement method because an abacus is needed to convert the measured bias to resistance [39].
(ii) Conductive AFM (C-AFM) that uses linear amplifier to probe current in the range from 100 pA to 100 nA.
(iii) Tunneling AFM (TUNA) which is a mode derived from the C-AFM and uses a low noise linear amplifier to probe current at lower range from 50 fA to around 100 pA [40].

From a general point of view the current measurements are obviously performed in contact mode. In this configuration the contact force between the AFM tip and the sample surface is a crucial parameter. Indeed, this force is determined as compromise between a high force to ensure reliable mechanical contact and a low force to avoid tip degradation (coating damage, tip radius increase, etc.).

Recently, in the attempt to find the best compromise for the contact force, a mode using Peak-Force technique named PF-TUNA was developed [41]. In this mode the current is probed in different configurations: Peak Current (current at the maximum contact force), Contact average Current (current mean value over the entire contact phase) and Cycle average Current (current mean value over the entire approach/retract process). Other advantage of this mode is to probe simultaneously surface topography, mechanical properties and electrical current. However, a comparison between the C-AFM and PF-TUNA revealed that higher contact force is required in PF-TUNA than in C-AFM to reach the same current value [42].

In some applications as solar cells or water-splitting photochemical cells, light is one of the main components of the system. Accordingly, to understand the influence of the active layer properties under illumination on the device performance, a light source was added to the AFM set-up to probe electrical current under dark and light conditions. Figure 9.3 represents a classical AFM set-up modified to provide sample backside illumination, thus allowing realization of electrical measurements (mainly C-AFM, KPFM and EFM) under light conditions [43]. Various light sources have been implemented to evaluate the measurement procedure: laser with accordable wavelength, reverse microscope and solar cell simulator. Independently on the light source, the light power density should be controlled accurately to avoid thermal effects that can induce abnormal carrier conduction or thermal expansion leading to an increase of the contact force between tip and surface [44].

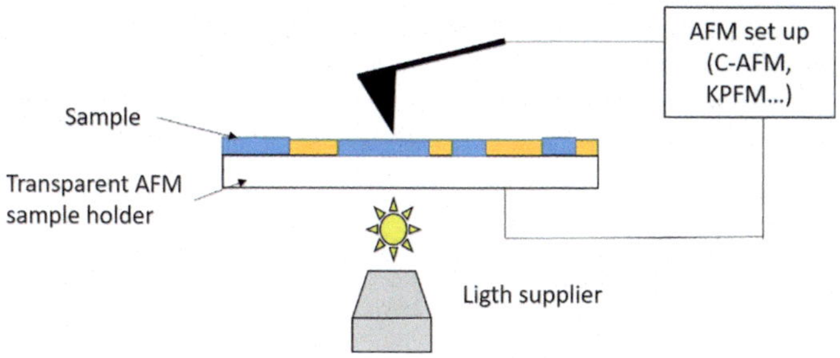

Fig. 9.3 Schematic diagram of AFM set-up for C-AFM/KPFM measurement under back side illumination

9.2.6 *KPFM Modelling*

The modelling of the electrostatic force and potential produced by interaction of the AFM tip and the sample surface faces the problem of strongly divergent field and requires multiscale approach. Therefore, simplifying hypotheses have been considered for the geometry. The aim of the modelling is multifold: first it serves as optimization tool for operating conditions and design of the instrument; second it may help identifying artefacts from useful signal and finally, it could help, in a reverse form to reach the charge distribution providing a given KPFM signal.

9.2.6.1 Electrostatic Force Modelling

The electrostatic force $\mathbf{F_e}$ between the AFM tip and sample surface can be expressed by the following equation:

$$\mathbf{F_e} = \frac{\varepsilon_0}{2} \iint_{Sonde} \|\mathbf{E}\|^2 \mathbf{dS} \tag{9.9}$$

with E the electric field, ε_0 is the dielectric permittivity and S in the sample surface.

According to the literature, different methods to compute the electrostatic force applied on the AFM tip were developed.

- Image charge model [45];
- Analytical model which computes the electrostatic force for each part of the AFM tip: sphere (tip apex), cone (tip main part) and cantilever [46]. Results provided by this approach present a good agreement with experimental ones for short tip-sample distances;
- Finite element model (FEM) reproducing the AFM tip in 2D axisymmetrical geometry [47, 48]. This model is typically used for interpretation of KPFM measurements;

– 3D FEM reproducing the real pyramidal AFM-tip shape [49]. This approach proposed recently, exhibits a good agreement with experimental results also for large tip-sample distances but requires longer computational times.

9.2.6.2 Surface Potential Modelling

In the current state, two different models were developed to reproduce KPFM measurements:

– Standard model. The electrostatic force is computed using one of the methods described above as a function of the bias applied on the tip V_{DC}. The relationship between the electrostatic force and the V_{DC} is of parabolic type with minimum occurring for the surface potential [50]. This model is commonly used even though it does not take into account the fact that the KPFM is not sensitive to image charge;
– Model proposed by Borowik et al., which considers the fact that the KPFM is insensitive to image charge [51]. The model accounts for side-capacitance and nonlinear effects taking place in the KPFM experiments. Concerning AM-KPFM, the surface potential V_{Surf} is expressed as follows:

$$V_{Surf} = -\frac{V_{DC}\left(F_{Q\text{-}V} - F_Q - F_V\right)}{2F_V} \tag{9.10}$$

For a given applied bias V_{DC}, $F_{Q\text{-}V}$ is the electrostatic force with charge Q and bias applied on tip, F_Q is the electrostatic force with charge Q (image charge effect) and F_V is the electrostatic force with applied bias V_{DC} on tip.

A similar model was proposed for FM-KPFM by Borowik et al. [51].

9.3 Local Charges Injection Mechanisms

Local charge injection and decay mechanisms, in thin dielectric layers are important issues due to their impact on the performances of a number of micro- and nano-devices. Indeed, even if the charge retention is exploited in non-volatile memories [52] for example, this phenomenon remains the main cause of failure in MicroElectroMechanical Systems [53] or CMOS devices with thin gate dielectric layers [54]. Consequently, charge injection and dynamics in thin active dielectric layers require characterization at local scale.

During the past decade, charge injection and retention were extensively studied locally using the electrical modes derived from AFM. First of all, the Electrostatic Force Microscopy (EFM) was employed [55–57]. However, the EFM sensitivity to image charge implies difficulties in the interpretation of the experimental results. Therefore, KPFM tends to be preferred to investigate charge injection and retention in thin dielectric films [58, 59].

From an experimental point of view, the charge injection and decay studies appear strongly influenced by the way charges are generated and by the measurement conditions. These issues related to the charging method and the KPFM measurement constrains will be detailed in the following part before providing some illustration of the KPFM potentiality for investigation of charge injection and decay mechanisms. In the last part the remaining bottlenecks and ongoing developments will be presented.

9.3.1 Local Injection Using Conductive AFM-Tip and Surface Potential Measurements

9.3.1.1 Methodology for Local Charges Injection Using AFM Tip

Charge retention in dielectrics was extensively studied at macroscopic scale by using Corona gas discharge for charging (ions deposition) and Kelvin probe technique for macroscopic surface potential measurements [60, 61]. In the early 90s, due to the AFM improvement, local charge injection using conductive AFM tip was developed. This approach was introduced by C. Schonenberger and based on triboelectrication as charging method [62]. In this configuration, the AFM tip is grounded and rubbed on the surface. Five years later, the charge injection in thin dielectric layers was performed using applied bias on a conductive AFM tip [63]. Two configurations are available for charge injection with AFM tip either in contact or in tapping mode (Fig. 9.4). In contact mode, a constant force $\mathbf{F_c}$ is maintained between the AFM tip and the sample surface during all charging process. In tapping mode, the conductive AFM tip oscillates at fixed amplitude/frequency close to the dielectric surface. For charging, a bias is applied on the AFM tip. This can be dc or ac bias whose characteristics influence strongly the charging process. After charging, the following step consists in probing the resulting surface potential KPFM (Fig. 9.4). The experimental results show that the surface topography is not influenced by the charges injection (Fig. 9.5a) whereas the surface potential is modified by the injected charges cloud (Fig. 9.5b). During the charging process, the AFM tip can be moved over the dielectric surface resulting in a contour for the injected charges cloud. This process named electrostatic patterning is illustrated on Fig. 9.5c, d. In this example, charges are generated in tapping mode after applying ac-bias on the AFM-tip. So, the charges injection using AFM tip appears a complex process due to the combination of injection mode (contact vs. tapping), patterning (local or surface charging) and applied bias (dc vs. ac). The relative merit of each configuration is summarized in Table 9.3.

According to the charge injection configuration three main approaches can be identified in the literature:

(i) Electro-triboelectrification which consists in applying DC-bias on the AFM tip brought in contact with dielectric layer and rubbed over it [64, 65];

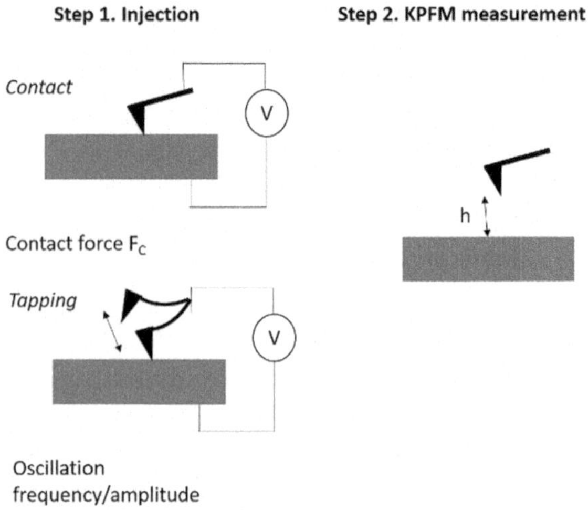

Fig. 9.4 Injection configuration scheme in contact and in tapping mode

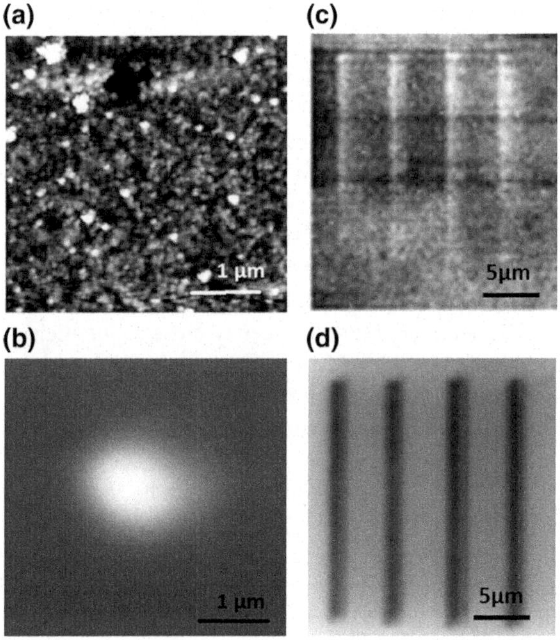

Fig. 9.5 **a** Topography and **b** resulting surface potential for dc-injection in contact in 40 nm-thick SiO_xN_y dielectric layer (2 min at 30 V). **c** Topography and **d** resulting surface potential for ac-injection in tapping mode in 40 nm-thick SiO_xN_y. Injected area 25 µm-long with bias pulse of 60 V

Table 9.3 Comparison of advantages and drawbacks of different injection conditions

	Injection mode		Pattern		Applied voltage	
	Contact	Tapping	Point	Contour	DC	AC
Advantages	Reliable injection conditions	Less tip damage	Simple shape of potential profile	Tunable dimensions	Reliable injection conditions	Amount of injected charge higher than DC
Drawbacks	Possible tip damage	Injection conditions roughly evaluated	Charged area depends on material	Possible tip damage	Amount of injected charge weaker than AC	Possible parasitic tip oscillations Charging mechanism

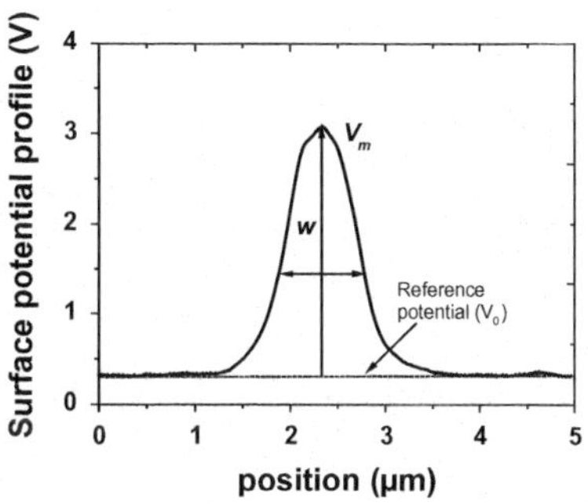

Fig. 9.6 Surface potential profile cross-section over contact point for punctual charge injection (2D surface potential map is depicted on Fig. 9.5b)

(ii) Charge injection in contact mode applying DC-bias on fixed AFM tip. This is mainly used to study punctual charge injection process with contact between the tip and the sample surface thus improving the injection configuration repeatability [66, 67];

(iii) Electrostatic patterning configuration using tapping mode and DC-bias to produce injected charge pattern. This configuration is used for applications in nanolithography for nano patterning [68].

For punctual charge injection configuration (Fig. 9.5b) the lateral cross-section of the potential profile exhibits a Gaussian shape (Fig. 9.6). Three parameters characterizing the injected charge can be extracted from this experimental profile:

(i) Full-Width at Half Maximum (FWHM) which reflects the lateral spreading of the injected charge;

(ii) Surface potential maximum V_m relative to the reference surface potential without charge;

(iii) Area under the potential profile I_s which is supposed to represents an image of the amount of injected charge. Indeed, when the charges are located on the surface (or close to surface compared to tip radius) the area under the potential profile, i.e. I_s, is proportional to the amount of injected charge [63].

Moreover, these parameters can be used to provide information on the process of charge release, in particular to distinguish charge surface spreading and charge drift in the insulation bulk.

9.3.1.2 Influence of the Experimental Conditions on the Charges Injection Process

Various studies emphasize that the experimental results provided by KPFM are very sensitive to the experimental conditions, such as nature of the tip, polarity and magnitude of the applied voltage, charging time, etc. These are parameters easy to control. So, they will not be involved in the analysis here. On the contrary, other subtler parameters have strong contribution to resulting surface potential as explained in the following.

- Environmental conditions
  The influence of environmental conditions (humidity or ambient gas) on KPFM measurement after local charge injection depends strongly on the dielectric material under study. High humidity level in the measurement chamber induces a water layer build-up on the dielectric surface which creates a meniscus with the AFM tip. This phenomenon has different impacts:

  (i) Increasing charge dissipation (mainly through lateral charge spreading) due to charge deposition in water layer instead of on the dielectric layer. In this case, the KPFM results reflect mainly charge dynamic in the water layer [7, 58];
  (ii) Local oxidation due to chemical reaction in the water layer induced by the strong electric field. This effect results in modification of the surface topography which in general is not altered by the applied voltage on the tip during topography measurement [69];
  (iii) No real impact mainly due to dielectric layer hydrophobic properties [70]. Moreover, Sridhara et al. reported that in dry conditions, the ambient gas may impact charge injection and decay [71]. Consequently, strict control of the environmental conditions is essential for gathering reliable and reproducible results.

- Apparent height effect
  When the amount of injected charge is large, the electrostatic force could parasite the AFM probe control during topography measurements. This induces a correlation between topography and surface potential mapping and an apparent height is observed on topography map as shown on Fig. 9.5c for high injection bias. Ziegler et al. demonstrate that this effect can be reduced by applying bias during topography measurement [72]. As the topography is reproduced during measurement

with KPFM in lift mode, this apparent height induces surface potential decrease which is not linked to the charge decay in the dielectric layer.

9.3.2 *Charges Injection and Decay in Thin Dielectric Layers*

9.3.2.1 Issue About Charges Injection Mechanisms

From an outlook of the literature, it can be realized that charges stabilization and release in thin dielectric layers were much more extensively studied than the injection mechanisms. The identification of mechanisms occurring during charges injection from metalized AFM-tip into dielectric layers is not straightforward. Indeed, the local character of mechanisms, the highly diverging nature of the electric field, its enhancement close to the tip and the small distance between tip and sample surface prevents the use of macroscopic models. However, this aspect is ill referenced in the literature. Up to now only few mechanisms were proposed to explain the charge injection in thin dielectric films by AFM. Historically, based on macroscopic approaches, injection by corona discharge was evoked as the main mechanism for dielectric charging by AFM [73]. However, recent theoretical calculations seconded by experimental demonstration, confirmed that corona discharge cannot develop in such small tip-to-sample distances [74]. Instead, the field electron emission enhanced by thermionic electron emission is proposed as the main mechanism.

Moreover, the influence of tip characteristics (geometry, work function of the coating…) is rarely considered. Sun et al. briefly addressed the influence of the material work function of the AFM tip on the charge injection, by comparing features from two tips with different coatings [75]. This was completed by Villeneuve-Faure et al., demonstrating that the work function of the metal coating of the tip influences carrier injection differently depending on the applied voltage polarity [14]. Indeed, electrons injection follows the Schottky barrier law whereas holes seem to be insensitive to the injection barrier height. Holes injection would be mainly driven by interface states. These aspects remain to be revised after deeper investigation.

9.3.2.2 Charges Injection and Decay

According to Morita et al., when charges are stored locally on an insulating material, two dissipation processes can be envisaged [59]:

(i) Surface charge spreading, which is a conservative mechanism for the charge quantity. Thus, the integrated intensity under the potential profile I_s is constant with time, whereas the maximum potential V_m decreases. Moreover, strong peak broadening is observed, i.e. increasing of FWHM;
(ii) Drift of charges in the bulk of the material with/without charge dissipation or recombination that appears conservative/non-conservative process for the

measured potential. Indeed, when charges penetrate into the bulk layer, their influence on the tip decreases with the distance from the surface. The integrated intensity I_s and the maximum potential V_m decrease with the same kinetics. When the two phenomena (surface diffusion and volume intake) are superimposed, the integrated intensity I_s and the maximum potential V_m decrease with different time constants. It is to note that already at the charging step, the charges spread: the lateral electric field radiated by the biased tip promotes charge lateral spreading whereas electric field at contact point controls the amount of injected charges [14].

Considering a particular application one can investigate different phenomena. The main failure mechanism in capacitive Radio Frequency MicroElectroMechanical Sytems (RF-MEMS) with electrostatic actuation is related to charging effect in the dielectric layer which induces stitching of the mobile membrane [53]. To mitigate this effect, one way is to provide dielectric layer favoring charge release, nevertheless keeping high insulating characteristics. One possibility is to create plasma processed dielectric layers with gradual variation of their electrical properties, for example amorphous silicon oxynitride layers SiO_xN_y (with only 4 at.% of N in the layer) [76]. In this case, the Si concentration is tuned in the layer by acting on the γ parameter which reflects the ratio of the partial pressures of gas precursors in the plasma ($\gamma = N_2O/SiH_4$). Figure 9.7 compares surface potential decay in SiO_xN_y layers with different structural characteristics. Results emphasize a strong influence of the material properties on the charge decay. Indeed, in the SiO_xN_y ($\gamma = 100$) layer, which is close in composition to thermal silica, the resistivity is high (22.0×10^{14} Ω m) and charges remain trapped in the dielectric layer. No broadening of the potential peak and only weak I_s decrease are observed. In the SiO_xN_y ($\gamma = 10$) layer, although the resistivity remains high (4.9×10^{14} Ω m) the charges spread laterally. The potential peak broadening is important. Moreover, an important decrease of the peak I_s area is observed. It is concluded that surface diffusion and volume intake mechanisms of charge occur simultaneously for this layer. Finally, in the SiO_xN_y ($\gamma = 5$) layer, which possesses slightly higher concentration of Si and consequently a bit lower resistivity (3.8×10^{14} Ω m) compared to SiO_xN_y ($\gamma = 10$) layer, high I_s decrease is observed alongside with weak potential peak broadening. Here, the main charge release mechanism is the volume one which is ascribed to high conduction during the injection process [67].

Concerning floating gate MOS memories, a thin dielectric layer with embedded nanoparticles [77] or quantum dots appears promising for the device performance in terms of charging and retention times [78]. Lwin et al. demonstrated that the work function of metallic nanoparticles has a strong influence on the charge decay in the device active layer [77]. Indeed, as shown on Fig. 9.8 the charge decay and spreading dynamics are quicker for Au nanoparticles than for Pt ones, with Au having lower work function compared to the Pt-work function.

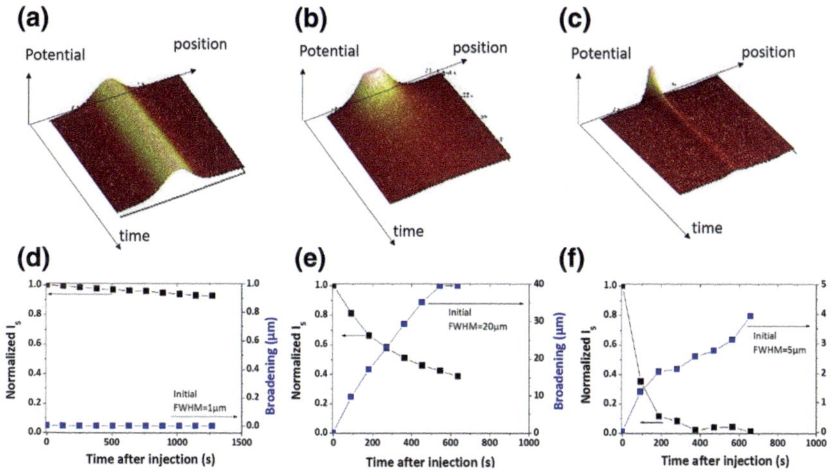

Fig. 9.7 Evolution of surface potential profile (**a**–**c**) and normalized I_s and broadening (**d**, **e**) as a function of time after injection for SiO_xN_y layers elaborated by plasma process with different gas ratio γ [76]: **a**, **d** $\gamma = 100$, **b**, **e** $\gamma = 10$ and **c**, **f** $\gamma = 5$. Charge injection in contact mode during 1 min at 25 V

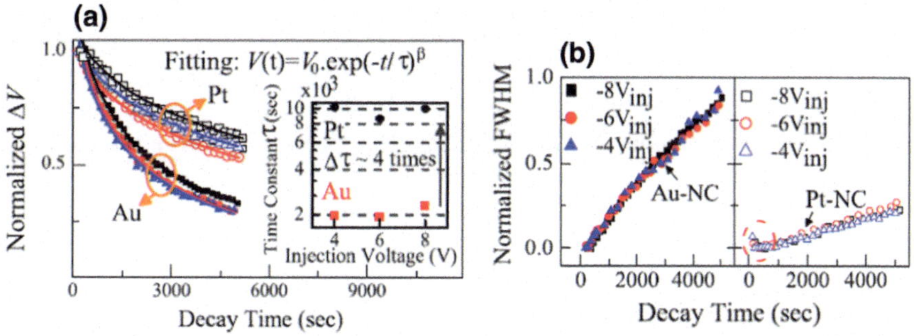

Fig. 9.8 Time dependence of normalized **a** maximum potential and **b** FWHM for sample with Au and Pt nanoparticles after charge injection of −4, −6 and −8 V during 10 s. The inset shows characteristic time constant extracted from the curve fitting. Reproduced from Lwin et al. [77]

9.3.2.3 From Surface Potential Measurement to Charge Density Determination

Following the evolution of KPFM profile in time after charge injection provides information about the charge dissipation processes but to identify the physical mechanisms, actual charge density and charge distribution are necessary. Different attempts to extract charge density from KPFM measurements after local charge injection are reported in the literature. However, the two main limitations to reach this goal are related to measurements of the real surface potential and to hypotheses about the in-depth charges cloud extension. The mainly adopted hypothesis in a lot of studies is that the charges are located on the dielectric surface [64, 65]. This hypothesis seems valid for charges injection using triboelectrification but appears questionable for contact/tapping mode of injection. Other hypotheses are proposed in this case like charge located at internal interface [66] or distributed in the entire volume of the dielectric layer [71].

Experimental results show that the measured surface potential is lower in intensity and broader in shape than the theoretical one mainly due to parasitic capacitance between the AFM probe and the sample. To overcome this issue, Xu et al. proposed to use PSF function (cf. part 2.2.2) to extract the real surface potential induced by the injected charge [79]. Under the hypothesis that charges are located on the surface, the authors concluded that the deconvoluted profile provides twice higher charge density with value closer to the macroscopic one.

To try to reproduce the real configuration, Palleau et al. investigated the influence of charges penetration in thin PMMA layers [48]. In this study charges are injected using electrostatic patterning and charge density is determined for various charges area dimensions. As shown in Table 9.4, for a fixed charge pattern, the charge density is influenced by hypothesis about their in-depth penetration. The more charges are considered to spread in the volume the lower is the density needed to reproduce KPFM signal.

Up to now, no one of the reported methods based on KPFM measurements has permitted to extract 3D charge density without hypothesis about the charge in-depth behavior. To face this problem approaches, based on Electrostatic Force Distance Curve (EFDC) have been recently proposed [15]. The method based on Force Distance Curve measurements appears promising to probe space charge in 3D in thin dielectric layers due to its sensitivity to charge localization [80]. However, the EFDC modeling needs to be improved and inverse method to be developed in order to ascribe unique charge density profile to each experimental EFDC.

Table 9.4 Charge density as function of pattern size and charge penetration depth p used in the modelling [48]

Pattern size (μm)	Charge density (C/m^2)			
	p = 1 nm	p = 10 nm	p = 50 nm	p = 100 nm
1	3.75×10^{-3}	3.6×10^{-3}	2.8×10^{-3}	1.9×10^{-3}
5	3.65×10^{-3}	3.5×10^{-3}	2.75×10^{-3}	1.75×10^{-3}
10	3.25×10^{-3}	3.05×10^{-3}	2.45×10^{-3}	1.6×10^{-3}

9.4 KPFM for Space Charge Probing in Semiconductor and Dielectric Materials

As shown previously, the versatility of the KPFM technique provides a guess for the fate of injected charges (lateral vs. in-depth spreading) in non-invasive way. However, sometimes the investigated structures are real devices (PN junctions, solar cells…) in which metal/dielectric, metal/semiconductor or semiconductor/semiconductor interfaces are present and should be considered. Palermo et al. demonstrated that the KPFM is a powerful technique to characterize thin films for electronic applications, however without extracting information about the amount of related charge [26]. This points to the need of methodology for estimation of the charge distributions using KPFM to investigate charge generated in processes under external stress (electric field, light…) In the following, methodologies developed to extract charge density profile from surface potential measurement are presented, before addressing their application to different devices.

9.4.1 KPFM Measurements on Bias Electronic Devices: Challenge and Bottleneck

During KPFM measurements over real biased electronic devices (organic/inorganic FET, CMOS, solar cells…), the bias applied on electrodes (Fig. 9.9) induces electrostatic forces which impact the measured topography. The apparent height effect discussed above decreases the KPFM sensitivity close to the electrodes [81]. A way to avoid this effect is to investigate charge behavior after bias application. This method is mainly valid for dielectric layers due to charge trapping effect. Another way is to exploit the Feed-forward compensation proposed by Ziegler et al. during KPFM measurements [72]. More recently, Bercu et al. proposed a modified AM-KPFM set-up which avoids error on surface topography due to biased electrodes [82]. The proposed system uses an external voltage source (for electrode biasing) and synchronized AFM tip which permits to apply bias on the electrode only during surface potential measurements.

9.4.2 Methodology for Charge Density Profile Determination from KPFM Measurements

To extract the space charge density profile $\rho(x)$ from the surface potential profile $V_s(x)$ measured by KPFM three approaches were explored in the literature based on solution of the Poisson's equation for electrostatics. The first approach, named Second Derivative Model (SDM) is the simplest one and relies on the hypothesis that the probed surface potential is representative of the potential distribution in the layer

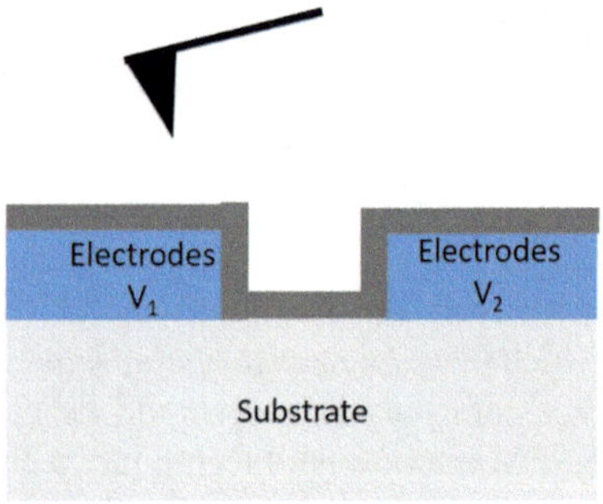

Fig. 9.9 KPFM configuration during measurement over real device under bias. In grey, dielectric or semiconductor active probed layer

volume. Here, the charge distribution is deduced from the second derivative of the measured surface potential (ε_0 is the vacuum permittivity, ε_r the relative permittivity of the dielectric material):

$$\rho(x) = -\varepsilon_0 \varepsilon_r \frac{d^2 V_S}{dx^2} \tag{9.11}$$

The spatial resolution in this approach is sensitive to the derivation step dx and to the noise level superimposed to the KPFM recorded profile [83]. Refining the derivation step is mandatory to improve spatial resolution. This method is also used, in two dimensions, to extract charge density profile for localized injected charge [84].

The second approach is based on the same hypothesis (charge distribution is obtained using 9.11), but prior to the derivation step a signal treatment is applied on the raw data to reduce noise impact [85]. The most powerful smoothing method for derivation approach is Savitzky and Golay (SG) one [86]. In this method a fixed degree polynomial function is used to fit the experimental data on a fixed amount of equally-spaced data points named "data point window". The fitting process is based on least squares polynomial regression. In the follow this method will be named SG-SDM. Using this method Faliya et al. optimized the technique in term of "data point window" and noise robustness [85].

The last method is based on Finite Element Model (FEM). The main advantage of this method is to reproduce the real sample geometry without assumption on the potential distribution in 3D [87]. The Poisson equation is solved in two dimensions in the dielectric layer and in the surrounding air box to determine the potential distribution V(x, z):

$$\frac{d^2V}{dx^2} + \frac{d^2V}{dz^2} = \frac{\rho(x,z)}{\varepsilon_0 \varepsilon_r} \tag{9.12}$$

In this model, a hypothesis on the shape of the charge density cloud is needed. The process followed to determine charge density is a sequential one: (i) Initial charge density profile is supposed; (ii) The surface potential map is computed using (9.12) and compared to the experimental one; (iii) If the difference is less than the KPFM resolution the charge density is determined, otherwise the initial charge density is modified and the surface potential is computed till convergence with experimental profile.

9.4.3 Applications to Dielectrics and Semiconductors

9.4.3.1 Charge Dynamic in Solid Electrolytes

The development of improved lithium-ion batteries needs understanding of the charge dynamic inside the solid polymer electrolyte. In this perspective, KPFM measurements appear very appropriate. In the last decade numerous studies focused on the charge dynamic in poly(ethylene oxide) (PEO), a ion conductor polymer, which is promising material for solid electrolyte applications. In this context, Martin et al. extracted charge density profile in PEO using the SDM method. However, due to poor signal/noise ratio the obtained results are not subject to interpretation [88]. Few years later, Faliya et al. proposed the SG-SDM method to overcome the influence of noise on the surface potential measurements [85]. Optimizing the SG smoothing process, the authors managed to extract space charge density in PEO (Fig. 9.10c) from noisy surface potential profiles (Fig. 9.10a). Following the same approach results were obtained for the PEO layer after applying external stress and charge oscillation in time (Fig. 9.10b, d). Note that in this last case, the influence of the electric field on topography is not compensated.

9.4.3.2 Charge Density in Thin-Films Transistors

KPFM was used to investigate charge density distribution in thin-films transistors. Due to mobile charges, which are present and transported continuously in the device, the KPFM measurements and interpretation of the obtained profiles in operating transistor devices appear challenging [25]. In this case the surface potential is measured by KPFM between source and drain with or without applied potential. The resulting surface potential profile is being interpreted using the SDM method to obtain two kinds of information. The first one is related to the investigation of different doped regions. Applying this methodology to a doped drain of a metal–oxide–silicon field effect transistor, Henning et al. distinguished relative changes in dopant concentration with lateral resolution of less than 100 nm [89]. They emphasize that, even though this method does not provide absolute dopant concentration, it is sensitive to changes in the dopant concentration, from 10^{15} to 10^{20} cm^{-3}. The same kind of information was obtained by Kryvchenko et al. applying bias between drain and source to accentuate the contrast in surface potential and to improve resolution (Fig. 9.11) [90]. However, as shown on Fig. 9.11b these authors do not reach quantitative values for the amount of charge. The second kind of information concerns process issues. As an example, Kryvchenko et al. investigated contact metal diffusion in In_2O_3 thin-film transistor

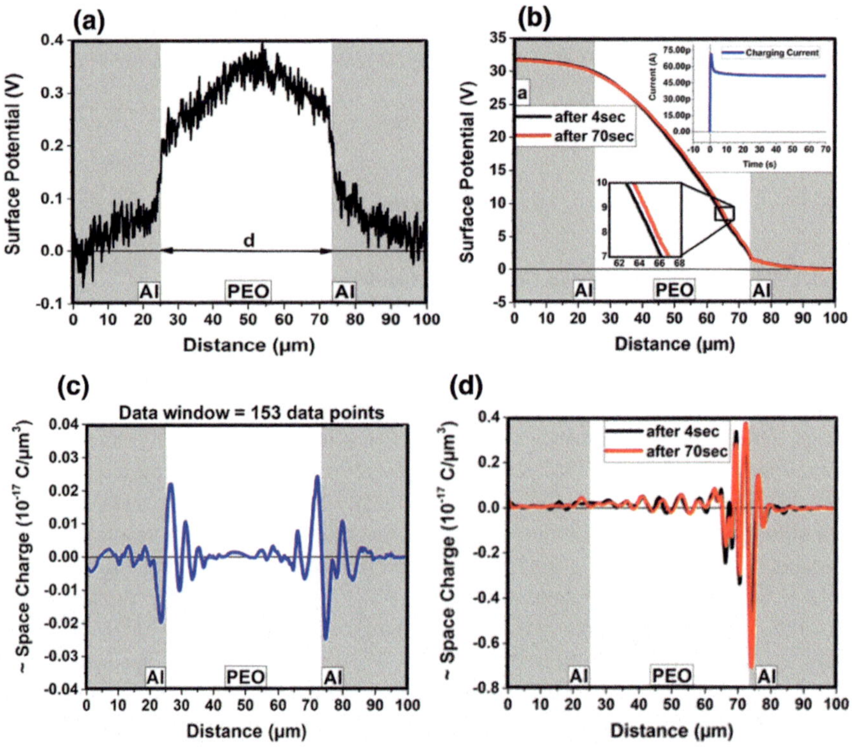

Fig. 9.10 Surface potential profile measured by KPFM over Al/PEO/Al structure **a** without applied bias and **b** 30 V (ground) applied on left (right) electrode. Resulting space charge density profile obtained using SG-SDM method **c** without applied bias and **d** 30 V (ground) applied on left (right) electrode. Reproduced from Faliya et al. [85]

devices [90]. As shown on Fig. 9.11a, surface potential close to the right electrode is poorly reproduced by Technology Computer-Aided Design (TCAD) software even without applied voltage on the drain. Indeed, charges are located on this electrode (Fig. 9.11b) which is due to contact metal diffusion.

9.4.3.3 Injection at Metal Dielectric Interface

KPFM measurements were used to investigate charge injection at metal/dielectric interface. As an example Okamoto et al. investigated the influence of $BaTiO_3$ degradation close to the anode for multilayer ceramic capacitor applications [91]. In another study, the charge injection at SiN_x/Al interface was investigated using FM-KPFM measurements after lateral Al-electrode polarization. Figure 9.12a represents the resulting surface potential with negative peak close to cathode and positive peak to anode. Due to the profile shape, a Gaussian distribution of the charges density at

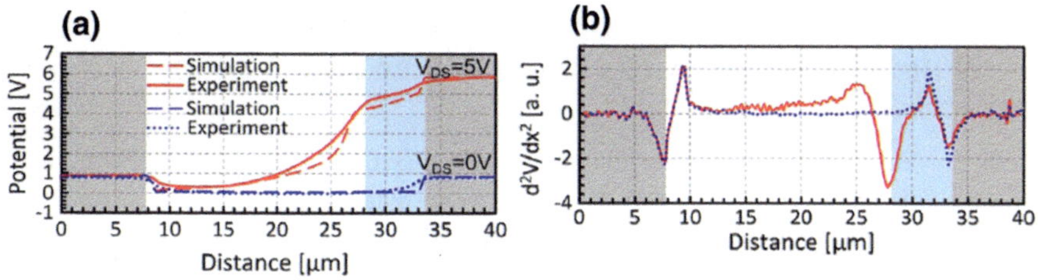

Fig. 9.11 **a** Comparison of surface potential profile measured by KPFM and simulated by TCAD software at $V_{DS} = 0$ V and $V_{DS} = 5$ V when $V_{GS} = 0$ V. **b** Calculated second derivative of the measured surface potential at $V_{DS} = 0$ V (dashed line) and $V_{DS} = 5$ V (solid line) represent the measured charge profiles. Reproduced with permission from Kryvchenko et al. [90]. Copyright (2016) American Chemical Society

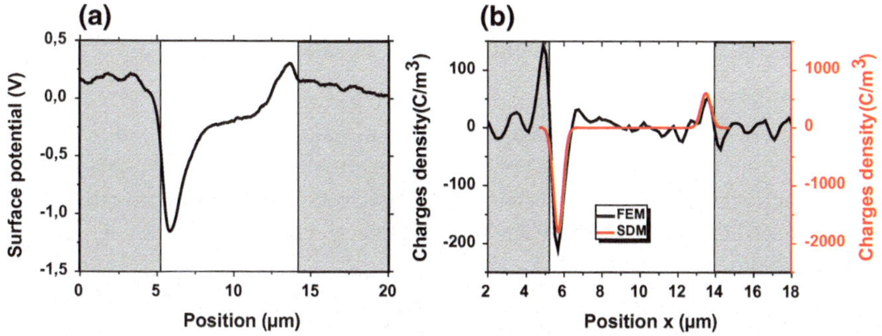

Fig. 9.12 **a** Surface potential profile probed by FM-KPFM and **b** charge density profiles extracted from the surface potential profile measured using the SDM and the FEM models

both electrodes is supposed. Thus, for positive and for negative charges, the density is expressed as:

$$\rho(x) = \rho_0 \exp\left(-\frac{(x - x_0)^2}{0.36W^2}\right), \tag{9.13}$$

where ρ_0 is the maximum value (positive for holes and negative for electrons), x_0 is the position at which the maximum density occurs and W is the FWHM of the charges cloud.

SDM and FEM methods were applied to the FM-KPFM profile showing that both methods provide the same shape for charge density (Fig. 9.11b). However, the amount of charges found is different [87]. Indeed, the authors demonstrate that the FEM method, which reproduces the real geometry of the device, provides more representative value for the charge density. Therefore, for quantitative charge density determination the FEM method should be preferred.

9.5 Nanoscale Opto-Electrical Characterization of Thin Film Based Solar Cells Using KPFM and C-AFM

Energy production is the defining challenge of the 21st century as the global demand is projected to more than double by 2050 [92]. An exponential increase of the production capacity from the solar energy is highly required. Two approaches have been proposed to achieve cheaper photovoltaic (PV) electricity [11]:

(i) Increasing the power conversion efficiency while keeping PV-materials costs the same (Si-wafer-based solar cells (1st generation) or high efficiency concepts ("all-Si" tandem cells for 3rd generation);
(ii) Developing low-cost, moderate efficiency PV-materials (thin-film PV, 2nd generation). Organic photovoltaic (OPV) are an important emerging technology, part of 2nd generation, which promise to provide solar-to-electric energy conversion in portable, light-weight packages, at extremely low cost;

Independently on the technology, PV or OPV, device performances are controlled by phenomena occurring at nanoscale (exciton formation by light, excitation dissociation at p/n junction, carriers transport in p or n layers/domains). So, charges creation under illumination and their transport in the active layer should be investigated at local scale. To that end, as shown in the following, three kinds of measurements are implemented:

(i) Current or surface potential mapping with or without illumination to localize the area where charges are created or collected by top/bottom electrode;
(ii) Localized current versus voltage measurements to investigate charge transport mechanisms and determine material properties as charge mobility;
(iii) Time resolved measurements to follow the current or surface potential evolution in time during light/dark transition and to investigate charge dynamics;

These techniques first developed for solar cells are now frequently applied for photoelectrochemical cells involved in water splitting applications [93].

9.5.1 *Mapping Measurements*

The influence of illumination on surface potential [94–96] or current [97–100] maps was extensively studied in the literature. Figure 9.13a, b show the evolution of surface potential map over PFB:F8BT blend film for different light wavelengths. At long wavelengths, light does not create excitons in the both materials due to the low absorption (Fig. 9.13a) and surface potential is not modified. According to results from absorbance spectroscopy, other wavelengths are chosen to excite only F8BT (370 nm) or for both materials inducing a decrease of their surface potential due to the photocharge effect.

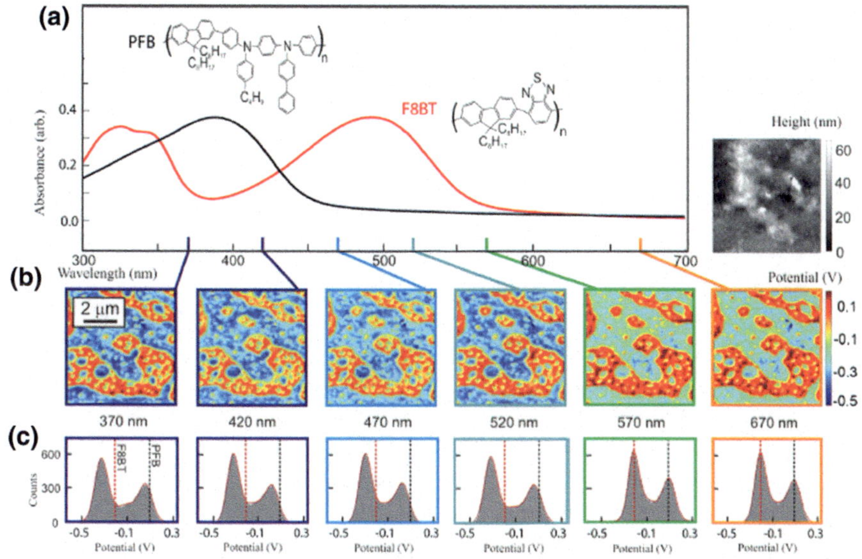

Fig. 9.13 Evolution of surface potential map of PFB:F8BT film illuminated under different wavelengths. Reprinted with permission from Luria [94]. Copyright 2012 American Chemical Society

9.5.2 Localized Current-Voltage Measurements

Concerning electrical measurements at local scale two kinds of measurements can be performed: current versus voltage and time resolved characterization. According to the literature, current versus voltage measurements are the most frequently performed ones. Direct interpretation of the obtained results provides information for the voltage threshold evolution as a function of light power [93] or for the current as a function of light wavelength or for minimum photon energy sufficient for the photoexcitation of mobile charge carriers [101]. However, deeper analysis of the results is not straightforward. The main limits are imposed by the used model and determination of the collection surface. Indeed, the electrical characteristics were mainly obtained on the basis of Space Charge Limited Current (SCLC) model to extract carrier mobilities [102]. However, many studies point out that the mobility determined using C-AFM measurements is of some orders of magnitude higher than the mobility obtained in macroscale way [103]. This difference is related to the tip-plane configuration during C-AFM measurements. Indeed, this configuration induces inhomogeneous electric field distribution with large field enhancement close to the tip compared to the one computed in plane-plane configuration. Reid et al. proposed a modified SCLC model to consider the actual electric field (and current density inhomogeneity) due to the tip-plane configuration [103]. This approach provides carrier mobility in the same order of magnitude as the C-AFM and macroscale measurements.

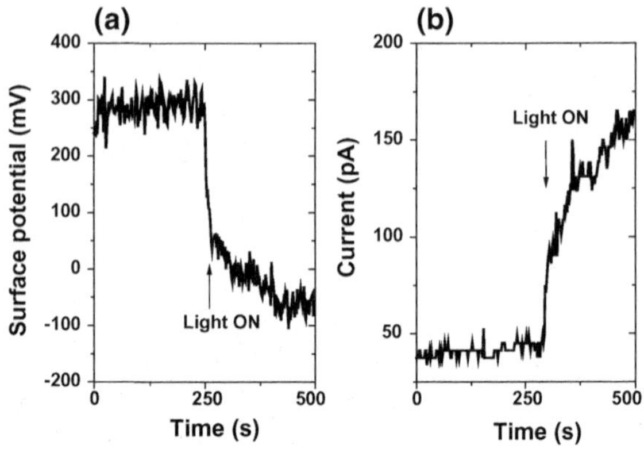

Fig. 9.14 Evolution of **a** photovoltage and **b** photo-current (polarization bias of 3 V) for 40 nm-thick P3HT:PCBM blend

Concerning determination of the collection surface some improvements have been proposed:

(i) Contact area is computed using the Hertz approach and corresponds to the mechanical contact area [104];
(ii) Effective surface is determined by fitting current-voltage experimental curve. This surface depends on several parameters like tip work function, contact force and dielectric thickness, and presents a broad range of values from 10 to 100 nm^2 [105];
(iii) SEM observations of the AFM tip after measurements [40]. Moeman et al. proposed an approach to estimate the radius of area in which the current is collected as function of the sample mechanical properties [106].

Time resolved electrical measurements consist in acquiring surface potential and/or current in time during or after illumination to probe the charge transport. Figure 9.14 shows surface potential and current evolution during light and dark cycle. After illumination the surface potential decreases whereas the current increases. The response time can be related to the device performances. Indeed, Coffey et al. reported that the response time is determined by the solar cell external quantum efficiency [107].

Current versus voltage and time resolved measurements provide complementary information on the electrical properties of devices. Indeed, combining these techniques Villeneuve-Faure et al. showed that in the P3HT:PCBM blend charges are trapped during light exposition which induces a decrease of the Schottky barrier to injection and a modification of the current versus voltage characteristics [108].

9.6 Conclusion and Overview

KPFM and C-AFM appear as powerful techniques to investigate physical phenomena related to charges injection and transport in the active layers of different devices. However, their performances remain limited by the instrumental set-up achievements and existing methodologies for interpretation of the obtained results. From experimental point of view the KPFM and C-AFM performances are mainly limited by the tip characteristics. Indeed, the AFM tips used for electrical measurements are silicon-tips with conductive coating and present greater tip radius than silicon one (from 25 nm (Pt-coating) to 125 nm (diamond-coating) vs. only around 5 nm for bare Si-tip). This implies worse spatial resolution either in topography or in electrical properties and higher parasitic capacitance (lower sensitivity). To improve performance new probes are under development as Si-doped conductive probes used for PF-KPFM or PF-TUNA modes.

The results interpretation appears strongly limited by the lack of appropriate modelling of the interaction between the AFM probe and the studied sample (electrostatic force, contact area…). Concerning space charge measurements by using KPFM, 2D-measurement methods are available and can be used to investigate charge injection and transport at the interfaces appearing in many devices. However, up to now no existing technique allows to probe localized charges in 3D without hypothesis on the charge distribution (shape, in-depth penetration…). These issues require mutual evolution of both the AFM instrument with new probing method really sensitive to the charge localization in the volume and the modelling of electrostatic interaction between the AFM tip and the sample (real tip geometry as example). Concerning charge dynamic probing by C-AFM, the drawbacks are mainly related to results interpretation. The used modelling approaches should be improved to consider the real sample geometry (and the related electric field heterogeneity) and the collection surface to extract the electrical properties of the studied devices.

References

1. A. Galembeck, C.A.R. Costa, M.C.V.M. da Silva, E.F. Souza, F. Galembeck, Scanning electric potential microscopy imaging of polymers: electrical charge distribution in dielectrics. Polymer **42**, 4845 (2001)
2. D.J. Lacks, R.M. Sankaran, Contact electrification of insulating materials. J. Phys. D: Appl. Phys. **44**, 453001 (2011)
3. T. Shinbrot, M. Rutala, H. Herrmann, Surface contact charging. Phys. Rev. E **96**, 032912 (2017)
4. G. Bersuker, P. Zeitzoff, G. Brown, H.R. Huff, Dielectrics for future transistors. Mater. Today **7**, 26–33 (2004)
5. D.J. Fonseca, M. Sequera, On MEMS reliability and failure mechanisms. Int. J. Qual. Stat. Reliab. **2011**, 820243 (2011)
6. B. Bhushan, Nanotribology and nanomechanics of MEMS/NEMS and BioMEMS/BioNEMS materials and devices. J. Microelectron. Eng. **84**, 387 (2007)
7. U. Zaghloul, G. Papaioannou, B. Bhushan, F. Coccetti, P. Pons, R. Plana, On the reliability of electrostatic NEMS/MEMS devices: review of present knowledge on the dielectric charging and stiction failure mechanisms and novel characterization methodologies. Microelectron. Reliab. **51**, 1810–1818 (2011)
8. U. Zaghloul, B. Bushan, P. Pons, G.J. Papaioannou, F. Coccetti, R. Plana, On the influence of environment gases, relative humidity and gas purification on dielectric charging/discharging processes in electrostatically driven MEMS/NEMS devices. Nanotechnology **22**, 035705 (2011)
9. W. Zhou, J.B. He, X.P. He, H.J. Yu, B. Peng, Dielectric charging induced drift in micro device reliability-a review. Microelectron. Reliab. **66**, 1–9 (2016)
10. C. Zhao, C. Zhou Zhao, S. Taylor, P.R. Chalker, Review on non-volatile memory with high-k dielectrics: flash for generation beyond 32 nm. Materials **7**, 5117–5145 (2014)
11. M.C. Scharber, N.S. Sariciftci, Efficiency of bulk-heterojunction organic solar cells. Progress in Polymer Science **38**, 1929–1940 (2013)
12. J. Le Rouzo, D. Duché, C. Ruiz, F. Thierry, M. Carlberg, G. Berginc, M. Pasquinelli, J.J. Simon, L. Escoubas, F.R. Flory, Specific tools for studying the optical response of heterogeneous thin film layers. J. Nanophotonics **11**, 016009 (2017)
13. M.Y. Li, C.H. Chen, Y. Shi, L.J. Li, Heterostructures based on two-dimensional layered materials and their potential applications. Mater Today **19**, 322–335 (2016)
14. C. Villeneuve-Faure, K. Makasheva, L. Boudou, G. Teyssedre, Charge injection in thin dielectric films by Atomic Force Microscopy: influence of geometry and material work function of the AFM tip on the injection process. Nanotechnology **27**, 245702 (2016)
15. C. Villeneuve-Faure, L. Boudou, K. Makasheva, G. Teyssedre, Atomic force microscopy developments for probing space charge at sub-micrometer scale in thin dielectric films. IEEE Trans. Dielectr. Electr. Insul. **23**, 713 (2016)
16. S. Diaham, F. Saysouk, M.-L. Locatelli, T. Lebey, Huge nanodielectric effects in polyimide/boron nitride nanocomposites revealed by the nanofiller size. J. Phys. D Appl. Phys. **48**, 385301 (2015)
17. T. Takada, Y. Hayase, Y. Tanaka, T. Okamoto, Space charge trapping in electrical potential well caused by permanent and induced dipoles for LDPE/MgO nanocomposite. IEEE Trans. Dielectr. Electr. Insul. **15**, 152–160 (2008)
18. G. Teyssedre, S.T. Li, K. Makasheva, N. Zhao, L. Milliere, C. Laurent, Interface tailoring for charge injection control in polyethylene. IEEE Trans. Dielectr. Electr. **24**, 1319–1330 (2017)
19. P. Van Rysselberghe, A note on work functions and chemical potentials. J. Chem. Phys. **21**, 1550 (1953)
20. M. Nonnenmacher, M.P. O'Boyle, H.K. Wickramasinghe, Kelvin probe force microscopy. Appl. Phys. Lett. **58**, 2921 (1991)
21. Y. Rosenwaks, R. Shikler, T. Glatzel, S. Sadewasser, Kelvin probe force microscopy of semiconductor surface defects. Phys. Rev. B **70**, 085320 (2004)

22. Y. Shen, D.M. Barnett, P.M. Pinsky, Simulating and interpreting Kelvin probe force microscopy images on dielectrics with boundary integral equations. Rev. Sci. Instrum. **79**, 023711 (2008)
23. R. Shikler, T. Meoded, N. Fried, B. Mishori, Y. Rosenwaks, Two-dimensional surface band structure of operating light emitting devices. J. Appl. Phys. **86**, 107 (1999)
24. H.O. Jacobs, P. Leuchtmann, O.J. Homan, A. Stemmer, Resolution and contrast in Kelvin probe force microscopy. J. Appl. Phys. **84**, 1168 (1998)
25. A. Liscio, V. Palermo, K. Mullen, P. Samori, Tip-sample interaction in Kelvin probe force microscopy: quantitative measurement of the local surface potential. J. Phys. Chem. C **112**, 17368 (2008)
26. V. Palermo, M. Palma, P. Samori, Electronic characterization of organic thin films by Kelvin probe force microscopy. Adv. Matter **18**, 145 (2006)
27. U. Zerweck, C. Loppacher, T. Otto, S. Grafström, L.M. Eng, Accuracy and resolution limits of Kelvin probe force microscopy. Phys. Rev. B **71**, 125424 (2005)
28. W. Meliz, J. Shen, A.C. Kummel, S. Lee, Kelvin probe force microscopy and its application. Surf. Sci. Rep. **66**, 1 (2011)
29. M. Bieletzki, T. Hynninen, T. Soini, M. Pivetta, C. Henry, A. Foster, F. Esch, C. Barth, U. Heiz, Topography and work function measurements of thin MgO(001) films on Ag(001) by nc-AFM and KPFM. Phys. Chem. Chem. Phys. **12**, 3203 (2010)
30. C. Barth, C.R. Henry, Kelvin probe force microscopy on surfaces of UHV cleaved ionic crystals. Nanotechnology **17**, 5155 (2006)
31. E. Strassburg, A. Boag, Y. Rosenwaks, Reconstruction of electrostatic force microscopy images. Rev. Sci. Instrum. **76**, 083705 (2005)
32. G. Cohen, E. Halpern, S.U. Nanayakkara, J.M. Luther, C. Held, R. Bennewitz, A. Boag, Y. Rosenwaks, Reconstruction of surface potential from Kelvin probe force microscopy images. Nanotechnology **24**, 295702 (2013)
33. S. Sadewasser, C. Leendertz, F. Streicher, M.C. Lux-Steiner, The influence of surface topography on Kelvin probe force microscopy. Nanotechnology **20**, 505503 (2009)
34. M. Lee, M. Lee, F.B. Prinz, Geometric artefact suppressed surface potential measurement. Nanotechnology **17**, 3728 (2006)
35. D. Ziegler, A. Stemmer, Force gradient sensitive detection in lift-mode Kelvin probe force microscopy. Nanotechnology **22**, 075501 (2011)
36. G. Li, B. Mao, F. Lan, L. Liu, Practical aspects of single-pass scan Kelvin probe force microscopy. Rev. Sci. Instrum. **83**, 113701 (2012)
37. C. Li, S. Minne, Y. Hu, J. Ma, J. He, H. Mittel, V. Kelly, N. Erina, S. Guo, T. Mueller, PeakForce Kelvin probe force microscopy. Bruker Application Note #140 (2013)
38. B. Pittenger, N. Erina, C. Su, Quantitative mechanical property mapping at the nanoscale with PeakForce QNM. Bruker Application Note #128 (2012)
39. P. Eyben, M. Xu, N. Duhayon, T. Clarysse, S. Callewaert, W. Vandervorst, Scanning spreading resistance microscopy and spectroscopy for routine and quantitative two-dimensional carrier profiling. J. Vac. Sci. Technol., B **20**, 471 (2002)
40. V. Yanev, T. Erlbacher, M. Rommel, A.J. Bauer, L. Frey, Comparative study between conventionnal macroscopic I-V techniques and advanced AFM based methods for electrical characterization of dielectrics at nanoscale. Microelectron. Eng. **86**, 1911 (2009)
41. C. Li, S. Minne, B. Pittenger, A. Mednick, M. Guide, T.-Q. Nguyen, Simultaneous electrical and mechanical property mapping at the nanoscale with PeakForce TUNA. Bruker Application Note #132 (2011)
42. U. Celano "Metrology and Physical Mechanisms in New Generation Ionic Devices" Springer thesis (2016)
43. R. Giridharagopal, G. Shao, C. Groves, D.S. Ginger1, New SPM techniques for analyzing OPV materials. Mater. Todays **13**, 50 (2010)
44. D.E. Sviridov, V.I. Kozlovsky, Toward reliable photoconductive atomic force microscopy measurements. J. Vac. Sci. & Technol. B **34**, 061801 (2016)

45. C. Riedel, A. Legria, R. Arinero, J. Colmenero, J.J. Saenz, Contrast inversion in electrostatic force microscopy imaging of trapped charges: tip–sample distance and dielectric constant dependence. Nanotechnology **22**, 345702 (2011)
46. M. Labardi, J. Barsotti, D. Prevosto, S. Capaccioli, C.M. Roland, R. Casalini, Extended model for the interaction of dielectric thin films with an electrostatic force microscope probe. J. Appl. Phys. **118**, 224104 (2015)
47. R. Arinero, C. Riedel, C. Guasch, Numerical simulations of electrostatic interactions between an atomic force microscopy tip and a dielectric sample in buried nano-particules. J. Appl. Phys. **112**, 114313 (2012)
48. E. Palleau, L. Ressier, L. Borowik, T. Melin, Numerical simulations for a quantitative analysis of AFM electrostatic nanopatterning on PMMA by Kelvin Force Microscopy. Nanotech. **21**, 225706 (2010)
49. A. Boularas, F. Baudoin, G. Teyssedre, C. Villeneuve-Faure, S. Clain, 3D modelling of electrostatic interaction between AFM probe and dielectric surface: impact of tip shape and cantilever contribution. IEEE Trans. Dielectr. Electr. Insul. **23**, 705 (2016)
50. D.S.H. Charrier, M. Kemerink, B.E. Smalbrugge, T. de Vries, R.A.J. Janssen, Real versus measured surface potentials in scanning Kelvin probe microscopy. ACSNano **2**, 622 (2008)
51. L. Borowik, K. Kushiaku, D. Theron, T. Melin, Calculating Kelvin force microscopy signals from static force field. Appl. Phys. Lett. **96**, 103119 (2010)
52. P. Normand, E. Kapetanakis, P. Dimitrakis, D. Skarlatos, D. Tsoukalas, K. Beltsios, A. Claverie, G. Benassayag, C. Bonafos, M. Carrada, N. Cherkashin, V. Soncini, A. Agarwal, Ch. Sohl, M. Amee, Effects of annealing conditions on charge storage of Si nanocrystal memory devices obtained by low-energy ion beam synthesis. Microelectron. Eng. **67–68**, 629 (2003)
53. A. Witvrouw, H.A.C. Tilmans, I. De Wolf, Materials issues in the processing, the operation and the reliability of MEMS. Microelectron. Eng. **76**, 245 (2004)
54. A. Kerber, E. Cartier, R. Degraeve, P.J. Roussel, L. Pantisano, T. Kauerauf, G. Groeseneken, H.E. Maes, U. Schwalke, Charge trapping and dielectric reliability of SiO_2–Al_2O_3 Gate stacks with TiN electrodes. IEEE Trans. Electron Dev. **50**, 1261 (2003)
55. J.E. Stern, B.D. Terris, H.J. Mamin, D. Rugar, Exposition and imaging of localized charge on insulator surfaces using force microscope. Appl. Phys. Lett. **53**, 2717 (1988)
56. F. Marchi, R. Dianoux, H.J.H. Smilde, P. Mur, F. Comin, J. Chevrier, Characterisation of trapped electric charge carriers behaviour at nanometer scale by electrostatic force microscopy. J. Electrostatics **6**, 538 (2008)
57. J. Lambert, G. de Loubens, C. Guthmann, M. Saint-Jean, Dispersive charge transport along the surface of an insulating layer observed by electrostatic force microscopy. Phys. Rev. B **71**, 155418 (2005)
58. C.A. Rezende, R.F. Gouveia, M.A. da Silva, F. Galembeck, Detection of charge distribution in insulator surfaces. J. Phys.: Condens. Matter. **21**, 263002 (2009)
59. S. Morita, Y. Sugawara, Microscopic contact charging and dissipation. Thin Solid Films **393**, 310 (2001)
60. R. Williams, M.H. Woods, High electric fields in silicon dioxide produced by corona charging. J. Appl. Phys. **44**, 1026 (1973)
61. I.D. Baikie, S. Mackenzie, P.J.Z. Estrup, J.A. Meyer, Noise and the Kelvin method. Rev. Sci. Instrum. **62**, 1326 (1991)
62. C. Schonenberger, Charge flow during metal-insulator contact. Phys. Rev B **45**, 3861 (1992)
63. M. Saint Jean, S. Hudlet, C. Guthmann, J. Berger, Charge dynamics and time evolution of contact potential studied by atomic force microscopy. Phys. Rev. B **56**, 15391 (1997)
64. N. Balke, P. Maksymovych, S. Jesse, I.I. Kravchenko, Q. Li, S.V. Kalinin, Exploring local electrostatic effects with scanning probe microscopy: implications for piezoresponse force microscopy and triboelectricity. ACS Nano **8**, 10229 (2014)
65. Y.S. Zhou, S. Wang, Y. Yang, G. Zhu, S. Niu, Z.-H. Lin, Y. Liu, Z.L. Wang, Manipulating nanoscale contact electrification by an applied electric field. Nano Lett. **14**, 1567 (2014)
66. Y. Han, Z. Huo, X. Li, G. Chen, X. Yang, D. Zhang, Y. Wang, T. Ye, M. Liu, Investigation of charge loss mechanism of thickness-scalable trapping layer by variable temperature Kelvin probe force microscopy. IEEE Electron Devices Lett. **34**, 870 (2013)

67. C. Villeneuve-Faure, K. Makasheva, C. Bonafos, B. Despax, L. Boudou, P. Pons, G. Teyssedre, Kelvin force microscopy characterization of charging effect in thin a-SiO_xN_y:H layers deposited in pulsed plasma enhanced chemical vapor deposition process by tuning the Silicon-environment. J. Appl. Phys. **113**, 204102 (2013)
68. R. Diaz, E. Palleau, D. Poirot, N.M. Sangeetha, L. Ressier, High-throughput fabrication of anti-counterfeiting colloid-based photoluminescent microtags using electrical nanoimprint lithography. Nanotechnology **25**, 345302 (2014)
69. M. Chiesa, R. Garcia, Nanoscale space charge generation in local oxidation nanolithography. Appl. Phys. Lett. **96**, 263112 (2010)
70. L. Ressier, V. Le Nader, Electrostatic nanopatterning of PMMA by AFM charge writing for directed nano-assembly. Nanotechnology **19**, 135301 (2008)
71. K. Sridhara, Ł. Borowik, D. Mariolle, N. Chevalier, J.-P. Colonna, Environment and baking influence on charge retention on silicon nitride charge trap layers. J. Appl. Phys. **111**, 023710 (2012)
72. D. Ziegler, N. Naujoks, A. Stemmer, Feed-forward compensation of surface potential in atomic force microscopy. Rev. Sci. Instrum. **79**, 063704 (2008)
73. S. Morita, Y. Sugawara, Y. Fukano, Atomic force microscope combined with scanning tunneling microscope. Jpn. J. Appl. Phys. **32**, 2983 (1993)
74. K. Makasheva, C. Villeneuve-Faure, C. Laurent, B. Despax, L. Boudou, G. Teyssedre, Dielectric charging by AFM in tip-to-sample space mode: overview and challenges in revealing the appropriate mechanisms. Nanotechnology **26**, 295704 (2015)
75. Z. Sun, X. Wang, W. Song, Q. Lei, A research on the surface charge with different coating tip by electrostatic force microscope. Annual report IEEE conference electrical insulation dielectric phenomena, vol. 295 (2013)
76. K. Makasheva, B. Despax, L. Boudou, G. Teyssedre, Dielectric layers for RF-MEMS switches: design and study of appropriate structures preventing electrostatic charging. IEEE Trans. Dielectr. Electr. Insul. **19**, 1195 (2012)
77. Z.Z. Lwin, K.L. Pey, C. Liu, Q. Liu, Q. Zhang, Y.N.K. Singh, S. Mahapatra, Localized charge trapping and lateral charge diffusion in metal nanocrystal-embedded High-j/SiO_2 gate stack. Appl. Phys. Lett. **99**, 222102 (2011)
78. J. Nishitani, K. Makihara, M. Ikeda, H. Murakami, S. Higashi, S. Miyazaki, Decay characteristics of electronic charged states of Si quantum dots as evaluated by an AFM/Kelvin probe technique. Thin Solid Films **508**, 190 (2006)
79. J. Xu, Y. Wu, W. Li, J. Xu, Surface potential modeling and reconstruction in Kelvin probe force microscopy. Nanotechnology **28**, 365705 (2017)
80. I. Alhossen, C. Villeneuve-Faure, F. Baudoin, F. Bugarin, S. Segonds, Sensitivity analysis of the electrostatic force distance curve using Sobol's method and design of experiments. J. Phys. D Appl. Phys. **50**, 035304 (2017)
81. P. Girard, Electrostatic force microscopy: principles and some applications to semiconductors. Nanotechnology **12**, 485 (2001)
82. N.B. Bercu, L. Giraudet, O. Simonetti, M. Molinari, Development of an improved Kelvin probe force microscope for accurate local potential measurements on biased electronic devices. J. Microscopy **267**, 272 (2017)
83. F. Mortreuil, C. Villeneuve-Faure, L. Boudou, K. Makasheva, G. Teyssedre, Charges injection phenomena at metal/dielectric interface investigated by Kelvin probe force microscopy. J. Phys. D **50**, 175302 (2017)
84. G. Lubarsky, R. Shikler, N. Ashkenasy, Y. Rosenwaks, Quantitative evaluation of local charge trapping in dielectric stacked gate structures using Kelvin probe force microscopy. Appl. Phys. Lett. **58**, 2921 (1991)
85. K. Faliya, H. Kliem, C.J. Dias, Space charge measurements with Kelvin probe force microscopy. IEEE Trans. Dielec. Electr. Insul. **24**, 1913 (2017)
86. A. Savitzky, M.J. Golay, Smoothing and differentiation of data by simplified least squares procedures. Anal. Chem. **36**, 1627 (1964)

87. C. Villeneuve-Faure, L. Boudou, K. Makasheva, G. Teyssedre, Methodology for extraction of space charge density profiles at nanoscale from Kelvin probe force microscopy measurements. Nanotechnology **28**, 505701 (2017)
88. B. Martin, H. Kliem, Transient formation of internal potentials in solid electrolytes. J. Appl. Phys. **107**, 076103 (2010)
89. A.K. Henning, T. Hochwitz, Two-dimensional surface dopant profiling in silicon using scanning Kelvin probe microscopy. J. Appl. Phys. **77**, 1888 (1995)
90. O. Kryvchenkova, I. Abdullah, J.E. Macdonald, M. Elliott, T.D. Anthopoulos, Y.-H. Lin, P. Igic, K. Kalna, R.J. Cobley, Nondestructive method for mapping metal contact diffusion in In_2O_3 thin-film transistors. ACS Appl. Mater. Interfaces **8**, 25631–25636 (2016)
91. T. Okamoto, S. Kitagawa, N. Inoue, A. Ando, Electric field concentration in the vicinity of the interface between anode and degraded $BaTiO_3$-based ceramics in multilayer ceramic capacitor. Appl. Phys. Lett. **98**, 072905 (2011)
92. R. Eisenberg, D.G. Nocera, Preface: overview of the forum on solar and renewable energy. Inorg. Chem. **44**, 6799 (2005)
93. S.-Y. Park, R. Elbersen, J. Huskens, H. Gardeniers, J.-Y. Lee, G. Mul, J. Heo, Characterization of opto-electrical enhancement of tandem photoelectrochemical cells by using photoconductive-AFM. Nanotechnology **28**, 295401 (2017)
94. J.L. Luria, N. Hoepker, R. Bruce, A.R. Jacobs, C. Groves, J.A. Marohn, Spectroscopic imaging of photopotentials and photoinduced potential fluctuations in a bulk heterojunction solar cell film. ACS Nano **6**, 9392 (2012)
95. E.J. Spadafora, R. Demadrille, B. Ratier, B. Grevin, Imaging the carrier photogeneration in nanoscale phase segregated organic heterojunctions by Kelvin probe force microscopy. Nano Lett. **10**, 3337 (2010)
96. K. Maturova, M. Kemerink, M.M. Wienk, D.S.H. Charrier, A.J. Janssen, Scanning Kelvin probe microscopy on bulk heterojunction polymer blends. Adv. Funct. Mater. **19**, 1379–1386 (2009)
97. D.A. Kamkar, M. Wang, F. Wudl, T.-Q. Nguyen, Single nanowire OPV properties of a fullerene-capped P3HT dyad investigated using conductive and photoconductive AFM. ACS Nano **6**, 1149 (2012)
98. D.C. Coffey, O.G. Reid, D.B. Rodovsky, G.P. Bartholomew, D.S. Ginger, Mapping local photocurrents in polymer/fullerene solar cells with photoconductive atomic force microscopy. Nanoletters **7**, 738 (2008)
99. O. Douhéret, A. Swinnen, M. Breselge, I. Van Severen, L. Lutsen, D. Vanderzande, J. Manca, High resolution electrical characterisation of organic photovoltaic blends. Micro Eng. **84**, 431(2007)
100. B.J. Leever, M.F. Durstock, Spatially resolved photocurrent mapping of operating organic photovoltaic devices using atomic force photovoltaic microscopy. Appl. Phys. Lett. **92**, 013302 (2008)
101. I. Beinik, M. Kratzer, A. Wachauer, L. Wang, Y.P. Piryatinski, G. Brauer, X.Y. Chen, Y.F. Hsu, A.B. Djurišić, C. Teichert, Photoresponse from single upright-standing ZnO nanorods explored by photoconductive AFM. Beilstein J. Nanotechnol. **4**, 208 (2018)
102. M. Dante, J. Peet, T.Q. Nguyen, Nanoscale charge transport and internal structure of bulk heterojunction conjugated polymer/fullerene solar cells by scanning probe microscopy. J. Phys. Chem. **112**, 7241 (2008)
103. O.G. Reid, K. Munechika, D.S. Ginger, Space charge limited current measurements on conjugated polymer films using conductive atomic force microscopy. Nanoletters **8**, 1602 (2008)
104. T.G. Ruskell, R.K. Workman, D. Chen, D. Sarid, S. Dahl, S. Gilbert, High resolution Fowler Nordheim field emission maps of thin silicon oxide layers. Appl. Phys. Lett. **68**, 1063 (1996)
105. U. Celano, T. Hantschel, G. Giammaria, R. Chintala, T. Conard, H. Bender, W. Vandervorst, Evaluation of the electrical contact area in contact-mode scanning probe microscopy. J. Appl. Phys. **117**, 214305 (2015)
106. D. Moerman, N. Sebaihi, S.E. Kaviyil, P. Leclere, R. Lazzaroni, O. Douheret, Towards a unified description of the charge transport mechanisms in conductive atomic force microscopy studies of semiconducting polymers. Nanoscale **6**, 10596 (2014)

107. D.C. Coffey, D.S. Ginger, Time-resolved electrostatic force microscopy of polymer solar cells. Nat. Mater. **5**, 735 (2006)
108. C. Villeneuve-Faure, D. Le Borgne, E. Bedel-Pereira, K.I. Moineau Chane-Ching, D. Hernandez-Maldonado, I. Seguy, Cross Kelvin force microscopy and conductive atomic force microscopy studies of organic bulk heterojunction blends for local morphology and electrical behavior analysis. J. Appl. Phys. **117**, 055501 (2015)

Chapter 10
Conductive AFM of 2D Materials and Heterostructures for Nanoelectronics

Filippo Giannazzo, Giuseppe Greco, Fabrizio Roccaforte, Chandreswar Mahata and Mario Lanza

Abstract Two-dimensional materials (2DM), such as the semimetal graphene, semiconducting MoS2 and insulating h-BN, are currently the object of wide interests for next generation electronic applications. Despite recent progresses in large area synthesis of 2DMs, their electronic properties are still affected by nano- or micro-scale defects/inhomogeneities related to the specific growth process. Electrical scanning probe methods, such as conductive atomic force microscopy (C-AFM), are essential tools to investigate charge transport phenomena in 2DMs with nanoscale resolution. This chapter illustrates some case studies of C-AFM applications to graphene, MoS2 and h-BN. Furthermore, the results of the nanoscale electrical characterization have been correlated to the behavior of macroscopic devices fabricated on these materials.

10.1 Introduction

The isolation of graphene (Gr) obtained by exfoliation of graphite represented the first experimental demonstration that two-dimensional (2D) materials can be stable

F. Giannazzo (✉) · G. Greco · F. Roccaforte
Consiglio Nazionale delle Ricerche, Institute for Microelectronics and Microsystems (CNR-IMM), Strada VIII, 5, 95121 Catania, Italy
e-mail: filippo.giannazzo@imm.cnr.it

G. Greco
e-mail: giuseppe.greco@imm.cnr.it

F. Roccaforte
e-mail: fabrizio.roccaforte@imm.cbr.it

C. Mahata · M. Lanza
Institute of Functional Nano & Soft Materials, Collaborative Innovation Center of the Ministry of Education, Soochow University, 199 Ren-Ai Road, Bld. 909, Office 3215, Suzhou 215123, China
e-mail: cmahata@suda.edu.cn

M. Lanza
e-mail: mlanza@suda.edu.cn

U. Celano (ed.), *Electrical Atomic Force Microscopy for Nanoelectronics*, NanoScience and Technology, https://doi.org/10.1007/978-3-030-15612-1_10

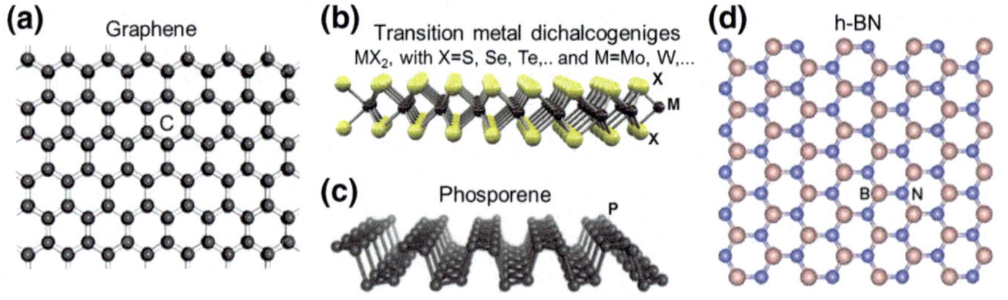

Fig. 10.1 Schematic representation of the lattice structure of monolayer of graphene (**a**), transition metal dichalcogenides (**b**), phosporene (**c**), hexagonal-boron nitride (**d**)

under ambient conditions after separation from the bulk crystal [1]. This opened also the way to the investigation of an entire class of layered materials occurring in nature, which are composed by the vertical stacking of 2D sheets bond by van der Waals (vdW) interaction [2]. 2D layered materials are currently the object of wide scientific interests, due to their unique electrical, optical, mechanical, and chemical properties, which make them attractive both from a basic and from a technological standpoint. Looking at the electronic properties, the 2D materials (2DMs) family includes both semimetals (such as Gr), semiconductors (such as some transition metal dichalcogenides [3] and phosporene [4]) and insulators (such as the hexagonal boron nitride) [5].

To date, graphene, the single atomic layer of sp^2 carbon (see Fig. 10.1a), has been the most widely studied 2DM, because of its excellent carrier mobility (up to 10^5 $cm^2\ V^{-1}s^{-1}$) [6, 7] and large electron mean free path [8–11], combined to optical transparency in the visible wavelength range and outstanding mechanical robustness and flexibility. These properties make Gr suitable for several applications, e.g. as a transparent conductive electrode in optoelectronic devices [12], as a channel material for high frequency field effect transistors (FETs) both on rigid and flexible substrates [13], and as active layer in environmental, chemical or biological sensors [14]. The main intrinsic limitation of Gr is the lack of a band gap, which hinders its application as a channel material for logic FETs [15].

Transition metal dichalcogenides (TMDs) are composed by the van der Waals stacking of MX_2 layers, where M is transition metal (Mo, W, Te,…) and X is a chalcogen (S, Se,…), as illustrated in Fig. 10.1b. Due to the proper bandgap (in the range from 1 to 2 eV, also depending on the layers' number), combined with a good stability under ambient conditions, semiconductor TMDs are very promising candidates as channel materials for digital electronics [16, 17]. The energy gap of TMDs comes, however, at the cost of a relatively low mobility. As an example, the upper theoretical limit for mobility of monolayer MoS_2 at room temperature is about 400 $cm^2\ V^{-1}\ s^{-1}$ [18], whereas the experimental values reported so far are in the range of few tens of $cm^2\ V^{-1}\ s^{-1}$ [19, 20].

Besides TMDs, also phosporene, the single layer of black phosphorous (see Fig. 10.1c), recently attracted the interest as semiconducting channel material for

FETs. A mobility of 286 $cm^2\ V^{-1}\ s^{-1}$ has been reported for few-layer phosphorene [4], whereas a value up to ~1000 $cm^2\ V^{-1}\ s^{-1}$ have been shown for multilayer phosphorene with ~10 nm thickness [21]. However, the main disadvantage of this material (as compared to TMDs) is its high chemical reactivity under ambient conditions, which can represent a serious concern for practical applications.

Finally, hexagonal boron nitride (h-BN), a single layer of B and N atoms arranged in a honeycomb lattice (see Fig. 10.1d), is an insulator with a band gap of 5.5–5.7 eV [5, 22] and a dielectric constant ranging from 2 to 4 [23]. Compared to traditional dielectrics (like SiO_2, HfO_2, TiO_2 and Al_2O_3), h-BN shows many advantages. Due to its uniform thickness and atomically flat surface free of dangling bonds it has been used as a substrate and/or as a cap for Gr and TMDs, in order to enhance the carrier mobility and electron mean free path of these 2DMs [7, 8, 24–28]. Hexagonal boron nitride performances as a dielectric have been also evaluated. Despite it holds a dielectric constant similar to that of SiO_2, its high crystalline quality layered structure may remarkably slow down the speed for defect formation, leading to superior device reliability [29]. Finally, its high thermal conductivity (about 20 times larger than that of SiO_2) [30], can improve the heat dissipation within the device, which will contribute to enhance its lifetime.

To date, 2DMs have been considered to replace one or more components of currently employed device structures, like the metal-insulator-semiconductor FET. The most inspiring recent advance in this direction has been the development of fully 2D metal-insulator-semiconductor devices by combining MoS_2 as channel material, h-BN as gate insulator, and Gr as source-drain and gate electrode [31]. Besides conventional FETs architectures, novel vertical device concepts [32] based on stacking of different 2DMs [33, 34] or on their integration with bulk semiconductors [35–37] or semiconductors heterostructures [38–41] have been also proposed in the last few years.

In many cases, the first demonstrators of these devices have been fabricated using 2DMs mechanically exfoliated from the parent bulk crystals. Although the samples obtained by this approach typically exhibit excellent electronic properties, the small size of the flakes makes this synthesis method not suitable for large scale electronics applications. In the last decade, many efforts have been devoted to produce 2DMs on large area by synthesis approaches compatible with current semiconductor devices technology, such as chemical vapour deposition (CVD), molecular beam epitaxy (MBE), atomic layer deposition (ALD).

As a matter of fact, the electronic properties of 2DMs are strongly dependent on the growth method, on the substrate where they are grown, as well as on post-growth processes required to integrate these materials for devices fabrication. As an example, peculiar nano- or micro-scale defects/inhomogeneities (e.g. grain boundaries, thickness fluctuations, wrinkles,…) can be typically found in large area 2DMs deposited by CVD. The density and distribution of these defects are related to the morphology and structural/chemical properties of the substrate and to the specific nucleation/growth mechanisms. Clearly, these local electrical and structural inhomogeneities are reflected in the macroscopic electrical behaviour of devices based on 2DMs.

From the above discussion, it is clear how nanoscale resolution electrical characterization of 2DMs is essential to get a complete understanding of their macroscopic electrical behaviour on a large area. In this context scanning probe methods, such as conductive atomic force microscopy (C-AFM), scanning capacitance microscopy (SCM), electrostatic force microscopy (EFM) and Kelvin probe force microscopy (KPFM), proved to be essential tools to understand the nature and electrical activity of defects/nonuniformities in 2DMs [42–46]. Furthermore, these techniques can provide a guide to solve growth problems and to understand the phenomena limiting the performances of electronic devices based on these systems. In particular, C-AFM allows performing high resolution current mapping and local current-voltage (*I*-*V*) characterization at the nanoscale [47]. Hence, it is the method of choice to investigate the mechanisms of current injection from contacts to 2DMs, the lateral homogeneity of conductivity of 2D semimetals (Gr) and semiconductors (TMDs), the conduction mechanisms across 2D insulators (h-BN) [48, 49]. Furthermore, it allows to investigate the vertical current injection across the van der Waals heterostructures of 2DMs or the heterojunctions between 2DMs and bulk semiconductors.

This chapter is organized as follows:

- Section 10.2 discusses the state-of-the-art synthesis methods for electronic-grade Gr, MoS_2 and h-BN on large area.
- Section 10.3 includes an overview of the 2DMs-based electronic devices currently under investigation.
- Section 10.4 presents some case studies of C-AFM applications to Gr, MoS_2 and h-BN, to illustrate the potentiality of this characterization method for 2DMs investigation. In some cases, the results of the nanoscale electrical characterization will be correlated to the behaviour of macroscopic devices fabricated on these materials.

10.2 Large Area Synthesis of Graphene, MoS_2 and h-BN for Electronics

Despite their short history, tremendous progresses have been done in the last years in the synthesis of 2DMs on large area for electronic applications.

To date, growth methods have reached the highest degree of maturity in the case of Gr. This is due to the simple (monoatomic) lattice structure and high thermodynamic stability of this 2D allotrope of carbon. From the historical point of view, the first approach to obtain Gr on wafer scale was the controlled graphitisation of hexagonal silicon carbide (4H– and 6H–SiC) by high temperature annealing [50–52]. Thermal processes at temperatures >1400 °C lead to preferential sublimation of silicon from SiC surface and to a rearrangement of the excess of carbon in a graphitic structure. Besides temperature, pressure in the growth chamber is a key parameter ruling the Si sublimation rate. The highest quality Gr is obtained by atmospheric pressure processes, which allow to increase the annealing temperature (above 1600 °C) maintaining a low Si evaporation. Clearly, the high temperature is beneficial for the surface mobility of the excess C atoms and their reorganization in the Gr lattice. The structural and electrical properties of Gr grown on SiC strongly depend also on the substrate orientation, i.e. the Si face (0001) [50, 53], the C face (0001) [54–56] or the non-polar faces (11–20) and (1–100) [57]. In particular, single or few-layers Gr films with very precise epitaxial orientation with respect to the substrate can be obtained on the (0001) face of SiC [50]. For this reason, Gr grown on SiC(0001) is commonly named epitaxial Gr (EG). These interface properties are the result of the peculiar growth mechanism of Gr on this SiC orientation, i.e. the formation of a carbon reconstruction of the (0001) face, the so-called C buffer layer, which acts as the precursor for Gr formation [54]. This C buffer layer is, in part, covalently bonded to SiC substrate, with the presence of a large density of positively charged Si dangling bonds. These are responsible of high n-type doping (up to 10^{13} cm^{-2}) and reduced carrier mobility (typically ~1000 cm^2 V^{-1} s^{-1}) of EG, as well as of a small Schottky barrier height (SBH) at EG/SiC(0001) interface [58, 59]. Intercalation with hydrogen or other atomic species (F, Li, Si, Ge, …) has been employed to decouple the C buffer layer from the substrate, resulting into a quasi-free-standing epitaxial graphene (QFSEG) [60]. Besides an improved in-plane carrier mobility, QFSEG shows an increase of the SBH for vertical current injection across the interface with SiC.

The main advantage of the growth on SiC is that it provides Gr directly on a semi-conducting or semi-insulating substrate ready for advanced electronics, metrology and sensing applications. However, due to the relatively high price of SiC and to the difficulty of detaching Gr from it, alternative substrates and growth methods have been also explored. Among these, chemical vapour deposition (CVD) from hydro-carbon precursors on catalytic substrates (typically metals, such us Ni, Cu, Pt, Ru,…) [61] is nowadays the most widely employed. Gr growth on catalytic metals is ruled by two different mechanisms, which depend on the solubility of carbon atoms in the metal at the growth temperature. In this context, Ni and Cu, which are the most used substrates for Gr CVD growth, exhibit a different behavior, due to the very different C solid solubility values in the two materials (0.6% for Ni at the melting temperature $T_{M,Ni} = 1455$ °C, and 0.0076% for Cu at $T_{M,Cu} = 1085$ °C). Hence, when a polycrystalline Ni substrate is exposed at high temperature (~900 to 1000 °C) to the hydrocarbon gas, C atoms resulting from its decomposition dissolve into the metal bulk. During the cooling down step, the reduced C solubility at lower temperature gives rise to the segregation of few layers of Gr on the Ni surface. The resulting graphitic film typically exhibits inhomogeneous thickness, with larger number of layers formed at Ni grain boundaries. On the other hand, in the case of Cu substrates, carbon dissolution/segregation are suppressed and Gr formation is a surface process occurs in three steps: (i) nucleation of Gr domains, (ii) their growth by diffusion and attachment of C ad-atoms, (iii) their merging with the formation of domain boundaries. Studies on the coating as a function of growth time have shown that, under proper conditions, Gr growth on Cu is a self-limiting process, resulting in monolayer Gr coverage on large fraction (>90%) of the surface, with the presence of thicker Gr films mainly in the proximity of the grain boundaries of polycrystalline Cu. Besides

Gr domains boundaries, wrinkles (i.e. corrugations of the Gr membrane) are other typical defects of CVD grown Gr on catalytic metals. They originate, in part, from the compressive strain experienced by the Gr membrane during the cooling down of the sample after the deposition, due to the large thermal expansion coefficient mismatch with respect to the substrate.

As a matter of fact, using CVD-grown Gr on metals for electronics applications commonly implies its transfer from the original substrate to an insulating or semiconducting one. Although this is a very versatile and widely used method, it suffers from some drawbacks related to Gr damage (cracks, folding, additional wrinkles,…) and residual polymer contaminations (typically PMMA) from the transfer procedure, as well as of possible adhesion problems between Gr and the substrate. Furthermore, it typically introduces undesired metal (Cu, Fe) contaminations [62] originating from the growth substrate and the typically used Cu etchants, that can be only partially mitigated by Gr delamination without Cu etching [63]. Several advances have been performed to achieve Gr with low cracks and wrinkles on arbitrary substrates. As an example, wetting the target substrate with low surface tension organic liquids (such as heptane or isopropyl alcohol) was demonstrated to improve Gr adhesion [64]. Furthermore, replacing PMMA with alternative supporting polymers for Gr transfer (such as ethylene-vinyl acetate) enabled a conformal transfer even onto uneven substrates, leaving less residues after removal [65].

The increasing interest in semiconductor transition metal dichalchogenides (TMDs) for digital electronics and optoelectronics motivated significant efforts in the synthesis of these materials on large area. To date, MoS_2 has been the most investigated member of the TMDs family. Although the best quality MoS_2 samples for electronics are still obtained by exfoliation from bulk molybdenite, many progresses have been made in the CVD growth of MoS_2 directly on insulating substrates (such as SiO_2 or sapphire). Vapors from sulfur powders or the H_2S gas are the commonly used precursors to deliver S to the growth chamber [66]. Mo can be directly deposited on the substrate in form of thin metallic (Mo) or metal-oxide (MoO_3) layer [67, 68] and then subjected to sulfurization. Alternatively, it can be delivered to the growth chamber starting from vapourized metal oxide powders or liquids precursors. The usual growth temperatures for MoS_2 are 700–1000 °C, depending on the deposition method. The inclusion of a metal catalyst such as Au is favorable for the film quality [69]. With plasma-enhanced chemical vapor deposition (PECVD), the growth temperature of MoS_2 can be lowered to 150–300 °C, making it possible to directly deposit MoS_2 even on plastic substrates [70]. Metal-organic CVD is a special case of CVD where organometallic precursors are used. Recently, this method has been applied for the growth of high quality MoS_2 thin films on 4-in. oxidized silicon wafers, using $Mo(CO)_6$ and $(C_2H_5)_2S$ gas precursors [71]. Another promising synthesis method for MoS_2 is atomic layer deposition (ALD), which allows the layer-by-layer growth by sequential and non-overlapping pulses of the precursors, separated by purges in between. In each pulse, molecules react with the surface in a self-limiting way. Hence, it is possible to grow materials with high-thickness precision. Despite its potential, ALD is still not so widespread for TMDs, mainly because of the lack of suitable precursors. Nevertheless, several groups used ALD also for the deposition of metals or metal oxides with controlled thickness to be further sulfurized ex situ by other techniques. Recently, ALD growth of monolayer to few-layer MoS_2 on large SiO_2/Si or quartz substrates using $MoCl_5$ (vaporized) and H_2S precursors has been reported [72].

In spite of these progresses, deposition of MoS_2 (as well as other TMDs) with high spatial homogeneity is still challenging. Large-area MoS_2 films are typically polycrystalline with the grain size commonly ranging from 1 to 10 micrometers. Grain boundaries between these domains introduce significant resistance contributions and degrade the overall conductivity of the TMD films.

Also in the case of h-BN ultra-thin insulator, mechanical exfoliation still provides the highest quality material [48, 49], but its use is reduced to research purposes, due to the small size and thickness fluctuations of the nano-sheets. For this reason, physical vapor deposition (PVD) and CVD methodologies have been explored to produce h-BN on large area. Within PVD family, magnetron sputtering has been used [73] to successfully synthesize monocrystalline and high-quality wafer-scale boron nitride films on a Ru/Al_2O_3 substrate. More recently, molecular beam epitaxy (MBE) has been proposed as a promising way to produce atomically thin h-BN stacks with high-crystalline quality. For instance, Nakhaie et al. [74] successfully synthesized h-BN thin films on polycrystalline Ni foils through MBE method. Additionally, MBE also has been used in the field of 2D heterostructures fabrication, for in situ growth of h-BN on TMDs [75]. CVD deposition of h-BN has been also investigated in the last years, using different kind of precursors for B and N atoms (including borazine or BCl_3/NH_3, BF_3/NH_3, B_2H_6/NH_3, and $B_{10}H_{14}/NH_3$). Besides the influence of different precursors, the catalyst substrate also has a very important impact on the growth of h-BN. Until now, many kinds of materials have served as substrate for CVD growth of h- BN, including metals (Ni [76], Cu [77],…), as well as insulators (such as sapphire [78]). Because Ni and Cu are cheap, accessible and show good catalytic effect, they have become the most widespread substrate materials for CVD growth of h-BN. Moreover, the relatively small lattice mismatch between h-BN and the (111) faces of Ni or Cu greatly improves the thickness uniformity of the h-BN thin films.

Similarly to the case of other 2DMs, GBs are common defects in polycrystalline h-BN grown by CVD [79, 80]. The binary composition of h-BN shows a very complicated configuration of GBs. Lattice distortions in h-BN at GB locations results in the apparition of in-gap states, which increase the local conductivity through the 2D layered dielectric stack. Scanning tunneling spectroscopy measurements and density functional theory calculations indicated that the band gap of h-BN with dislocation GBs is lower than in defect-free locations [81]. Consequently, the local leakage current increase at the GBs of h-BN must be considered during the design of devices with h-BN as a dielectric. h-BN films with very large domains [82] will be necessary to achieve optimal device performances. On the other hand, the presence of GBs in dielectrics could be even useful in different configurations, as they may allow some GB-driven additional properties, such as resistive switching [83].

Similarly to the case of Gr, wrinkles are typically present in CVD-grown h-BN on catalytic metals (e.g. copper). Also for h-BN, the origin of the wrinkles is related to the large temperatures required by the CVD growth process (above 750 °C) and to the compressive strain experienced by the h-BN film due to the thermal expansion coefficient mismatch with the growth substrate. These local corrugations of the h-BN membrane are typically maintained after transfer to another substrate. From an electrical point of view, wrinkles introduce a gap between the h-BN and the substrate, corresponding to an effective increase of the dielectric thickness at that location. The more insulating nature of the wrinkled sites could reduce the amount of charge stored in a capacitor and induce threshold voltage inhomogeneities in FETs.

Another kind of defect that can be found in CVD-grown h-BN is the presence of cracks. These can be generated during the CVD growth process if the substrate is not fully covered with at least one layer of h-BN. Moreover, they can be produced during a transfer process from the growth substrate to another substrate. As a matter of fact, cracks are extremely harmful defects in 2D insulators used as dielectric. In fact, an atomic scale crack within a dielectric sandwiched by two electrodes can produce the dielectric breakdown (BD) of the whole structure, and the irreversible failure of the device.

10.3 Overview of 2D Materials-Based Electronic Devices

10.3.1 Graphene FETs for High Frequency Electronics

Due to its excellent carrier mobility, Gr has been considered since the first studies as channel material for high frequency FETs. The poor ON/OFF current ratio, originating from the lack of a bandgap in Gr electronic band-structure, represents a minor issue for radio frequency (RF) applications, where fast modulation of the current in the device operated in the ON-state is the primary requirement [13, 15]. Current and power amplification are two key functions of RF transistors, and the main figures of merit associated to them are the cut-off frequency f_T (i.e. the frequency for which current gain is reduced to unity) and the maximum oscillation frequency f_{MAX} (i.e. the frequency for which power gain is reduced to unity).

Figure 10.2 illustrates some of the milestones in the development of Gr-based FETs specifically for RF applications. The first example of top-gated Gr FET, fabricated using monolayer Gr exfoliated from HOPG and SiO_2 as a gate dielectric, was reported in 2007 [84]. The first Gr-based FET capable of RF operation, with a cut-off frequency $f_T = 14.7$ GHz for a gate length $L_G = 500$ nm, was reported one year later, in 2008 [85]. The development of advanced synthesis methods (e.g. epitaxial growth of Gr on SiC or CVD on catalytic metals) provided high quality Gr on large area for devices fabrication. In 2010, arrays of RF Gr-FETs with very promising RF performances ($f_T = 100$ GHz) were fabricated at IBM using epitaxial Gr grown on SiC wafers [86]. Ultra-scaled transistors with interesting RF performances have been also demonstrated using transferred CVD Gr [87]. In 2012, a record cut-off frequency $f_T = 427$ GHz has been reported for Gr-FETs with self-aligned gates (L_G

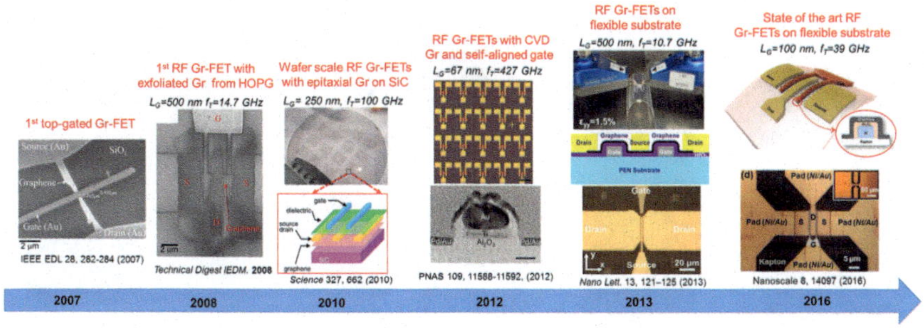

Fig. 10.2 Milestones in the development of Gr FET for RF applications on rigid and on flexible substrates

= 67 nm) [88]. As mentioned before, besides the cut-off frequency f_T, also f_{MAX} must be considered when evaluating the high frequency performances of RF devices. While similar values of f_T and f_{MAX} are commonly measured in RF transistors fabricated with conventional semiconductors, Gr-FETs typically show poor f_{MAX} as compared to f_T. This can be ascribed, in part, to the poor saturation of the output characteristics, also descending from the missing bandgap. These limitations of Gr FETs in terms of power gain performances recently stimulated the investigation of novel Gr-based device concepts for high frequency applications, as discussed later on in this section.

In the last few years, the possibility of fabricating RF transistors also on flexible plastics substrates, by exploiting the excellent mechanical properties of Gr, has been intensively explored [89, 90]. The most recent progresses in this field are also highlighted in Fig. 10.2.

It should be noted that the above discussed performances have been achieved by integration of state-of-the-art large area Gr within the fabrication flow of RF transistors. As a matter of fact, both epitaxial Gr on SiC and transferred CVD Gr exhibit peculiar defects/inhomogeneities which are responsible of a degradation of Gr conductivity with respect to the ideal values. Hence, further improvements in the cut-off frequency will require specifically addressing the microscopic mechanisms responsible of Gr mobility degradation in real devices. Section 10.4.1 will show some examples of nanoscale electrical investigations identifying the role played by some peculiar features on the electrical behaviour of Gr, and their impact on transistors performances.

10.3.2 2D-Semiconductors FETs for Digital Electronics

To overcome Gr limitations arising from the lack of a bandgap, new solutions have been explored inside the wide family of 2DMs. In particular, 2D semiconductors, such as MoS_2 and phosporene, have been considered for digital applications. Figure 10.3

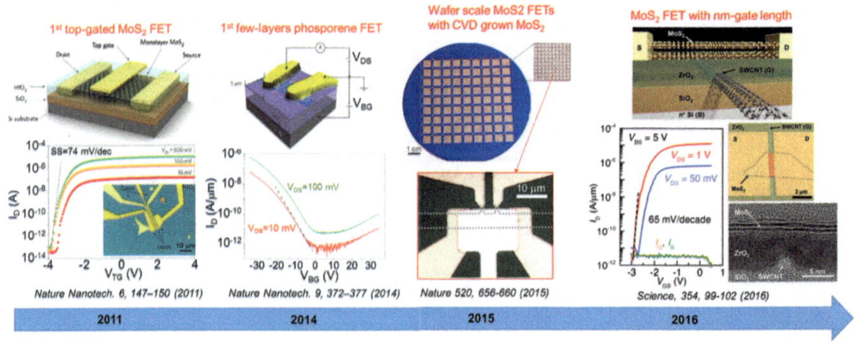

Fig. 10.3 Milestones in the development of field effect transistors with a 2D semiconductors channel

illustrates some of the milestones in the development of FETs with 2D semiconductors channels. The first example of top-gated FET fabricated with exfoliated monolayer MoS_2 was reported in 2011 by Radisavljevic et al. [16]. This first demonstrator showed already very interesting performances, i.e. large I_{on}/I_{off} ratio ($>10^7$) and small subthreshold swing (SS $= 74$ mV dec^{-1}), approaching the desired requirements of FETs for CMOS digital circuits. The high-k top gate dielectric (HfO_2) played a key role in this device, as it allowed the improvement of electron mobility by effectively screening charged impurities. The first FET with few layers phosporene channel (obtained by exfoliation of black phosphorous) was reported in 2014 [21]. This device showed p-type behaviour, with I_{ON}/I_{OFF} current ratio of 10^5 and saturation of the output characteristics. Progress in CVD deposition of MoS_2 on large area insulating substrates allowed wafer scale fabrication of arrays of single layer MoS_2 FETs [71], as shown in Fig. 10.3.

The ultimate thin body of TMDs can be very beneficial for the scaling prospects of lateral FETs for CMOS applications, as discussed in many simulation works [17, 91–93]. As an example, Liu et al. [92] predicted that MoS_2 FETs can meet the requirements of the ITRS [94] down to a minimum channel length of 8 nm. For such aggressively reduced geometries, the low mobility of MoS_2 is not a real issue, because transport can be considered as almost ballistic for channel length below 10 nm. Although most of these predictions are based on simulations, some experimental work has been also reported, where the challenges of channel length scaling in TMDs transistors down to the nanometric limit started to be addressed. As an example, in 2016, Desai et al. [95] demonstrated a MoS_2 FET with the gate represented by a single wall carbon nanotube (L_G = 1 nm), that showed a nearly ideal SS = 65 mV/dec (see Fig. 10.3).

In spite of these great promises, several challenges still need to be faced to exploit the potentialities of MoS_2 and other TMDs for next generation CMOS technology. Besides the issues related to the lattice defects (such as chalcogen vacancies) [96–98] and impurities [99] commonly present in these compound materials, some critical processing steps need to be developed. These include the fabrication of low resistance source/drain contacts [100, 101] and doping [102–104]. Both exfoliated and CVD grown MoS_2 are typically unintentionally n-type doped [105]. Furthermore, most of the metals exhibit a Fermi level pinning in the upper part of the MoS_2 band gap, which results in a low Schottky barrier height (SBH), typically ranging from ~25 to ~300 meV [100]. The origin of the unintentional n-type doping and of the Fermi level pinning in pristine MoS_2 is still matter of debate, and many authors tried to explain these effects in terms of defects/inhomogeneities of MoS_2 [106] or by the presence of foreign impurities. These properties have important implications for the development of a CMOS technology based on MoS_2. In fact, while n-type FETs with an electron accumulation channel can be easily obtained with unintentionally doped MoS_2, selective-area p-type doping of MoS_2 under the source/drain electrodes and/or the use of unconventional contacts forming a low Schottky barrier for holes injection are required to obtain p-type FETs working in the channel inversion regime. To date, several approaches have been explored in the literature to reach this goal, including the use of high work-function MoO_x (with $2 < x < 3$) contacts [107] and pre-treatments of MoS_2 with oxygen plasmas in the source and drain areas [108].

10.3.3 Vertical Transistors Based on 2D-Materials Heterostructures

Besides the conventional lateral FET, novel device architectures based on van der Waals (vdW) heterostructures obtained by the stacking of 2DMs [2, 33] have been introduced. These devices rely on quite different working principles than lateral FETs, and mainly exploit vertical current transport across the 2DM interfaces. In the following subsections, two relevant examples of these devices, i.e. the tunnelling field effect transistor (TFET) [34, 109], the band-to-band tunnelling transistor [110], will be described.

10.3.3.1 Tunnelling Field Effect Transistor

Britnell and co-workers first reported a TFET with the vertically stacked heterostructure composed of Gr and thin h-BN [34]. The basic principle of this vertical transistor is the quantum tunneling between the two Gr electrodes separated by the thin h-BN barrier (see Fig. 10.4a, b). Current tunneling was controlled by tuning Gr density of states and the associated barrier by the external gate voltage. As the transit time associated to tunneling is very low, the TFET can be potentially suitable for high speed operation. On the other hand, as a result of the direct tunneling mechanism, this device suffers from very low current density (in the order of 10–100 pA/μm^2, as shown in Fig. 10.4c), making it not useful for practical applications. Starting from the same idea, Georgiou et al. [109] reported a Gr vertical FET with WS_2 layers as barrier (see Fig. 10.4d). Due to the smaller band gap of WS_2, current transport between

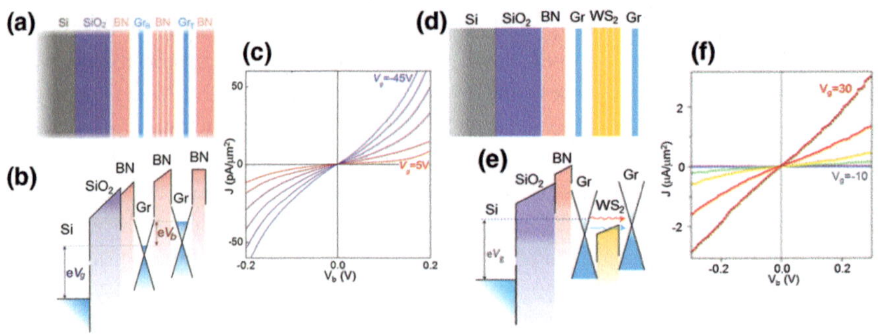

Fig. 10.4 **a** Schematic cross section and **b** energy band diagram of a Gr/h-BN/Gr tunneling field effect transistor. **c** Current density J versus interlayer bias V_b for different gate bias applied to the Si back-gate. Images adapted with permission from [34]. **d** Schematic cross section and **e** energy band diagram of a Gr/WS_2/Gr tunneling field effect transistor. **f** Current density J versus interlayer bias V_b for different gate bias applied to the Si back-gate. Images adapted with permission from [109]

the two Gr layers occurs by tunneling or thermionic emission (see Fig. 10.4e). As compared to the Gr/h-BN/Gr prototype, this device exhibits much higher ON current (in the order of 1 $\mu A/\mu m^2$, as shown in Fig. 10.4f) and ON/OFF current ratio up to 10^6.

10.3.3.2 Band-to-Band Tunneling Vertical Transistor

Van der waals heterojunctions of different TMDs offer the possibility of realizing ultrasharp semiconductor heterostructures with tailored band alignment, to implement novel device concepts. As an example, MoS_2/WSe_2 stacks have been demonstrated to give rise to a n-p semiconductor heterostructure with a type II band alignment, that can be used to realize Esaki diodes, working on the principle of band-to-band resonant tunneling.

Roy et al. [111] first experimentally demonstrated interlayer band-to-band tunneling in vertical MoS_2/WSe_2 vdW heterostructures using a dual-gate device architecture, where the electric potential and carrier concentration of MoS_2 and WSe_2 layers were independently controlled by the two symmetric gates (see schematic in Fig. 10.5a and optical microscopy in Fig. 10.5b). Notably, the atomically thin MoS_2 and WSe_2 layers resulted in a weak electrostatic screening. As a result, degenerate n-type doping of MoS_2 and p-type doping of WSe_2 could be induced by properly biasing the two gates (see Fig. 10.5c). Under these conditions, the device output characteristics (see Fig. 10.5d) exhibit a negative differential resistance (NDR) behavior, with the peak associated to the band-to-band-tunneling (BTBT) phenomena at the heterointerface.

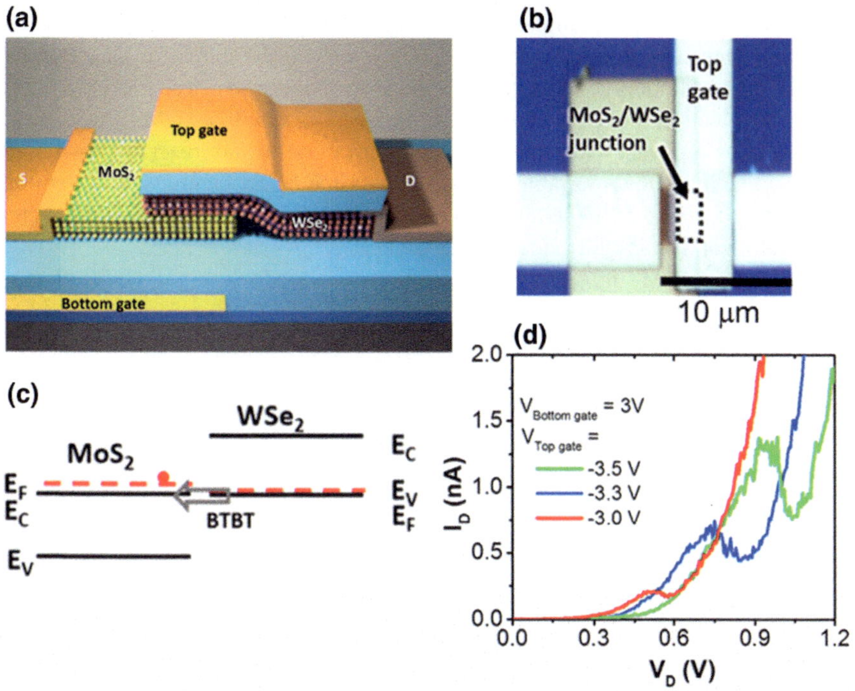

Fig. 10.5 Schematic representation (**a**) and optical microscopy (**b**) of a MoS_2/WSe_2 heterojunction diode, where the electric potential and carrier concentration of MoS_2 and WSe_2 layers are independently controlled by two gates. **c** Energy band diagram of the heterostructure with the MoS_2 and WSe_2 driven into n^+ and p^+ doping by positive biasing of the bottom gate and negative biasing of the top gate, respectively. **d** Output characteristics of the device in these biasing conditions showing characteristic negative differential resistance (NDR) peaks. Images adapted with permission from [111]

10.3.4 Transistors Based on 2D Materials Heterojunctions with Bulk Semiconductors

The development of new device concepts based on all-2D vdW heterostructures has been one of the main driving forces for the research on 2DMs in the last few years. However, producing entire families of 2DMs and their heterostructures over large areas with high electronic quality remains an outstanding challenge, limiting the progress of all-2D vdW heterostructures for future applications. On the other hand, the integration of 2DMs with bulk (3D) semiconductors [112] presents the advantage of combining the functional properties of these novel materials with the well-assessed electronic quality of 3D substrates. Hence, 2D/3D heterostructures currently represent the most viable approach toward the exploitation of 2DMs for advanced electronics/optoelectronics applications.

In this section, some examples of transistors arising from the integration of 2DM with semiconductors will be introduced. They include the transistor based on the field effect modulation of the Gr/semiconductor Schottky barrier (Barristor) [35], and the hot electron transistor (HET) with the base made with a 2DM [36, 37, 113]. These devices typically show high ON/OFF current ratios, not reachable by lateral Gr FETs, that make them suitable for logic and switching applications. Furthermore, some of these device concepts, like the HET with a Gr base, are especially targeted for operating at ultra-high frequencies up to THz.

10.3.4.1 Gate Modulated Schottky Barrier Transistor (Barristor)

The Barristor device concept is based on the tunability of the Schottky barrier height of a Gr contact with a semiconductor by an external electric field. Clearly, a nearly ideal interface between Gr and the semiconductor, without interface states responsible of Fermi level pinning, is required to achieve an efficient field effect modulation of the Schottky barrier height. The first Barristor was demonstrated by transferring CVD graphene onto hydrogen-passivated Si, thus obtaining a nearly ideal Schottky diode behavior both with n- and p-type Si [35]. Figure 10.6a illustrates a cross-sectional schematic of a Gr/Si Barristor, and Fig. 10.6b shows the band diagram for the Gr/n-Si Schottky junction for $V_g > 0$. The output characteristics of the device (for different V_g values) are reported in Fig. 10.6c and the modulation of the Gr/n-Si Schottky barrier height with the gate bias is shown in the insert of the same figure. A current ON/OFF ratio of $\sim 10^5$ under forward bias was achieved, which is suitable for digital applications. The complementary p-type Barristor was obtained by field effect modulation of the Gr/p-Si Schottky barrier height. Furthermore, by combining the complementary p- and n-type devices, logic circuits such as an inverter and a half-adder were demonstrated [35].

10.3.4.2 Graphene/SiC MESFET for Logic Applications

As discussed in the Sect. 10.3.2, epitaxial Gr on silicon carbide was first employed for the fabrication of high performance RF transistors [86]. More recently, some research groups evaluated the monolithic integrating of both RF and logic devices in the same SiC wafer. The idea was to exploit the high in-plane mobility of Gr for high frequency devices, and the peculiar electrical properties of EG/SiC(0001) interface, as well as the wide bandgap of SiC to implement logic devices. In particular, Hertel et al. [114] demonstrated a MESFET-like transistor on SiC for logic applications, where the source and drain Ohmic contacts were made with monolayer epitaxial graphene and the Schottky gate contact obtained by quasi-free-standing bilayer Gr (QFBLG) obtained by selective-area H_2 intercalation at Gr/SiC(0001) interface (see schematic in Fig. 10.7a). The current flow in the n-type SiC conductive channel was modulated by the Gr Schottky gate and by a back-contact onto p-type SiC substrate controlling the width of the space charge region. Figure 10.7b shows the transfer

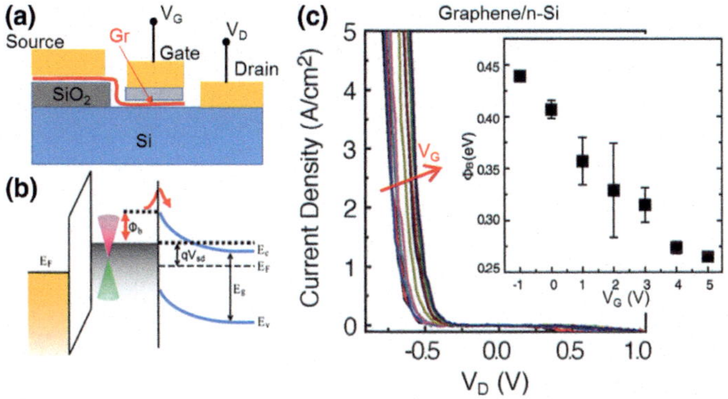

Fig. 10.6 **a** Schematic cross-section and **b** band-diagram for a gate modulated Gr/Schottky barrier transistor (Barristor). **c** Output characteristics at different gate bias values of the Gr/n-Si Barristor. In the insert, the modulation of the Gr/n-Si Schottky barrier with the gate bias. Images adapted with permission from [35]

characteristics I_D-V_{TG} of the transistor for different values of the backgate bias V_{BG}, showing how the device can work in the normally-ON conditions for V_{BG} from −5 to −6.5 V, and in the normally-OFF conditions V_{BG} above −6.5 V. This device has been used to efficiently implement logic circuits [115], such as an inverter (Fig. 10.7c) or a NAND logic gate.

10.3.4.3 Hot Electron Transistor for THz Applications

The hot electron transistor (HET) is a three-terminal (i.e., emitter, base and collector) unipolar vertical device where the ultra-thin base layer is sandwiched between two thin insulating barriers, i.e., the emitter-base and base-collector barriers. For a sufficiently high forward bias between the base and the emitter, electrons are injected into the base by Fowler-Nordheim tunneling through the barrier or by thermionic emission above the barrier, depending on the barrier height and thickness. A key aspect for the HET operation is that the injected electrons (hot electrons) have a higher energy compared to the Fermi energy of the electrons thermal population (cold electrons) in the base. Ideally, for a base thickness lower than the scattering mean free path of hot-electrons, a large fraction of these hot electrons can traverse the base ballistically, i.e. without losing energy, and finally overcome the base-collector barrier to reach the collector electrode.

Although this device concept was introduced more than 50 years ago, the successful demonstration of high-performance HETs has been limited by the difficulty to scale the base thickness below the electron mean free path of the carriers. In this context, 2DMs, in particular Gr and TMDs, can represent ideal candidates to fabricate the base of HETs, since they maintain excellent conduction properties and structural integrity down to single atomic layer thickness, allowing one to overcome the base

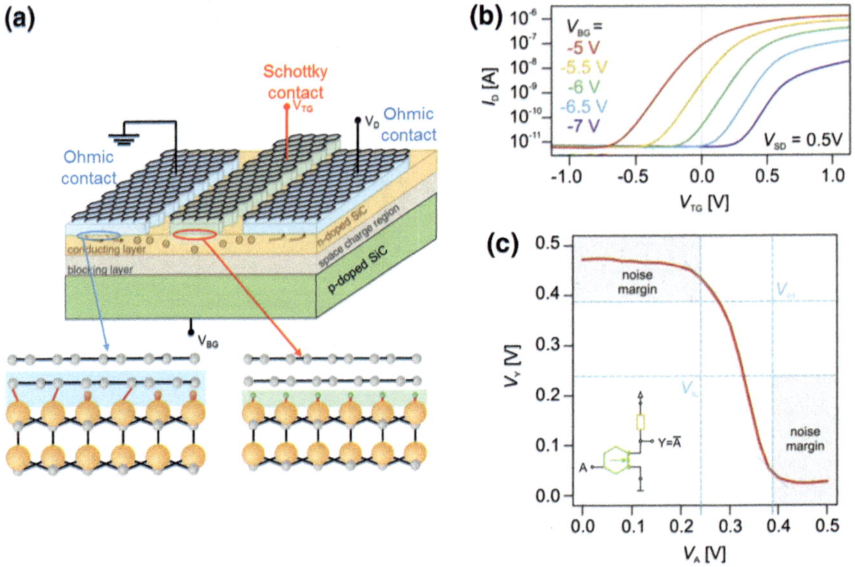

Fig. 10.7 **a** Schematic cross section of an epitaxial Gr/SiC MESFET, with monolayer epitaxial Gr source/drain Ohmic contacts and the gate Schottky contact made by Quasi-Free-Standing Bilayer Gr by selective area H_2 intercalation. **b** Transfer characteristics I_D-V_{TG} of the transistor for different values of the backgate bias V_{BG}. **c** Implementation of an inverter using the epitaxial Gr/SiC MESFET. Images adapted with permission from [114, 115]

scalability issue. Figure 10.8 illustrates some of the milestones in the development of HETs with 2DMs base. The first experimental prototypes of GBHETs were reported by Vaziri et al. [36] and by Zeng et al. [113] in 2013. Those demonstrators were fabricated on Si wafers using a CMOS-compatible technology and were based on metal/insulator/Gr/SiO_2/n^+-Si stacks, where n^+-doped Si substrate worked as the emitter, a few nm thick SiO_2 as the E-B barrier, a thicker high-k insulator (Al_2O_3 or HfO_2) as the B-C barrier and the topmost metal layer as the collector. In spite of the wide modulation of I_C as a function of V_{BE}, these first prototypes suffered from a high threshold voltage and a very poor injected current density (in the order of $\mu A/cm^2$) due to the high Si/SiO_2 barrier, hindering their practical application.

Besides Gr, monolayer MoS_2 has been also considered as the base material. As an example, Torres et al. [37] demonstrated an HET device based on a stack of ITO/HfO_2/MoS_2/SiO_2/n^+-Si, where the n^+-doped Si substrate worked as the emitter, thermally-grown SiO_2 as the emitter-base tunneling barrier, a monolayer of CVD-grown MoS_2 as the base, the HfO_2 layer as the base-collector barrier and the topmost ITO as the collector electrode. This device showed an improved current gain with respect to the previously described Gr-base HET prototypes, mainly due to the lower conduction band offset between MoS_2 and HfO_2, with respect to the cases of Gr/HfO_2 and Gr/Al_2O_3. In spite of this, the collector current density of these devices was still poor (in the order of $\mu A/cm^2$), due to the high E-B barrier between Si and SiO_2.

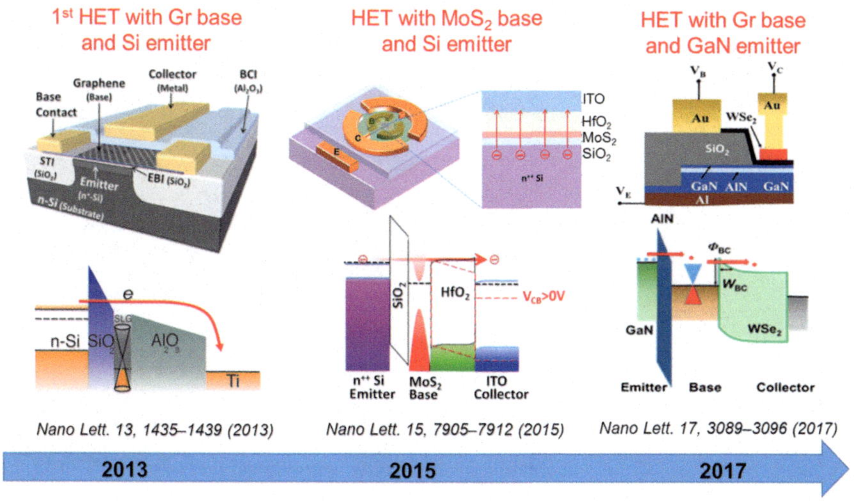

Fig. 10.8 Milestones in the development of hot electron transistors with a 2D materials (Gr, MoS_2) base

More recently, GaN/AlGaN or GaN/AlN heterostructures have been considered as emitter/emitter-base barriers for HETs [38–41], due to the presence of high density 2DEG at the interface and to the high structural quality of the barrier layer. In particular, Zubair et al. [41] fabricated a Gr-base HET with a GaN/AlN/Gr/WSe_2/Au stack, where exfoliated WSe_2 layer (forming a vdW heterojunction with Gr) was adopted as the base-collector barrier layer.

10.4 C-AFM Applications to 2D Material and Devices: Case Studies

10.4.1 Nanoscale Mapping of Transport Properties in Graphene and MoS_2

As discussed in the Sect. 10.2, 2DMs typically present peculiar nanoscale inhomogeneities (layer thickness fluctuations, wrinkles, cracks, contaminations, etc.) related to the synthesis method, to the growth substrate and to the processes eventually needed to transfer them to the final substrate for electronic applications. These inhomogeneities ultimately determine the macroscopic electrical behavior of 2DM- based devices. In the case of Gr and semiconductor TMDs residing on an insulating or semi-insulating substrate, C-AFM has been employed to probe the lateral homogeneity of transport properties, by performing current mapping using the configuration illustrated in Fig. 10.9a. Macroscopic contacts are deposited on Gr or TMD. During the

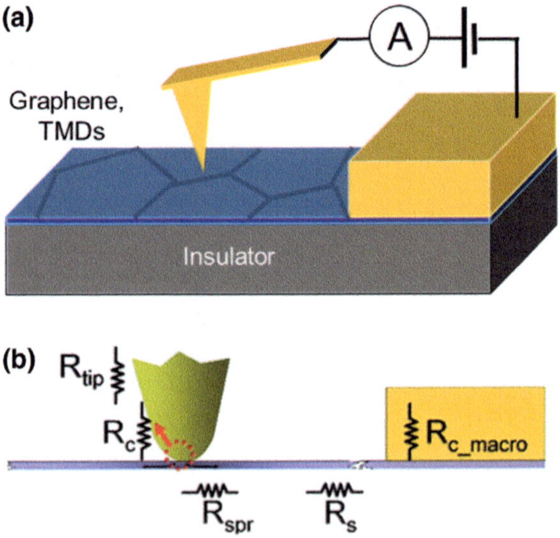

Fig. 10.9 **a** Schematic of the C-AFM experimental setup for lateral current mapping on Gr or TMDs. **b** Representation of the different contributions to the measured resistance *R*

scan of the 2DM surface with a metal-coated tip, a DC bias is applied to the macroscopic contact, and the lateral current flow inside the 2DM between this contact and the nanometric tip is measured by a current amplifier connected to the tip.

As illustrated schematically in Fig. 10.9b, the resistance *R* measured at a fixed DC bias can be expressed as the series of several contributions, i.e.,

$$R = R_{\text{tip}} + R_{\text{c}} + R_{\text{spr}} + R_{\text{s}} + R_{\text{c\_macro}}, \tag{10.1}$$

where R_{tip} is the tip resistance, R_c is the tip/2DM contact resistance, R_{spr} is the spreading resistance encountered by the current to spread from the tip to 2DM, R_s is the series resistance and $R_{\text{c\_macro}}$ is the macroscopic contact resistance. Among the five terms in (10.1), R_{tip}, $R_{\text{c\_macro}}$ and R_{S} are independent on the local 2DM properties. R_{tip} can be evaluated by C-AFM measurements on metal structures, whereas the series resistance R_s of the 2DM and the macroscopic contact resistance $R_{\text{c\_macro}}$ can be independently measured using properly fabricated macroscopic test patterns (e.g., transmission line model test structures) on the 2DM. On the other hand, the R_c and R_{spr} contributions in (10.1) capture the information on the local electrical properties of the 2DM with a lateral resolution comparable with the tip radius. In particular, R_{spr} is proportional to the local resistivity of the 2DM. The contact resistance R_c depends on the specific mechanism of current injection from the tip to the 2DM, which can be deduced from the behaviour of the local *I-V* curves collected by C-AFM. As an example, in the case of rectifying *I-V* curves, current injection is typically ruled by thermionic emission or thermionic field emission mechanisms, and R_c depends on the local Schottky barrier between the tip and 2DM. On the other hand, for Ohmic *I-V* curves, field emission from the tip to the 2DM is the dominant current injection mechanism, and R_c depends both on the local tip/2DM Schottky barrier and on the local doping of the 2DM.

The following subsections will illustrate C-AFM applications to local current mapping in CVD grown Gr transferred to SiO_2, in epitaxial Gr grown on semi-insulating SiC (0001), and in thin MoS_2 films mechanically exfoliated from the bulk crystal onto SiO_2.

10.4.1.1 Current Mapping in CVD-Grown Monolayer Gr Transferred onto SiO_2

Grain boundaries (GB) between Gr domains and wrinkles of the Gr membrane are two main extended defects in CVD-grown monolayer on copper. GB are peculiar of the 2D polycrystalline structure of this materials, which originates from the growth and merging of Gr domains randomly nucleated on the Cu surface. Wrinkles are, in part, present already in the as-grown material, but can be introduced also during the transfer process. Several authors employed electrical AFM and STM to evaluate the electrical properties of these extended defects [116–119].

Figure 10.10a, b show the topography and the corresponding C-AFM current map of CVD Gr transferred onto SiO_2/Si. These results are from [116]. Wrinkles are clearly visible in the topographic image as bright lines. Their directions are not related to each other. In addition to wrinkles, small corrugations of the Gr membrane (ripples) distributed along arrays of curved lines are visible in the AFM image as well, and they are highlighted with wavy dashed lines in Fig. 10.10a. Generally, the current drops when AFM tip goes across the wrinkles, while it is rather constant and high on the Gr flat areas. As in the case of the flat Gr sheet, the current does not drop significantly along the ripples. The topography and the corresponding current map on a zoomed area containing wrinkles are shown in Fig. 10.10c, d, respectively. Two wrinkles are denoted with $W1$ and $W2$. As can be seen from the current map, the current drops mostly on the wrinkles, and especially along the wrinkle $W1$. Here the current drops by more than 50% compared to the current on the flat Gr sheet, while at some point of $W1$ the current drops by around 90%. Bright lines inclined at around 45° in the topographic image correspond to Gr ripples. In the highest parts of the ripples denoted with arrows in Fig. 10.9b, the current falls down but not more than 10%.

Two representative local I-V curves collected close to the wrinkle (1) and on the wrinkle (2) are shown in Fig. 10.10c. In the bias voltage range ±0.1 V both curves are approximately linear, indicating that the point contact between the AFM tip and Gr can is Ohmic. The lower slope of the I-V curve (2) corresponds to a higher "local" resistance on the wrinkle. Under the used measurement conditions [116], the tip/Gr contact resistance was the main contribution to the local resistance. As the nanoscale contact between the metallic tip and semimetallic Gr can be considered as point contact separating two large conductors [120], R_c can be approximated as $R_C \propto \rho / r_c^2$, being ρ the Gr resistivity and r_c the contact radius. Hence, the locally increased resistance on the wrinkles can be ascribed to a locally higher resistivity.

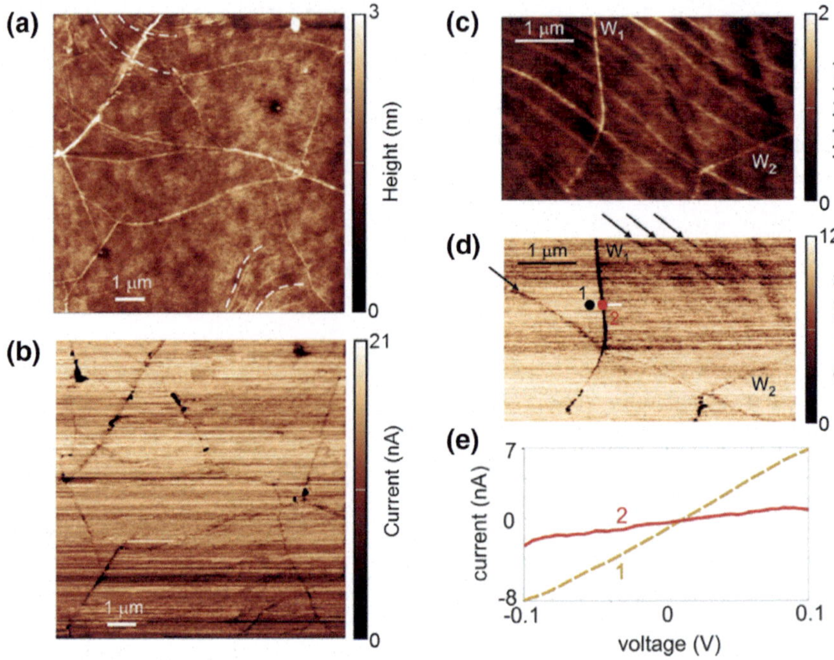

Fig. 10.10 **a** Topography and **b** C-AFM current map of CVD Gr transferred onto SiO_2/Si. Zoom-in images of the morphology (**c**) and current (**d**) in a region with wrinkles and ripples. **e** Local *I-V* curves in a planar Gr region nearby a wrinkle (1) and on top of the wrinkle (2). Images adapted with permission from [116]

One explanation for this effect is that a local strain in graphene wrinkles increases the scattering for charge carriers [121]. As a result, the transmission through the strained part of Gr is lowered and the overall conductivity falls down. At the same time, straining and deformation of Gr can lead to local perturbation of Gr lattice, resulting also into a decreased conductivity.

A similar degradation of local Gr conductivity has been observed for grain boundaries in CVD Gr [117, 118]. It should be observed that, since wrinkles and grain boundaries are randomly oriented, the overall effect of the local conductivity degradation at these defects' locations is the reduction of the average Gr conductivity.

10.4.1.2 Current Mapping in Epitaxial Graphene on Silicon Carbide

Epitaxial Gr was grown by high temperature (1600 °C) sublimation of 4H–SiC (0001) in inert gas (Ar) at atmospheric pressure [122, 123]. The 4H–SiC substrates used for this study had 8° off-axis miscut angle in the [11–20] direction. After the growth the, SiC showed a typical stepped morphology, with steps oriented perpendicularly to the original miscut direction all over the sample. After deposition of a macroscopic

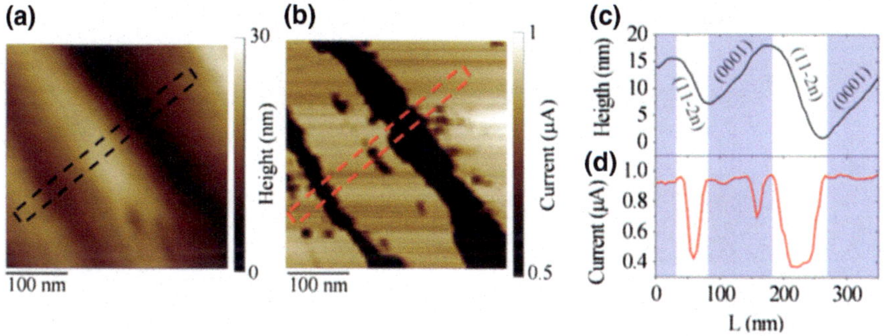

Fig. 10.11 Surface morphology (**a**) and the corresponding current map (**b**) on few layers of EG grown on 4H–SiC (0001), 8°-off. Line-profiles of the height (**c**) and of the current (**d**) along the indicated directions in the maps. Adapted and reprinted with permission from [122]

Ohmic contact onto Gr, C-AFM current mapping was carried out as illustrated in Fig. 10.9a by a Pt-coated Si tip. The current-voltage (*I-V*) characteristics measured in this configuration exhibit a linear behaviour, indicating an Ohmic contact between the tip and epitaxial Gr.

Figure 10.11a, b show the surface morphology and the corresponding current map measured on the EG sample [124]. The line-profiles of the height and of the current along the indicated directions in the maps are depicted in Fig. 10.11c, d, respectively. The as-grown Gr resides on a surface exposing (0001) basal plane terraces (with an average width of 300 nm) separated by (11-2*n*) facets. This characteristic topography of the SiC substrate results from the evolution of atomic steps present in the virgin sample due to the 8° miscut angle of the SiC wafer with respect to the growth axis (0001). In Fig. 10.9b, it can be observed a significant decrease from ~1 to ~0.4 μA, i.e., by more than a factor of 2, in the local current measured on the (11-2*n*) facets compared to the values measured on the (0001) basal plane terraces. As already mentioned, the contact between the tip and Gr is Ohmic and the tip/Gr contact resistance R_c was minimized by properly setting the contact force between the tip and epitaxial Gr [125]. Hence, the decrease of the local current on (11-2*n*) facets corresponds to a local increase of Gr resistance.

From a structural point of view, it has been observed by atomic resolution STEM analyses on cross-sectioned samples that the carbon buffer layer is locally delaminated from the (11-2*n*) facets [122]. As the presence of this interfacial layer is the main responsible of the high n-type doping of epitaxial Gr on the basal plane (0001), its local delamination from (11-2*n*) facets can account for a local reduction of electrons density.

To understand how these nanoscale electrical inhomogeneities translate in the macroscopic electrical behavior of epitaxial Gr devices, top gated field effect transistors with the channel length in the direction parallel ([1–100]) or perpendicular ([11–20]) to substrate steps have been fabricated [123]. The cross-section schematic of the FET is illustrated in Fig. 10.12a. Long channel devices, with gate length L_G

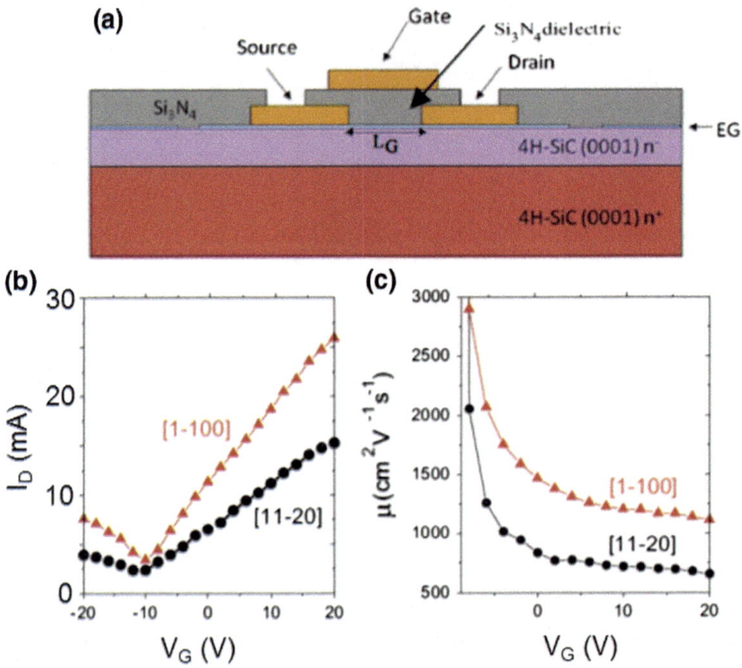

Fig. 10.12 **a** Cross-section schematics of a top-gated epitaxial Gr field effect transistor. **b** Drain current vs gate bias for Gr FETs with the channel length along the [11–20] direction, i.e. orthogonal to the SiC steps, and the [1–100] direction, i.e. parallel to the SiC steps. **c** Calculated mobility μ in the electron branch for FETs with the channel length along [11–20] and [1–100] directions. Adapted with permission from [123]

from 5 to 20 μm and channel width $W = 100$ μm, were used to maximize the importance of the channel resistance with respect to the contact resistance on the source-drain current. For both channel orientations, the drain current versus gate bias (I_D-V_G) characteristics (Fig. 10.12b) exhibit an ambipolar behavior, with the minimum conductivity (neutrality point) shifted to negative gate bias (−10 V), consistently with an average n-type doping of epitaxial Gr. Interestingly, a significantly higher drain current is measured in the [1–100] direction with respect to the [11–20] direction. The effective channel mobility μ in the electron branch was also evaluated and is depicted in Fig. 10.12c. The mobility values for the two orientation are very similar (2900 cm^2 V^{-1} s^{-1}) at V_G approaching the neutrality point, whereas $\mu_{[11-20]} = 830$ cm^2 V^{-1} s^{-1} and $\mu_{[1-100]} = 1470$ cm^2 V^{-1} s^{-1} were obtained at $V_G = 0$ V.

The observed channel resistance anisotropy in macroscopic FETs can be explained in terms of the nanoscale conductance inhomogeneity in Fig. 10.11. In fact, the macroscopic channel resistance measured in the direction perpendicular to the substrate steps can be expressed as the series combination (i.e. the sum) of the local resistances of (0001) terraces and (11-2n) facets, whereas the channel resistance in the direction parallel to the steps can be expressed as the parallel combination. As a matter of fact, the channel resistance resulting from the series combination is higher than the one resulting from the parallel combination.

10.4.1.3 Nanoscale Inhomogeneity of the Schottky Barrier in MoS_2

As discussed in the Sect. 10.3.2, the study of the metal/MoS_2 Schottky barrier currently represents a key topic for the development of prototypes of thin film transistors based on this 2D semiconductor. In this context, C-AFM has been employed to investigate the current injection from a nanoscale metal contact, i.e., the a Pt-coated AFM tip, to the surface of MoS_2 multilayers exfoliated on a SiO_2 substrate [126]. A macroscopic contact consisting of a metal film with circular holes of radius $L \gg r_{tip}$ was deposited on the MoS_2 flakes, as shown in the schematic of Fig. 10.13a. Arrays of local *I-V* curves were collected in the central region of this structure. All the curves exhibit an asymmetric behaviour with respect to bias inversion, consistently with the expected Schottky behaviour for the tip/MoS_2 contact. A representative forward bias $I\text{-}V_{tip}$ characteristic from this set of measurements is reported in Fig. 10.13b. In the semilog plot, current exhibits a more than two decades linear increase (from 1×10^{-10} to 5×10^{-8} A) with V_{tip}, followed by a saturation. In order to extract the SBH Φ_B and the ideality factor n, the thermionic emission law was applied to fit the $I\text{-}V_{tip}$ curves in the forward bias regime,

$$I = AA^*T^2 \exp\left(-\frac{q\Phi_B}{kT}\right)\exp\left[\frac{q(V_{tip} - IR)}{nkT}\right] \quad (10.2)$$

where q is the electron charge, k is the Boltzmann constant, T the absolute temperature ($T = 300$ K), $A = \pi r_{tip}^2$ the tip contact area, and A^* the Richardson constant. At small forward biases the weight of the resistive term R in the exponential factor of (10.2) is negligible and $\ln(I)$ depends linearly on V_{tip}. By linear fitting of the forward bias characteristic in the low voltage region, $\Phi_B = 307$ meV and $n = 1.61$ have been determined from the intercept and the slope of the fit, respectively. For higher bias values R causes a deviation of current from linearity and its saturation on the semilog scale.

By performing the same analysis on a full set of 100 $I\text{-}V_{tip}$ characteristics collected on a 0.5 μm × 0.5 μm area, a map of the local SBHs has been determined. The map and the histogram of the Φ_B values for pristine MoS_2 is reported in Fig. 10.13c, f, respectively. An average SBH of 300 meV with a standard deviation of 24 meV has been estimated from the distribution in Fig. 10.13f. Clearly, these SBH values are much lower than the ideal one expected according to the Schottky-Mott theory $\Phi_B = W_{Pt} - \chi_{MoS2} \approx 1.3$ eV (with W_{Pt} the Pt work function and χ_{MoS2} the MoS_2 electron affinity), consistently with the commonly reported Fermi level pinning for most of the metals in the upper part of the MoS_2 band gap [100].

The Schottky barrier of metal contacts with MoS_2 can be tuned by O_2 plasma pre-functionalization of MoS_2 surface [108]. Figure 10.13d reports the SBH map

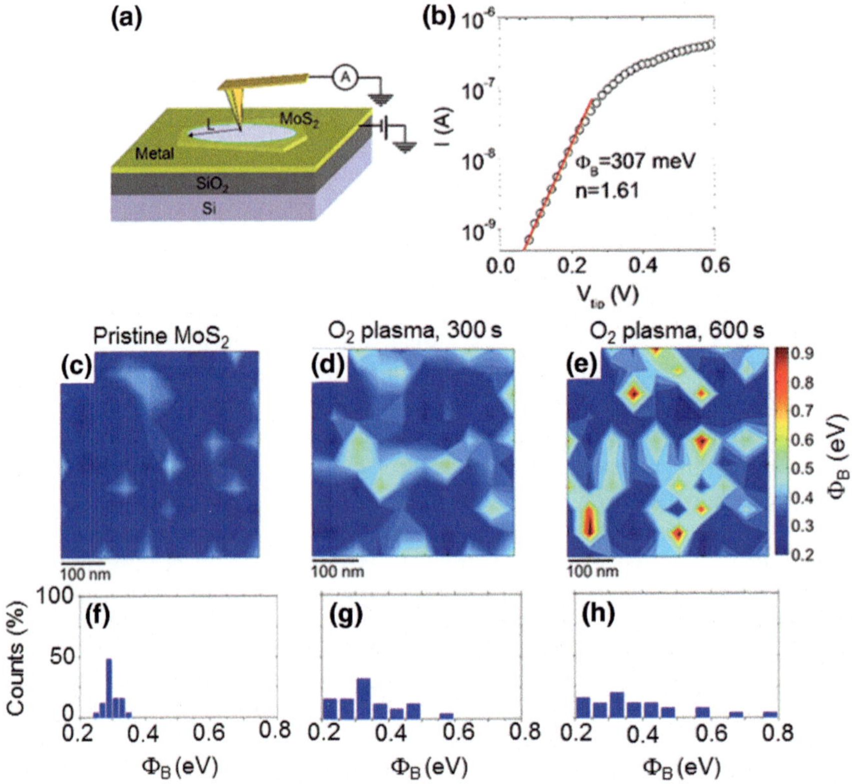

Fig. 10.13 **a** Schematics of the experimental setup for C-AFM measurements on MoS_2. **b** Representative forward bias I-V_{tip} characteristic and fit with the thermionic emission law to extract the SBH and ideality factor. 2D maps and histograms of the local Φ_B values on pristine MoS_2 (**c**, **f**), after O_2 plasma for 300 s (**d**, **g**) and 600 s (**e**, **h**). Adapted and reprinted with permission from [108, 126]

after 300 s soft plasma treatment, which results in a broader SBH distribution, with Φ_B ranging from 0.21 to 0.58 eV, as shown by the histogram of the Φ_B values (see Fig. 10.13g). Finally, Fig. 10.13e shows the SBH map after 600 s soft plasma treatment. In this case, the SBH distribution extends from ~0.2 to ~0.9 eV (see Fig. 10.13h). Hence, it includes both regions with low barrier for electrons and regions with low barrier for holes, being the Schottky barrier for holes $\Phi_{B,h} = E_g - \Phi_B$.

To understand the implications on devices electrical characteristics of the nanoscale modification of MoS_2 surface properties induced by the O_2 plasma, back-gated FETs were fabricated with source and drain contacts deposited on pristine MoS_2 (see schematic in Fig. 10.14a) or on areas selectively exposed to O_2 plasma functionalization for 600 s (see schematic in Fig. 10.14d). Figure 10.14b shows the transfer characteristics $I_D - V_G$ for different drain bias values (V_{DS} = 1, 2, and

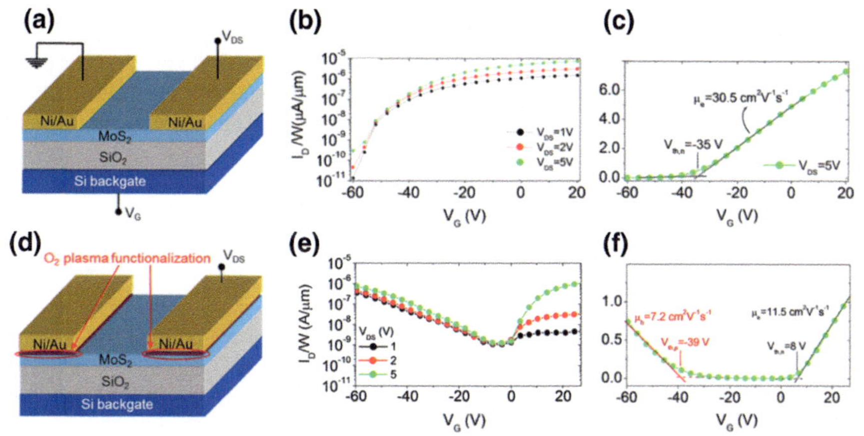

Fig. 10.14 Schematic illustrations of back-gated field-effect transistors (FETs) with source and drain contacts deposited on pristine MoS_2 (**a**) or on areas selectively exposed to O_2 plasma for an optimal time of 600 s (**d**). I_D-V_G characteristics for a pristine MoS_2 FET on semilog (**b**) and linear scale (**c**). I_D- V_G characteristics for a FET with O_2 functionalized contact areas on semilog (**e**) and linear scale (**f**). Adapted and reprinted with permission from [108]

5 V) measured on a FET with channel length $L = 10\ \mu m$ fabricated on a pristine MoS_2 flake. The n-type transistor behavior typically reported for MoS_2 FETs can be observed, with a monotonic increase of I_D over more than 5 decades in the considered gate bias range. The exponential increase of I_D with V_G at high negative bias values is due to thermionic emission of electrons above the Schottky barrier at Ni/MoS_2 interface. A linear plot of the transfer characteristic (measured with $V_{DS} =$ 5 V) is also reported in Fig. 10.14c. By fitting the linear region in the $I_D - V_G$ curve, the threshold voltage for electron accumulation in the channel ($V_{th,n} = -35$ V) was obtained as the intercept with the horizontal axis, and the field-effect mobility ($\mu_e \approx$ 30 cm^2 V^{-1} s^{-1}) was evaluated. Figure 10.14e shows the transfer characteristics I_D $- V_G$ (for $V_{DS} = 1, 2$, and 5 V) of a MoS_2 transistor with the same channel length and thickness as those of the pristine one but with the source and drain contacts deposited on plasma O_2 functionalized regions. For $V_{DS} = 1$ and 2 V, the transfer characteristics exhibit a pronounced p-type behavior, with the hole current branch (for $V_G < 0$) significantly higher than the electron current branch (for $V_G > 0$), whereas for V_{DS} $= 5$ V an ambipolar behavior is observed, with I_{on}/I_{off} current ratio of ~10^3 for both the electrons and holes branches. The linear scale I_D-V_G characteristic in Fig. 10.14f exhibits two threshold voltages: one for electron accumulation ($V_{th,n} \approx 8$ V) and the other for hole inversion channel formation ($V_{th,p} \approx -39$ V). Furthermore, from the slope of I_D-V_G characteristics in the two branches, the field-effect mobilities for electrons ($\mu_e = 11.5$ cm^2 V^{-1} s^{-1}) and holes ($\mu_h = 7.2$ cm^2 V^{-1} s^{-1}) have been evaluated.

The key to achieve this ambipolar behavior in O_2 functionalized FETs is represented by the coexistence of regions with low SBH for electrons and regions with low SBH for holes within the same source and drain contact areas, as demonstrated in the Schottky barrier height map of Fig. 10.13e. In particular, for a positively biased drain contact ($V_{DS} > 0$), the injection of electrons from the source to the accumulation channel (for $V_G > 0$) occurs through regions with lower SBH for electrons, whereas the injection of holes from the drain to the inversion channel (for $V_G < 0$) is allowed by the regions with low SBH for holes.

The possibility to have both n- and p-type current transport in a single device structure is an important step towards the implementation of a CMOS technology with MoS_2.

10.4.2 Vertical Current Injection Through 2D/3D or 2D/2D Materials Heterojunctions

10.4.2.1 Quasi-Free-Standing-Epitaxial-Graphene on Silicon Carbide

Epitaxial graphene grown on Si-face silicon carbide (SiC) is a versatile system allowing the fabrication of lateral field effect transistors for RF applications, as well as novel device concepts by exploiting the tunability of Gr/SiC contact from Ohmic to Schottky via hydrogen intercalation.

Here we present a micro- and nanoscale electrical characterization of quasi-free-standing-bilayer graphene (QFBLG) grown by thermal decomposition of on-axis 4H–SiC at 1680 °C under Ar at atmospheric pressure followed by hydrogen intercalation at 850 °C in H_2 flux. Macroscopic Schottky diodes were first fabricated by deposition of large diameter ($d = 400$ μm) Au contacts on QFBLG followed by O_2 plasma etching for lateral isolation. The Schottky barrier height Φ_B of these diodes were determined by the standard electrical methods, i.e. *C-V* measurements and *I-V* analyses [127]. Interestingly, different barrier height values have been obtained from the $1/C^2$ versus *V* plot under reverse bias polarization ($\Phi_{B,CV} = 1.57$ eV, see Fig. 10.15a) and by fitting the forward bias *I-V* curve with the thermionic emission model ($\Phi_{B,IV} = 0.9$ eV, see Fig. 10.15a).

This kind of discrepancy between $\Phi_{B,CV}$ and $\Phi_{B,IV}$ has been observed also in traditional metal/semiconductor Schottky contacts, and it is commonly ascribed to the Schottky barrier height inhomogeneity, i.e. to the presence of nanoscale low barrier regions embedded in a high barrier background. In fact, while *C-V* measurements under reverse polarization are more sensitive to the high Schottky barrier background, the presence of low barrier areas (even of small size and density) has a strong impact on the current conduction under forward bias.

In order to get insight on the nature and the density of these low-barrier paths in the case of QFBLG/SiC interface, *I-V* measurements were performed using the C-AFM tip on Au/QFBLG/SiC diodes [128] with different sizes (from 10 to 0.5 μm diameter), as illustrated in the schematic of Fig. 10.15c. Typical AFM images of diodes with $d = 10$ μm and $d = 1$ μm are reported in the inserts of Fig. 10.15d,

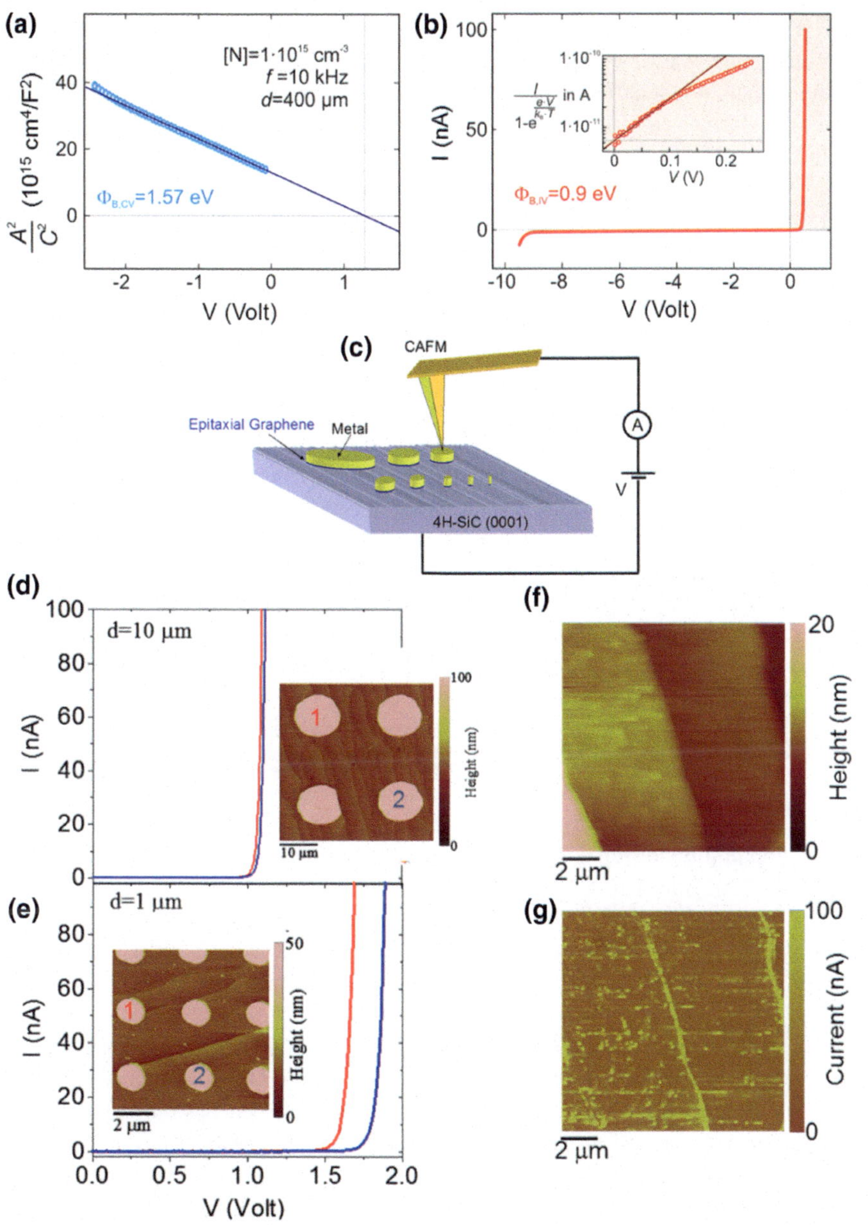

Fig. 10.15 Macroscopic *C-V* and *I-V* characterization of QFBLG/SiC Schottky contacts. **a** $1/C^2$ versus *V* plot under reverse bias polarization and **b** I-V curve on the diode. **c** Schematic representation of the experimental setup for I-V characterization of variable size QFBLG/SiC Schottky contacts. Representative I-V characteristics measured on Au capped QFBLG/SiC diodes with 10 μm (**d**) and 1 μm diameter (**e**). AFM morphologies of the contacts are shown in the inserts. C-AFM morphology (**f**) and current map (**g**), on uncapped QFBLG/SiC diodes. Images adapted from [127, 128]

e. The average SiC terrace width was ~3–6 μm, as determined from morphological measurements. While several SiC terraces are included in the contact area in the case of the 10 μm contacts (Fig. 10.15d), the 1 μm contacts are typically fully residing inside SiC terraces and some of them intercept only one step (Fig. 10.15e).

In Fig. 10.15d, e are reported representative forward bias *I-V* characteristics measured on the Au-capped QFBLG contacts to SiC with 10 and 1 μm diameters. The SBH values were extracted from the linear region of the log(*I*)-*V* curves using the thermionic emission model. While similar SBH values (~1 eV) were obtained from the *I-V* characteristics measured on the two representative 10 μm contacts, higher and significantly different SBH values are obtained for the two representative 1 μm contacts, i.e. ~1.55 eV for the contact inside the terrace and ~1.4 eV for the contact on the step.

These measurements revealed the critical role of steps in reducing the *I-V* measured effective SBH and suggest a different electrical behaviour of the QFBLG interface on terraces and on steps. Finally, Fig. 10.15f, g report nanoscale resolution morphology and current maps on the uncapped QFBLG contacts onto SiC. Current maps reveal a higher current flowing through the QFBLG-SiC interface on step edges than on terraces. This indicates that edges, provide low SBH paths, which are responsible for the reduced average Schottky barrier on large contacts. The different properties of the QFBLG contacts on the edges and planar terraces can be explained in terms of the different crystallographic planes exposed by the SiC substrate, i.e. the (0001) polar face on terraces and non-polar face (11–20) at the edges.

The average p-type doping of QFBLG has been ascribed to piezoelectric polarization from the (0001) polar SiC surface [129]. On the other hand, such a p-type doping is expected to be locally reduced along (11–20) non polar faces of SiC [57]. Hence, using a Schottky-Mott model to describe the QFBLG/SiC interface, a lower p-type doping on the step edges is expected to result in a reduced SBH with respect to planar (0001) surface.

10.4.2.2 Graphene Junctions with AlGaN/GaN Heterostructures

As discussed in the Sect. 10.3.4.3, Gr junctions with AlN/GaN or AlGaN/GaN heterostructures represent a key building block of hot electron transistors for THz applications. To date, the most used approach to fabricate the Gr base in these devices has been transferring of monolayer Gr grown by CVD on copper to the surface of the emitter-base barrier layer. However, transferred CVD Gr exhibits some peculiar in homogeneites (wrinkles, cracks,…), that have an impact not only on the in-plane current transport but also on the vertical current injection across Gr heterojunctions with semiconductors. While cracks density in the Gr membrane have been minimized by adopting proper transfer procedures [40], wrinkles remain typical defects of this system. Here, C-AFM measurements elucidate the role of wrinkles in the current injection through Gr/AlGaN/GaN heterojunctions [40]. The experimental setup for C-AFM-based vertical current measurements is schematically illustrated in Fig. 10.16a. In this configuration, a macroscopic ohmic contact was fabricated onto

AlGaN. The local current flowing vertically from Gr to the 2DEG at AlGaN/GaN interface is measured by applying a positive bias between the nanoscale tip scanned onto Gr and this macroscopic contact. A typical current-voltage (I-V_{tip}) characteristic measured in this configuration is reported in Fig. 10.7b. It exhibits a rectifying behavior, with negligible current at negative bias values and current onset at positive ones. As demonstrated by Fisichella et al. [38], this behavior can be fitted by thermionic current emission above the Gr/AlGaN/GaN Schottky barrier. Figure 10.16c, d show a typical morphology and the corresponding vertical current map measured with the tip scanned on the Gr membrane in a region including wrinkles. The height profile and the corresponding current profile extracted from the morphological and electrical maps are also reported in Fig. 10.16e, left and right axis, respectively. The current measured on the Gr wrinkle is ~35% lower with respect to the planar Gr regions. Figure 10.16f schematically illustrates a model of the vertical current path in the planar (i) and wrinkled (ii) Gr regions. As the wrinkle height is few nanometers (~2.5 nm in the present case), direct current flow from the tip contact on top of the wrinkle to the AlGaN/GaN 2DEG would imply tunneling across this few nm-thick vacuum barrier. Hence, the least resistive path is along the wrinkle walls up to the planar region, and then by thermionic emission above the AlGaN barrier layer. This additional path along the wrinkles walls adds a resistance term R_{wr} to the thermionic emission expression, i.e. $I \propto \exp[(qV - IR_{wr})/nkT]$, and can justify the local decrease of the injected current at higher bias values.

10.4.2.3 2D Transition Metal Dichalcogenides Heterojunctions

As discussed in the Sect. 10.3.3.2, heterostructures of TMDs with a type II band alignment, such as MoS_2 and WSe_2, allows the implementation of band-to-band tunnelling transistors with a sharp negative differential resistance peaks. The first examples of this device concept (such as the one illustrated in Fig. 10.5) were fabricated by exfoliation of MoS_2 and WSe_2 flakes and sequential transfer to obtain the heterostructure. However, this fabrication method is laborious and not scalable. Furthermore, the crystalline flakes comprised in the heterostructure are randomly misoriented and residual contaminations can be trapped at the interface. All these factors reduce the layers coupling and the efficiency of BTBT. More recently some attempts to fabricate MoS_2/WSe_2 heterostructures by CVD methods have been explored, with the aim to produce these vdW heterostructures on large area and to avoid the drawbacks of the pick-up and transfer method. As an example, Lin et al. [130] recently reported the growth of MoS_2/WSe_2 heterostructures onto epitaxial Gr on SiC by sequentially depositing monolayers of the two TMDs employing oxide powder vapourization or MOCVD. In particular, WSe_2 was first grown on tri-layer EG at 950 °C. Following this first growth step, the surface coverage of the WSe_2 on EG was typically >60%, with a lateral size of 2 μm for WSe_2 domains (as illustrated by the AFM image in Fig. 10.17a). Subsequently, a second ex situ growth of MoS_2 on WSe_2/EG was performed at 750 °C. The MoS_2 domains are smaller (~300 nm) and typically grow on WSe_2 starting from domain's edges, as illustrated by the AFM image in Fig. 10.17b).

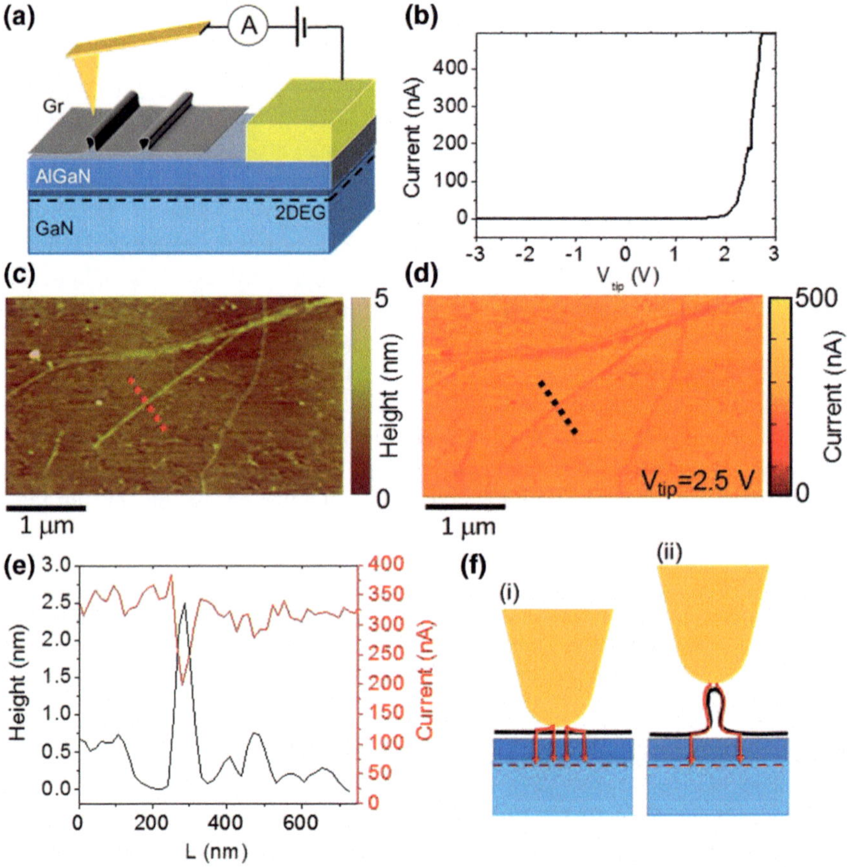

Fig. 10.16 **a** Schematic illustration of the C-AFM setup for vertical current measurements in Gr/AlGaN/GaN heterojunctions. **b** Typical current-voltage (I-V_{tip}) characteristic measured in this configuration. C-AFM morphology (**c**) and corresponding vertical current map (**d**) measured with the tip scanned on the Gr membrane in a region including wrinkles. **e** Height line profile (left vertical axis) and current line profile (right vertical axis) extracted from the morphological and electrical maps. **f** Model of the vertical current path in the planar (i) and wrinkled (ii) Gr regions. Adapted and reprinted with permission from [40]

Scanning tunnelling microscopy/spectroscopy on these heterostructures showed a Moirè pattern associated to a fixed misorientation angle of ~1.9° between MoS_2 and the underlying WSe_2 layer. While the mechanical stacking technique leads to a variety of rotation angles between layers, the direct growth of vdW layers appears to have a strict rotational alignment, which may be critical for achieving optimal coupling between the layers.

Given the small size of the MoS_2/WSe_2 areas, C-AFM was employed as a method of choice to investigate vertical current transport across the heterostructure [130]. In particular, local I-V measurements through the MoS_2/WSe_2/EG and WSe_2/EG heterostructures were carried out at room temperature according to the schematic in Fig. 10.17c. Representative *I-V* curves collected on the two systems are reported in Fig. 10.17d. The *I-V* characteristics WSe_2/EG heterojunction exhibit a slight rectifying behavior, associated to the p-n junction between p-type WSe_2 and n-type EG on SiC. On the other hand, the curves measured on the MoS_2/WSe_2/EG show a current peak (at $V_{peak} \approx 1.1$ V) followed by a valley (with a peak-to-valley current ratio of 1.9) and, finally, an exponential current increase. Such negative-differential-resistance behavior, previously demonstrated in dual-gated MoS_2/WSe_2 devices fabricated by layers transfer [111], is here observed in a much easier configuration, indicating a better electronic coupling of the directly grown layers. The main challenge to exploit this approach for large area device applications will be to develop laterally uniform heterostructures by van der Waals epitaxy.

10.4.3 Electrical Characterization of h-BN as Two Dimensional Dielectric: Lateral Inhomogeneity, Reliability and Dielectric Breakdown

As discussed in the introduction, h-BN is a dielectric 2D layered material with a considerably high band gap of 5.5–5.7 eV [5, 22] and dielectric constant of ~3 [23], which has recently attracted a lot of interest for its use in different electronic devices

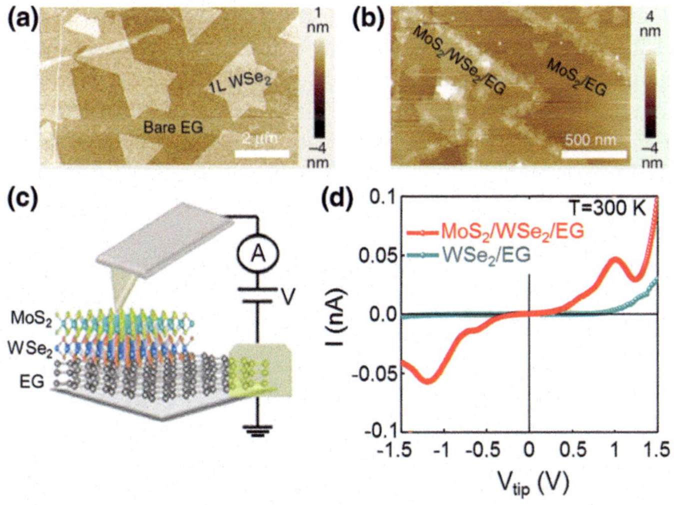

Fig. 10.17 AFM morphology of **a** single layer WSe_2 and **b** MoS_2/WSe_2 stacks deposited on epitaxial graphene (EG) on SiC. **c** Schematic of the C-AFM set-up for the *I-V* measurement in this layered system. **d** Experimental *I-V* traces for the WSe_2/EG p-n junction and for the MoS_2/WSe_2/EG interface showing a negative differential resistance (NDR) peak. Images adapted with permission from [130]

[83] mainly for three reasons. (i) The first one is its intrinsic good performance as dielectric. As it will be explained in the following subsections, monolayer h-BN (0.33 nm thick) can resist electrical fields better than 2 nm HfO_2. (ii) h-BN presents an ideal interaction with Gr and other 2DMs [131], which at the same time show a very bad interface full of local defects with high-k dielectrics and SiO_2 [132–135]. (iii) h-BN holds several properties wanted in electronic devices, such as a high thermal conductivity [136] which is beneficial for slowing down the dielectric breakdown process. Therefore, using h-BN to replace traditional dielectrics in Gr and TMD based devices is an interesting approach.

The electrical properties of ultra-thin h-BN dielectric films are strongly dependent on the synthesis method. In the following subsections, mechanically exfoliated h-BN from bulk crystals and CVD-grown h-BN on copper substrates will be considered. C-AFM applications to evaluate the nanoscale lateral homogeneity and the current transport properties h-BN produced by these different methods will be illustrated. Furthermore, C-AFM permitted to investigate the reliability and breakdown mechanism of h-BN films on local scale and to compare the results with measurements on macroscopic devices (capacitors) with h-BN dielectric.

10.4.3.1 Nanoscale Homogeneity and Variability

In the case of h-BN obtained by mechanical exfoliation from the bulk crystal, thickness inhomogeneity is the main source of variability. Britnell et al. [48] used C-AFM for nanoscale resolution mapping of the tunneling current through h-BN flakes exfoliated onto a graphite substrate, as illustrated in the schematic of Fig. 10.18a. Figure 10.18b, c show the AFM topography and the corresponding current map in a region including the conductive graphite substrate, 2 and 4 layers of h-BN. The height profile in Fig. 10.18b indicates that h-BN terraces are extremely flat. Furthermore, little variations of the current was observed in Fig. 10.18c within regions of the same thickness, suggesting lack of pinholes or other conductive defects.

Lee et al. [49] used local I-V curves measured by C-AFM to investigate current transport mechanisms through mechanically exfoliated h-BN nanosheets with different thicknesses. Figure 10.18d reports the I-V curves for h-BN stacks with thicknesses from 1 to 3 layers. At low bias they follow the direct tunneling (DT) model, as shown by the linear dependence of the current on the tip bias:

$$I(V) = \frac{A_{\text{eff}}\sqrt{m\Phi_B}q^2V}{h^2d}\exp\left[-\frac{4\pi\sqrt{m\Phi_B}d}{h}\right] \tag{10.3}$$

For larger thickness ($N > 4$), the DT conduction at low bias becomes negligible, and Fowler-Nordheim (FN) tunneling becomes the dominating transport mechanism at high electric fields, as demonstrated by the linear behavior of the $\log(I/V^2)$ versus $1/V$ curves in Fig. 10.18e.

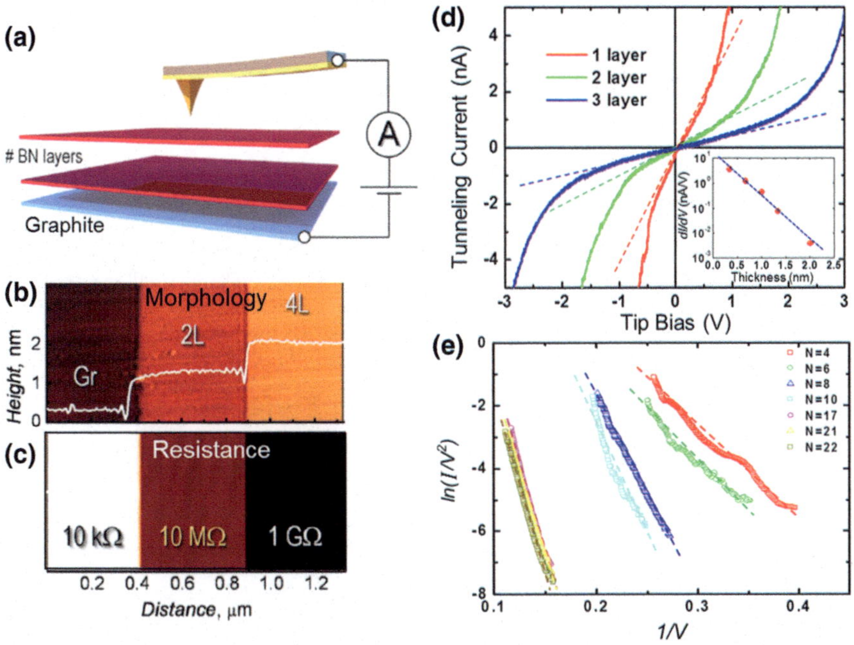

Fig. 10.18 **a** Setup for C-AFM measurements on thin h-BN stacks exfoliated on a conductive substrate (graphite). Morphology (**b**) and current map (**c**) measured on a sample region including bare graphite, two layers (2L) and four layers (4L) of h-BN. The average values of the differential resistance on each terrace are also reported. Images adapted with permission from [48]. **d** Tunneling current versus tip bias measured for 1, 2 and 3 layers of h-BN. The linear behavior at low bias is consistent with direct tunneling (DT). The insert shows a semilog plot of the differential conductance versus h-BN thickness, whose linear behavior further supports DT. **e** $\ln(I/V^2)$ versus $1/V$ plots of the current-voltage characteristics measured for h-BN stacks including from 4 to 22 layers. The linear behavior is consistent with Fowler-Nordheim tunneling. Images adapted with permission from [49]

$$\ln\left[\frac{I(V)}{V^2}\right] = \ln\left[\frac{A_{\mathrm{eff}} q^3 m}{8\pi\, h \Phi_{\mathrm{B}} d^2 m^*}\right] - \frac{8\pi\sqrt{2m^*}\Phi_{\mathrm{B}}^{3/2} d}{3hqV} \tag{10.4}$$

In the case of CVD-grown h-BN, conduction properties are affected by several kinds of morphological and structural inhomogeneities, including thickness fluctuations, grain boundaries (GBs), wrinkles and cracks. Figure 10.19a shows a representative SEM image of CVD-grown single layer h-BN on Cu [137], where some of these typical inhomogeneities (wrinkles, multilayer islands, as well as copper substrate steps) are indicated. Figure 10.19b reports a C-AFM current map collected on the h-BN/Cu stack by applying a DC bias of 1 V between the tip and Cu. A reduced current contrast can be observed in multilayer h-BN islands and wrinkles with respect to monolayer h-BN regions. Wrinkles appear as long insulating lines in the current image because they introduce a gap between the h-BN and the conductive substrate, resulting in an effective increase of the dielectric thickness at those

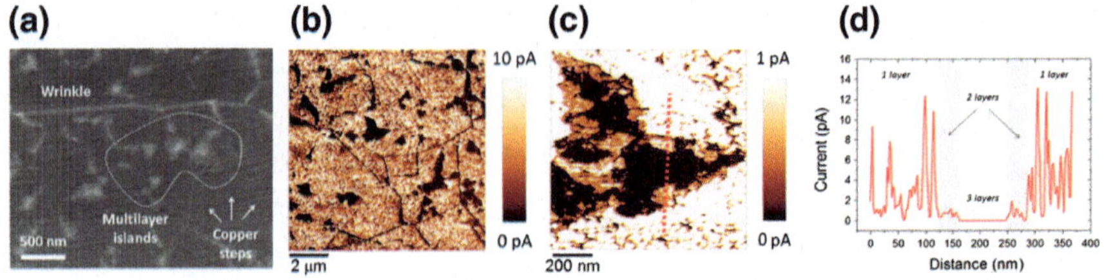

Fig. 10.19 **a** SEM image of CVD-grown single-layer h-BN on Cu. **b** C-AFM current map collected on the h-BN/Cu stack by applying 1 V. In both cases, the wrinkles, insulating (multilayer) islands, and copper steps can be easily observed. **c** C-AFM current map displaying a multilayer island of the h-BN/Cu stack. **d** Line-profile extracted from the current map, displaying the tunneling current through different amounts of layers. Images adapted with permission from [137]

locations. Figure 10.19c, d show a higher resolution current map in a multilayer h-BN island. A line-profile across this region is reported in Fig. 10.19d, showing the significant decrease of the current values from 1 layer to 3 layers h-BN. Hence, C-AFM mapping is a powerful tool for the quantification of the percentage of area for a specific thickness in CVD grown h-BN.

However, Fig. 10.19c shows that, even for the same h-BN thickness all wrinkle-free locations show deviations on the tunneling current. This can be better observed in Fig. 10.20a, which shows 94 *I*-*V* curves collected at different random locations of the h-BN stack. The study shows a continuous group of *I*-*V* curves that expands progressively. The current conduction through monolayer BN films was modeled using tunneling conduction by considering both intrinsic and defect assisted tunneling. Wentzel-Kramer- Brillouin (WKB) method was used to calculate the tunneling probability for monolayer BN. On the other side 3D Poisson equation was solved to calculate electric field across stack which consist of defect and charge trapping (CT) contributions. Final tunneling current calculations for monolayer, bilayer, and trilayer were presented in Fig. 10.20a. A close fitting with experimental data further confirmed that current from h-BN films measured in C-AFM originated from electron tunneling [137] across monolayer (~82%), bilayer and trilayer regions of the h-BN stack. However, the distribution of the *I*-*V* curves does not follow a stepped nature, but just show a continuous variation. This is an indication that other defects present in the h-BN may produce a certain degree of variability.

This hypothesis was corroborated by Pan et al. [138] who collected cross-sectional transmission microscope (XTEM) images of multilayer h-BN stacks grown by CVD on Cu (see Fig. 10.21). As it can be observed, the h-BN shows excellent layered structure and sharp interface with the Cu substrate (and also with the top Ti electrode). The layered structure is interrupted by some local lattice distortions, which can be as narrow as just one single atom.

In view of the results in Figs. 10.18, 10.19, 10.20 and 10.21 one would say that exfoliated h-BN is more homogeneous than CVD grown h-BN. However, such affirmation would not be properly demonstrated. First of all, Fig. 10.18d only shows one *I*-*V* curve per thickness, without any kind of variability. More *I*-*V* curves for

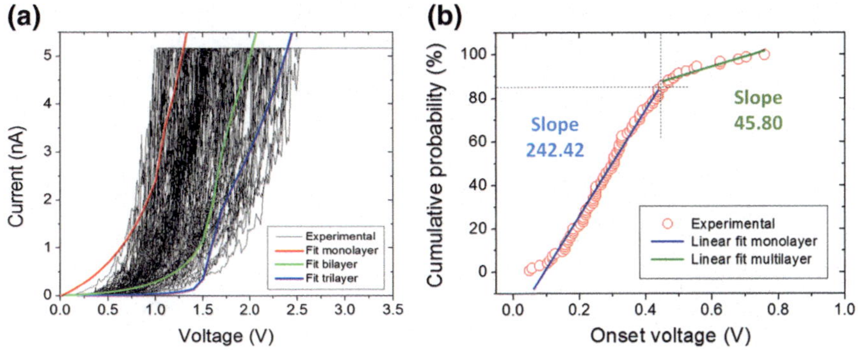

Fig. 10.20 **a** 94 *I-V* curves collected at different random locations on the h-BN/Cu stack. The voltage at which the C-AFM current saturation (5 nA) is reached shows deviation between 1 and 2.5 V. The plot also shows the calculated tunneling current for monolayer, bilayer, and trilayer BN films. **b** Statistical distribution plot of the onset voltages [$V(I = 10\ \text{pA})$] extracted from (**a**), where two groups of locations can be distinguished: monolayer (~82%) and multilayer (~18%). Images adapted with permission from [137]

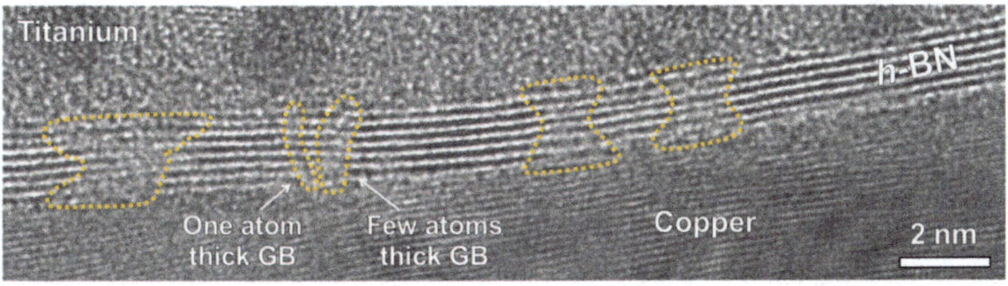

Fig. 10.21 Cross-sectional TEM image of Ti/thin h-BN/Cu RRAM device showing defective paths (GBs) through the h-BN. Images adapted with permission from [138]

each thickness need to be plotted in the same graph in order to see if the distribution is continuous like in Fig. 10.20a or stepped. Moreover, the current maps in Fig. 10.18c are plotted in a black/white current scale that covers 5 orders of magnitude (10 KΩ–1 MΩ), which makes impossible to distinguish inhomogeneities within each map (thickness). In order to analyze the current inhomogeneities within the flakes of exfoliated h-BN the experiments in Fig. 10.18 [48, 49] need to be repeated statistically, and the data need to be plotted in the proper scales.

Another important feature that may induce electrical inhomogeneity in CVD-grown h-BN stacks is the poly-crystallinity of the substrate where it is grown. h-BN stacks grown on Cu do not show large variation of the tunneling current from one metallic grain to another. However, if the h-BN is grown on Pt foils, large variability has been detected [139] Fig. 10.22a, b show the topographic and current maps simultaneously measured on the surface of h-BN grown on a polycrystalline Pt substrate by applying a bias of -2 V to the Pt substrate with respect to the grounded C-AFM tip. Strategically a large enough scan area of $80 \times 80\ \mu\text{m}^2$ was selected so that the selected

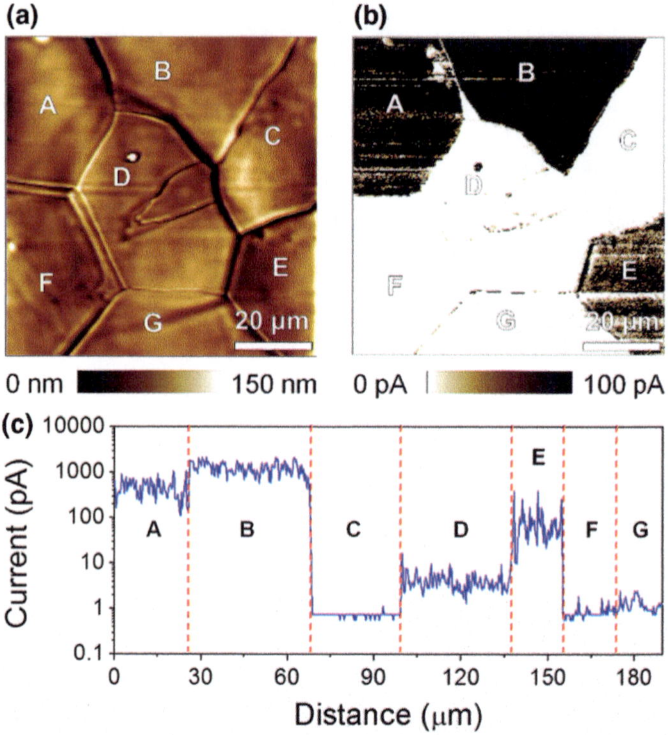

Fig. 10.22 C-AFM characterization of as-grown h-BN/Pt. **a** Topographic and **b** current maps simultaneously collected on h-BN grown on a polycrystalline Pt substrate, under a bias of −2 V (applied to the substrate, tip grounded). **c** Assembly of cross sections collected at the different grains of the current map in (**b**). Images adapted with permission from [139]

area can contain several Pt grains to study the effect of local inhomogeneities on the performance of electronic devices. The current map in Fig. 10.22b clearly shows the sharp variation of conductivity from grain to grain; the grain boundaries of the Pt foil can be seen in the topographic map shown in Fig. 10.22a. The differences within each grain in Fig. 10.22b can be better distinguished in Fig. 10.22c, which shows the current of each grain in logarithmic scale. In [139] it was demonstrated that the conductivity variations are related to thickness fluctuations from one grain to another. These thickness fluctuations are related to the different catalytic activities of the different Pt grains, due to their different crystallographic orientation at their surface (e.g. 100, 111). Interestingly, this phenomenon does not take place when growing the h-BN on other metals (Cu, Ni).

10.4.3.2 Reliability and Dielectric Breakdown

For a dielectric material, reliability is the ability of keeping unaltered its insulating properties during device operation [140]. In particular, thin insulating films employed in the state-of-the-art ultra-scaled technologies usually have to withstand large electric fields [141], which can lead to high leakage currents, degradation phenomena and, ultimately, dielectric breakdown. Typically, the electrical field can generate different types of defects in the microstructure of the dielectric, leading to the apparition of stress induced leakage currents (SILC) through it [142]. In some materials with high densities of native defects (such as high-*k* dielectrics) strong leakage current fluctuations during the degradation process are commonly observed, which are a consequence of charge trapping and de-trapping in these defects [143]. When the density of defects prohibitively increases, many partially formed defective paths can propagate across the dielectric, leading to a remarkable increase of the leakage current at very low voltages: this is the onset of the soft breakdown [144]. Finally, if the electrical stress still persists one of these conductive paths becomes dominant, forming a defect-related conductive filament that physically connects the top and bottom electrodes. This situation is called hard dielectric breakdown (BD), and it usually produces the failure of the device [141].

Recent investigations have reported important differences between the degradation process of h-BN dielectric stacks and traditional oxide dielectrics. Hattori et al. [29] used C-AFM to monitor the dielectric breakdown of mechanically exfoliated multilayer h-BN nanosheets with different thicknesses, showing that the failure occurs layer-by-layer until complete physical breakdown of the h-BN stack (generation of a hole by material removal). The dielectric strength measured out-of-plane in multilayer h-BN is ~12 MV/cm [145].

Ji et al. [137] compared the performance of a traditional 3D dielectric (2 nm HfO_2) and a monolayer h-BN sheet (which hold the same equivalent oxide thickness), by collecting local *I-V* curves under identical ramped voltage stress (RVS) using C-AFM. Figure 10.23a shows a sequence of 45 *I-V* curves collected on a single location of 2 nm thick HfO_2 dielectric film, by applying a bias ramp from 0 to 3 V. The results indicate that HfO_2 films suffer from charge trapping/detrapping, SILC and premature BD within the 45 RVS. Figure 10.23b shows a sequence of 112 *I-V* curves measured on a single location of monolayer h-BN grown on copper by ramping the tip bias from 0 to 3 V. h-BN showed almost unaltered conduction during the 112 RVS. This finding is in fact quite exciting, as this is an almost ideal behavior. The superior stability of the monolayer h-BN compared to the 2 nm thick HfO_2 film can be better observed in Fig. 10.23c, showing the behavior of the current onset voltage of individual *I-V* curves versus the *I-V* curve number.

The degradation and dielectric breakdown of h-BN were also studied from local *I-V* analyses using higher driving voltages [137]. Figure 10.24a shows a sequence of 19 *I-V* curves collected on a single position in a bilayer region of h-BN on copper. The 1st and 7th *I-V* curves used ramped voltages from 0 to 5 V (to break a h-BN layer), whereas all the rest only employed ramping voltage from 0 to 3 V to characterize the conduction. The 1st *I-V* curve shows typical currents of bilayer h-BN, whereas from

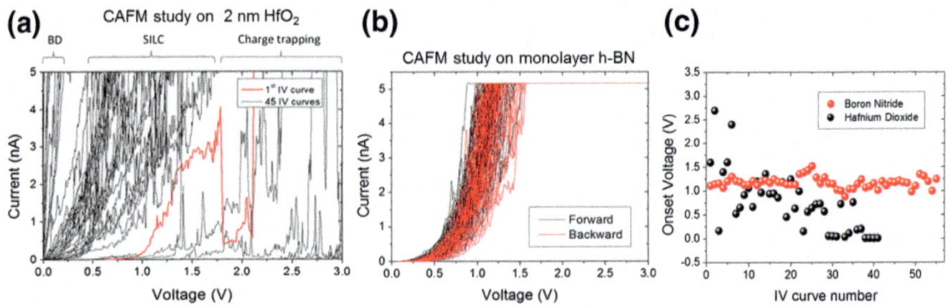

Fig. 10.23 **a** Sequence of 45 *I-V* curves collected with the C-AFM on a single location of a 2 nm thick HfO_2 dielectric film by ramping the tip bias from 0 to 3 V. The plot clearly displays three typical phenomena in the degradation of HfO_2: charge trapping/detrapping, SILC and Ohmic conduction (BD). **b** Sequence of 112 *I-V* curves measured in a single location of monolayer h-BN grown on copper by ramping the tip bias from 0 to 3 V (black lines) and backward from 3 to 0 V (red lines). **c** Onset voltage of the *I-V* curves measured on the two dielectric systems as a function of the curve number, showing a very stable behavior in the case of h-BN. Images adapted with permission from [137]

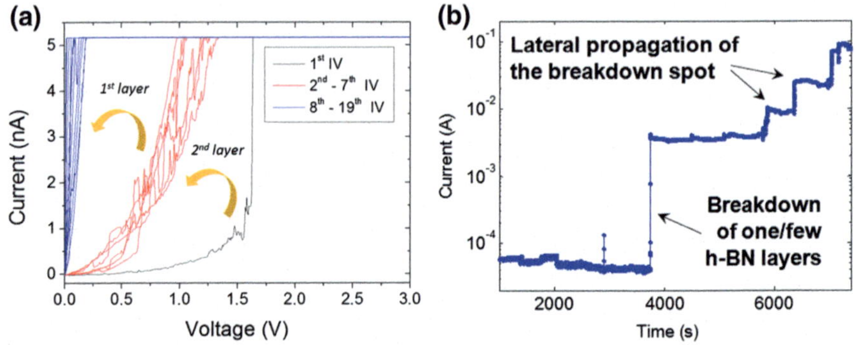

Fig. 10.24 **a** C-AFM study of breakdown of a bilayer region of CVD h-BN on copper. The 1st and 7th *I-V* curves used ramped voltages from 0 to 5 V (to break a h-BN layer), whereas all the rest only employed ramping voltage from 0 to 3 V to characterize the conduction. Image adapted with permission from [137]. **b** Device level study of breakdown performed on metal/multilayer h-BN/metal stack by measuring the current versus time (*I-t*) curves. Image adapted with permission from [148]

the 2nd to 7th *I-V* curves the conduction is similar to that of monolayer h-BN [49]. Finally, from the 8th to the last *I-V* curve, the conduction is mainly Ohmic, indicating that the complete BD is reached. The clear separation between the three groups of *I-V* curves indicates that the BD of the bilayer h-BN film occurs in a layer-by-layer mode.

It is worth noting that such jump was not preceded by the large current fluctuations typically observed in 3D dielectrics [146], and only small fluctuations were recorded. Hattori et al. [29] explained this unusual degradation mechanism of h-BN in terms of the unique anisotropic speed for defect formation in 2D layered dielectrics [29].

In fact, unlike 3D insulators (i.e. HfO_2, Al_2O_3 and TiO_2) where bonding is spatially isotropic, 2D layered h-BN stacks exhibit in-plane covalent bonding and Van der Waals forces between adjacent layers. This is expected to generate different speeds for defect formation in-plane and out-of-plane.

The behavior of the *I*-*V* characteristics for bilayer h-BN reported in Fig. 10.24a was explained as follows. The total current measured in the *I*-*V* curves can be described as $I = J(t_{OX}) \times A_{eff}$ [147], where J is the current density, t_{OX} is the thickness of the dielectric and A_{eff} is the effective area through which J flows. In-plane propagation of defects does not modify the effective thickness of the h-BN stack, and just produce a linear increase of area for current flow (A_{eff}). On the contrary, the formation of defects in adjacent layers implies a reduction of t_{OX}, which strongly alters the value of J. Due to the exponential dependence between J and t_{OX} in most conduction mechanisms though a dielectric, the current increase produced by the formation of out-of-plane defects is larger than that of in-plane ones. Hence, the sudden current shifts displayed in Fig. 10.24a should be related to the BD of a layer within the h-BN stack, while the small current fluctuations within each current step should be related to the formation of in-plane defects. This behavior was also later demonstrated in metal/multilayer h-BN/metal device level via current versus time (*I*-*t*) curves, as shown in Fig. 10.24b [148].

The *I*-*V* curves and *I*-*t* curves shown in Fig. 10.24 are able to monitor the degradation of one/layer, as this is a phenomenon that takes place in a much shorter time scale. Ranjan et al. [149] measured *I*-*t* curves with higher temporal resolution by placing the C-AFM tip on the surface of 5–7 layers thick h-BN grown by CVD, and observed characteristic random telegraph noise (RTN) signals at different biases. This observation indicates that the degradation of the h-BN stack takes place by trapping and de-trapping of charges, in a similar way than that of traditional dielectrics. However, the layered structure plays an important role in the overall dielectric breakdown process (as shown in Fig. 10.24).

One shear difference between exfoliated h-BN and CVD-grown h-BN is the morphological changes produced by the breakdown event. As mentioned, the BD event in multilayer exfoliated h-BN produce material removal and dramatic sample destruction, as shown in Fig. 10.25a [29]. However, this does not happen in CVD-grown multilayer h-BN stacks. In such samples the BD is accompanied by the formation of a hillock, as shown in Fig. 10.25b, similar to what happens in ultra-thin SiO_2 and high-*k* dielectric films [150]. The reason should be the larger density of native defects in the multilayer h-BN stack. Ranjan et al. [149] observed the formation of small pits in CVD-grown h-BN, but never a dramatic degradation like that in exfoliated samples. This discrepancy between the observations in [149, 150] should be related to the environmental conditions (air or ultra high vacuum) and polarity of the RVS used. But in any case, Ranjan et al. [149] never reported in CVD-grown h-BN a dramatic material removal like the one observed in exfoliated samples. The most surprising observation is that monolayer h-BN does not show any kind of hillock formation after hard BD, as shown in Fig. 10.25c. This observation has been attributed to the high thermal conductivity of monolayer h-BN [150].

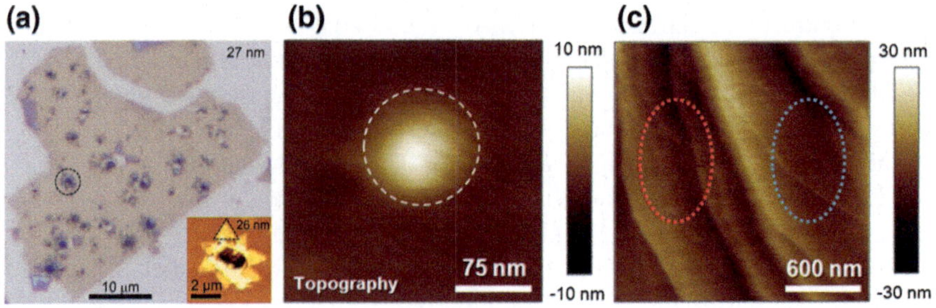

Fig. 10.25 **a** Optical microscopy image of an exfoliated multilayer h-BN flake after a series of breakdown tests with the C-AFM tip. One of the fractured areas is highlighted with the broken circle. The inset presents a typical AFM topographical image of a fractured fragment. Image adapted with permission from [29]. **b** Topography image of a hillock formed during the BD in CVD multilayer h-BN. **c** Topography image of CVD monolayer h-BN after ramped voltage stress (in the blue and red circles regions), showing the absence BD induced hillocks. Image adapted with permission from [150]

Finally, it should be highlighted that, while exfoliated h-BN has never shown any form of resistive switching (RS), several groups demonstrated the presence of bipolar RS in CVD-grown multilayer h-BN stacks, both at the device level [151–153] and at the nanoscale [149]. Probably the reason is that the BD event in CVD-grown h-BN takes place at lower energies, leading to less damage that allows recuperation. In fact, h-BN not only shows bipolar resistive switching, but also threshold type RS (i.e. self-recovery after the bias is switched off), a behavior that (for the same conditions) high-k dielectrics do not show. This behavior also indicates the higher reliability of h-BN compared to high-k dielectrics, and it may enable the use of h-BN in volatile RS applications, such as selector for RRAM and/or electronic synapse.

10.5 Summary

This chapter provided an overview of C-AFM applications to 2D materials for next generation micro- and nano-electronic devices. The state-of-the-art growth methods for electronic quality Gr, MoS_2 and h-BN have been initially discussed, as the morphological and electrical properties of these materials strongly depend on the synthesis approach. Furthermore, the main device concepts based on 2D materials, their van der Waals heterostructures, as well as on heterojunctions of 2D materials with 3D (i.e. bulk) semiconductors have been illustrated, discussing the evolution of these technologies in the last years and the open challenges for future industrial applications of these devices.

Finally, a number of relevant case studies of C-AFM on 2D materials has been reported:

(i) the lateral homogeneity of current transport in Gr grown by CVD or by thermal decomposition of SiC;
(ii) the Schottky barrier homogeneity of MoS_2;
(iii) the vertical current injection through 2D/3D or 2D/2D materials heterojunctions;
(iv) the conduction mechanisms, lateral homogeneity and reliability of h-BN obtained by exfoliation from bulk crystals or by CVD on metals.

The results of the nanoscale electrical characterization were correlated to measurements in fabricated devices, thus providing an insight in the phenomena limiting the performances of 2D materials—based devices.

Acknowledgements The authors want to acknowledge these colleagues for useful discussions: E. Schilirò, S. Di Franco, P. Fiorenza, R. Lo Nigro, I. Deretzis, A. La Magna, G. Nicotra, (CNR-IMM, Catania, Italy). This work has been supported, in part, by the Flag-ERA project "GraNitE: Graphene heterostructures with Nitrides for high frequency Electronics".

References

1. K.S. Novoselov, A.K. Geim, S.V. Morozov, D. Jiang, Y. Zhang, S.V. Dubonos, I.V. Grigorieva, A.A. Firsov, Science **306**, 666 (2004)
2. A.K. Geim, I.V. Grigorieva, Van der Waals heterostructures. Nature **499**, 419–425 (2013)
3. Q.H. Wang, K.K.-Zadeh, A. Kis, J.N. Coleman, M.S. Strano, Nat. Nanotechnol. **7**, 699 (2012)
4. H. Liu, A.T. Neal, Z. Zhu, Z. Luo, X. Xu, D. Tománek, P.D. Ye, Phosphorene: an unexplored 2D semiconductor with a high hole mobility. ACS Nano **8**, 4033–4041 (2014)
5. K. Watanabe, T. Taniguchi, H. Kanda, Direct-bandgap properties and evidence for ultraviolet lasing of hexagonal boron nitride single crystal. Nat. Mater. **3**, 404–409 (2004)
6. K.I. Bolotin, K.J. Sikes, J.H. Hone, L. Stormer, P. Kim, Temperature-dependent transport in suspended graphene. Phys. Rev. Lett. **101**, 096802 (2008)
7. C.R. Dean, A.F. Young, I. Meric, C. Lee, L. Wang, S. Sorgenfrei, K. Watanabe, T. Taniguchi, P. Kim, K.L. Shepard, J. Hone, Boron nitride substrates for high-quality graphene electronics. Nat. Nanotechnol. **5**, 722–726 (2010)
8. A.S. Mayorov, R.V. Gorbachev, S.V. Morozov, L. Britnell, R. Jalil, L.A. Ponomarenko, P. Blake, K.S. Novoselov, K. Watanabe, T. Taniguchi, A.K. Geim, Micrometer-scale ballistic transport in encapsulated graphene at room temperature. Nano Lett. **11**, 2396–2399 (2011)
9. F. Giannazzo, V. Raineri, Graphene: synthesis and nanoscale characterization of electronic properties. Riv. del Nuovo Cimento **35**, 267–304 (2012)
10. S. Sonde, F. Giannazzo, C. Vecchio, R. Yakimova, E. Rimini, V. Raineri, Role of graphene/substrate interface on the local transport properties of the two-dimensional electron gas. Appl. Phys. Lett. **97**, 132101 (2010)
11. F. Giannazzo, S. Sonde, R. Lo Nigro, E. Rimini, V. Raineri, Mapping the density of scattering centers limiting the electron mean free path in graphene. Nano Lett. **11**, 4612–4618 (2011)
12. F. Bonaccorso, Z. Sun, T. Hasan, A.C. Ferrari, Graphene photonics and optoelectronics. Nat. Photonics **4**, 611–622 (2010)
13. L. Liao, X. Duan, Graphene for radio frequency electronics. Mater. Today **15**, 328–338 (2012)
14. F. Schedin, A.K. Geim, S.V. Morozov, E.W. Hill, P. Blake, M.I. Katsnelson, K.S. Novoselov, Detection of individual gas molecules adsorbed on graphene. Nat. Mater. **6**, 652–655 (2007)
15. F. Schwierz, Graphene transistors. Nat. Nanotech. **5**, 487–496 (2010)
16. B. Radisavljevic, A. Radenovic, J. Brivio, V. Giacometti, A. Kis, Single-layer MoS_2 transistors. Nat. Nanotechnol. **6**, 147 (2011)

17. Y. Yoon, K. Ganapathi, S. Salahuddin, How good can monolayer MoS_2 transistors be? Nano Lett. **11**, 3768–3773 (2011)
18. K. Kaasbjerg, K.S. Thygesen, K. Jacobsen, W, Phonon-limited mobility in n-type single-layer MoS_2 from first principles. Phys. Rev. B **85**, 115317 (2012)
19. B. Radisavljevic, A. Kis, Mobility engineering and a metal–insulator transition in monolayer MoS_2. Nat. Mater. **12**, 815–820 (2013)
20. M.S. Fuhrer, J. Hone, Measurement of mobility in dual-gate MoS_2 transistor. Nat. Nanotech. **8**, 146–147 (2013)
21. L. Li, Y. Yu, G.J. Ye, Q. Ge, X. Ou, H. Wu, D. Feng, X.H. Chen, Y. Zhang, Black phosphorus field-effect transistors. Nat. Nanotech. **9**, 372–377 (2014)
22. L. Song, L.J. Ci, H. Lu, P.B. Sorokin, C.H. Jin, J. Ni, A.G. Kvashnin, D.G. Kvashnin, J. Lou, B.I. Yakobson, P.M. Ajayan, Large scale growth and characterization of atomic hexagonal boron nitride layers. Nano Lett. **10**, 3209–3215 (2010)
23. K.K. Kim, A. Hsu, X.T. Jia, S.M. Kim, Y.M. Shi, M. Dresselhaus, T. Palacios, J. Kong, Synthesis and characterization of hexagonal boron nitride film as a dielectric layer for graphene devices. ACS Nano **6**, 8583–8590 (2012)
24. M.W. Iqbal, M.Z. Iqbal, M.F. Khan, M.A. Shehzad, Y. Seo, J.H. Park, C. Hwang, J. Eom, High-mobility and air-stable single-layer WS_2 field-effect transistors sandwiched between chemical vapor deposition-grown hexagonal BN films. Sci. Rep. **5**, 10699 (2015)
25. N. Petrone, T. Chari, I. Meric, L. Wang, K.L. Shepard, J. Hone, Flexible graphene field-effect transistors encapsulated in hexagonal boron nitride. ACS Nano **9**, 8953–8959 (2015)
26. T. Chari, I. Meric, C. Dean, K. Shepard, Properties of self-aligned short-channel graphene field-effect transistors based on boron-nitride-dielectric encapsulation and edge contacts. IEEE Trans. Electron Devices **62**, 4322–4326 (2015)
27. M.S. Bresnehan, M.J. Hollander, M. Wetherington, M. LaBella, K.A. Trumbull, R. Cavallero, D.W. Snyder, J.A. Robinson, Integration of hexagonal boron nitride with quasi-freestanding epitaxial graphene: toward Wafer-Scale, high-performance devices. ACS Nano **6**, 5234–5241 (2012)
28. E. Kim, T.H. Yu, E.S. Song, B. Yu, Chemical vapor deposition-assembled graphene field-effect transistor on hexagonal boron nitride. Appl. Phys. Lett. **98**, 262103 (2011)
29. Y. Hattori, T. Taniguchi, K. Watanabe, K. Nagashio, Layer-by-layer dielectric breakdown of hexagonal boron nitride. ACS Nano **9**, 916–921 (2015)
30. A. Lipp, K.A. Schwetz, K. Hunold, Hexagonal boron nitride: fabrication, properties and applications. J. Eur. Ceram. Soc. **5**, 3–9 (1989)
31. T. Roy, M. Tosun, J.K. Kang, A.B. Sachid, S.B. Desai, M. Hettick, C.C. Hu, A. Javey, Field-effect transistors built from all two-dimensional material components. ACS Nano **8**, 6259–6264 (2014)
32. F. Giannazzo, G. Greco, F. Roccaforte, S.S. Sonde, Vertical transistors based on 2D materials: status and prospects. Crystals **8**, 70 (2018)
33. H. Wang, F. Liu, W. Fu, Z. Fang, W. Zhoue, Z. Liu, Two-dimensional heterostructures: fabrication, characterization, and application. Nanoscale **6**, 12250–12272 (2014)
34. L. Britnell, R.V. Gorbachev, R. Jalil, B.D. Belle, F. Schedin, A. Mishchenko, T. Georgiou, M.I. Katsnelson, L. Eaves, S.V. Morozov, N.M.R. Peres, J. Leist, A.K. Geim, K.S. Novoselov, L.A. Ponomarenko, Field-effect tunneling transistor based on vertical graphene heterostructures. Science **335**, 947–950 (2012)
35. H. Yang, J. Heo, S. Park, H.J. Song, D.H. Seo, K.-E. Byun, P. Kim, I. Yoo, H.-J. Chung, K. Kim, Graphene barristor, a triode device with a gate-controlled Schottky barrier. Science **336**, 1140–1143 (2012)
36. S. Vaziri, G. Lupina, C. Henkel, A.D. Smith, M. Ostling, J. Dabrowski, G. Lippert, W. Mehr, M.C. Lemme, A graphene-based hot electron transistor. Nano Lett. **13**, 1435–1439 (2013)
37. C.M. Torres, Y.W. Lan, C. Zeng, J.H. Chen, X. Kou, A. Navabi, J. Tang, M. Montazeri, J.R. Adleman, M.B. Lerner, Y.L. Zhong, L.J. Li, C.D. Chen, K.L. Wang, High-current gain two-dimensional MoS_2 -base hot-electron transistors. Nano Lett. **15**, 7905–7912 (2015)

38. G. Fisichella, G. Greco, F. Roccaforte, F. Giannazzo, Current transport in graphene/AlGaN/GaN vertical heterostructures probed at nanoscale. Nanoscale **6**, 8671–8680 (2014)
39. F. Giannazzo, G. Fisichella, G. Greco, A. La Magna, F. Roccaforte, B. Pecz, R. Yakimova, R. Dagher, A. Michon, Y. Cordier, Graphene integration with nitride semiconductors for high power and high frequency electronics. Phys. Status Solidi A **214**, 1600460 (2017)
40. F. Giannazzo, G. Fisichella, G. Greco, E. Schilirò, I. Deretzis, R. Lo Nigro, A. La Magna, F. Roccaforte, F. Iucolano, S. Lo Verso, S. Ravesi, P. Prystawko, P. Kruszewski, M. Leszczyński, R. Dagher, E. Frayssinet, A. Michon, Y. Cordier, Fabrication and characterization of graphene heterostructures with nitride semiconductors for high frequency vertical transistors. Phys. Status Solidi A **215**, 1700653 (2018)
41. A. Zubair, A. Nourbakhsh, J.-Y. Hong, M. Qi, Y. Song, D. Jena, J. Kong, M. Dresselhaus, T. Palacios, Hot electron transistor with van der Waals base-collector heterojunction and high-performance GaN emitter. Nano Lett. **17**, 3089–3096 (2017)
42. F. Giannazzo, P. Fiorenza, V. Raineri, Carrier transport in advanced semiconductor materials, in *Applied Scanning Probe Methods*, ed. by B. Bhushan, H. Fuchs, M. Tomitori vol 8–10 (Springer, Heidelberg, 2008)
43. F. Giannazzo, V. Raineri, E. Rimini, Transport properties of graphene with nanoscale lateral resolution, in *Scanning Probe Microscopy in Nanoscience and Nanotechnology* 2, ed. by B. Bhushan (Springer Berlin Heidelberg, 2011)
44. F. Giannazzo, S. Sonde, V. Raineri, E. Rimini, Irradiation damage in graphene on SiO_2 probed by local mobility measurements. Appl. Phys. Lett. **95**, 263109 (2009)
45. F. Giannazzo, S. Sonde, V. Raineri, E. Rimini, Screening length and quantum capacitance in graphene by scanning probe microscopy. Nano Lett. **9**, 23 (2009)
46. J. Eriksson, R. Pearce, T. Iakimov, C. Virojanadara, D. Gogova, M. Andersson, M. Syväjärvi, A. Lloyd Spetz, R. Yakimova, The influence of substrate morphology on thickness uniformity and unintentional doping of epitaxial graphene on SiC. Appl. Phys. Lett. **100**, 241607 (2012)
47. F. Giannazzo, G. Fisichella, G. Greco, P. Fiorenza, F. Roccaforte, in *Conductive Atomic Force Microscopy: Applications in Nanomaterials*, ed. by M. Lanza (WILEY-VCH Verlag, Weinheim 2017), Ch. 7, pp. 163–186. ISBN: 978-3-527-34091-0
48. L. Britnell, R.V. Gorbachev, R. Jalil, B.D. Belle, F. Schedin, M.I. Katsnelson, L. Eaves, S.V. Morozov, A.S. Mayorov, N.M.R. Peres, A.H.C. Neto, J. Leist, A.K. Geim, L.A. Ponomarenko, K.S. Novoselov, Electron tunneling through ultrathin boron nitride crystalline barriers. Nano Lett. **12**, 1707–1710 (2012)
49. G.H. Lee, Y.J. Yu, C.G. Lee, C. Dean, K.L. Shepard, P. Kim, J. Hone, Electron tunneling through atomically flat and ultrathin hexagonal boron nitride. Appl. Phys. Lett. **99**, 243114 (2011)
50. C. Berger, Z. Song, X. Li, X. Wu, N. Brown, C. Naud, D. Mayou, T. Li, J. Hass, A.N. Marchenkov, E.H. Conrad, P.N. First, W.A. De Heer, Electronic confinement and coherence in patterned epitaxial graphene. Science **312**, 1191–1196 (2006)
51. C. Virojanadara, M. Syvajarvi, R. Yakimova, L.I. Johansson, A.A. Zakharov, T. Balasubramanian, Homogeneous large-area graphene layer growth on 6H-SiC(0001). Phys. Rev. B **78**, 245403 (2008)
52. K.V. Emtsev, A. Bostwick, K. Horn, J. Jobst, G.L. Kellogg, L. Ley, J.L. McChesney, T. Ohta, S.A. Reshanov, J. Röhrl, E. Rotenberg, A.K. Schmid, D. Waldmann, H.B. Weber, Th Seyller, Towards wafer-size graphene layers by atmospheric pressure graphitization of silicon carbide. Nat. Mater. **8**, 203–207 (2009)
53. C. Vecchio, S. Sonde, C. Bongiorno, M. Rambach, R. Yakimova, E. Rimini, V. Raineri, F. Giannazzo, Nanoscale structural characterization of epitaxial graphene grown on off-axis 4H-SiC (0001). Nanoscale Res. Lett. **6**, 269 (2011)
54. K.V. Emtsev, F. Speck, Th Seyller, L. Ley, J.D. Riley, Interaction, growth, and ordering of epitaxial graphene on SiC{0001} surfaces: a comparative photoelectron spectroscopy study. Phys. Rev. B **77**, 155303 (2008)

55. G. Nicotra, I. Deretzis, M. Scuderi, C. Spinella, P. Longo, R. Yakimova, F. Giannazzo, A. La Magna, Interface disorder probed at the atomic scale for graphene grown on the C face of SiC. Phys. Rev. B **91**, 155411 (2015)
56. C. Bouhafs, A.A. Zakharov, I.G. Ivanov, F. Giannazzo, J. Eriksson, V. Stanishev, P. Kühne, T. Iakimov, T. Hofmann, M. Schubert et al., Multi-scale investigation of interface properties, stacking order and decoupling of few layer graphene on C-face 4H-SiC. Carbon **116**, 722–732 (2017)
57. M. Ostler, I. Deretzis, S. Mammadov, F. Giannazzo, G. Nicotra, C. Spinella, Th Seyller, A. La Magna, Direct growth of quasi-free-standing epitaxial graphene on nonpolar SiC surfaces. Phys. Rev. B **88**, 085408 (2013)
58. F. Varchon, R. Feng, J. Hass, X. Li, B. Ngoc Nguyen, C. Naud, P. Mallet, J.-Y. Veuillen, C. Berger, E.H. Conrad, L. Magaud, Electronic structure of epitaxial graphene layers on SiC: effect of the substrate. Phys. Rev. Lett. **99**, 126805 (2007)
59. S. Sonde, F. Giannazzo, V. Raineri, R. Yakimova, J.-R. Huntzinger, A. Tiberj, J. Camassel, Electrical properties of the graphene/4H-SiC (0001) interface probed by scanning current spectroscopy. Phys. Rev. B **80**, 241406(R) (2009)
60. C. Riedl, C. Coletti, T. Iwasaki, A.A. Zakharov, U. Starke, Quasi-free-standing epitaxial graphene on sic obtained by hydrogen intercalation. Phys. Rev. Lett. **103**, 246804 (2009)
61. X. Li, W. Cai, J. An, S. Kim, J. Nah, D. Yang, R. Piner, A. Velamakanni, I. Jung, E. Tutuc, S.K. Banerjee, L. Colombo, R.S. Ruoff, Large-area synthesis of high-quality and uniform graphene films on copper foils. Science **324**, 1312–1314 (2009)
62. G. Lupina, J. Kitzmann, I. Costina, M. Lukosius, C. Wenger, A. Wolff, S. Vaziri, M. Östling, I. Pasternak, A. Krajewska, W. Strupinski, S. Kataria, A. Gahoi, M.C. Lemme, G. Ruhl, G. Zoth, O. Luxenhofer, W. Mehr, Residual metallic contamination of transferred chemical vapor deposited graphene. ACS Nano **9**, 4776–4785 (2015)
63. G. Fisichella, S. Di Franco, F. Roccaforte, S. Ravesi, F. Giannazzo, Microscopic mechanisms of graphene electrolytic delamination from metal substrates. Appl. Phys. Lett. **104**, 233105 (2014)
64. H.H. Kim, S.K. Lee, S.G. Lee, E. Lee, K. Cho, Wetting-assisted crack- and wrinkle-free transfer of wafer-scale graphene onto arbitrary substrates over a wide range of surface energies. Adv. Funct. Mater. **26**, 2070–2077 (2016)
65. J.-Y. Hong, Y.C. Shin, A. Zubair, Y. Mao, T. Palacios, M.S. Dresselhaus, S.H. Kim, J. Kong, A rational strategy for graphene transfer on substrates with rough features. Adv. Mater. **28**, 2382–2392 (2016)
66. Y. Yu, C. Li, Y. Liu, L. Su, Y. Zhang, L. Cao, Controlled scalable synthesis of uniform, high-quality monolayer and few-layer MoS_2 Films. Sci. Rep. **3**, 1866 (2013)
67. Y.-C. Lin, W. Zhang, J.-K. Huang, K.-K. Liu, Y.-H. Lee, C.-T. Liang, C.-W. Chu, L.-J. Li, Wafer-scale MoS_2 thin layers prepared by MoO_3 sulfurization. Nanoscale **4**, 6637–6641 (2012)
68. Y. Zhan, Z. Liu, S. Najmaei, P. Ajayan, J. Lou, Large-area vapor-phase growth and characterization of MoS_2 atomic layers on a SiO_2 substrate. Small **8**, 966–971 (2014)
69. I. Song, C. Park, M. Hong, J. Baik, H.-J. Shin, H. Choi, Patternable large-scale molybdenium disulfide atomic layers grown by gold-assisted chemical vapor deposition. Angew. Chem. Int. Ed. **53**, 1266–1269 (2014)
70. C. Ahn, J. Lee, H.-U. Kim, H. Bark, M. Jeon, G. Ryu, Z. Lee, G. Yeom, K. Kim, J. Jung, Y. Kim, C. Lee, T. Kim, Low-temperature synthesis of large-scale molybdenum disulfide thin films directly on a plastic substrate using plasma-enhanced chemical vapor deposition. Adv. Mater. **27**, 5223–5229 (2015)
71. K. Kang, S. Xie, L. Huang, Y. Han, P. Huang, K. Mak, C.-J. Kim, D. Muller, J. Park, High-mobility three-atom-thick semiconducting films with wafer-scale homogeneity. Nature **520**, 656–660 (2015)
72. A. Valdivia, D. Tweet, J. Conley Jr., Atomic layer deposition of two dimensional MoS_2 on 150 mm substrates. J. Vac. Sci. Technol. A **34**, 021515 (2016)

73. P. Sutter, J. Lahiri, P. Zahl, B. Wang, E. Sutter, Scalable synthesis of uniform few-layer hexagonal boron nitride dielectric films. Nano Lett. **13**, 276–281 (2013)
74. S. Nakhaie, J.M. Wofford, T. Schumann, U. Jahn, M. Ramsteiner, M. Hanke, J.M.J. Lopes, H. Riechert, Synthesis of atomically thin hexagonal boron nitride films on nickel foils by molecular beam epitaxy. Appl. Phys. Lett. **106**, 213108 (2015)
75. A.T. Barton, R. Yue, S. Anwar, H. Zhu, X. Peng, S. McDonnell, N. Lu, R. Addou, L. Colombo, M.J. Kim, R.M. Wallace, C.L. Hinkle, Transition metal dichalcogenide and hexagonal boron nitride heterostructures grown by molecular beam epitaxy. Microelectron. Eng. **147**, 306–309 (2015)
76. Y.M. Shi, C. Hamsen, X.T. Jia, K.K. Kim, A. Reina, M. Hofmann, A.L. Hsu, K. Zhang, H. Li, Z.Y. Juang, M.S. Dresselhaus, L.J. Li, J. Kong, Synthesis of few-layer hexagonal boron nitride thin film by chemical vapor deposition. Nano Lett. **10**, 4134–4139 (2010)
77. K.K. Kim, A. Hsu, X.T. Jia, S.M. Kim, Y.M. Shi, M. Hofmann, D. Nezich, J.F. Rodriguez-Nieva, M. Dresselhaus, T. Palacios, J. Kong, Synthesis of monolayer hexagonal boron nitride on Cu foil using chemical vapor deposition. Nano Lett. **12**, 161–166 (2012)
78. A. Ismach, H. Chou, D.A. Ferrer, Y.P. Wu, S. McDonnell, H.C. Floresca, A. Covacevich, C. Pope, R. Piner, M.J. Kim, R.M. Wallace, L.G. Colombo, R.S. Ruoff, Toward the controlled synthesis of hexagonal boron nitride films. ACS Nano **6**, 6378–6385 (2012)
79. Y.Y. Liu, X.L. Zou, B.I. Yakobson, Dislocations and grain boundaries in two-dimensional boron nitride. ACS Nano **6**, 7053–7058 (2012)
80. A.L. Gibb, N. Alem, J.H. Chen, K.J. Erickson, J. Ciston, A. Gautam, M. Linck, A. Zettl, Atomic resolution imaging of grain boundary defects in monolayer chemical vapor deposition-grown hexagonal boron nitride. J. Am. Chem. Soc. **135**, 6758–6761 (2013)
81. Q.C. Li, X.L. Zou, M.X. Liu, J.Y. Sun, Y.B. Gao, Y. Qi, X.B. Zhou, B.I. Yakobson, Y.F. Zhang, Z.F. Liu, Grain boundary structures and electronic properties of hexagonal boron nitride on Cu(111). Nano Lett. **15**, 5804–5810 (2015)
82. G.Y. Lu, T.R. Wu, Q.H. Yuan, H.S. Wang, H.M. Wang, F. Ding, X.M. Xie, M.H. Jiang, Synthesis of large single-crystal hexagonal boron nitride grains on Cu–Ni alloy. Nat. Commun. **6**, 6160 (2015)
83. F. Hui, C. Pan, Y. Ji, Y. Shi, M. Lanza, On the use of two dimensional hexagonal boron nitride as dielectric. Microelectron. Eng. **163**, 119–133 (2016)
84. M.C. Lemme, T.J. Echtermeyer, M. Baus, H. Kurz, A graphene field-effect device. IEEE Electron Device Lett. **28**, 282–284 (2007)
85. I. Meric, N. Baklitskaya, P. Kim, K.L Shepard, RF performance of to p-gated, zero-bandgap graphene field-effect transistors. Technical Digest IEDM (2008), pp. 1–4
86. Y.M. Lin, C. Dimitrakopoulos, K.A. Jenkins, D.B. Farmer, H.Y. Chiu, A. Grill, P. Avouris, 100-GHz transistors from wafer-scale epitaxial graphene. Science **327**, 662 (2010)
87. Y. Wu, K.A. Jenkins, A. Valdes-Garcia, D.B. Farmer, Y. Zhu, A.A. Bol, C. Dimitrakopoulos, W. Zhu, F. Xia, P. Avouris, Y.-M. Lin, State-of-the-art graphene high-frequency electronics. Nano Lett. **12**, 3062–3067 (2012)
88. R. Cheng, J. Bai, L. Liao, H. Zhou, Y. Chen, L. Liu, Y.-C. Lin, S. Jiang, Y. Huang, X. Duan, High-frequency self-aligned graphene transistors with transferred gate stacks. PNAS **109**, 11588–11592 (2012)
89. N. Petrone, I. Meric, J. Hone, K.L. Shepard, Graphene field-effect transistors with gigahertz-frequency power gain on flexible substrates. Nano Lett. **13**, 121–125 (2013)
90. W. Wei, E. Pallecchi, S. Haque, S. Borini, V. Avramovic, A. Centeno, A. Zurutuza, H. Happy, Mechanically robust 39 GHz cut-off frequency graphene field effect transistors on flexible substrates. Nanoscale **8**, 14097 (2016)
91. K. Alam, R. Lake, Monolayer MoS_2 transistors beyond the technology road map. IEEE Trans. Electron Dev. **59**, 3250–3254 (2012)
92. L. Liu, Y. Lu, J. Guo, On monolayer MoS_2 field-effect transistors at the scaling limit. IEEE Trans. Electron Dev. **60**, 4133–4139 (2013)
93. K. Majumdar, C. Hobbs, P.D. Kirsch, Benchmarking transition metal dichalcogenide MOS-FET in the ultimate physical scaling limit. IEEE Electron Dev. Lett. **35**, 402–404 (2014)

94. International Technology Roadmap for Semiconductors (ITRS, 2012); http://www.itrs.net/
95. S.B. Desai, S.R. Madhvapathy, A.B. Sachid, J.P. Llinas, Q. Wang, G.H. Ahn, G. Pitner, M.J. Kim, J. Bokor, C. Hu, H.-S.P. Wong, A. Javey, MoS_2 transistors with 1-nanometer gate lengths. Science **354**, 99–102 (2016)
96. J. Hong, Z. Hu, M. Probert, K. Li, D. Lv, X. Yang, L. Gu, N. Mao, Q. Feng, L. Xie, J. Zhang, D. Wu, Z. Zhang, C. Jin, W. Ji, X. Zhang, J. Yuan, Z. Zhang, Exploring atomic defects in molybdenum disulphide monolayers. Nat. Commun. **6**, 6293 (2015)
97. Z. Yu, Y. Pan, Y. Shen, Z. Wang, Z.-Y. Ong, T. Xu, R. Xin, L. Pan, B. Wang, L. Sun, J. Wang, G. Zhang, Y.W. Zhang, Y. Shi, X. Wang, Towards intrinsic charge transport in monolayer molybdenum disulfide by defect and interface engineering. Nat. Commun. **5**, 5290 (2014)
98. Y. Guo, D. Liu, J. Robertson, Chalcogen vacancies in monolayer transition metal dichalcogenides and Fermi level pinning at contacts. Appl. Phys. Lett. **106**, 173106 (2015)
99. R. Addou, S. McDonnell, D. Barrera, Z. Guo, A. Azcatl, J. Wang, H. Zhu, C.L. Hinkle, M. Quevedo-Lopez, H.N. Alshareef, L. Colombo, J.W.P. Hsu, R.M. Wallace, Impurities and electronic property variations of natural MoS_2 crystal surfaces. ACS Nano **9**, 9124–9133 (2015)
100. S. Das, H.-Y. Chen, A.V. Penumatcha, J. Appenzeller, High performance multi-layer MoS_2 transistors with scandium contacts. Nano Lett. **13**, 100–105 (2013)
101. R. Kappera, D. Voiry, S.E. Yalcin, B. Branch, G. Gupta, A.D. Mohite, M. Chhowalla, Phase-engineered low-resistance contacts for ultrathin MoS_2 transistors. Nat. Mater. **13**, 1128–1134 (2014)
102. J. Suh, T.-E. Park, D.-Y. Lin, D. Fu, J. Park, H.J. Jung, Y. Chen, C. Ko, C. Jang, Y. Sun, R. Sinclair, J. Chang, S. Tongay, J. Wu, Doping against the native propensity of MoS_2: degenerate hole doping by cation substitution. Nano Lett. **14**, 6976–6982 (2014)
103. H. Fang, M. Tosun, G. Seol, T.C. Chang, K. Takei, J. Guo, A. Javey, Degenerate n-doping of few-layer transition metal dichalcogenides by potassium. Nano Lett. **13**, 1991–1995 (2013)
104. A. Nipane, D. Karmakar, N. Kaushik, S. Karande, S. Lodha, Few-layer MoS_2 p-type devices enabled by selective doping using low energy phosphorus implantation. ACS Nano **10**, 2128–2137 (2016)
105. H. Qiu, T. Xu, Z. Wang, W. Ren, H. Nan, Z. Ni, Q. Chen, S. Yuan, F. Miao, F. Song, G. Long, Y. Shi, L. Sun, J. Wang, X. Wang, Hopping transport through defect-induced localized states in molybdenum disulphide. Nat. Commun. **4**, 2642 (2013)
106. S. McDonnell, R. Addou, C. Buie, R.M. Wallace, C.L. Hinkle, Defect-dominated doping and contact resistance in MoS_2. ACS Nano **8**, 2880–2888 (2014)
107. S. Chuang, C. Battaglia, A. Azcatl, S. McDonnell, J.S. Kang, X. Yin, M. Tosun, R. Kapadia, H. Fang, R.M. Wallace, A. Javey, MoS2 p-type transistors and diodes enabled by high work function MoOx contacts. Nano Lett. **14**, 1337–1342 (2014)
108. F. Giannazzo, G. Fisichella, G. Greco, S. Di Franco, I. Deretzis, A. La Magna, C. Bongiorno, G. Nicotra, C. Spinella, M. Scopelliti, B. Pignataro, S. Agnello, F. Roccaforte, Ambipolar MoS_2 transistors by nanoscale tailoring of Schottky Barrier using oxygen plasma functionalization. ACS Appl. Mater. Interfaces. **9**, 23164 (2017)
109. T. Georgiou, R. Jalil, B.D. Belle, L. Britnell, R.V. Gorbachev, S.V. Morozov, Y.-J. Kim, A. Gholinia, S.J. Haigh, O. Makarovsky, L. Eaves, L.A. Ponomarenko, A.K. Geim, K.S. Novoselov, A. Mishchenko, Vertical field-effect transistor based on graphene-WS_2 heterostructures for flexible and transparent electronics. Nat. Nanotechnol. **8**, 100–103 (2013)
110. D. Sarkar, X. Xie, W. Liu, W. Cao, J. Kang, Y. Gong, S. Kraemer, P.M. Ajayan, K. Banerjee, A subthermionic tunnel field-effect transistor with an atomically thin channel. Nature **526**, 91–95 (2015)
111. T. Roy, M. Tosun, X. Cao, H. Fang, D.-H. Lien, P. Zhao, Y.-Z. Chen, Y.-L. Chueh, J. Guo, A. Javey, Dual-gated MoS_2/WSe_2 van der Waals tunnel diodes and transistors. ACS Nano **9**, 2071–2079 (2015)
112. D. Jariwala, T.J. Marks, M.C. Hersam, Mixed-dimensional van der Waals heterostructures. Nat. Mater. **16**, 170–181 (2017)

113. C. Zeng, E.B. Song, M. Wang, S. Lee, C.M. Torres, J. Tang, B.H. Weiller, K.L. Wang, Vertical graphene-base hot electron transistor. Nano Lett. **13**, 2370 (2013)
114. S. Hertel, D. Waldmann, J. Jobst, A. Albert, M. Albrecht, S. Reshanov, A. Schöner, M. Krieger, H.B. Weber, Tailoring the graphene/silicon carbide interface for monolithic wafer-scale electronics. Nat. Commun. **3**, 957 (2012)
115. S. Hertel, M. Krieger, H.B. Weber, Monolithic circuits with epitaxial graphene/silicon carbide transistors. Phys. Status Solidi RRL **8**, 688–691 (2014)
116. B. Vasic, A. Zurutuza, R. Gajic, Spatial variation of wear and electrical properties across wrinkles in chemical vapour deposition graphene. Carbon **102**, 304–310 (2016)
117. M. Ahmad, H. An, Y.S. Kim, J.H. Lee, J. Jung, S.-H. Chun, Y. Seo, Nanoscale investigation of charge transport at the grain boundaries and wrinkles in graphene film. Nanotechnology **23**, 285705 (2012)
118. L. Tapaszto, P. Nemes-Incze, G. Dobrik, K. Jae Yoo, C. Hwang, L. Biro, Mapping the electronic properties of individual graphene grain boundaries. Appl. Phys. Lett. **100**, 053114 (2012)
119. J.C. Koepke, J.D. Wood, D. Estrada, Z.-Y. Ong, K.T. He, E. Pop et al., Atomic-scale evidence for potential barriers and strong carrier scattering at graphene grain boundaries: a scanning tunneling microscopy study. Nano Lett. **7**, 75–86 (2013)
120. B. Nikolic, P.B. Allen, Electron transport through a circular constriction. Phys. Rev. B **60**, 3963–3969 (1999)
121. M.I. Katsnelson, A.K. Geim, Electron scattering on microscopic corrugations in graphene. Philos. T. Roy. Soc. A **366**, 195–204 (2008)
122. G. Nicotra, Q.M. Ramasse, I. Deretzis, A. La Magna, C. Spinella, F. Giannazzo, Delaminated graphene at silicon carbide facets: atomic scale imaging and spectroscopy. ACS Nano **7**, 3045–3052 (2013)
123. F. Giannazzo, I. Deretzis, G. Nicotra, G. Fisichella, Q.M. Ramasse, C. Spinella, F. Roccaforte, A. La Magna, High resolution study of structural and electronic properties of epitaxial graphene grown on off-axis 4H-SiC (0001). J. Cryst. Growth **393**, 150–155 (2014)
124. F. Giannazzo, I. Deretzis, A. La Magna, F. Roccaforte, R. Yakimova, Electronic transport at monolayer-bilayer junctions in epitaxial graphene on SiC. Phys. Rev. B **86**, 235422 (2012)
125. M. Nagase, H. Hibino, H. Kageshima, H. Yamaguchi, Local conductance measurements of double-layer graphene on SiC substrate. Nanotechnology **20**, 445704 (2009)
126. F. Giannazzo, G. Fisichella, A. Piazza, S. Agnello, F. Roccaforte, Nanoscale inhomogeneity of the Schottky barrier and resistivity in MoS_2 multilayers. Phys. Rev. B **92**, 081307(R) (2015)
127. F. Giannazzo, S. Hertel, A. Albert, A. La Magna, F. Roccaforte, M. Krieger, H.B. Weber, Electrical nanocharacterization of epitaxial graphene/silicon carbide Schottky contacts. Mater. Sci. Forum **778–780**, 1142–1145 (2014)
128. F. Giannazzo, S. Hertel, A. Albert, G. Fisichella, A. La Magna, F. Roccaforte, M. Krieger, H.B. Weber, Electrical properties of the hydrogen intercalated epitaxial graphene/SiC interface investigated by nanoscale current mapping. Mater. Sci. Forum **821–823**, 929–932 (2015)
129. J. Ristein, S. Mammadov, Th Seyller, Origin of doping in quasi-free-standing graphene on silicon carbide. Phys. Rev. Lett. **108**, 246104 (2012)
130. Y.-C. Lin, R.K. Ghosh, R. Addou, N. Lu, S.M. Eichfeld, H. Zhu, M.-Y. Li, X. Peng, M.J. Kim, L.-J. Li, R.M. Wallace, S. Datta, J.A. Robinson, Atomically thin resonant tunnel diodes built from synthetic van der Waals heterostructures. Nat. Communications **6**, 7311 (2015)
131. G. Giovannetti, P.A. Khomyakov, G. Brocks, P.J. Kelly, J. van den Brink, Substrate-induced band gap in graphene on hexagonal boron nitride: Ab initio density functional calculations. Phys. Rev. B **76**, 73103 (2007)
132. D.B. Farmer, H.Y. Chiu, Y.M. Lin, K.A. Jenkins, F. Xia, P. Avouris, Utilization of a buffered dielectric to achieve high field-effect carrier mobility in graphene transistors. Nano Lett. **9**, 4474–4478 (2009)
133. S. Kim, J. Nah, I. Jo, D. Shahrjerdi, L. Colombo, Z. Yao, E. Tutuc, S.K. Banerjee, Realization of a high mobility dual-gated graphene field-effect transistor with Al_2O_3 dielectric. Appl. Phys. Lett. **94**, 062107 (2009)

134. I. Meric, C.R. Dean, A.F. Young, N. Baklitskaya, N.J. Tremblay, C. Nuckolls, P. Kim, K.L. Shepard, Channel length scaling in graphene field-effect transistors studied with pulsed current-voltage measurements. Nano Lett. **11**, 1093–1097 (2011)
135. S. McDonnell, B. Brennan, A. Azcatl, N. Lu, H. Dong, C. Buie, J. Kim, C.L. Hinkle, M.J. Kim, R.M. Wallace, HfO_2 on MoS_2 by atomic layer deposition: adsorption mechanisms and thickness scalability. ACS Nano **7**, 10354–10361 (2013)
136. L. Lindsay, D.A. Broido, Enhanced thermal conductivity and isotope effect in single-layer hexagonal boron nitride. Phys. Rev. B: Condens. Matter Mater. Phys. **84**, 155421 (2011)
137. Y.F. Ji, C.B. Pan, M.Y. Zhang, S.B. Long, X.J. Lian, F. Miao, F. Hui, Y.Y. Shi, L. Larcher, E. Wu, M. Lanza, Boron nitride as two dimensional dielectric: reliability and dielectric breakdown. Appl. Phys. Lett. **108**, 012905 (2016)
138. C. Pan, Y. Ji, N. Xiao, F. Hui, K. Tang, Y. Guo, X. Xie, F.M. Puglisi, L. Larcher, E. Miranda, L. Jiang, Y. Shi, I. Valov, P.C. McIntyre, R. Waser, M. Lanza, Coexistence of grain-boundaries-assisted bipolar and threshold resistive switching in multilayer hexagonal boron nitride. Adv. Funct. Mater. **27**, 1604811 (2017)
139. F. Hui, W. Fang, W.S. Leong, T. Kpulun, H. Wang, H.Y. Yang, M.A. Villena, G. Harris, J. Kong, M. Lanza, Electrical homogeneity of large-area chemical vapor deposited multilayer hexagonal boron nitride sheets. ACS Appl. Mater. Interfaces. **9**, 39895–39900 (2017)
140. C.H. Ho, S.Y. Kim, K. Roy, Ultra-thin dielectric breakdown in devices and circuits: a brief review. Microelectron. Reliab. **55**, 308–317 (2015)
141. R. Degraeve, B. Kaczer, G. Groeseneken, Degradation and breakdown in thin oxide layers: mechanisms, models and reliability prediction. Microelectron. Reliab. **39**, 1445–1460 (1999)
142. D.J. Dimaria, E. Cartier, Mechanism for stress-induced leakage currents in thin silicon dioxide films. J. Appl. Phys. **78**, 3883–3894 (1995)
143. L. Vandelli, A. Padovani, L. Larcher, G. Bersuker, Microscopic modeling of electrical stress-induced breakdown in poly-crystalline hafnium oxide dielectrics. IEEE Trans. Electron Devices **60**, 1754–1762 (2013)
144. E. Miranda, J. Sune, R. Rodriguez, M. Nafria, X. Aymerich, Soft breakdown fluctuation events in ultrathin SiO_2 layers. Appl. Phys. Lett. **73**, 490–492 (1998)
145. Y. Hattori, T. Taniguchi, K. Watanabe, K. Nagashio, Anisotropic dielectric breakdown strength of single crystal hexagonal boron nitride. ACS Appl. Mater. Interfaces **8**, 27877–27884 (2016)
146. L. Aguilera, M. Porti, M. Nafria, X. Aymerich, Charge trapping and degradation of $HfO_2/SiO_2/$ MOS gate stacks observed with enhanced CAFM. IEEE Electron Device Lett. **27**, 157–159 (2006)
147. M. Lanza, M. Porti, M. Nafría, X. Aymerich, E. Whittaker, B. Hamilton, UHV CAFM characterization of high-k dielectrics: effect of the technique resolution on the pre- and post-breakdown electrical measurements. Microelectron. Reliab. **50**, 1312–1315 (2010)
148. F. Palumbo, X. Liang, B. Yuan, Y. Shi, F. Hui, M.A. Villena, M. Lanza, Bimodal dielectric breakdown in electronic devices using chemical vapor deposited hexagonal boron nitride as dielectric. Adv. Electron. Mater. **4**, 1700506 (2018)
149. A. Ranjan, F.M. Puglisi, N. Raghavan, S.J. O'Shea, K. Shubhakar, P. Pavan, A. Padovani, L. Larcher, K.L. Pey, Random telegraph noise in 2D hexagonal boron nitride dielectric films. Appl. Phys. Lett. **112**, 133505 (2018)
150. L. Jiang, Y. Shi, F. Hui, K. Tang, Q. Wu, C. Pan, X. Jing, H. Uppal, F. Palumbo, G. Lu, T. Wu, H. Wang, M.A. Villena, X. Xie, P.C. McIntyre, M. Lanza, Dielectric breakdown in chemical vapor deposited hexagonal boron nitride. ACS Appl. Mater. Interfaces **9**, 39758–39770 (2017)
151. F.M. Puglisi, L. Larcher, C. Pan, N. Xiao, Y. Shi, F. Hui, M. Lanza, 2D h-BN based RRAM devices. Int. Electron Devices Meeting (IEDM) Tech. Dig., 34.8.1–38.8.4 (2016)
152. C. Pan, E. Miranda, M.A. Villena, N. Xiao, X. Jing, X. Xie, T. Wu, F. Hui, Y. Shi, M. Lanza, Model for multi-filamentary conduction in graphene/hexagonal-boron-nitride/graphene based resistive switching devices. 2D Mater. **4**, 025099 (2017)
153. F. Hui, M.A Villena, W. Fang, A.-Yu Lu, J. Kong, Y. Shi, X. Jing, K. Zhu, M. Lanza, Synthesis of large-area multilayer hexagonal boron nitride sheets on iron substrates and its use in resistive switching devices. 2D Mater. (2018). https://doi.org/10.1088/2053-1583/aac615

Chapter 11
Diamond Probes Technology

Thomas Hantschel, Thierry Conard, Jason Kilpatrick and Graham Cross

Abstract The superior properties of diamond being the hardest, best thermally conductive, high chemical inert and low friction material makes it very attractive for use as a tip material in scanning probe microscopy (SPM). The commercial availability of micromachined Si probes at the beginning of the 1990s triggered soon the interest and need for different tip coatings such as diamond which was first wanted for increasing the tip lifetime. Although first reports on diamond growth from the wafer phase were first reported in the 1980s, it took until the early 1990s before first applications using diamond grown by chemical vapor deposition (CVD) appeared on the market. Therefore, the development of fabrication processes for diamond tips, especially for electrically conductive ones, required also substantial efforts on the development of the diamond coating knowhow itself. As commercial probe companies considered diamond probes as specialty probes with a small market size in the early days, it explains well why most diamond tip innovations were established by universities and research centers.

T. Hantschel (✉) · T. Conard
IMEC, Kapeldreef 75, 3001 Leuven, Belgium
e-mail: thomas.hantschel@imec.be

T. Conard
e-mail: thierry.conard@imec.be

J. Kilpatrick
Conway Institute of Biomedical and Biomolecular Research,
University College Dublin, Dublin, Ireland
e-mail: jason.kilpatrick@ucd.ie

J. Kilpatrick · G. Cross
Adama Innovations, CRANN, Trinity College Dublin, Dublin, Ireland
e-mail: graham.cross@tcd.ie

G. Cross
Trinity College, School of Physics, Dublin 2, Ireland

U. Celano (ed.), *Electrical Atomic Force Microscopy for Nanoelectronics*,
NanoScience and Technology, https://doi.org/10.1007/978-3-030-15612-1_11

11.1 Introduction

The superior properties of diamond being the hardest, best thermally conductive, high chemical inert and low friction material makes it very attractive for use as a tip material in scanning probe microscopy (SPM). The commercial availability of micromachined Si probes at the beginning of the 1990s triggered soon the interest and need for different tip coatings such as diamond which was first wanted for increasing the tip lifetime. Although first reports on diamond growth from the wafer phase were first reported in the 1980s, it took until the early 1990s before first applications using diamond grown by chemical vapor deposition (CVD) appeared on the market. Therefore, the development of fabrication processes for diamond tips, especially for electrically conductive ones, required also substantial efforts on the development of the diamond coating knowhow itself. As commercial probe companies considered diamond probes as specialty probes with a small market size in the early days, it explains well why most diamond tip innovations were established by universities and research centers.

Figure 11.1 shows some milestones in the history of electrical diamond probes. In 1995, De Wolf and Vandervorst [1] introduced the scanning spreading resistance microscopy (SSRM) method [first being called nano-spreading resistance probe (nano-SRP)] for carrier profiling and showed with a crude hand-crafted probe (Fig. 11.1a) still operated in point-contact mode that only diamond is hard enough to withstand the required GPa pressure range. This probe consisted of a sharpened three-sided diamond crystal (~100 nm tip radius) implanted with boron and glued to a Ti shaft and was originally developed for scanning tunneling microscopy (STM). For using it as a cantilever probe, De Wolf et al. [1] fixed two Pt wires and mirrors to these tips. Although these diamond probes allowed for first SSRM measurements, the heavy and stiff probes caused a lot of sample damage and allowed only for one-dimensional measurement. The first boron doped diamond films deposited on Si AFM probes (Fig. 11.1e) were reported by Niedermann et al. in 1996 [2]. They had sharp diamond crystals protruding from the apex which gave the high resolution in AFM measurements. They were still suffering from multiple point contacts due to several diamond grains making contact with the sample surface at the same time. Similar coated diamond tips (CDT) with a slightly higher aspect ratio were the first commercially offered diamond probes (by the company Nanosensors, Germany) (Fig. 11.1b). Although these microfabricated tips allowed for the first two-dimensional (2D) SSRM measurements, they are typically breaking off close to the apex during scanning when exposed to SRRM typical forces in the μN range as illustrated in Fig. 11.2 (note that monocrystalline Si is a very brittle material). Most AFM users do not notice this effect as it happens mostly at the beginning of the measurements and the electrical contact in the SSRM measurement is then provided by a sharp nanocrystal sticking out from the sidewall of the broken-off tip. Note that CDTs are today included in the probe portfolio of most leading probe suppliers and are mainly used where the resolution requirements are less stringent.

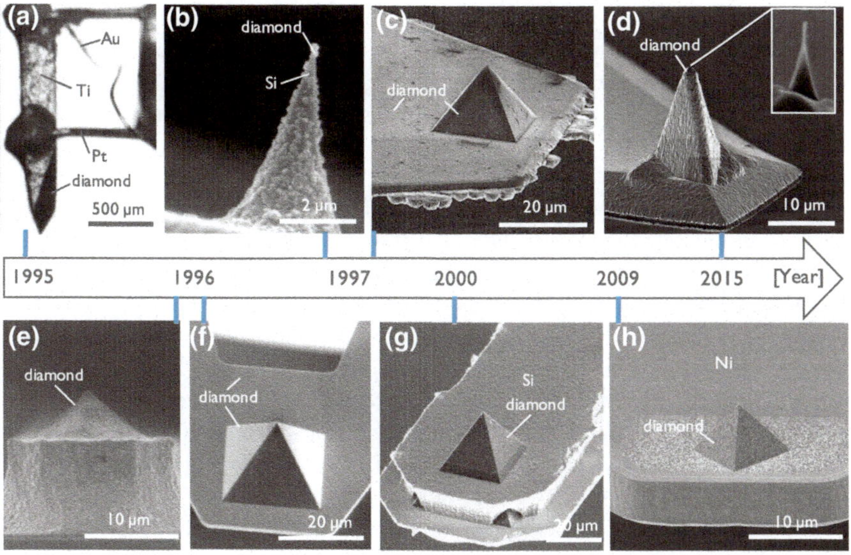

Fig. 11.1 History of diamond probes for SSRM application: **a** Handmade diamond probe by IMEC, **b** first commercial coated diamond probe (CDT) offered by Nanosensors, **c** monolithic diamond probe (University of Kassel), **d** first commercial plasma etched diamond probe (Adama Innovations), tip apex diameter of zoom-in is 5 nm, **e** first CDT by CSEM, **f** first full diamond tip (FDT) probe by CSEM, **g** first high conductive FDT by IMEC with diamond pyramid on Si cantilever, **h** nanometer resolution FDT by IMEC with diamond tip on metal cantilever

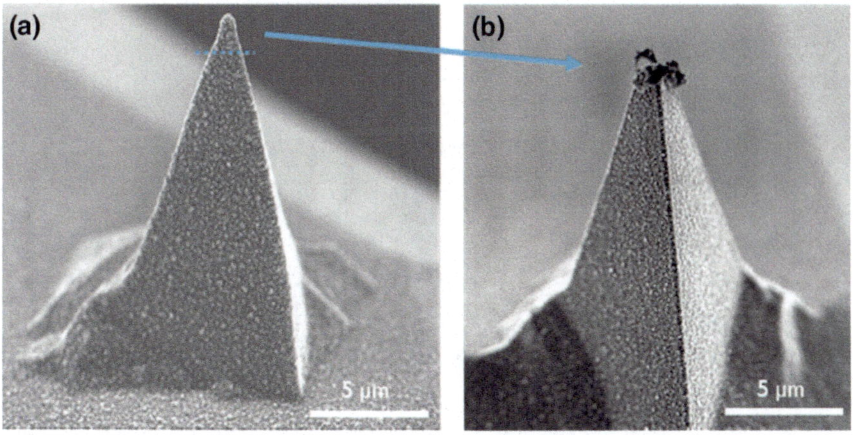

Fig. 11.2 Apex of diamond coated Si tip (CDT) (**a**) is often breaking off in SSRM (**b**)

Pyramidal diamond tips, commonly referred to as full diamond tips (FDT), were developed to overcome the problem of multiple point contacts and to avoid tip breaking. In 1996, Niedermann et al. [2] (Fig. 11.1f) and shortly afterwards Oesterschulze et al. [3] (Fig. 11.1c) demonstrated the first tip prototypes but still without measurements. It still took a few years of process improvement before the first highly conductive diamond tips with a base width of $40 \times 40\ \mu m^2$ integrated into Si cantilevers were reported by Hantschel et al. [4] for high-resolution SSRM measurements (Fig. 11.1g). The introduction of metal cantilever probes with a smaller tip base width of $7 \times 7\ \mu m^2$ [5] and improvements of the diamond process were crucial to obtain sub-1 nm electrical resolution in SSRM measurements of device structures (Fig. 11.1h). Today, the increased materials complexity in combination with confined volumes in sub-10 nm technology nodes are resulting in new challenges for pyramidal tips as illustrated in Fig. 11.3a where the low aspect ratio can lead to materials smearing artifacts in SSRM measurements. Therefore, diamond tips with a higher aspect ratio and ultra-high sharpness have more recently been developed (and commercialized by Adama Innovations) representing basically a CDT with a nanoscopic diamond tip on top (Fig. 11.1d) which are overcoming the smearing effects of low-aspect-ratio pyramidal tips as illustrated in Fig. 11.3b. The following two sections discuss these two currently highest performing electrical diamond tip configurations in more detail: pyramidal molded tip probes and plasma etched tip probes. Note that many other alternative diamond tip configurations have been reported and are being used in some cases but they are not being widely used for routine high-conductive electrical AFM measurements. Among them there are a few promising configurations which the interested reader can follow up. For example, the single crystal diamond needle tips as reported by Obratzov et al. [6] where a several micrometers thick blanket diamond film is oxygen plasma etched into conically shaped diamond needles which are then manually attached to Si cantilever probes . Unfortunately, this approach is limited to undoped tips so far. Moldovan et al. [7] reported oxide mold sharpened ultra-nanocrystalline diamond (UNCD) tips with a higher aspect ratio than the traditional pyramidal tips but boron doped UNCD suffers from lower conductivity than NCD and MCD and is therefore not preferred for SSRM.

11.2 Molded Diamond Tip Probes

When the development of nano-SRP (later on called SSRM) in the mid-1990s called for the urgent fabrication of conducting diamond tips, the application of so-called tip molding—whereby creating first an inverted pyramid by anisotropic Si etching, subsequent mold filling and final mold removal—seemed an obvious choice for their fabrication as this approach was already well established for manufacturing pyramidal silicon nitride AFM tips [8]. However, the low-pressure low-temperature diamond CVD synthesis was just being developed itself during that time and hence the required diamond thin films were not widely available yet. Therefore, a Ph.D. project

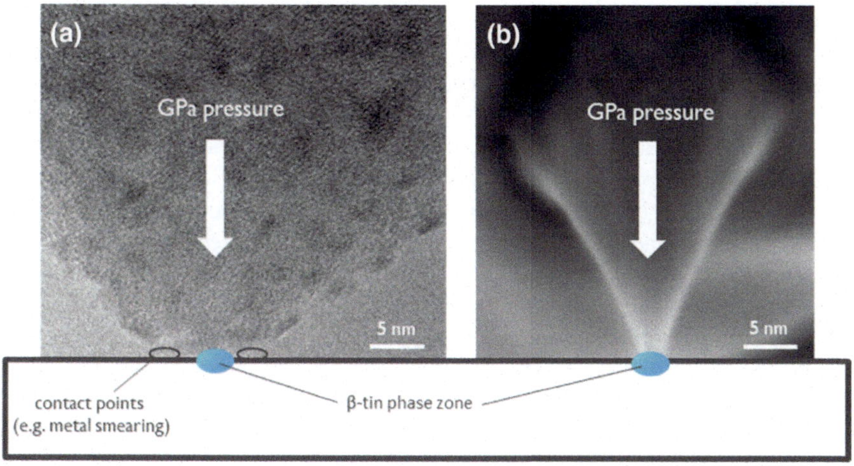

Fig. 11.3 Comparison of contact point zone for low-aspect-ratio FDT (**a**) and high-aspect-ratio plasma etched tip (Adama Innovations) (**b**). Although the same nanoscopic electrical resolution is observed, additional contact points are often found for FDTs leading also to materials smearing effects in confined volume device structures

was launched at the time to establish molded diamond tips for application in SSRM. The major aim was to achieve mechanically stable tips (without tip breakage as typically observed for coated diamond coated tips), with nanometer scale resolution and high conductivity. The full diamond tips (FDT) established in this work proved crucial for the breakthrough of the SSRM method for the carrier profiling of most advanced micro-/nano-electronics devices. In the following decade, their wafer scale fabrication using common 200-mm Si wafer fab technology represented another major step towards the routine use of SSRM in the semiconductor industry. This section covers the FDT probe fabrication and characterization, discusses several alternative tip configurations based on the molding approach, and discusses important (but often overlooked) probe storage aspects for metal-based cantilever probes. The FDT probe application is covered extensively in Chap. 2.

11.2.1 Basic Fabrication Process

Figure 11.4 shows the basic process scheme for molded diamond tip probes. First, molds with a pyramidal shape are formed on top of 200 mm diameter (100)-Si wafer substrates using anisotropic Si etching in 30% KOH at 70 °C whereby SiO_2 serves as an etch mask (Fig. 11.4a). Although the base width of the inverted pyramid can be chosen over a wide range, a smaller base width is preferred as a sharper apex is easier to obtain with a smaller base width. Therefore, we reduced the base width from originally 40×40 to $7 \times 7\ \mu m^2$. Next, in the so-called seeding step, the

Si substrate is covered with diamond nanoparticles which serve as nucleation sites during heteroepitaxial diamond growth. For this the Si substrate is immersed into the seeding dispersion containing diamond nanoparticles dissolved in ethanol. The potential of the diamond nanoparticles (referred to as zeta-potential) should be high (~20 to 40 mV) to obtain a high seeding density on the substrate surface ($\geq 10^{10}$ cm^{-2}) as shown by Williams et al. [9]. Tsigkourakos et al. [10] showed that the seeding parameters, in particular the seed dispersion concentration, strongly impact the final quality of the diamond pyramids and must be optimized. The seeded substrate surface is shown in the zoom-in SEM image of Fig. 11.4a. As the conventional diamond seeds are undoped and hence non-conductive, it can lead to a non-conductive tip when the seed particle is deposited at the apex. Therefore, two groups explored high boron doped seed fabrication [11, 12] and most recently boron doped detonation nanodiamond seeds were demonstrated [13]. Next, the seeded Si substrate is coated with an about 1 μm thick boron doped MCD layer by an sp3 model 655 hot-filament CVD (HFCVD) system using a CH_4/H_2 gas ratio of 1.5%, a pressure of 25 Torr, a trimethylboron (TMB)/CH_4 gas flow ratio of 0.4% and a substrate temperature of ~850 °C (Fig. 11.4b). Note that in this step the highest possible boron doping of the diamond layer is wanted for achieving the maximum electrical conductivity (minimum resistivity) without sacrificing in tip wear resistance too much. A detailed boron doping [14] and diamond film wear [15] study has been carried out for this showing that a resistivity of about 2–5 $\times$ 10^{-3} Ω cm can be obtained with the well calibrated doping process whereby the film wear rate is only about 3-times higher than for low doped films but there is still high enough intra grain cohesion ensuring the mechanical integrity of the diamond tip.

After diamond growth the diamond tip area is lithographically patterned (Fig. 11.4c) and then structured by reactive ion etching (RIE) in a O_2:SF_6 plasma using Al as an etch mask (Fig. 11.4d). For probe prototyping the diamond tips were integrated into Si cantilevers [4]; the preferred option of most commercial probe manufacturers; but for high compatibility with microelectronics fab manufacturing using substrate sizes of ≥150 mm in diameter (100-mm diameter Si wafers are mostly used by AFM probe companies whereby high wafer thickness uniformity is crucial for high cantilever thickness uniformity), the integration into 5–6 μm thick metal cantilevers was chosen whereby Ni became the material of choice (Fig. 11.4e). This approach allows for surface micromachining using Ni electroplating which was first demonstrated by Rasmussen et al. [16] whereby cantilever and holder chip are both made out of electroplated Ni. The drawback of this method is the need to underetch the probe chip entirely (=etching away the whole Si wafer material) to release the probes from the wafer which is incompatible with fab processing. Therefore, we established a procedure based on selective underetching of a perforated holder chip [17] (Fig. 11.4f) and demonstrated a rather unconventional peel-off approach [18] (Fig. 11.4g) whereby only the cantilever is locally underetched in KOH and then the probe is peeled away from the wafer surface like a post-it paper. During the peel-off step, the adhesion between the Si substrate and the peel-off layer is gradually overcome. Small bridges still hold the holder membrane in place after peel-off. These bridges break when the probe is removed from the wafer. Different peel-off layers

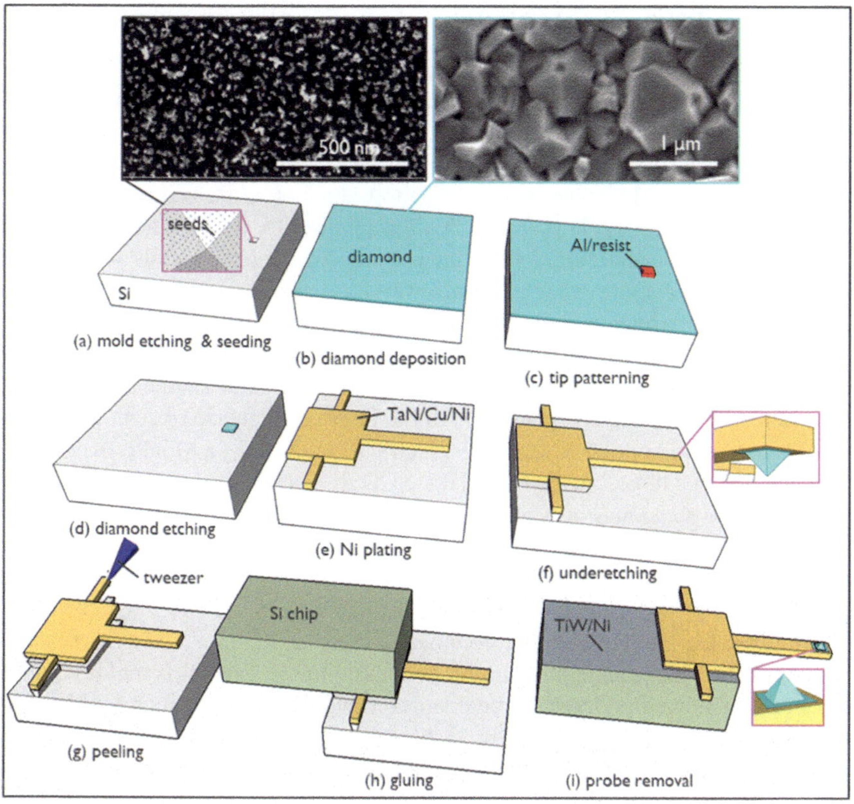

Fig. 11.4 Basic FDT fabrication scheme based on tip molding and cantilever peel-off

were explored and Ti:W and TaN became the standard thin film materials in our process. Although the probe peel-off is a manual step done under an optical microscope, it is very reproducible and is already routinely employed in our FDT processing since 2 decades. In the final step, a metalized probe holder chip of 1.6 mm in width and 3.4 mm in length is fixed to the peeled-off holder chip membrane for convenient probe handling (Fig. 11.4h). The connection between holder chip membrane and probe chip is established by a silver-based epoxy. Alternatively, a lead-free soldering procedure was also established and works as well [19]. In the final step, the probe is manually removed from the wafer using a pair of tweezers (Fig. 11.4i). Note that this approach is a manual step but we showed that this assembly step can in principle also be done by a robotic procedure [20]. Our experience with manual peel-off and gluing showed that a trained and skilled person can do this with an about ±5 to 10 μm of accuracy.

11.2.2 Probe Process Variations

Although the classical pyramidal tip configuration is still mostly used for our SSRM measurements, we demonstrated several alternative tip configurations. For example, the so-called tip-on-tip configuration was developed whereby a small diamond pyramid, for higher sharpness, is placed on top of a large truncated pyramid which serves as a pedestal [21]. This approach allows indeed for a very small apex of the mold but needs e-beam lithography at the bottom of the mold which is a more complicated process. A more straightforward way for a sharper molded tip apex is the low-temperature oxidation of the Si mold as first demonstrated for silicon nitride tips by Akamine [22] and applied to molded UNCD tips by Moldovan et al. [7]. Unfortunately, our experiments with NCD and MCD filling of oxide sharpened molds did not produce superior tip sharpness due to the larger crystal size of such films (and UNCD not being an option due to lower electrical conductivity). Moldovan et al. [7]. demonstrated that three-sided pyramidal UNCD tips can be made by molding when taking (311)-Si substrates instead of the commonly used (100)-Si wafers. Note that this three-sided tip shape has the advantage that the apex is always a single point whereas four-sided pyramids often have a knife-shaped apex as a result of small non-symmetrical effects during the process (e.g. lithography mask, etch mask). (311)-Si wafer substrates are however not commonly used in 200-mm diameter Si-wafer fabs. Another disadvantage of diamond pyramidal tips in general is that it is rather bulky and positioned underneath the cantilever and therefore the region of interest (ROI) is during scanning not directly visible which hinders the rapid tip positioning towards the ROI. As SSRM is mostly applied to nanoscopic devices, direct apex visibility is highly desired. For this, we have developed so-called in-plane diamond tips (Fig. 11.5) whereby the tip is in plane with the cantilever and is lithographically shaped [23]. This overcomes the need for anisotropic Si etching and allows for a high degree of freedom to define a desirable tip shape. Although FDTs are superior over CDTs in terms of spatial resolution, they suffer from a limited electrical conductivity resulting into an overall lower dynamic range which is due to the lower active doping of the interfacial diamond layer used for molded tips. To overcome this disadvantage, we established the concept of overcoated diamond tips (ODT) [24] (Fig. 11.6) which combines the advantages of FDTs (mechanical stable tip) and CDTs (high electrical conductivity). In this tip configuration, first a Si mold is defined by anisotropic etching (Fig. 11.6a, b) and is then filled with boron doped diamond (similar as in the basic process) (Fig. 11.6c); this is followed by diamond patterning (Fig. 11.6d) and local underetching of this diamond tip (Fig. 11.6e), and an overcoating of the diamond tip with a thin boron doped diamond layer on top of the interfacial diamond layer (Fig. 11.6f). This additional diamond coating step is done in an upside-own manner. Finally, the ODT structure is bonded to the end of a metal cantilever. Variations of the classic molding process have been reported also by other groups; for example, Oesterschulze et al. [25] first lithographically patterned and etched a diamond filled mold area in such a way that a triangular shaped diamond tip points away from the diamond cantilever allowing for direct tip visibility while scanning. Beuret et al. [26]

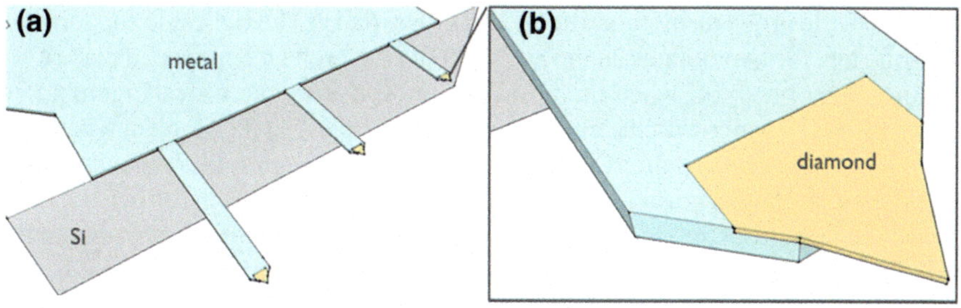

Fig. 11.5 Concept for in-plane diamond tips attached to metal cantilevers (**a**) allow for direct visibility of ROI during scanning. Zoom-in (**b**) illustrates the tapered shape

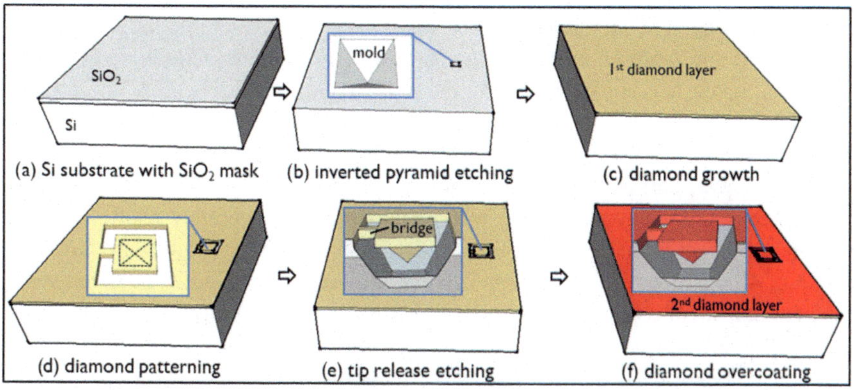

Fig. 11.6 Adapted process scheme for fabrication of overcoated diamond tips (ODT)

reported on an interesting double-molding concept whereby a Si tip is first coated with a silicon nitride layer, the core Si tip material is then removed, the formed mold is then filled with diamond and finally the outside silicon nitride layer is removed again. Although this is a more complex approach, it illustrates very well that diamond tip molding allows for many different shapes, geometries and dimensions.

11.2.3 Fabrication Results

The FDT probe process is running in a state-of-the-art Si wafer manufacturing line (using initially 150-mm and today 200-mm wafers) which allows for a high uniformity of important parameters such as tip sharpness and cantilever thickness. Figure 11.7a shows an optical microscopy images of a fabricated 200-mm FDT probe wafer containing about 1000 probes. The zoom-in (Fig. 11.7b) illustrates that two lines of probes are always facing each other for convenient holder chip mounting; some probe membranes have already been removed from the wafer. Figure 11.7c shows a single probe membrane after peeling. The flap on the right side supports the peel-off step, the two bridges on the left side hold the probe membrane in place after peeling. They break off when the probe is removed from the wafer. Figure 11.7d shows the probe wafer with the glued holder chips in place at the moment when they are picked up from the wafer using a pair of tweezers. Figure 11.7e shows a low-magnification 3D view of a mounted probe after removal from the wafer whereby inspection images of optical and SEM microscopy were merged with the actual probe design for a more realistic probe view. Note the straight walls of the holder chip which ensures a tight grip during probe handling by tweezers (conventional Si probes have 54° slopes from anisotropic etching which relatively easily slip away from the tweezer tip). The probe chip is metalized with a thin layer of Ti:W + Ni which ensures that the electrical current can flow from it to the holder membrane (via the silver epoxy connection) further along the Ni cantilever into the boron doped diamond tip and from its apex into the sample. The probe underetch and the peel-off area are clearly distinguishable by the red Cu and gray TaN color. The zoom-in (Fig. 11.7f) shows that the cantilever end has slanted edges for better visibility of the scanned area and helps to avoid that contact is being made to the sample surface by the cantilever corners. For a rigid connection between the diamond pyramid and Ni cantilever, the diamond area is larger than the diamond pyramid itself. As can be seen from Fig. 11.7e, each probe has three cantilevers with a width of 50 μm and a length of 465, 305, and 225 μm respectively allowing us to select the desired spring constant, and hence force range, for the specific type of electrical AFM measurement. The nominal spring constants are 3, 11, and 27 N/m. Note the good alignment between holder membrane and holder chip which is routinely achieved with the manual mounting step. The pyramidal diamond tip is just placed a few micrometers away from the cantilever end for proper contacting of the tip to the sample in AFM. For selecting a specific cantilever, the others are bent slightly away using a pair of tweezers.

Figure 11.8a, b shows SEM images with more details of the diamond tip and its apex. Note the perfectly molded tip shape with the sharp apex and the clearly visible structure of the boron doped diamond grains. For assessing the geometrical sharpness of the diamond tips in a more quantitative way, we developed a procedure for mounting the tips directly onto a Cu grid for inspection by transmission electron microscopy (TEM). From the 13 tips inspected in this way, an average tip radius of 14.9 nm was obtained ranging from 5.0 to 32.0 nm. The diamond tip apex shown in Fig. 11.8c, d has a tip radius of 5.5 nm. The higher spatial resolution observed in actual SSRM measurements can be explained by nanoscopic protrusion on the apex and the smaller size (~3-times for tips varying between 5 and 50 nm [27]) of the so-called beta-tin pockets formed at GPa pressures underneath the contact point. A smaller tip radius is reducing the risk for forming a multiple-tip electrical contact.

Two factors which strongly improved the tip performance are the switch over from NCD to MCD films and the better control over the seeding of the inverted pyramids on wafer scale. Despite a 3-times lower boron incorporation into the highest doped MCD films, its resistivity is nearly only half of that of the highest doped NCD film. Figure 11.9 shows top and cross-section through such an NCD and MCD film

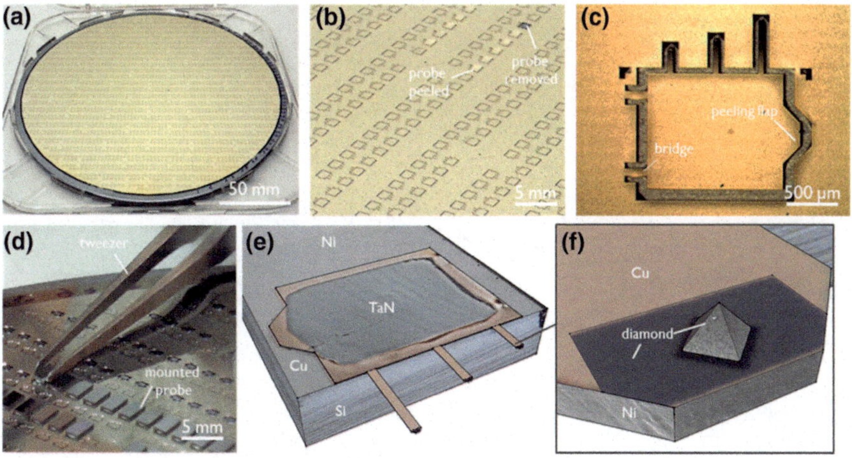

Fig. 11.7 Fabricated probe wafer (**a**) with detailed view of peel-off area (**b**) and probe membrane (**c**); probes are removed from wafer after holder chip mounting (**d**); view of merged probe design and both optical and SEM inspection images (**e, f**)

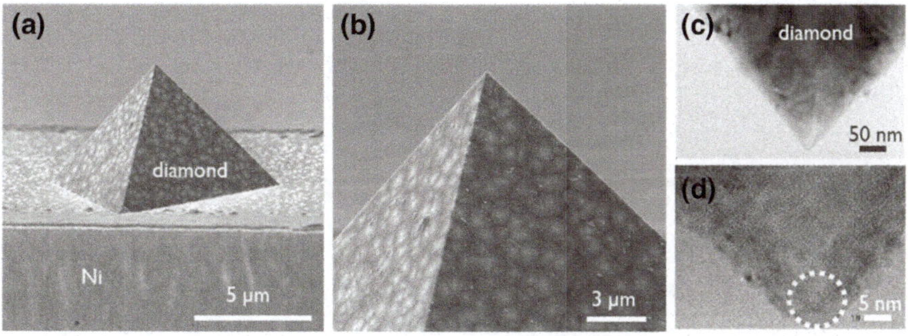

Fig. 11.8 Zoom-in view of diamond pyramid tip (**a, b**) with detailed TEM view of apex area (**c, d**)

illustrating the higher presence of grain boundaries in NCD compared to MCD films. We showed in a detailed study that the wear rate of such MCD films is much lower than that of NCD films [15]. In another study we showed that an average diamond nanoparticle size of $\leq$20 nm and a seeding density of ~5E10 cm^{-2} is preferred to achieve a diamond film with a high vertical electrical conductivity [28]. Figure 11.10 shows an SSRM image of the interfacial diamond side illustrating that the undoped diamond nanoparticles (white color) remain non-conductive even after being exposed to high temperatures of ~850 °C for 10 h during diamond growth; the replacement of undoped by doped seeds was demonstrated [11, 12] but a steady supply of this material for wafer-scale seeding is still to be realized. Figure 11.11 shows an inverted pyramidal pit which was seeded with our optimized seeding process [10] which was further developed into the so-called spin-zero approach for high seeding uniformity on wafer scale [29].

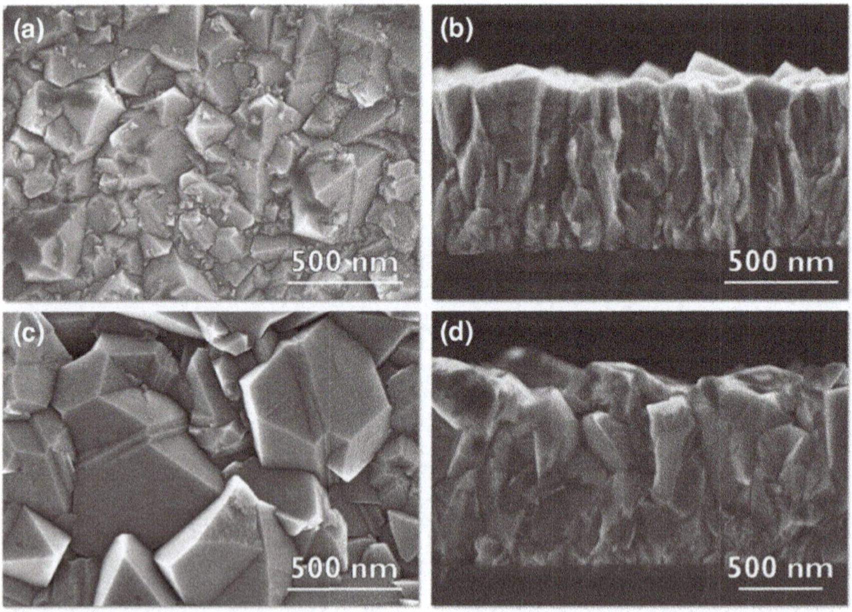

Fig. 11.9 SEM inspection of top and cross-section of 1 μm thick high boron doped NCD (**a**, **b**) and MCD (**c**, **d**) film

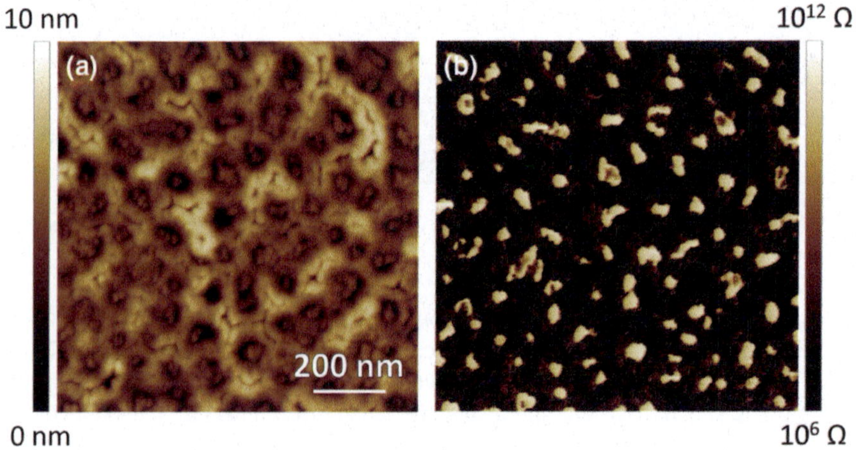

Fig. 11.10 AFM topography (**a**) and SSRM image (**b**) of interfacial diamond side illustrating that the undoped diamond nanoparticles embedded into the high boron doped diamond film (dark color) remain non-conductive (white color) even after prolonged diamond growth

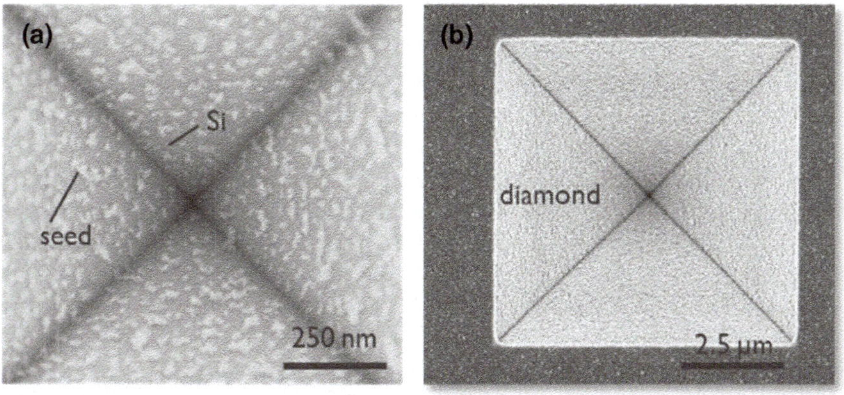

Fig. 11.11 Optimal seeded apex of inverted pyramid (**a**) and 100 nm of diamond growth into inverted diamond pyramid seeded with optimized procedure yielding a closed film (**b**)

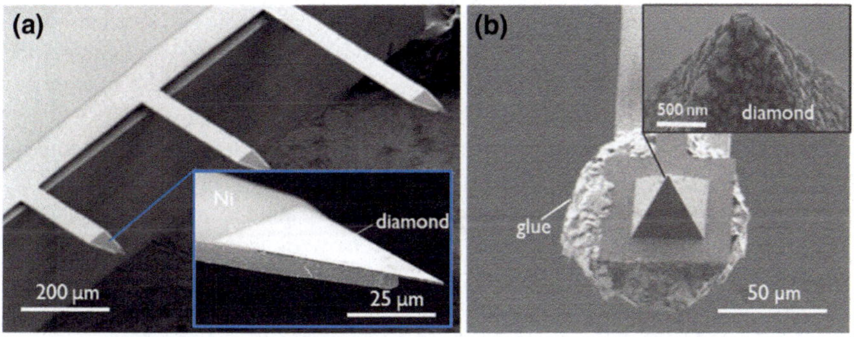

Fig. 11.12 Fabricated in-plane diamond tips on Ni cantilever (**a**), ODT structure glued to a Ni cantilever (**b**); tip structure received 2 diamond coatings, one before and one after molding

Figure 11.12 shows alternative diamond tip structures which we have fabricated with the molding approach: in-plane diamond tips with a tapered (Fig. 11.12a), and ODT tips made by the procedure described above and manually attached to the cantilever end (Fig. 11.12b).

11.3 Probe Characterization

Fabricated probes are extensively tested in a Bruker Icon PT AFM system with SSRM application module. Figure 11.13a shows an SSRM image taken with an FDT probe of a special Si p-type calibration structures resembling a staircase of doped regions with different doping levels ranging between 4E16 and 3E20 cm^{-3}. As can be seen in the line profile (Fig. 11.13b), all doping levels can be clearly identified. Figure 11.13c shows the calibration curves of 4 different FDT probes which are extracted from such SSRM staircase images by plotting the average measured resistance values against the known resistivity values of each stair. The curves show a lower but unsaturated slope for high doping levels which enables us to measure high dopant areas, which are most important in devices. Previously fabricated FDT probes [5] based on NCD showed a completely flattened slope for the high doping levels. This NCD saturation for high doping levels is caused by the series tip resistance which becomes dominant over the spreading resistance for high-doped areas. MCD growth gives a higher dynamic range and is therefore clearly preferred over NCD growth. We explain this strongly improved electrical behavior for MCD films on the interfacial side (being the active probing layer for FDT) by a process whereby boron incorporation into the diamond matrix is strongly suppressed at the early growth stages because of the presence of O from the chamber walls which forms stable B-O compounds. This happens during a constant time interval until the O has been flushed out from the chamber and is influenced by the growth rate which is ~2.5 times lower for our MCD than NCD process (80 vs. 185 nm/h). It results into a boron concentration ratio $[B]_{bulk}/[B]_{interfacial\ side}$ of ~12.5 for NCD and only ~1.5 for MCD; and hence the MCD growth mode is substantially less affected which is also confirmed by our electrical results. A further reduction in growth rate is limited for our HFCVD system as it affects the wire carburization process (e.g. very low CH_4/H_2 ratios leading to poor wire carburization). The presence of a load lock system is therefore the preferred option for high-doped BDD electrode layers. Figure 11.14a shows a dedicated test structure which we use for the evaluation of the electrical tip resolution; it consists of a 0.55 nm wide SiO_2 layer sandwiched in between two doped Si layers. De Wolf [30] reported that the electrical tip resolution is equal or smaller than the SiO_2 thickness if the maximum resistance of the SiO_2 layer is at least twice the resistance of the doped Si layers. The SiO_2 layer is becoming invisible for blunter tips. Based on measurements of this structure, Fig. 11.14b shows the tip yield for four MCD tip wafers (20 probe tips measured per wafer) with a typical 80–95% yield meaning that nearly every tip is working. This is a strong improvement in comparison to our previous process with lower seeding density and NCD growth with a typical yield of 30–45% meaning that only 1 out of 3 tips is working.

11.3.1 Probe Storage Considerations

SPM probes are commonly shipped and stored in silicone mat boxes because they adhere well to the mats and can be easily removed from it again. Furthermore, there are no visible residues on the probe back and the boxes can be used for a long time. A study by Lo et al. [31] showed however that probes stored in such boxes can be contaminated by monolayers of silicone oils (poly(dimethylsiloxane), PDMS). In their experiments, silicone oils released from this kind of packaging material formed a contamination layer on substrates stored on these mats in just a few days. Although such a monolayer does not necessarily harm the functionality of a pure Si probe,

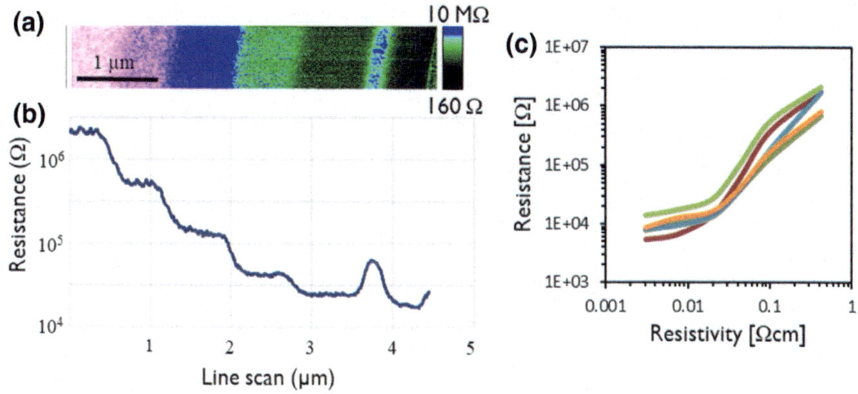

Fig. 11.13 Figure 11.7 SSRM image of p-type doped Si staircase structure (**a**); extracted line profile (**b**); calibration curve of four FDT probes (**c**)

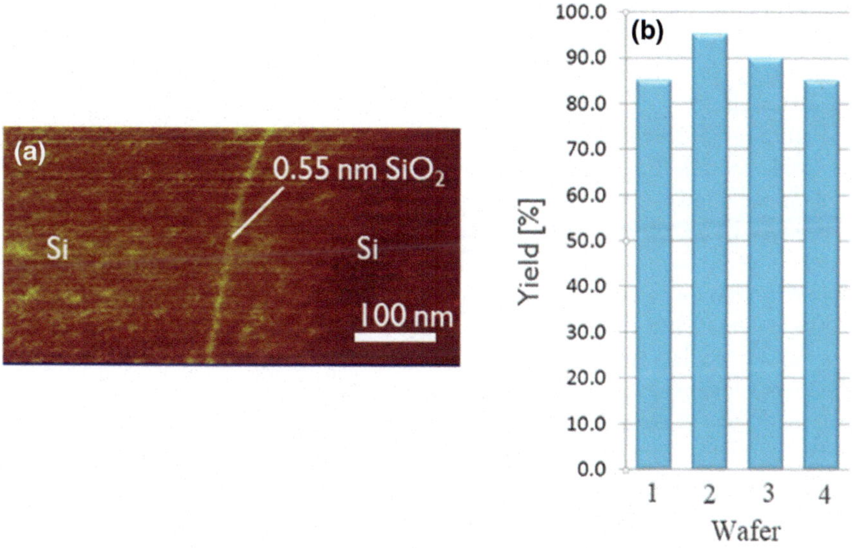

Fig. 11.14 SSRM measurement of 0.55 nm wide SiO_2 structure (**a**), wafer tip yield of FDT (**b**)

SPM probes which contain metals as a thin coating or when cantilever or tip are made from metal can be dramatically affected by this. Figure 11.15 shows three macroscopically contaminated metal containing probes which were stored for a few months in silicone mat boxes. SEM inspection showed that bubble- and sometimes also needle-like structures contaminated the tip or cantilever area if certain metals were involved. The bubble-structure was mostly observed, a needle-like growth was found in some cases. All probes in a storage box (typically 100 probes per box) were contaminated and looked nearly identical. Time-of-flight secondary ion-mass

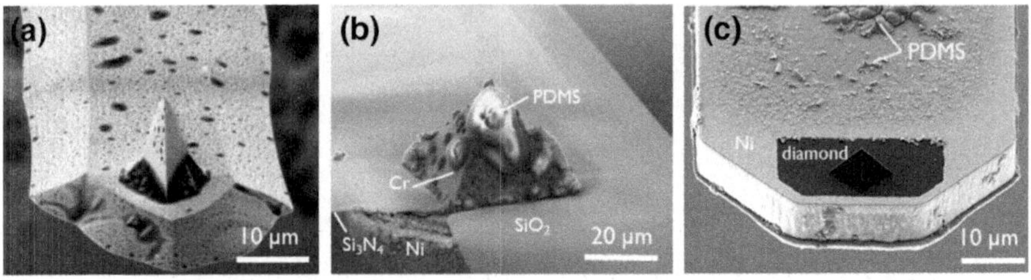

Fig. 11.15 Metal containing probes which were contaminated with PDMS after storage in silicone mat boxes; CoCr coated Si probe (**a**), Ni cantilever with metal tip (**b**), and Ni cantilever with integrated diamond tip (**c**)

spectrometry (TOF-SIMS) measurements taken above the tip area contained strong silicone oil signals indicating that the probes were indeed contaminated by PDMS. Different types of probes were inspected by SEM which were stored in silicone mat boxes for more than a year. We observed macroscopic probe contamination only in cases where the probes were made completely out of metal or were covered by a thin metal layer. No PDMS droplets were found on pure Si probes and on the diamond tip area itself.

Figure 11.15a shows a CoCr coated silicon probe which is coated by small silicone oil droplets. Interestingly, PtIr coated silicone probes stored in the same box were not contaminated macroscopically. It turned out that some materials are more sensitive to this type of contamination than others. Strong contamination was observed on all probes covered by CoCr, Cr, Cu, Ni and NiCr. No macroscopic contamination was visible in SEM on areas of Si, SiO_2, Si_3N_4, Pt, and PtIr. In case of Ti:W and Au, some areas showed PDMS droplets whereas others did not. Figure 11.15b illustrates well that PDMS droplets are formed only on certain materials. As can be seen, Ni and Cr areas were covered by PDMS droplets, SiO_2 and Si_3N_4 areas were not. This behavior was found for all contaminated probes. The mechanism of this contamination can be understood as a form of selective are deposition (SAD) where silicone oils from the mats become versatile and are selectively redeposited onto certain metallic areas; an effect which is recently being exploited in nanoelectronics technology for self-patterning purposes. Figure 11.15c illustrates that the metal cantilever of an FDT probe can be contaminated as well.

Figure 11.16 illustrates the evolution of probe contamination on full metal probes [32] over time which were stored for one year in a silicone mat box under cleanroom conditions. 15 probes were inspected by SEM at the beginning of the test, after 3 months and after one year. All probes showed the same contamination level during the test. As can be seen from Fig. 11.16a, the initially clean probe was strongly contaminated by silicone oil droplets after 3 months (Fig. 11.16b). After one year, the metal tip was buried completely under a big droplet (Fig. 11.16c). The same effect was observed for our Si cantilever probes with integrated metal pyramids.

Lo et al. [31] presented a probe cleaning procedure which uses a mixture of H_2SO_4 and H_2O_2 which is too aggressive for our metal-cantilever probes. Therefore, we

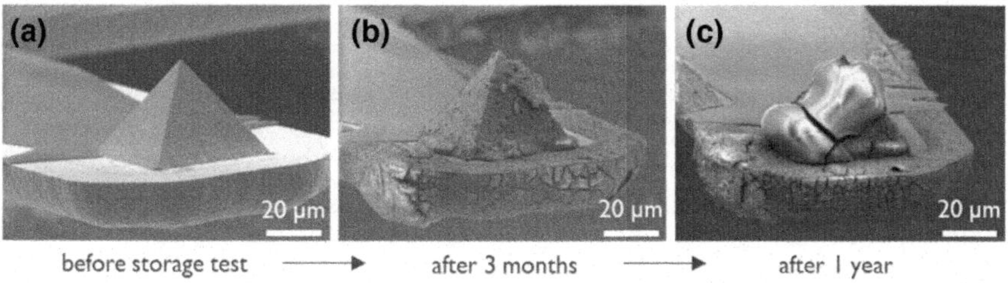

Fig. 11.16 Evolution of probe contamination in silicone mat box for one year

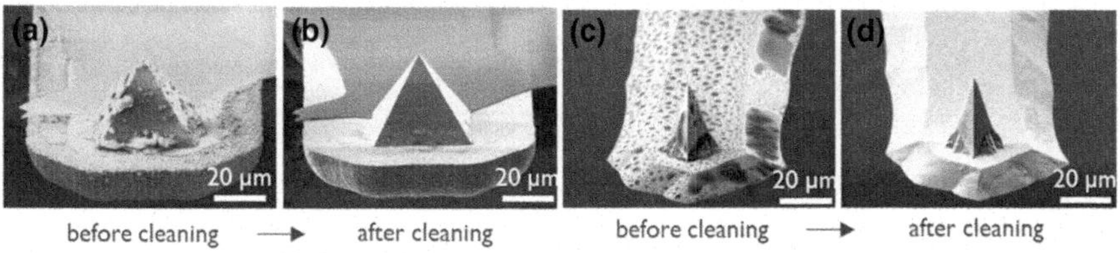

Fig. 11.17 Before and after 5 min H_2O cleaning: full metal probes (**a**, **b**), CoCr coated Si probes (**c**, **d**)

looked for ways to remove the PDMS droplets from the probe tips. We observed that the PDMS droplets can be removed from the probe by a short H_2O rinse. Figure 11.17 shows a contaminated probe before cleaning and after a 5 min rinse in H_2O. As can be seen, the macroscopic contamination was removed completely. Figure 11.17 illustrates that simple H_2O rinsing does also work for Si probes coated with CoCr.

Further tests showed that about 95% of the contamination is already removed from the tip after 10 s of H_2O rinse. This means that the PDMS droplets are not dissolved in H_2O but are rather stripped off from the probe when it is dipped into H_2O. TOF-SIMS measurements were done on strongly contaminated probes before and after H20 cleaning. Figure 11.18 presents the normalized intensities (to Ni+) of three PDMS ions measured on a full metal probe before and after the cleaning step (note that the intensities are shown in logarithmic scale). From this figure it is clear that a large fraction of the PDMS was removed by the H_2O cleaning step. The contamination level after cleaning was only about 10%. The PDMS residues are most likely monolayers sticking to the surface which cannot be removed by probe dipping and rinsing in H_2O.

Our experiments illustrate that metal containing SPM probes can be strongly contaminated by PDMS droplets if they are stored in certain silicone mat boxes. As this happens often unpredictably and can destroy a large batch of probes, our metal-cantilever probes are not stored in silicon mat boxes but are placed onto a post-it like adhesive which did not show the effect of macroscopic probe contamination.

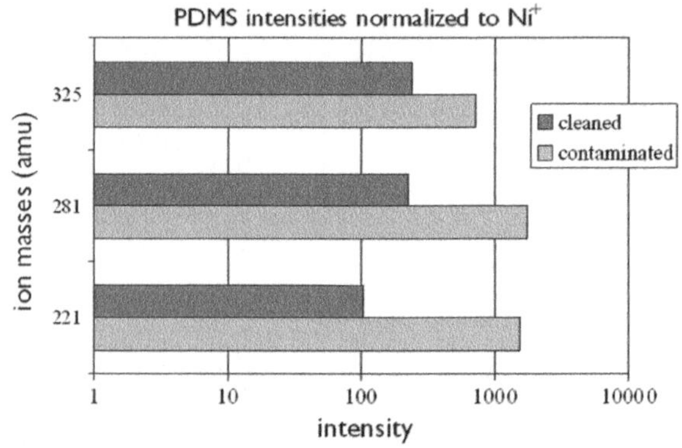

Fig. 11.18 TOF-SIMS measurement of contaminated metal probe before and after H_2O rinse. Note that the intensities are shown in logarithmic scale. A large fraction of the PDMS was removed by the cleaning step

11.4 Conclusions on Molded Diamond Probes

In the last 20 years, molded diamond probes have been an enabler for the SSRM method to become the leading method to electrically measure nanoelectronics device structures on the nanometer scale. The probe fabrication process steadily improved over the years and today a sub-nanometer resolution is routinely achieved. Although the electrical probing challenges directly ahead of us for sub-10 nm technology nodes are calling for high-aspect-ratio tips as discussed in the next section, the low-aspect-ratio classical pyramidal tips will still be needed for example where high-aspect-ratio tips do easily break off like in the slicing step of Scalpel AFM.

11.5 Plasma Etched Single Crystal Doped Diamond Probes

AFM provides several powerful techniques for nanometer scale characterization of the local properties of nanoelectronics such as charge, potential, conductivity, and capacitance. The availability of a wear-resistant, highly conductive, high-resolution probe is a critical component for successful and reliable measurements of these properties at length scales relevant to industry. Here we discuss the fabrication and performance of plasma etched single crystal doped diamond probes (PE-SC-DDPs) for high-resolution electrical metrology applications. We demonstrate that PE-SC-DDPs with tip radii <5 nm can be manufactured at wafer scale with superior electrical and mechanical properties enabling quantitative and consistent imaging, overcoming limitations of traditional AFM probes. Principle limitations as encountered by industry include: Trade-offs between resolution and reliability where probes exhibit significant degradation and imaging artefacts during operation, and significant performance variation from probe to probe. These limitations are particularly highlighted for those electrical modes requiring operation with constant tip-sample contact (e.g. C-AFM, TUNA, SSRM etc.), as these experience high lateral and normal forces at the tip apex. The unique combination of mechanical strength, conductivity and high-resolution offered by wafer scale fabricated of PE-SC-DDPs enables unique access to parameter spaces previously unobtainable using traditional AFM probes. These probes offer reliable and highly reproducible metrology tools for the characterization of present and future generations of nanoelectronic devices.

11.5.1 Manufacturing

The integration of diamond into AFM probe manufacturing, traditionally on 100 mm silicon wafers, poses significant manufacturing challenges: (1) at what stage in the process is diamond grown on the silicon wafer? and (2) how can the diamond be reliably shaped with nanoscale precision at wafer scale in order to optimize performance? There are two main process that have been used to create diamond probes: tip-mold-bonding [8, 33] and CVD diamond grown on silicon [2]. Both processes have significant trade-offs in terms of process complexity, yield, and functionality of the finished device. However, all processing techniques have a common challenge in accurately and reliably forming the diamond at the apex of the tip into shapes optimized for various applications with nanoscale precision. This processing challenge can be divided into two main areas: (1) handling and processing very fragile through-etch wafers, and (2) performing lithography techniques on relatively tall AFM tips with complex geometries.

When wafers of AFM probes are produced, a significant portion of the wafer is selectively removed, reducing the mechanical integrity of the wafer and making it susceptible to breakages when handling. Most traditional wafer processing equipment, including lithography tools, are unsuitable for the handling of finished AFM probe wafers and will shatter the wafer. Traditional lithography requires vacuum chucks, high RPM wafer spinners, pouring or spraying of viscous photosensitive polymers, robot wafer handlers, fast-moving XY stages, amongst other things. Without the use of this equipment, it would be next to impossible to form an etch mask at the very apex of the AFM tip to further refine it into a more advanced tip structure. Even if one could get past the challenges of wafer breakage with lithography processing equipment, it would still be extremely challenging to put nanometer-scale etch masks at the very apex of the relatively tall tips with complex geometry with any consistency across a wafer with high fidelity. This is due to AFM tips standing proud of the plane of the wafer on the order of 10–25 μm tall with tip apex lateral and height variations on the order of *microns*. Photomask exposure of such wafers would form etch masks poorly aligned with the tip apex and/or variations in the size of the mask due to nonuniform tip heights changing the focus of incident exposing UV light. These micron-scale X, Y, and Z variations inherent in AFM tip processing necessitate a tip-by-tip lithography technique with nanometer precision and accuracy if one wishes to perform nanometer-scale refinement on top of the existing micron-scale technology of silicon AFM probe manufacturing.

Ion beam implantation and plasma etching are clean and simple manufacturing techniques that can be realized without the need for spin coating photoresist, UV exposure or wet etching [34]. The precision of the focused ion beam (FIB) equipped with a *XYZ* stage allows for masking control on the nanometer scale despite complex substrate geometries. By combining beam shaping with depth of focus control, accurate masking of a single crystal diamond facet on a sloped surface at the tip apex can be reliably and accurately achieved. The gallium ion implanted region of diamond is then used as a hard mask during plasma etching to enable the desired tip shape to be etched into the diamond crystal. This manufacturing technique utilizes the FIB primarily as an ion implantation source and not to intentionally mill the diamond surface [35].

Current FIBs can also be integrated with pattern recognition and scripting capability in order to build recipes that can step an AFM probe wafer from tip-to-tip and focus energy at the very apex and overcome the X, Y, and Z variation inherent in the base silicon AFM probe wafers. When tailored correctly, one can optimize the ion dose to be part of the tip-refinement process. With state-of-the-art FIB tools an ion beam can be focused on every tip in a wafer, depositing the required ion dose at the exact location required, within a few hours. The plasma etch is then performed at the wafer scale and can be completed within a few minutes. This allows a degree of scale-up that can reliably mass produce high precision diamond probes at wafer scale with ample capacity to meet current and future market demands. A scheme showing the steps in the manufacture of PE-SC-DDPs using these techniques is shown in Fig. 11.19.

By adjusting key parameters such as the gallium mask shape and depth, and the plasma chamber pressure and bias (which allows combinations of anisotropic and isotropic etching), precise control over tip shape and tip radius can be achieved (Fig. 11.20). Figure 11.20b shows a PE-SC-DDPs with a radius of <15 nm. Plasma etching allows the precise control the tip radius of such probes, enabling a trade-off between lateral resolution and contact resistance (inversely proportional to tip radius) to be chosen. An example of the precision of this manufacturing technique is shown in Fig. 11.20c where a tip radius of 2 nm is achieved. Probes of this sharpness are often required for atomic resolution metrology applications but their susceptibility of mechanical damage at the tip apex makes them a poor choice contact mode (e.g. C-AFM, TUNA, SSRM etc.) based mapping of nanoelectronics. Flared or critical dimension tips (Fig. 11.20d) are generally used for the profiling of sidewalls of nanoelectronic devices [36, 37] and are an example of manufacturing shape control with mask and plasma etch techniques. This undercut etching is a phenomenon, widely reported in the 311 plane of Si, which can be transferred to all diamond planes under the appropriate etch chamber conditions [38]. Figure 11.20e shows high aspect ratio (HAR) probes suitable to the profiling of FinFET and nanowire structures at the N10, N7 and N5 nodes. These probes are manufactured through a complex process of anisotropic etching using a small and precisely controlled gallium implanted mask. In addition to the ability to precisely control the tip shape at the apex the cantilever properties can also be engineered to yield probes with spring constants of 2.8, 40, 150, 350 N/m and even 3500 N/m. Whilst the lower spring constants are standard in the industry the ability to selectively tune the diamond thickness allows unprecedented access to high spring constants for nanoindentation and tomography applications.

Here we have demonstrated the ability to precisely shape diamond that nanometer length scales for the formation of tip shaped relevant to AFM metrology of nanoelectronic devices. The precise nature of the mask and plasma etch technique combined with the scalability of the process to wafer scale manufacturing represents a paradigm shift in the production of precision PE-SC-DDPs for nanoelectronics applications. PE-SC-DDPs manufactured using this process have demonstrated high performance electrical data across a wide range of characterization techniques from the atomic to the whole device scale.

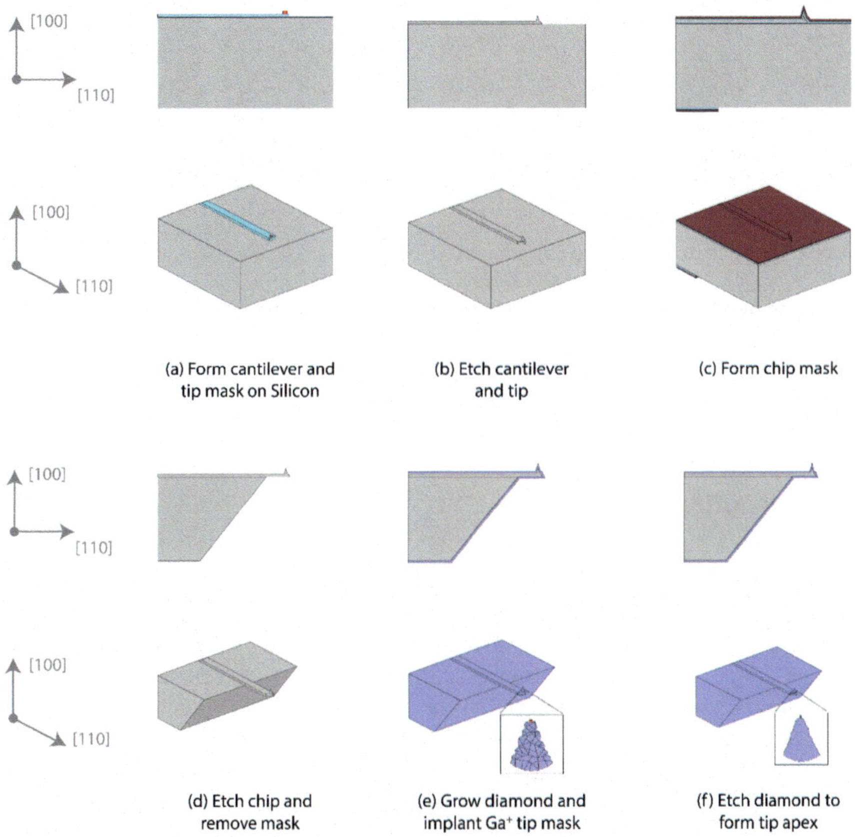

Fig. 11.19 Scheme showing the steps in the manufacture of PE-SC-DDPs. **a** Form cantilever and tip mask on Silicon, **b** Etch cantilever and tip, **c** Form chip mask, **d** Etch chip and remove mask, **e** Grow diamond and implant Ga^+ tip mask, and **f** Etch diamond to form tip apex

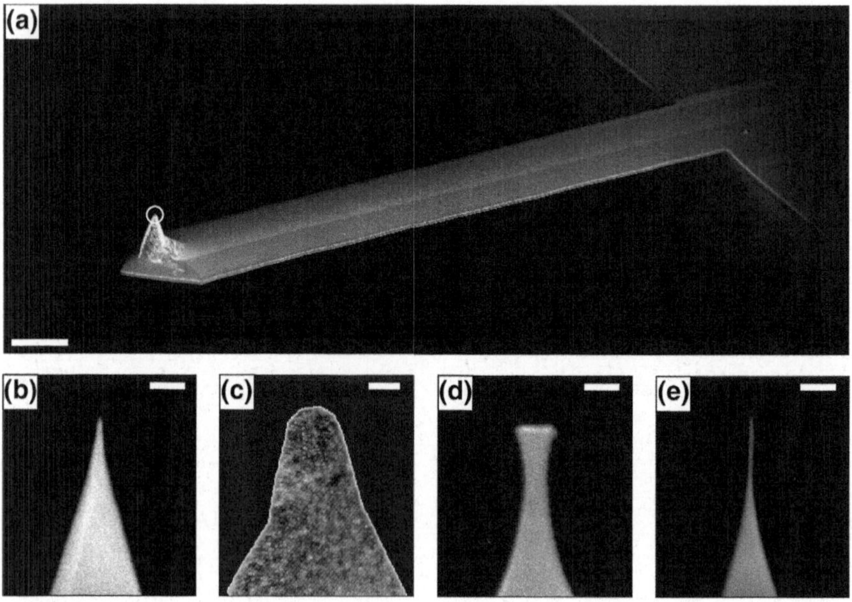

Fig. 11.20 Images of PE-SC-DDPs demonstrating precise control of tip shape and dimensions for specific applications. **a** Low magnification SEM image of 225 μm long silicon cantilever coated with ~1 μm of CVD diamond. Images **b–d** are located at the tip apex circled in (**a**). **b** SEM image of an Apex Sharp tip shape with a radius of <15 nm. **c** TEM image of a Super Sharp tip shape with a radius of ~2 nm. Note the presence of diamond lattice fringes at this resolution. **d** SEM image of an undercut CD-AFM probe. **d** SEM image of a high aspect ratio pillar probe. Images **b**, **c** and **e** correspond to Adama Innovations AD-40-AS, AD-40-SS and AD-40-P10 probes respectively. Scale bar is 20 μm in (**a**), 50 nm in (**b**), (**d**) and (**e**) and 2 nm in (**c**)

11.6 Applications

PE-SC-DDPs offer high versatility and are suitable for a wide range of demanding measurements of local electrical properties of nanoelectronic devices. Here we show a wide range of examples including: Measurements of tunneling current for individual surface atoms, conductivity profiling of carbon nanotubes and nanowires, solar cells, and metrology of high aspect ratio FinFET and nanowire type structures. The ability to precisely control tip shape and end radius of a conductive single crystal diamond structure allows a unique combination of electrical and mechanical properties to be realized in a novel measurement tool. Unprecedented reliable high-resolution measurements with a high degree of reproducibility become possible that are largely free from tip wear and damage induced imaging artefacts.

11.6.1 Scanning Tunneling Microscopy—Atomic Resolution Imaging

PE-SC-DDPs with an end radius of <5 nm, manufactured at wafer scale, are an ideal candidate for the profiling of the local electronic structure of materials at the atomic scale. Scanning Tunneling Microscopy (STM) is a technique which can measure the local density of electronic states at the sample surface with sub-atomic resolution using a very sharp conducting tip. STM techniques have been used to measure the presence of individual dopant atoms [39] and even the shape of electronic bonds within individual molecules adsorbed on a surface [40–42]. The technique is generally operated under vacuum due to the sensitivity of the measurement to surface contamination, but researchers have successfully extended its operation to both air [43] and liquid [44] environments. By applying a bias to the conductive tip and positioning the tip apex extremely close to the sample surface, quantum mechanical tunneling occurs whereby electrons pass across the gap between the probe and sample at a rate proportional to the density of states of the atom scanned over by the tip. By measuring the tunneling current, a map of the sample surface structure is constructed, and even chemical species identification of atoms beneath the tip can be achieved (with additional tunneling spectroscopy modes) [45]. Figure 11.21 shows example STM images obtained with PE-SC-DDPs in vacuum and air environments. These images show clear atomic resolution in both cases. Special tip conditioning is often required in STM due to poor sharpness and/or contamination. PE-SC-DDPs do not require any such conditioning and these probes can be operated continuously without degradation in performance, in some cases for months. This is a direct demonstration of the sharpness of the tip that can be formed using plasma etch manufacturing techniques.

11.6.2 Conductive Atomic Force Microscopy

The use of sharp conductive PE-SC-DDPs allows for the profiling of the conductivity of samples with spatial resolution and reliability that far surpasses traditional metal or diamond coated silicon probes or all metal probes. This unique combination of resolution, conductivity and robustness in a commercially mass-produced probe is particularly useful for the mapping of current flow through nanoscale circuit elements such as nanotubes and nanowires. Figure 11.22a demonstrates the ability to map the current flowing through an array of carbon nanotubes deposited on a silicon surface. In this experiment a bias is applied to the sample and the current passing through the network of nanotubes to the AFM probe is mapped. This allows the electrical conduction pathways across the surface to be resolved with nanometer resolution as well as mapping the resistance of each nanotube in the network. Figure 11.22b demonstrates conductivity mapping of an array of biased silver nanowires on a silicon surface. This data reveals the presence of individual grain boundaries within a single

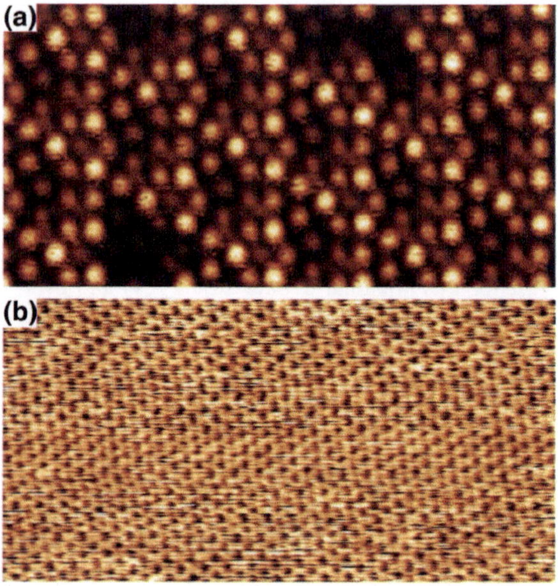

Fig. 11.21 Atomic resolution STM in vacuum and air. **a** 7 × 7 reconstructed silicon 111 measured via simultaneous AFM and STM in vacuum (14 × 7 nm). Data courtesy of the John Boland Lab, CRANN, Trinity College Dublin. Acquired using an AD-40-SS Adama Innovations probe. **b** STM image of graphite in air (15 × 7 nm). Acquired using an Asylum Research Cypher AFM in STM mode using a 80 N/m SS Adama Innovations probe

nanowire and the local change in resistance associated with these features [46, 47]. This type of information is crucial to the construction of nanoelectronic circuits using these types of elements.

In addition to spatial mapping of current flowing through materials deposited on an insulating or semiconducting surface, the conductivity profile of devices with complex three-dimensional architectures can also be assessed. Figure 11.23 shows a conductive AFM image of the back surface field (BSF) of a solar cell using a ResiScope [49] which provides 10 orders of magnitude of measurement range (from 100 fA to 1 mA) within a single measurement. Here the bands of conductivity are distinct and resolved with high spatial resolution. A 30 nm wide conductive band (light blue) is clearly resolved with high fidelity.

Conductive AFM can further interrogate samples by measuring the current versus voltage response (IV curves) either at discrete locations or mapped at high resolution across a surface [50]. The robustness of sharp conducting diamond probes further allows the mechanical exfoliation of surfaces in situ allowing for tomographical mapping of electrical properties as a function of milled depth in a process commonly referred to as scalpel AFM [51, 52].

11.6.3 Scanning Capacitance Microscopy

Scanning Capacitance Microscopy (SCM) enables mapping of dopants in a semiconductor device by measurement of changes in capacitance as a conductive probe

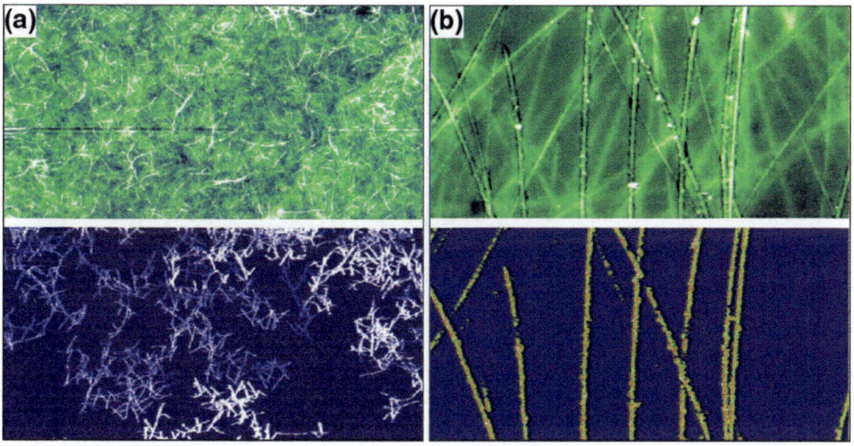

Fig. 11.22 Conductive AFM images of nanotubes and nanowires. **a** Conductivity mapping of an array of carbon nanotubes on a silicon surface connected to a gold electrode (8.5 × 15 μm). Topography (top) and Current (bottom). Sample courtesy: Matteo Palma, Queen Mary University London. Acquired using a Bruker Dimension Icon AFM in Peakforce-TUNA mode using a AD-40-AS Adama Innovations probe. **b** Conductivity mapping of silver nanowires on a silicon surface (5 × 2.5 μm). Acquired using a NT-MDT Spectrum Instruments Solver Nano AFM in Spreading Resistance Imaging mode [48] using a AD-2.8-SS Adama Innovations tip

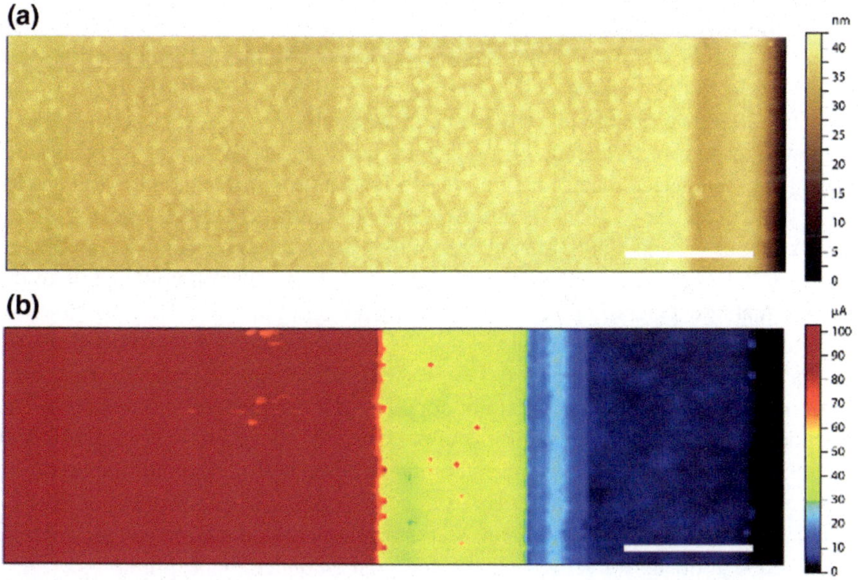

Fig. 11.23 Conductive AFM imaging of a back-surface field (BSF) solar cell using PE-SC-DDPs. **a** Topography. **b** Current. Acquired using a CSInstruments Nano-Observer AFM with a ResiScope using a AD-2.8-AS Adama Innovations probe. Scale bar is 500 nm

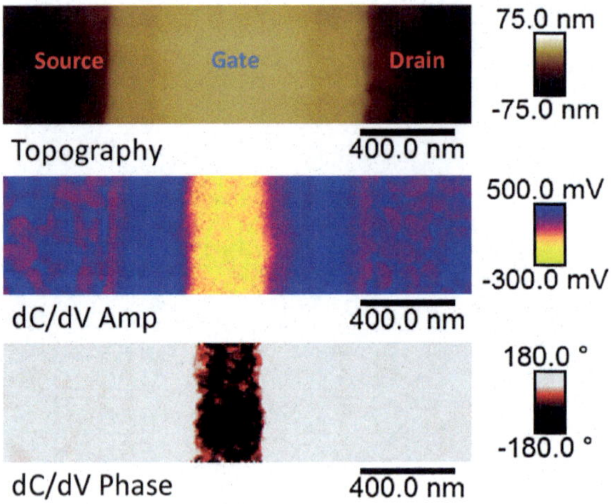

Fig. 11.24 Scanning Capacitance Microscopy (SCM) of a pnp transistor (top surface) with 250 nm effective gate length showing Topography (top), *dC/dV* amplitude (middle) and *dC/dV* phase channel (bottom). Acquired using a Bruker Dimension Icon AFM in SCM mode using a AD-2.8-AS Adama Innovations probe

is scanned across a surface in contact mode. Here the tip, native oxide and semiconductor surface form a tiny Metal Oxide Semiconductor (MOS) transistor where the tip acts as the gate [53]. By probing the voltage dependence of the capacitance of this MOS system the dopant type and density can be accurately mapped. In SCM, a modulated bias (*dV*) is applied to the sample to alternatively cause depletion (or accumulation) of electrons and accumulation (or depletion) of holes in the area directly beneath the probe at microsecond timescales. The depletion (or accumulation) that occurs below the tip is directly proportional to the dopant density and applied bias [54]. This causes a change in capacitance of MOS transistor which is detected as a change in resonance frequency of a gigahertz oscillator applying an AC bias to the probe [55]. The detected amplitude of *dC/dV* is then proportional to the local carrier density and the phase of *dC/dV* is indicative of the dopant polarity (n-type or p-type). Since changes in capacitance are detected with a dielectric layer between the tip and sample this technique does not require the removal or penetration of native oxide layers which are generally present under ambient conditions.

Figure 11.24 shows an example of a SCM measurement of the profile of a PNP memory junction using conductive PE-SC-DDPs. The different transistor regions, p-doped source and drain, and medium doped n-type channel are well resolved. The p-n junction resolution in this sample is approximately 30 nm. Here the measurement obtains both the topographic profile of the sample along with the dopant concentration (from *dC/dV* amplitude) and polarity (from *dC/dV* phase). The robustness of PE-SC-DDPs allows for many measurements to be made without changing the capacitance of the probe allowing for accurate and reliable SCM data to be acquired.

11.6.4 Scanning Spreading Resistance Microscopy

Scanning Spreading Resistance Microscopy was introduced as a high resolution extension of spreading resistance profiling (SRP) by Vandervorst [56] and De Wolf [1, 57] at IMEC in 1994. SSRM is used to map dopant profiles, diffusion and defects in two and three dimensions in semiconducting materials by measuring the current flow (proportional to the local carrier density) due to a biased probe applied to the sample surface. By using a logarithmic amplifier SSRM measurements can cover a range of ~10^{15} to 10^{20} atoms/cm^3 dopant densities in a single measurement. Critically this technique relies on the application of high force between the probe and sample such that the spreading resistance rather than the contact resistance is measured. The high forces required for stable SSRM operation in air (where the probe must penetrate through any native oxide and establish a β-tin phase transformation of the Si surface to ensure low contact resistance) were a driving force for the development and introduction to the market of diamond coated silicon probes [58, 59] and molded diamond probes [4, 60]. PE-SC-DDPs have recently demonstrated high resolution reproducible dopant mapping using SSRM. Due to the sharp, well defined tip interacting with the sample the force required is greatly reduced to achieve the high pressures required. This reduction in applied load leads to a higher degree of control during imaging which results in improved signal-to-noise ratio and increased lateral resolution compared to traditional coated diamond probes.

A demonstration of the dynamic range attainable with SSRM using PE-SC-DDPs is shown in Fig. 11.25, where a silicon sample with a p-doped staircase [61] is measured. Here all dopant steps are clearly resolved with good signal-to-noise illustrating the high sensitivity, lateral resolution and dynamic range of the conductive PE-SC-DDPs (10^{15} to above 10^{20} atoms/cm^3) for SSRM measurements. Figure 11.25c, d show a direct comparison of the SSRM resistance measured across the p-doped staircase with Secondary Ion Mass Spectrometry (SIMS) data. Here the profile measured with SSRM corresponds well (inversely proportional) to the SIMS data with small changes in dopant concentration fully resolved despite measuring over multiple orders of magnitude in concentration.

11.6.5 Measurement of High Aspect Ratio Structures

The imaging of high aspect ratio structures for industrial metrology of nanoelectronics via AFM remains challenging [62] as device elements continue to shrink and increase in geometric complexity [63]. This is inherently due to the imaging mechanism in AFM being a convolution of the sample topography and the shape of the tip apex [36]. Special care must be taken to mitigate the introduction of artifacts when scanning high aspect ratio structures through control of tip-sample interaction geometry and/or through post processing of data with deconvolution algorithms [64, 65]. PE-SC-DDPs can be tailored to this application by etching tip shapes with a

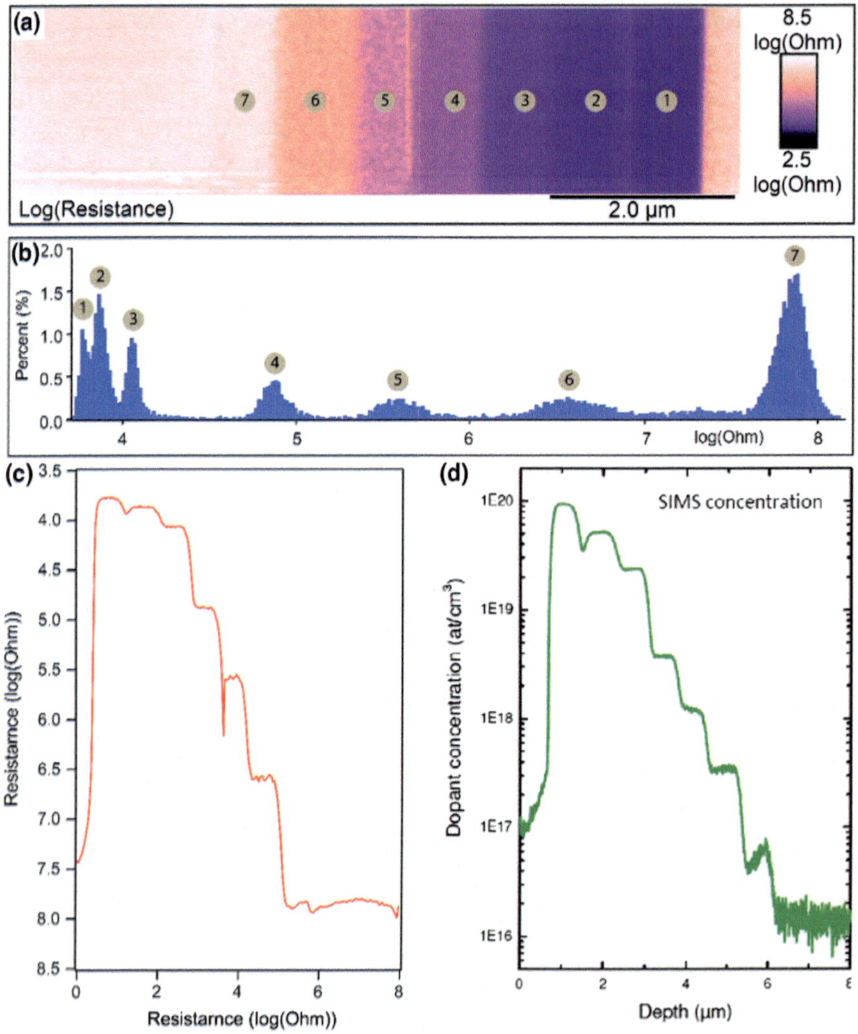

Fig. 11.25 Scanning Spreading Resistance Mode (SSRM) imaging of a p-doped staircase dopant profile. **a** SSRM log resistance image of the surface with the 7 dopant bands labelled. **b** Bearing analysis of (**a**) with labelled bands of dopant concentration. **c** Mean SSRM profile of (**a**). **d** SIMS profile of the sample from the manufacturer [33]. Acquired using Bruker Dimension Icon AFM in SSRM mode using a AD-40-AS Adama Innovations probe

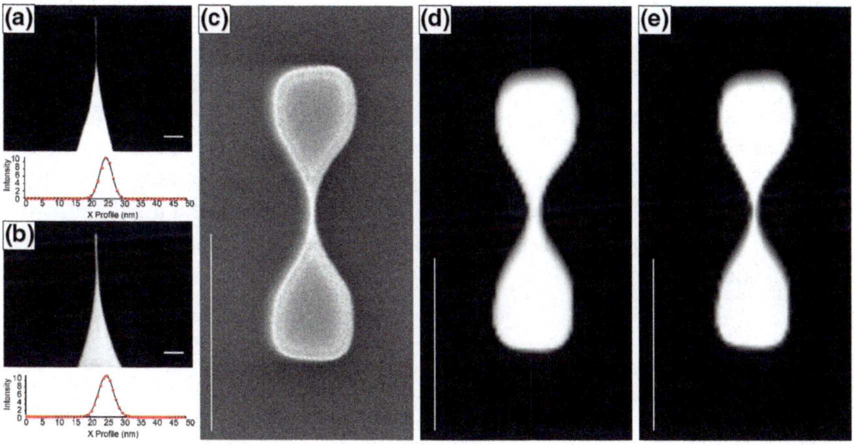

Fig. 11.26 High Aspect Ratio PE-SC-DDPs produced through a plasma etch process. **a**, **b** SEM images and pixel intensity profiles for 3.8 and 4.6 nm probe diameter (FWHM). Scale bars = 50 nm. **c** SEM image of a high aspect ratio FinFET type test structure. Minimum width = 12.5 nm. **d** AFM topography image of (**c**). Minimum width = 20.3 nm. **e** Data from (**d**) after correction for probe shape using a deconvolution algorithm. Minimum width = 12.1 nm. Scale bars in (**c**–**e**) = 400 nm. Acquired using an Asylum Research Cypher AFM in AC mode using a AD-40-P10 Adama Innovations probe

small diameter pillar with high aspect ratios (Fig. 11.26). In addition to the geometrical challenges of probe shape and scale, sidewall interactions between the probe and sample (e.g. capillary adhesion and Van der Waals forces) must often be overcome by employing the higher bending modulus offered by single crystal diamond combined with the low surface energy of diamond [66] and an ability to control the hydrophobicity through the chemistry of the diamond surface termination [67, 68]. Ultimately there will always be a resolution trade-off between sidewall interactions and probe diameter.

Examples of 100 nm tall PE-SC-DDPs for advanced high aspect ratio metrology are shown in Fig. 11.26a with a 25:1 aspect ratio probe with a 3.8 nm diameter and Fig. 11.26b with a 20:1 aspect ratio probe with a 4.6 nm diameter. These probes are highly doped (~1%) with boron to allow electrical measurements and/or applying a bias to effectively compensate tip-sample electrostatic interactions. A ~10 nm width FinFET type test structure provided by CEA-LETI [69] is first measured by SEM (Fig. 11.26c) and then by AFM using a high aspect ratio diamond probe (Fig. 11.26d). Since the probe has a finite diameter it is necessary to apply a deconvolution algorithm to the data in order to extract the true morphology of the structure [70]. Here the known probe geometry is input into the algorithm in order to apply deconvolution of the probe shape from the measured data [71] yielding the result shown in Fig. 11.26e. Here the gate width measured via SEM is 12.5 nm which corresponds well to the deconvoluted AFM measured gate width of 12.1 nm.

Whilst high aspect ratio probes made from silicon [72], carbon nanotubes [73, 74] and high density carbon (HDC)/diamond-like carbon (DLC) [75] have been available in the market for some time, PE-SC-DDPs offer a unique combination of mechanical strength, electrical conductivity and the ability to specify the hydrophobicity and surface energy of the probe through selection of the surface termination chemistry [76]. This unique combination of advantageous properties combined with a well-controlled and high yielding means of production is likely to be a highly attractive target for metrology of the complex three-dimensional structures on the roadmap of the semiconductor industry.

11.7 Conclusions and Outlook

Plasma based manufacturing techniques have been applied to precisely control the tip shape of doped diamond coated silicon probes with nanoscale precision at the wafer scale. The degree of control enabled by this scalable technique allows for a variety of tip shapes to be selectively manufactured, ranging from standard Apex Sharp and Super Sharp, to undercut CD probes for sidewall profiling and high aspect ratio probes for the measurement of the complex metrology challenges of industry now and into the future. PE-SC-DDPs have significantly altered the landscape of tools available for the characterization of nanoelectronic devices both in the research environment and for industrial applications.

The applications examples presented here are illustrative of the types of measurements enabled with this new technology, providing access to the metrology tools required for quality assurance as well as research and development for future generations of devices. The mass fabrication of high-fidelity sharp diamond probes with electrical conductivity is a unique enabling technology that is essential for the further development of nanoelectronics devices. These probes not only allow for the measurement of devices at length scales that were previously not able to be characterized by conventional metal coated silicon or diamond coated silicon probes, but they also greatly improve confidence in the data obtained as the probe to probes variation is tightly controlled through the plasma etched manufacturing process. This combined with the low wear and high resilience of sharp diamond probes allows the same probe to be used for many measurements and for statistical analysis of process control to be obtained.

Quantum sensing based on isolated solid-state spins promises to be an entire new paradigm for nanoelectronic measurement by sharp diamond probes. Nitrogen vacancy (NV) structures are stable, optically active point defects in the diamond lattice hosting an isolated single spin (or an ensemble). These structures have demonstrated the ability to probe a wide array of phenomena at unprecedented resolution and sensitivity including: Magnetic fields, electric fields, strain, temperature and pressure [77]. High resolution magnetometry has been demonstrated through optical readout of a scanned, rigidly mounted, single spin NV defect contained in a monolithic diamond nanostructure [78]. Combined topographic and electro-optical mode imaging was achieved by gluing a 100 nm diamond nanoparticle to an AFM cantilever that hosted an ensemble of NV spin centers [79]. We have recently shown that the ion beam based manufacturing methods used to produce PE-SC-DDPs can also introduce NV defects of controlled density in sub-micrometer scale spatial patterns [80]. Explorations to marry these techniques to allow for the mass fabrication NV diamond probes is currently ongoing and could lead to the commercial availability new probes that offer a new landscape of metrology for nanoelectronic devices.

Acknowledgements Thierry Conard is acknowledged for TOFSIMS measurements. Menelaos Tsigkourakos is thanked for diamond seeding, growth and SSRM support. W. Kulisch († November 2018) is thanked for his pioneering diamond research and molded tip support in this work.

References

1. P. De Wolf, J. Snauwaert, L. Hellemans, T. Clarysse, W. Vandervorst, M. D'Olieslaeger, D. Quaeyhaegens, Lateral and vertical dopant profiling in semiconductors by atomic force microscopy using conducting tips. J. Vac. Sci. Technol., A **13**, 1699–1704 (1995)
2. P. Niedermann, W. Hänni, N. Blanc, R. Christoph, J. Burger, Chemical vapor deposition diamond for tips in nanoprobe experiments. J. Vac. Sci. Technol. A **14**, 1233–1236 (1996)
3. E. Oesterschulze, W. Scholz, C. Mihalcea, D. Albert, B. Sobisch, W. Kulisch, Fabrication of small diamond tips for scanning probe microscopy application. Appl. Phys. Lett. **70**, 435 (1997). https://doi.org/10.1063/1.118173
4. T. Hantschel, P. Niedermann, T. Trenkler, W. Vandervorst, Highly conductive diamond probes for scanning spreading resistance microscopy. Appl. Phys. Lett. **76**, 1603–1605 (2000)
5. T. Hantschel, C. Demeulemeester, P. Eyben, V. Schulz, O. Richard, H. Bender, W. Vandervorst, Conductive diamond tips with sub-nanometer electrical resolution for characterization of nanoelectronics device structures. Phys. Status Solidi A **206**(9), 2077–2081 (2009). https://doi.org/10.1002/pssa.200982212
6. A.N. Obraztsov, P.G. Kopylov, B.A. Loginov, M.A. Dolganov, R.R. Ismagilov, N.V. Savenko, Single crystal diamond tips for scanning probe microscopy. Rev. Sci. Instrum. **81**, 013703 (2010). https://doi.org/10.1063/1.3280182
7. N. Moldovan, Z. Dai, H. Zeng, J.A. Carlisle, T.D.B. Jacobs, V. Vahdat, D.S. Grierson, J. Liu, K.T. Turner, R.W. Carpick, Advances in manufacturing of molded tips for scanning probe microscopy. J. Microelectromech. Syst. **21**, 431 (2012)
8. T.R. Albrecht, S. Akamine, T.E. Carver, C.F. Quate, Microfabrication of cantilever styli for the atomic force microscope. J. Vat. Sci. Technol. A **8**, 3386 (1990)
9. O.A. Williams, O. Douheret, M. Daenen, K. Haenen, E. Osawa, M. Takahashi, Enhanced diamond nucleation on monodispersed nanocrystalline diamond. Chem. Phys. Lett. **445**(4–6):255–258 (2007)
10. M. Tsigkourakos, T. Hantschel, K. Arstila, W. Vandervorst, Diamond nano-particle seeding for tip moulding application. Diam. Relat. Mater. **35**, 14–18 (2013)
11. S. Heyer, W. Janssen, S. Turner, Y. Lu, W.S. Yeap, J. Verbeeck, K. Haenen, A. Krueger, Toward deep blue nano hope diamonds: heavily boron-doped diamond nanoparticles. ACS Nano **8**, 5757–5764 (2014)
12. M. Tsigkourakos, T. Hantschel, C. Bangerter, W. Vandervorst, Electrical probing of B-doped diamond seeds embedded into the interfacial layer of a conductive diamond film. Phys. Status Solidi A, 1–6 (2014). https://doi.org/10.1002/pssa.201431215
13. A. Afandi, A. Howkins, I.W. Boyd, R.B. Jackman, Nanodiamonds for device applications: an investigation of the properties of boron-doped detonation nanodiamonds. Sci. Rep. **8**, 3270 (2018)

14. J. Zimmer, T. Hantschel, G. Chandler, W. Vandervorst, M. Peralta, Boron doping in hot filament MCD and NCD diamond films. Mater. Res. Soc. Symp. Proc. **1203** (1203-J12-01) (2010)
15. J.G. Buijnsters, M. Tsigkourakos, T. Hantschel, F.O.V. Gomes, T. Nuytten, P. Favia, H. Bender, K. Arstila, J.-P. Celis, W. Vandervorst, Effect of boron doping on the wear behavior of the growth and nucleation surfaces of micro- and nanocrystalline diamond films. ACS Appl. Mater. Interfaces **8**, 26381–26391 (2016). https://doi.org/10.1021/acsami.6b08083
16. J.P. Rasmussen, P.T. Tang, C. Sander, O. Hansen, P. Møller, M.J. Beeson, Fabrication of an all-metal atomic force microscope probe, in *Proceedings of the Transducers 1997, Chicago, USA* (June 1997), pp. 463–466
17. S. Koelling, T. Hantschel, W. Vandervorst, Conductive diamond probes with electroplated holder chips. Microelectron. Eng. **84**, 1178–1181 (2007)
18. T. Hantschel, S. Slesazeck, W. Vandervorst, The peel-off probe: a cost-effective probe for electrical atomic force microscopy. Proc. SPIE **4175**, 50 (2000)
19. T. Hantschel, U. Pape, S. Slesazeck, P. Niedermann, W. Vandervorst, Mounting of moulded AFM probes by soldering. Proc. SPIE **4175**, 62 (2000)
20. G. Gunther Reinhart, D. Jacob, M. Fouchier, Automated assembly of holder chips to AFM probes. Proc. SPIE **4568**, Microrobotics and Microassembly III (8 Oct 2001). https://doi.org/10.1117/12.444140
21. T. Hantschel, T. Trenkler, W. Vandervorst, A. Malave, D. Buechel, W. Kulisch, E. Oesterschulze, Tip-on-tip: a novel AFM tip configuration for the electrical characterization of semiconductor devices. Microelectron. Eng. **46**, 113 (1999)
22. S. Akamine, C.F. Quate, Low temperature thermal oxidation sharpening of microcast tips. J. Vac. Sci. Technol. B **10**, 2307 (1992)
23. T. Hantschel, T. Clarysse, A. Ajaykumar, F. Seidel, M. Tsigkourakos, T. Nuytten, K. Paredis, P. Eyben, B. Majeed, D.S. Tezcan, D.S. Tezcan, W. Vandervorst, Diamond nanoprobes for electrical probing of nanoelectronics device structures. Microelectron. Eng. **121**, 19 (2014)
24. T. Hantschel, M. Tsigkourakos, J. Kluge, T. Werner, L. Zha, K. Paredis, P. Eyben, T. Nuytten, Z. Xu, W. Vandervorst, Overcoated diamond tips for nanometer-scale semiconductor device characterization. Microelectron. Eng. **141**, 1 (2015)
25. E. Oesterschulze, A. Malave, U.F. Keyser, M. Paesler, R.J. Haug, Diamond cantilever with integrated tip for nanomachining. Diam. Relat. Mater. **11**, 667–671 (2002)
26. C. Beuret, T. Akiyama, U. Staufer, N.F. de Rooij, P. Niedermann, W. Hanni, Conical diamond tips realized by a double-molding process for high-resolution profilometry and atomic force microscopy applications. Appl. Phys. Lett. **76**, 1621 (2000)
27. P. Eyben, W. Vandervorst, D. Alvarez, M. Xu, M. Fouchier, in *Scanning Probe Microscopy*, vol. 31, ed. by S. Kalinin, A. Gruverman (Springer, New York, 2007)
28. M. Tsigkourakos, T. Hantschel, D.K. Simon, T. Nuytten, A.S. Verhulst, B. Douhard, W. Vandervorst, On the local conductivity of individual diamond seeds and their impact on the interfacial resistance of boron-doped diamond films. Carbon **79**, 103 (2014)
29. M. Tsigkourakos, T. Hantschel, S.D. Janssens, K. Haenen, W. Vandervorst, Spin-seeding approach for diamond growth on large area silicon-wafer substrates. Phys. Status Solidi A **209**, 1659 (2012)
30. P. De Wolf, Ph.D. thesis, University of Leuven, Belgium, 1998, 120
31. Y.-S. Lo, N.D. Huefner, W.S. Chan, P. Dryden, B. Hagenhoff, T.P. Beebe, Organic and inorganic contamination on commercial AFM cantilevers. Langmuir **15**, 6522 (1999)
32. T. Hantschel, T. Trenkler, M. Xu, W. Vandervorst, The fabrication of a full metal AFM probe and its applications for Si and InP device analysis. Proc. of SPIE **3875**, 20–31 (1999)
33. P. Niedermann, R.F. Christoph, *Components manufacturing method micromechanical having a diamond portion consisting at least of a tip, and micromechanical components having at least one diamond tip*. FR:2739494:A1 (1997)
34. L. Li, I. Bayn, M. Lu, C.-Y. Nam, T. Schröder, A. Stein, N.C. Harris, D. Englund, Nanofabrication on unconventional substrates using transferred hard masks. Sci. Rep. **5**, 7802 (2015)
35. W. McKenzie, J. Pethica, G. Cross, A direct-write, resistless hard mask for rapid nanoscale patterning of diamond. Diam. Relat. Mater. **20**, 707–710 (2011)

36. Y. Martin, H. Kumar Wickramasinghe, Method for imaging sidewalls by atomic force microscopy. Appl. Phys. Lett. **64**, 2498–2500 (1994)
37. G. Dai, H. Wolff, F. Pohlenz, H.-U. Danzebrink, G. Wilkening, Atomic force probe for sidewall scanning of nano- and microstructures. Appl. Phys. Lett. **88**, 171908 (2006)
38. M. Shikida, K.-I. Nanbara, T. Koizumi, H. Sasaki, M. Odagaki, K. Sato, M. Ando, S. Furuta, K. Asaumi, in *Transducers '01 Eurosensors XV*, ed. by E. Obermeier (Springer Berlin Heidelberg, 2001), pp. 648–651
39. M. Usman, J. Bocquel, J. Salfi, B. Voisin, A. Tankasala, R. Rahman, M.Y. Simmons, S. Rogge, L.C.L. Hollenberg, Spatial metrology of dopants in silicon with exact lattice site precision. Nat. Nanotechnol. **11**, 763–768 (2016)
40. L. Gross, F. Mohn, N. Moll, P. Liljeroth, G. Meyer, The chemical structure of a molecule resolved by atomic force microscopy. Science **325**, 1110–1114 (2009)
41. L. Gross, N. Moll, F. Mohn, A. Curioni, G. Meyer, F. Hanke, M. Persson, High-resolution molecular orbital imaging using a p-wave STM tip. Phys. Rev. Lett. **107**, 086101 (2011)
42. D. Ebeling, Q. Zhong, S. Ahles, L. Chi, H.A. Wegner, A. Schirmeisen, Chemical bond imaging using higher eigenmodes of tuning fork sensors in atomic force microscopy. Appl. Phys. Lett. **110**, 183102 (2017)
43. R.J. Colton, S.M. Baker, R.J. Driscoll, M.G. Youngquist, J.D. Baldeschwieler, W.J. Kaiser, Imaging graphite in air by scanning tunneling microscopy: role of the tip. J. Vac. Sci. Technol., A **6**, 349–353 (1988)
44. O. Marti, M. Amrein, *STM and SFM in Biology*. (Academic Press, London, 2012)
45. D. Stiévenard, in *Stress and Strain in Epitaxy*, ed. by M. Hanbücken, J.-P. Deville (Elsevier, Amsterdam, 2001), pp. 243–286
46. T.-H. Kim, X.G. Zhang, D.M. Nicholson, B.M. Evans, N.S. Kulkarni, B. Radhakrishnan, E.A. Kenik, A.-P. Li, Large discrete resistance jump at grain boundary in copper nanowire. Nano Lett. **10**, 3096–3100 (2010)
47. A. Bietsch, B. Michel, Size and grain-boundary effects of a gold nanowire measured by conducting atomic force microscopy. Appl. Phys. Lett. **80**, 3346–3348 (2002)
48. Spreading Resistance Imaging, https://www.ntmdt-si.com/resources/spm-principles/atomic-force-microscopy/contact-afm/spreading-resistance-imaging
49. ResiScope II AFM Electrical characterization, https://www.csinstruments.eu/resiscope
50. NanoElectrical Lab, https://www.bruker.com/products/surface-and-dimensional-analysis/atomic-force-microscopes/nanoelectrical-lab.html
51. A. Schulze, T. Hantschel, A. Dathe, P. Eyben, X. Ke, W. Vandervorst, Electrical tomography using atomic force microscopy and its application towards carbon nanotube-based interconnects. Nanotechnology **23**, 305707 (2012)
52. K. Atamanuk, J. Luria, B.D. Huey, Direct AFM-based nanoscale mapping and tomography of open-circuit voltages for photovoltaics. Beilstein J. Nanotechnol. **9**, 1802–1808 (2018)
53. N. Nakagiri, T. Yamamoto, H. Sugimura, Y. Suzuki et al., Application of scanning capacitance microscopy to semiconductor devices (1997)
54. M.L. O'Malley, G.L. Timp, S.V. Moccio, J.P. Garno, R.N. Kleiman, Quantification of scanning capacitance microscopy imaging of the pn junction through electrical simulation. Appl. Phys. Lett. **74**, 272–274 (1999)
55. Scanning Capacitance Microscopy—SCM, https://blog.brukerafmprobes.com/guide-to-spm-and-afm-modes/scanning-capacitance-microscopy-scm/
56. W. Vandervorst, M. Meuris, Method for resistance measurements on a semiconductor element with controlled probe pressure. 5369372 (1994)
57. P. De Wolf, Two-dimension carrier profiling of semiconductor structures with nm resolution (1998)
58. W. Hänni, N. Blanc, R. Christoph, J. Burger, Chemical vapor deposition diamond for tips in nanoprobe experiments. J. Vac. Sci. Technol. A Vac. Surf. Films (1996)
59. P. De Wolf, T. Clarysse, W. Vandervorst, L. Hellemans, P. Niedermann, W. Hänni, Cross-sectional nano-spreading resistance profiling. J. Vac. Sci. Technol. B Microelectron. Nanometer Struct. Process. Meas. Phenom. **16**, 355–361 (1998)

60. M. Fouchier, P. Eyben, D. Alvarez, N. Duhayon, M. Xu, S. Brongersma, J. Lisoni, W. Vandervorst, in *Smart Sensors, Actuators, and MEMS*, vol. 5116 (International Society for Optics and Photonics, 2003), pp. 607–617
61. Bruker AFM Probes—SSRM-SMPL-P, https://www.brukerafmprobes.com/a-3835-ssrm-smpl-p.aspx
62. N.G. Orji, R.G. Dixson, 3D-AFM measurements for semiconductor structures and devices, in *Metrology and Diagnostic Techniques for Nanoelectronics*, vol. 109 (2017)
63. R. Courtland, Transistors could stop shrinking in 2021. IEEE Spectr. **53**, 9–11 (2016)
64. J.P. Pelz, R.H. Koch, Tip-related artifacts in scanning tunneling potentiometry. Phys. Rev. B: Condens. Matter **41**, 1212–1215 (1990)
65. D. Keller, Reconstruction of STM and AFM images distorted by finite-size tips. Surf. Sci. **253**, 353–364 (1991)
66. J.S. Villarrubia, Algorithms for scanned probe microscope image simulation, surface reconstruction, and tip estimation. J. Res. Natl. Inst. Stand. Technol. **102**, 425–454 (1997)
67. G.J. Germann, G.M. McClelland, Y. Mitsuda, M. Buck, H. Seki, Diamond force microscope tips fabricated by chemical vapor deposition. Rev. Sci. Instrum. **63**, 4053–4055 (1992)
68. M.C. Salvadori, W.W.R. Araújo, F.S. Teixeira, M. Cattani, A. Pasquarelli, E.M. Oks, I.G. Brown, Termination of diamond surfaces with hydrogen, oxygen and fluorine using a small, simple plasma gun. Diam. Relat. Mater. **19**, 324–328 (2010)
69. S. Miyake, Tribological improvements of polished chemically vapor deposited diamond films by fluorination. Appl. Phys. Lett. **65**, 1109–1111 (1994)
70. CEA Leti, http://www.leti-cea.com
71. D. Nečas, P. Klapetek, Gwyddion: an open-source software for SPM data analysis. Cent. Eur. J. Phys. **10**, 181–188 (2012)
72. Y. Wang, D.W. van der Weide, Microfabrication and application of high-aspect-ratio silicon tips. J. Vac. Sci. Technol. B Microelectron. Nanometer Struct. Process. Meas. Phenom. **23**, 1582–1584 (2005)
73. H. Dai, J.H. Hafner, A.G. Rinzler, D.T. Colbert, R.E. Smalley, Nanotubes as nanoprobes in scanning probe microscopy. Nature **384**, 147 (1996)
74. C.V. Nguyen, Q. Ye, M. Meyyappan, Carbon nanotube tips for scanning probe microscopy: fabrication and high aspect ratio nanometrology. Meas. Sci. Technol. **16**, 2138 (2005)
75. J. Foucher, P. Filippov, C. Penzkofer, B. Irmer, S.W. Schmidt, in *SPIE Advanced Lithography*, vol. 8681 (International Society for Optics and Photonics, 2013), pp. 86811I-86816
76. A.V. Sumant, D.S. Grierson, J.E. Gerbi, J. Birrell, U.D. Lanke, O. Auciello, J.A. Carlisle, R.W. Carpick, Toward the ultimate tribological interface: surface chemistry and nanotribology of ultrananocrystalline diamond. Adv. Mater. **17**, 1039–1045 (2005)
77. R. Schirhagl, K. Chang, M. Loretz, C.L. Degen, Nitrogen-vacancy centers in diamond: nanoscale sensors for physics and biology. Annu. Rev. Phys. Chem. **65**, 83–105 (2014)
78. C.L. Degen, Scanning magnetic field microscope with a diamond single-spin sensor. Appl. Phys. Lett. **92**, 243111 (2008)
79. J.-P. Tetienne, A. Lombard, D.A. Simpson, C. Ritchie, J. Lu, P. Mulvaney, L.C.L. Hollenberg, Scanning nanospin ensemble microscope for nanoscale magnetic and thermal imaging. Nano Lett. **16**, 326–333 (2016)
80. D. McCloskey, D. Fox, N. O'Hara, V. Usov, D. Scanlan, N. McEvoy, G.S. Duesberg, G.L.W. Cross, H.Z. Zhang, J.F. Donegan, Helium ion microscope generated nitrogen-vacancy centres in type Ib diamond. Appl. Phys. Lett. **104**, 031109 (2014)

Chapter 12
Scanning Microwave Impedance Microscopy (sMIM) in Electronic and Quantum Materials

Kurt A. Rubin, Yongliang Yang, Oskar Amster, David A. Scrymgeour and Shashank Misra

Abstract Scanning Microwave Impedance Microscopy (sMIM) is a sensitive electrical measurement technique which can characterize local static and temporal variations of electrical permittivity, and conductivity of materials and devices as well as for failure analysis. It is being used to characterize dielectrics, semiconductors and their doping response, and metals. Measurements can be made at room temperature down to cryogenic temperatures where quantum effects become important. Leveraging near-field electrical interactions between a probe and the sample, sMIM can measure and image electrical properties and operation at the nanoscale to micron scale by incorporation into an atomic force microscope. sMIM is being applied to a wide range of industrial and scientific applications to improve fundamental and functional understanding and operational performance of advanced, exploratory and quantum electronic devices and materials and their fabrication.

12.1 Introduction

Scanning Microwave Impedance Microscopy (sMIM) is an electrical mode applied at the micron to nano scale through integration with Scanning probe microscopy (SPM). The technique provides a direct measurement of local permittivity and conductivity of materials [1]. It is effective in measuring a wide range of material types, metallics [2], semiconductors [3, 4], insulators and dielectrics [5, 6]. sMIM can be implemented in ambient conditions but is also compatible in cryogenic and ultra-high magnetic field environments including UHV [7], making it a very useful technique for investigating electronic materials and exploring quantum materials.

K. A. Rubin (✉) · Y. Yang · O. Amster
PrimeNano, Inc., 4701 Patrick Henry Dr., Bldg. 8, Santa Clara, CA 95054, USA
e-mail: KurtARubin@gmail.com

D. A. Scrymgeour · S. Misra
Sandia National Laboratories, P.O. Box 5800, Albuquerque, NM 87185, USA

Present Address
K. A. Rubin · O. Amster
KLA, One Technology Drive, Milpitas, CA 95035, USA

U. Celano (ed.), *Electrical Atomic Force Microscopy for Nanoelectronics*,
NanoScience and Technology, https://doi.org/10.1007/978-3-030-15612-1_12

Imaging using microwave frequencies integrated with a SPM is a near-field technique which provides particular benefits for measuring electronic materials and investigating quantum phenomena with high lateral resolution. sMIM has a relatively long-range interaction that is particularly useful for measuring subsurface materials or those requiring encapsulation, are layered systems, or cannot be directly contacted without changing the state of the sample under investigation. It enables investigation of materials for quantum electronics having properties that manifest on small length scales and at extreme environmental conditions, very low temperature and high magnetic field. Quantum edge effects and conductivity vary from edge to center of films and very small length scales that would otherwise be difficult to resolve. The electronic information from microwave impedance microscopy can be measured at the sub-micron scale at which new properties emerge from collections of many electrons. Examples of these phenomena include phase-separation during metal-insulator transitions [8], local carrier concentration [1], phase transition materials [9], and ferroelectric response [10].

In this chapter we present the general working principal of sMIM, as well as modeling the sMIM response from dielectric and semiconductor materials, illustrating some practical applications as well as providing a survey of the state of the art related to electronic and quantum materials.

12.2 Theory of Operation—sMIM

12.2.1 Working Principal of sMIM

The development of scanning near-field microwave microscopy has a long history stretching back to the 1960s. Several comprehensive review articles and book chapters are available for the interested reader [11–13]. Generally, the entire field of near-field microwave microscopy can be divided between waveguide and tip-based approaches, the latter being of current interest in nanoscale characterization and therefore the focus of this chapter. As its name indicates, this is a near-field technique where the oscillator (tip) is brought close to the sample and probed with evanescent waves, thus breaking the classical Abbe's limit to resolution, which enables nanoscale measurement of feature sizes approximately 10^6 times smaller than the microwave wavelength. There are a multitude of different flavors of scanning probe-based near-field scanning microwave microscopy, with some common variants including scanning microwave microscopy (SMM) [14–16], microwave impedance microscopy (MIM) [17], near-field scanning microwave microscopy (NSMM) [18–20], and non-linear dielectric microscopy (NDSM) [21]. SMM is a related technique that also uses microwave interactions to probe electrical properties based on a network analyzer as the RF source. NDSM uses a lock-in technique at microwave frequencies with a resonator integrated near the probe tip assembly to improve sensitivity. This chapter will focus on one variant, scanning microwave impedance microscopy (sMIM). For

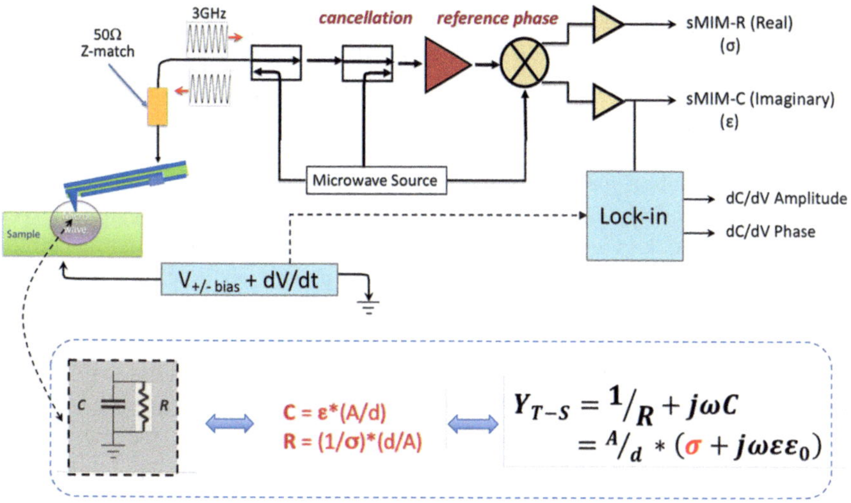

Fig. 12.1 Schematic of RF electronics and equivalent circuit of tip-sample, $Z_{tip\text{-}sample}$ lumped element model. The lumped element model can be expressed in terms of the system admittance showing the relation to permittivity and conductivity. In practice, the admittance depends upon the geometry of the tip and sample. The probe interface to the coaxial shielded probes create the probe-sample interface. The schematic includes a bias voltage sweep for measuring C-V responses and a lock-in to apply a modulated bias voltage for carrier modulation

simplicity, we will refer to the other varieties, aside from sMIM, as general near-field scanning microwave microscopy, keeping in mind that all implementations are examining the same basic physical phenomena—interaction of near-field microwaves with the sample.

In a sMIM measurement, microwaves are coupled through a custom shielded waveguide AFM cantilever to the probe, where they interact as evanescent waves with the portion of the sample immediately under the tip. Since the technique operates in the near-field, the spatial resolution of the microscope is determined by the size of the tip apex and not by diffraction that limits far-field type approaches. A fraction of the microwaves are reflected and the amplitude and phase (or equivalently, the real and imaginary parts) of the reflection are determined by the local electrical properties of the sample. For a linear sample the permittivity and conductivity determine the reflection, while for non-linear samples like doped semiconductors, the tip-bias-dependent-depletion-layer structure contributes significantly. By incorporating very sensitive detection electronics, lateral resolutions orders of magnitude smaller than the microwave wavelength can be achieved with good sensitivity in the aF range [22].

Figure 12.1 shows a schematic diagram of sMIM hardware that can be used for measuring nonlinear samples with a lock-in detection and with a bias voltage for measuring C-V response. A typical microwave system configuration includes an atomic force microscopes (AFM) integrated with the RF electronics which output to the AFM controller input channels, the probe interface module with matching electronics, and the shielded probe. The sMIM electronics couples a 3 GHz microwave signal to the AFM probe tip via a 50-ohm impedance interface module and a micro-fabricated transmission line in a custom-designed shielded metal SPM probe. The reflected microwave signal is amplified and demodulated by the sMIM electronics providing two analog signals representing the Real and Imaginary components of the complex reflection coefficient into the SPM controller analog to digital converter and it is synchronously captured with the topography signal during scanning. With the addition of a lock-in amplifier, sMIM can provide the dC/dV amplitude and phase signals correlating with the doping concentration and carrier type, respectively, of a doped semiconductor analogous to scanning capacitance microscopy (SCM). It is worth noting that the reflection has both an amplitude and phase. Alternatively, one can look at the in-phase reflection and the out-of-phase component of the reflection, often referred to as the Real and Imaginary part of the reflected signal.

Since the size of the probe is much smaller than the wavelength of the microwaves, the tip can be treated as a lumped element with tip admittance Z_{tip} and the electrical information of the sample is contained in the impedance between the tip and the ground Z_{tip}-sample. An equivalent circuit of the tip and the sample with ΔZ_{tip}-sample in parallel Z_{tip}. Two orthogonal sMIM outputs are proportional to the real and imaginary part of sample admittance $1/\Delta Z_{tip}$-sample. As the tip scans across the sample, encountering variations in dielectric constant ε or conductivity σ, the tip-sample interaction changes ΔZ_{tip}-sample, and this change is detected with the RF electronics.

12.2.2 Contrast Mechanism

Contrast arises from changes to the electrical interaction between the sample and probe tip, effectively modifying the tip-sample impedance. A microwave signal generator operating at GHz frequencies excites the tip which interacts with the sample. The reflected microwave signal is amplified and demodulated in quadrature to produce two output signals corresponding to the Real and Imaginary parts of the dielectric constant of the sample. A cancellation circuit enables a large dynamic range. The local permittivity and conductivity (ε, σ) can be derived from the real and imaginary parts of this impedance. The probe-sample interface is not a perfect impedance match to the input signal. The mis-match in impedance results in a reflected signal that is sensed by the RF electronics. The changes in the reflected wave represents the change in load as the probe is moved across the sample. The resulting perturbations captured in the reflected signal are a result of local variations of the sample impedance.

As the tip moves across a sample surface changes in the local material permittivity or conductivity are seen as perturbations of the probe impedance (the load). Provided the phase of the reflection is suitably calibrated, it can be shown that the Real part of the reflection is sensitive to the local conductivity and the Imaginary part of the reflection is sensitive to the permittivity. The signals can be collected as a map (image) of the local variations in the two images associated with the Re and Im output, σ and ε, respectively.

In a simplified way, an AFM probe in close proximity to a sample surface can be considered as a tiny leaky capacitor with an impedance. The reflection as illustrated in a typical sMIM image can be thought of as a small echo from a tiny parallel plate capacitor (probe-sample interface). The reflection depends on Z_{tip}, which depends on the capacitance, C, and the resistance, R, of this tiny parallel plate capacitor. As C and R change the reflection changes. So, as we move across the sample surface and build up an image of the reflection, we are imaging variations in C and R that depend on permittivity and conductivity.

12.2.3 Technique Benefits

If we consider the use of RF near-field from the measurement point of view, one can categorize the benefits of the technique into core responses: (1) single electrode measurement, (2) measure buried structures, (3) monotonic with the permittivity, and (4) monotonic with doping concentration.

12.2.4 Single Electrode Measurement

sMIM operates at RF frequencies high enough that the method can be thought of as a single electrode transport measurement. This single-electrode approach can be used with materials where it is difficult to introduce ground return path of the kind used for conductive AFM. An example of this approach is the Tselev liquid cell [23]. The configuration separates the probe, which is kept dry, from the liquid. It takes advantage of the ability of the microwaves to penetrate a very thin membrane so the sample can be imaged while eliminating complexities associated with probe-liquid interactions. The single-electrode approach can be helpful when applied to many material systems including ferroelectrics and energy materials. Another example, measuring buried charge, is discussed in the next section.

12.2.5 Measuring Buried Structures

Microwave impedance microscopy can leverage the advantages of long-range coupling to the sample as do other high frequency techniques. The measurements rely on the capacitive coupling to the sample, which requires no electrode to the sample, in comparison to lower frequency techniques, such as conductive AFM or scanning spread resistance microscopy, scanning kelvin probe microscopy, and scanning electrostatic force microscopy. This greatly reduces sample preparation complexity. In

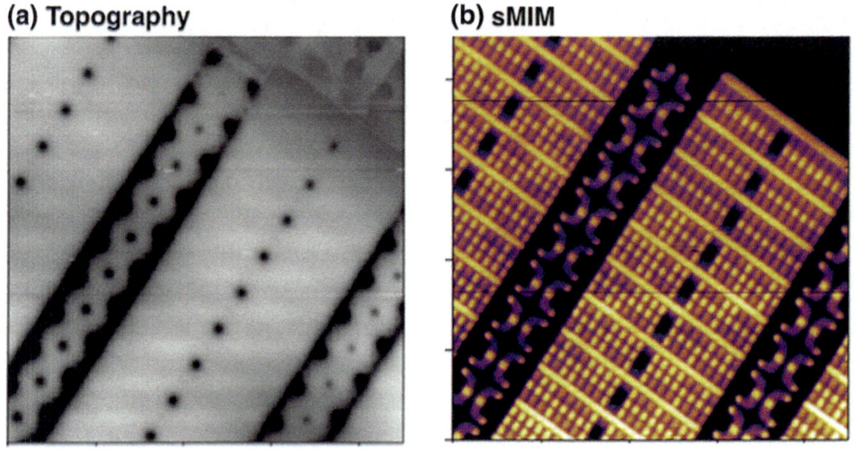

Fig. 12.2 **a** Topography of a backside polished FPGA non-volatile memory array. **b** sMIM image of stored charge imaged a specific location in the memory array under ~100 nm of Si. The measurement used only the probe tip as a single electrode with no bias or grounding required. Data collected by PrimeNano

conductive AFM, high tip-sample contact resistance can greatly compromise the signal and the measurement is difficult or not possible for a sample with native oxide. This problem is substantially decreased at microwave frequencies. sMIM can detect structures under the surface, which is important for studying buried circuits or a two-dimensional electron gas in a quantum well.

The ability to measure the presence or absence of charge is another attribute of sMIM which is useful for device physics, failure analysis, as well as basic studies. Figure 12.2 illustrates a FPGA that was previously programmed and then thinned from the backside. Polishing resulted in a shallow residual silicon thickness ramp ranging from 100s of nm's to 0 nm thickness. Figure 12.2b shows there was a strong sMIM signal response from stored charge buried under ~100 nm Si thickness. The figure indicates how it is possible to detect which memory cells had been programmed with charge, which enables connecting with the 1's and 0's of the programmed data. In contrast, Fig. 12.2a shows the topography of the same surface. The rather contrast-free topographic response above the buried charge illustrates how it was not possible to detect evidence of the buried stored charge by using topography.

Figure 12.3 shows imaging of a buried SiO_2 oxide pattern. The SiO_2 is readily observed through the 35 nm thick Si_3N_4 layer. Capacitive contrast imaging can be accomplished for structures buried even deeper. Generally, wider features can be imaged more deeply than narrower features. Figure 12.8 below discusses an additional example of measurement of buried structures, which may also be relevant towards future experimental quantum device fabrication [24].

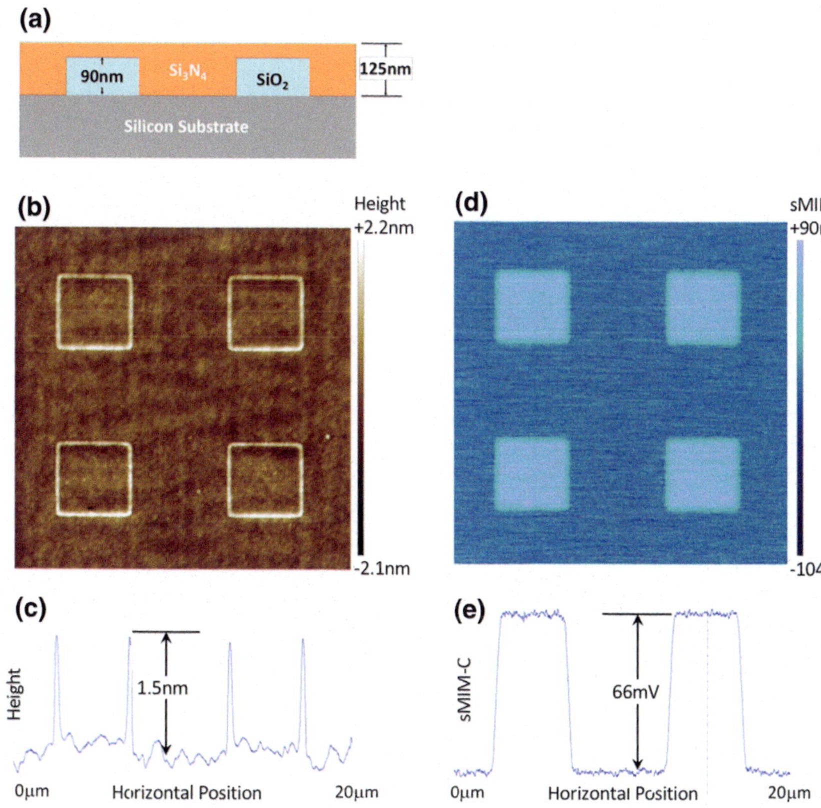

Fig. 12.3 **a** Drawing of a buried patterned oxide structure. The SiO_2 oxide surface of 90 nm tall SiO_2 pillars is buried 35 nm below the top of the 125 nm thick Si_3N_4 layer. **b** Topography of Si_3N_4 surface after planarization. The image was flattened. **c** Average height profile across buried SiO_2 features. A residual 1.5 nm feature remains at the edge of the SiO_2 region after planarization. The surface of the Si_3N_4 above the SiO_2 layer sits 0.15 nm taller than the surface of the Si_3N_4 layer between the SiO_2 regions. **d** sMIM-C signal from the identical region in (**b**). The sMIM-C signal measured with the tip retracted 500 nm away from the surface is subtracted from the in-contact sMIM-C signal to remove background and drift effects. The image was flattened and filtered with a 5-point median filter. **e** Average sMIM-C profile across the SiO_2 and Si_3N_4 regions. A strong 66 mV contrast sMIM-C signal is seen from the SiO_2 region buried 35 nm below the surface of the Si_3N_4. Data collected by PrimeNano

12.2.6 Monotonic with Permittivity

Figure 12.4 shows sMIM-C measured on various bulk dielectrics with dielectric constants from approximately 2–500. The measured values are shown as black circles, after subtracting for background, drift and normalizing the values with respect to the probe radius. The graph shows a clear linear relationship between sMIM-C and the log of the permittivity. The red squares shown in Fig. 12.4 are from a model

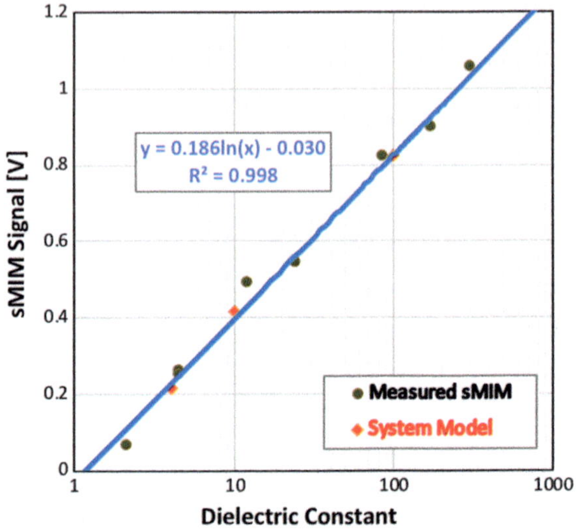

Fig. 12.4 Modeled admittance versus dielectric value (in red) linearly scaled and overlayed with experimentally measured sMIM signal versus dielectric value (in blue) for a set of bulk crystal dielectric samples. Data collected by PrimeNano

that originates with a finite-element calculation of the tip-sample admittance for the conical geometry of the sMIM probe. The origins of the log(ε) dependence can be seen in analytical models for spherically terminated conical tips above and in contact with linear materials.

12.2.7 Monotonic with Log Doping Concentration

For sMIM measurements on non-linear materials, such as doped semiconductors, the tip-sample bias influences the tip-sample impedance, or, more conveniently, the reciprocal of the tip-sample impedance, the tip-sample admittance, $Y_{\text{tip-sample}}$. As with linear samples, the sMIM signals are still proportional to the imaginary and real parts of $Y_{\text{tip-sample}}$, the capacitance and conductance below the tip-sample interface, but the capacitance and conductance now depend on the local permittivity and conductivity of the sample under the tip as well as on the geometry of the depletion layer. The depletion-layer geometry, in turn, depends on the tip-sample potential difference and on the doping level of the semiconductor. Analytical solutions exist for one dimensional geometries, and these can be used to model the results from macroscopic parallel-plate metal-oxide-semiconductor structures. A lumped-element approximation for an sMIM tip on an oxide-coated semiconductor and expressions from a delta depletion model for depletion-layer thickness can be used to examine trends.

Figure 12.5a shows the classic parallel plate model for describing a two-terminal metal-oxide-semiconductor (MOS) device. Since depletion-layer geometry has a strong impact on sMIM signals and the depletion-layer geometry varies with tip-sample voltage and with doping, varying the tip-sample voltage is a way to characterize semiconductor materials and devices, particularly the local doping level under

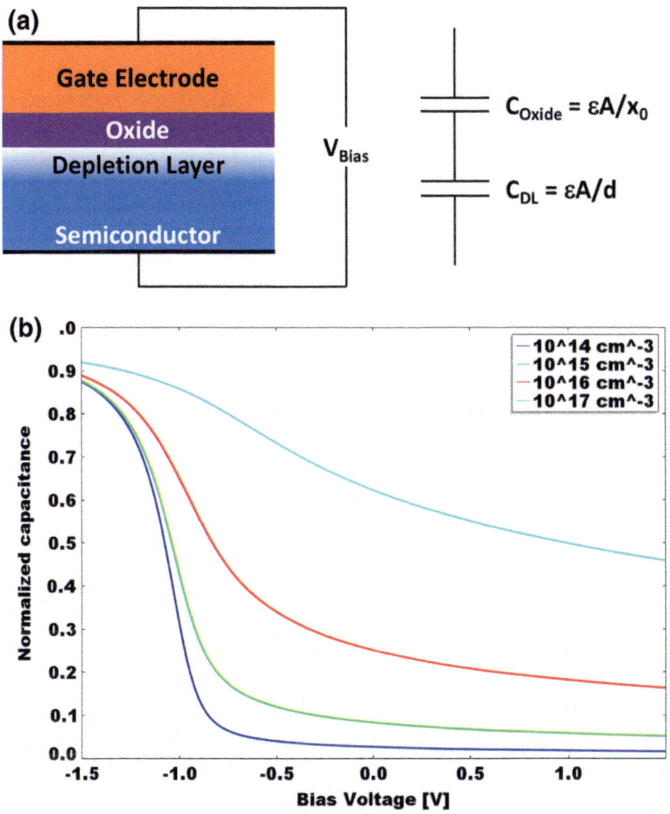

Fig. 12.5 **a** Schematic of the classical MOS capacitor configuration. The gate electrode is analogous to the sMIM probe contacting the sample surface. The MOS capacitor can be treated as two series capacitors, one with fixed capacitance from the oxide layer in series with a depletion layer capacitance that varies with applied voltage. **b** Capacitance-voltage curves from parallel-plate model simulated with COMSOL illustrates sMIM sensitivity to semiconductor doping level. The oxide thickness was 10 nm and the doping was *p*-type. The model shows the relationship of the capacitance (and analogously, the sMIM measurement) measurement on the depletion-layer thickness for various doping concentrations. Data collected by PrimeNano

the tip (or electrode in the case of patterned samples with electrodes present). This is similar to capacitance vs. voltage curves from macroscopic samples commonly used to characterize semiconductor materials and test structures. Figure 12.5b presents the classical MOS capacitor parallel plate model, here numerically generated for a range of doping concentration levels. But the 1-d delta-depletion and finite-element models are incomplete since they do not incorporate the geometries for AFM probe-sample interactions.

Similar to what was observed in the measurements of linear dielectrics shown in Fig. 12.4 above, where the sMIM signal is proportional to log(ε), experimental data measured with a sMIM AFM tip on doped semiconductors show sMIM sig-

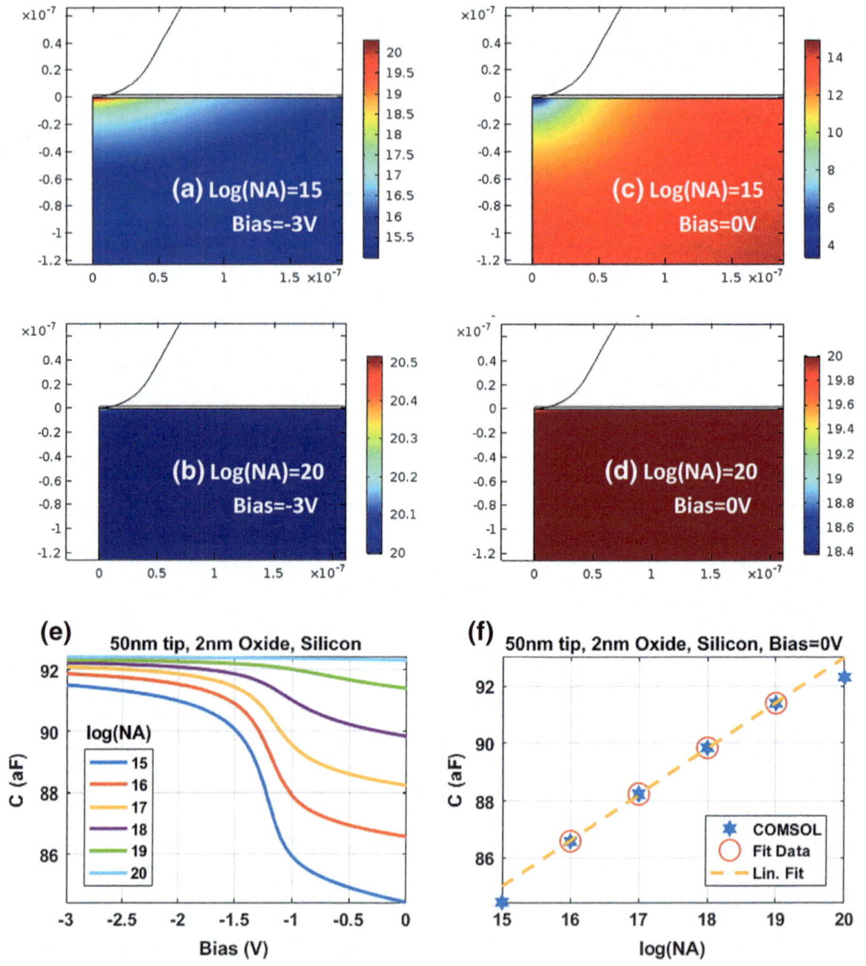

Fig. 12.6 **a–d** FEM predictions of the majority carrier hole density in the presence of marked biases on an sMIM probe for the marked p-type doping densities in silicon. Many more such simulations led to **e** FEM predictions of C-V curves with the doping specified by the legend. **f** Calibration of the probe tip's capacitance over the various doped samples as a function of their doping density. Data collected by PrimeNano

nals varying linearly with log([doping concentration]). To confirm the origins of the log([doping concentration]) behavior, finite-element modeling was used to assess the depletion-layer geometry for a conical tip, and how this geometry varies for both doping and applied gate (i.e. tip) voltage. Figure 12.6a shows results for one doping level.

Finite element modelling also allows calculation of the tip-sample capacitance for each doping level and gate voltage, resulting in C-V curves for the geometry of an sMIM probe on an oxide coated semiconductor as shown in Fig. 12.6e. Experimental data presented below in this article resemble the model results, suggesting that key critical physics are accounted for by the models. Figure 12.6f shows that the capacitance seen and measured by sMIM is linear in log doping over several orders of magnitude for doping of practical importance, enabling the possibility of calibrating sMIM results to invert for doping density.

12.3 Characterization of Nanomaterials with sMIM

12.3.1 Dielectric Materials

Some of the earliest work with near-field microwave microscopy was in the area of mapping dielectric constant variation within a sample. This included work to rapidly assess combinatorial material libraries showing compositional variation of dielectric constant and loss tangent in the doped $Ba_xSr_{1-x}TiO_3$ library [25], correlated scanning electron microscopy (SEM) and microwave microscopy of a the complex microstructures related to a diffusion couple of $CaTiO_3$ and $MgTiO_3$ detailing the variations in the local dielectric properties and the phase evolution during solidification [26], and variation of dielectric constant due to secondary phases in the $Ba(Mg_{1/3}Ta_{2/3})O_3$ and $Ba_2Ti_9O_{20}$ [27]. More recent tip-based work include the examination of both linear and nonlinear dielectric constants of $BaTiO_3$–$CoFe_2O_4$ (BTO–CFO) as a function of temperature and electric fields [28]. One of the first instances of near-field microwave imaging realized with scanning probe was by K. Lai et al. demonstrating contrast on semiconducting materials which will be discussed in the next section [29]. Proper calibration techniques utilizing sMIM or equivalent techniques allow dielectric constant mapping with well characterized systems/films and tips [30]. For example, high resolution dielectric characterization of mineral inclusions in rock was performed with sMIM where a calibrated dielectric constant map was correlated to elemental mapping collected with scanning electron microscopy coupled with energy dispersive X-ray (SEM/EDX) spectroscopy [5].

12.3.2 Semiconducting Materials

Due to significant commercial interest in understanding nanoscale dopant distribution in modern integrated circuits, near-field microscopy in general, and sMIM in particular, has proved to be a logical technique for examining doping concentrations. Efforts have been made to study semiconducting samples using scanning capacitance microscopy (SCM) [31], but the technique suffers from some limitations, including a necessary backplane electrode and a non-monotonic signal response which has limited the quantitative imaging of such structures [32, 33]. Near-field microwave measurements overcome both of these weaknesses. For example, Fig. 12.7a shows the monotonic signal response of sMIM as it is scanned over an epi-layer staircase

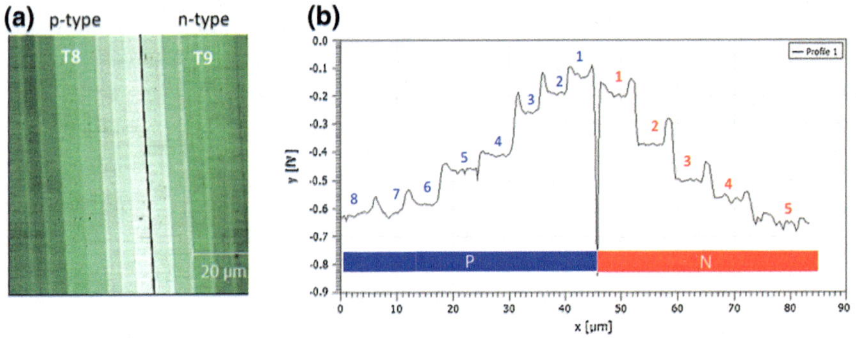

Fig. 12.7 Details of stair step sample **a** MIM-C image of IMEC staircase T8 & T9 bonded together face-to-face forming a continuous region with both p-type and n-type doping. The graph in **b** is the average profile of the doped steps. Data collected by PrimeNano

silicon wafer (IMEC, Belgium, part number T8_3 and T9) bonded together, face-to-face, to form a single region with both p-type and n-type doping. The graph in Fig. 12.7b is the average profile MIM values for the doping levels. Each step represents a different progressive doping level of the calibration reference.

The first probe near-field microwave examination of doped silicon was in 2003 utilizing an STM to image 10^{18} atoms/cm^3 boron doped silicon adjacent to undoped regions and showing clear sub-nanometer resolution [34]. In the development of the sMIM technique by K. Lai et al. imaging contrast was demonstrated between implanted n-type regions in background p doping [1]. Similar results were shown in related scanning probe based near-field microwave imaging of doped semiconductors demonstrating a clear monotonic response of dC/dV signals and a calibrated C-V curves on both *n* and *p* doped silicon with doping concentrations ranging from 10^{15} to 10^{20} atoms/cm^3 [3]. sMIM was used to map of doping concentrations in both n- and p-type staircase samples and commercial SRAM samples with both n- and p- channels, lightly doped drains, and p-epi and n-well regions [35]. They were also able to distinguish subtle regions of contrast attributed to thin dopant ions that penetrated through the protective layers during heavy implantation steps which was not seen in other studies. Several groups have published works focused on improving the accuracy of measuring doping concentrations [36, 37], as well as calibration of measured capacitance values showing ~0.1 fF resolution on test structures [3, 30, 38, 39]. Similar examination of variable dopant density staircase samples in GaAs with regions of Si n-doped GaAs with ranges from 3.6×10^{16} cm^{-3} to 5.5×10^{18} cm^{-3} could be distinguished with probe-based near-field microwave techniques, as well as the local examination of a p-n junction GaAs [40].

Recently, near-field microwave microscopy has been used to non-destructively examine buried, atomically-thin, highly phosphorous doped silicon nanostructures fabricated via the hydrogen resist lithography [41–43]. These so-called δ-layer nanostructures can be as small as a single phosphorous dopant up to features of several microns with dopant densities as high as ~10^{14} cm^{-2}, far in excess of the metal-

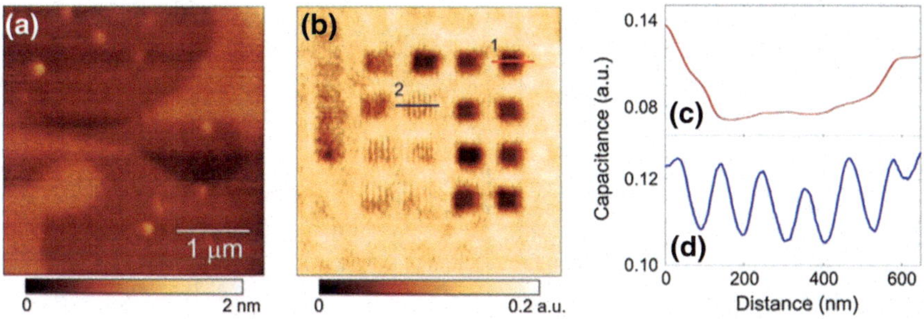

Fig. 12.8 **a** sMIM topography image and corresponding **b** sMIM capacitive image of buried resolution test structure in same orientation in (**a**). An averaged line cut across **c** line 1 (32 nm grating) showing no resolved lines and **d** line 2 (100 nm line spacing) showing resolved lines from (**b**). Reprinted from [44], 2017, with permission from Elsevier

insulator transition in Si [44]. Gramse et al. examined such structures with variable dopant density and distances from the silicon surface [43]. By examining a series of nanostructures patterned at different depths they were able to clearly distinguish 3 different doping density levels at both 4 and 9 nm from the surface, and through modeling and measurements to establish that as few as 10^3 atoms in a 25×25 nm^2 can be distinguished at 1 nm from the sample surface.

Scrymgeour et al. examined the resolution of the sMIM technique by creating and examining a series line pairs composed of ~10 nm wide δ-layer lines via hydrogen resist lithography [24]. The lines were arranged with variable pitches varying from 4 nm (fully depassivated surface) up to spacing as wide as 136 nm at a depth of approximately 30 nm below the surface. The sMIM capacitive image of the resolution tester is shown in Fig. 12.8b. The tip was capable of resolving individual lines at pitches greater than 88 nm. It was observed during wear studies that even after 28 h of continuous scanning which blunted the tip to a 1 μm plateau, it was possible to readily resolve 136 nm line pair, indicating that resolution of sMIM is ultimately more complicated than simply tip size.

The high resolution offered by scanning probe based near-field microwave microscopy is ideal for studying nanostructured semiconductors. Being able to apply biases directly to the tip helps to illuminate carrier dynamics within the nanowire that are ordinarily not observable. The p-n junction in a GaAs nanowire was examined by near-field microwave microscopy with dc tip biases ranging from 0 to 7 V which obviated the need for external electric connection to nanowire [45]. An apparent width of 1 μm was measured for the depletion region along with an observable shift in the width of the depletion region into the p-type region with increasing negative tip biases. Studies of a hexagonal GaN nanowire with near-field microwave microscopy combined with scanning gate microcopy show significant lateral variations in free carrier concentrations, and differences in electronic properties between nanowire facets and vertices attributed to band bending at the vertices [47].

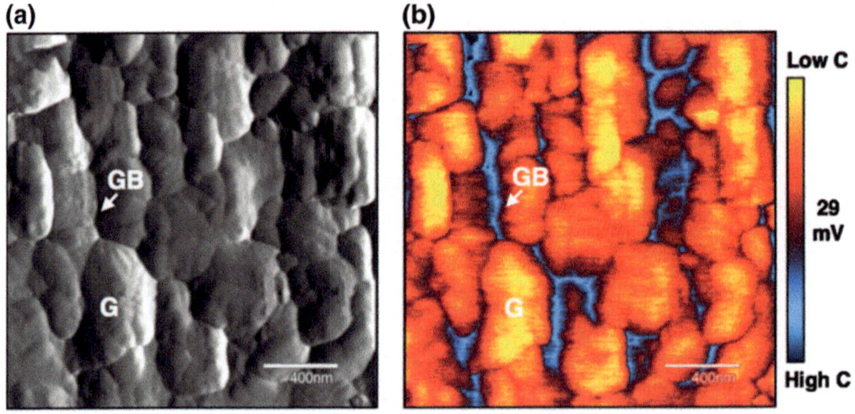

Fig. 12.9 **a** Deflection (derivative of topography) image and **b** corresponding capacitance image for an sMIM scan. For this measurement, the tip and the sample were both electrically grounded. The sample's rms roughness was 22 nm. Sample grain (G) and grain boundary (GB) positions have been marked. The scale bar is 400 nm. Reprinted from [48], with permission of AIP Publishing

Other semiconducting materials with variable electrical properties arising from material inhomogeneity such as grain boundaries or multiple phases have been studied with near-field microwave microscopy. One of the first studies was performed in an operational photovoltaic Cu(In,Ga)Se_2 (CIGS) sample looking at the material in both the dark and light state, revealing state-dependent contrast associated with light-generated charge reducing the depleted regions of the material [47]. Recently, as shown in Fig. 12.9, sMIM was used to image the surfaces of photovoltaic CdTe films annealed with $CdCl_2$ to show grain boundary carrier depletion that was not present for unannealed bulk films [48]. Similarly, nanoribbons of In_2Se_3 were shown to be inhomogeneous due to the presence of multiple local phases with resistivity values of over 6 orders of magnitude variation [49].

12.3.3 2D Materials

There is a lot of excitement behind 2D materials related to their novel mechanical, transport, and electrical behaviors which can be vastly different from their 3D varieties [50, 51]. Scanning probe based near-field microwave techniques are well suited for investigating 2D materials and their novel electronic properties. Of special interest to sMIM is the direct study of their electrical properties which can be profoundly influenced by inadvertent modification through either oxidation or damage associated with electrode fabrication. Near-field scanning microscopy techniques and sMIM in particular are ideal for studying these materials because they can per-

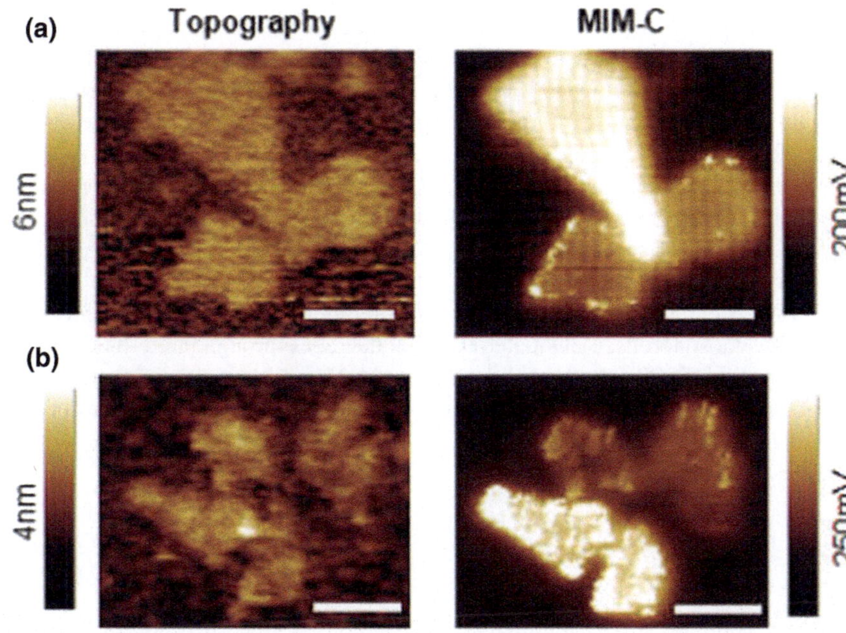

Fig. 12.10 Electrical connection between overlapped graphene sheets imaged with microwave microscopy (**a**, **b**). Examples of topographic and MIM images, respectively, of graphene sheet. The MIM-C images clearly display the electrically connected pieces as the same color. All scale bars are 200 nm. Reprinted with permission from [52]. Copyright 2009 American Chemical Society

form direct electrical characterization without the need of external electrodes or other sample preparations.

One of the first studies was of both exfoliated pristine and chemically derived graphene with conductivities varying over several orders of magnitude [52]. A strong contrast was observed in the microwave capacitive image related to sample conductivity, with lower quality higher defect samples have low and inhomogeneous response while the pristine graphene had strong and uniform response proportional to examined graphene area. The microwave imaging was also able to verify the electrical contact between overlapping graphene pieces as shown in Fig. 12.10.

Exfoliated nanoribbons of Bi_2Se_3 examined by microwave impedance microscopy shows that the layers only become highly conductive at thicknesses greater than three quintuple layers (two Bi and three Se atomic sheets bonded together) [53]. A thickness-dependent dielectric constant in nanoflakes of In_2Se_3 was directly measured with sMIM and confirmed against similar literature results measured with parallel plate capacitors geometry showing the strength of minimal sample preparation before measurement [54].

Mesoscale imperfections in vapor transport grown MoS_2 atomic layers where studied by sMIM, which was able to provide direct nanoscale characterization of the dendritic ad-layers and grain boundaries defects [55]. More recent work examined

MoS_2 and doped WSe_2 by near-field scanning microwave microscopy in combination with applied tip biases to explore electronic structure and local effects of surface contaminants [57]. Scanning microwave impedance microscopy has been used to study the thermal oxidation of WSe_2 nanosheets [57]. In near-field microwave measurements correlated to local Raman measurements, the authors were able to show that thermal oxidation starts at the edge of the sheets and propagates laterally toward the middle. The technique has sufficient resolution to indicate non-uniform conductivities with the nanoribbons attributed to partial oxidation to room temperature conductive sub-stoichiometric WO_{3-x}.

Finally, recent work has been performed studying the stability in air of encapsulated black phosphorous, a 2D material which degrades through photo-oxidation to phosphoric acid [58, 59]. Both works show via sMIM and other near-field microwave techniques, the time dependence of the electrical properties related to their decomposition in air and the effectiveness of different encapsulation methods [58, 59].

12.3.4 Quantum Materials

Because scanning microwave impedance microscopy (sMIM) is useful in mapping the spatial dependence of metallic behavior, it has also proven useful in the study of quantum materials, which can have spatially distinct metallic and non-metallic regions. Quantum materials are defined by the emergence of electronic and magnetic phases that are not perturbative extensions of the behavior found in simple metals, insulators, and magnets [60, 61].

Reductively, quantum materials come in two flavors [62]. The first of these contains some long-range order characterized by a local order parameter [63] (Fig. 12.11). Tuning a material parameter, like doping or strain, can induce a quantum phase transition into/out of the ordered phase [64]. Real samples often have extrinsic and intrinsic inhomogeneities, resulting in the *local* value of the tuning parameter being either below or above the critical value. This can result in different parts of the sample locally having a zero or a non-zero order parameter. Should the ordered phase have a different metallicity than the other phases in the phase diagram, which is common, a spatial map of the microwave response can be used to reveal whether a given region of the sample is ordered [65].

The second flavor of quantum materials contains a non-local kind of symmetry in the quantum wave function, resulting in so-called topological order [66]. The boundary between a topological material and a trivial material (including vacuum) supports protected in-gap states (Fig. 12.12). These in-gap states are protected from local sources of noise or dissipation by the topological order, and have been the subject of tremendous interest for application to low-dissipation transport [67] and low-decoherence quantum information [68]. In many interesting cases, both the topological state and the boundary are exposed near the surface, as is the case with a 2D insulator, with the 1D boundary state being metallic, or a 3D insulator confined to a 1D wire, with 0D in-gap states at the ends of the wire.

The spatial dependence of metallic response can reveal important details about both the nature of the two domains in a quantum material near a phase transition, or the mechanism driving a quantum phase transition. Patchy metallic and insulating islands can suggest a dopant inhomogeneity-driven metal-to-insulator transition [69, 70]. In cases where the quantum phase transition is driven by strain, these domains will instead be quasi-periodic 1D stripes. Data near a phase transition has also been used to reveal the nature of the phases themselves. For example, the enhanced conductivity between domains in manganites have been used to intuit the charge ordering on both sides of a ferroelectric/anti-ferroelectric phase transition [8, 9, 71].

A particularly compelling demonstration of the power of near-field microwave imaging in understanding phase transitions both in and out of equilibrium has been accomplished for the metal-to-insulator transition in VO_2 [72, 73]. Vanadium dioxide has a well-known structural phase transition between a high symmetry tetragonal phase and a broken symmetry monoclinic phase below 68 °C (Fig. 12.13a). The low-temperature phase is a strongly correlated Mott insulator (Fig. 12.13c), while the high temperature phase is a correlated metal. Spatial maps of the amplitude of the reflected microwave signal show an increase in metallic stripe-like metallic features when warming towards the transition out of the insulating phase, eventually giving way to a grainy metallic phase above the transition (Fig. 12.13b). The 1D quasiperiodic nature of the pattern is stereotypical of a strain-driven inhomogeneity near a phase transition.

As quantum materials attract greater attention for applications, there is a growing need to examine what happens to quantum phases in conditions that approximate device operation, which are well suited to scanning microwave techniques. In contrast, these conditions are often a challenge for traditional material characterization tools, which need large (homogeneous) samples or long averaging times to get good signal-to-noise ratios. Figure 12.13d, e show a simple vanadium dioxide device that

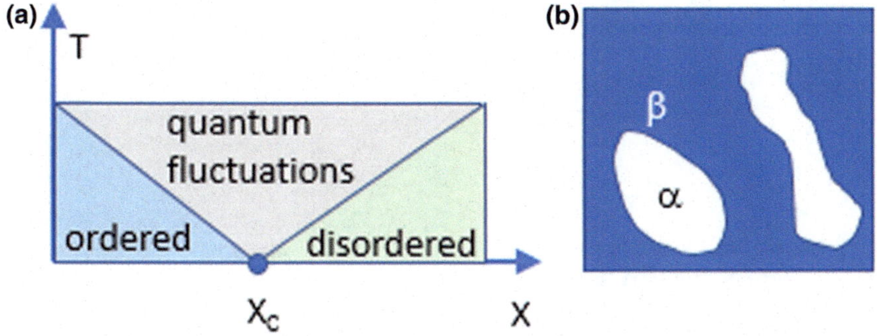

Fig. 12.11 **a** Cartoon illustration of a quantum phase transition with tuning parameter X and critical point X_C. **b** Real space map of a slightly inhomogeneous sample, where different regions α and β have different local values of X, meaning some regions of the sample are in the ordered phase and others are not. A simple instantiation would be X being doping, the ordered phase being a correlated insulator, and the disordered phase being metallic

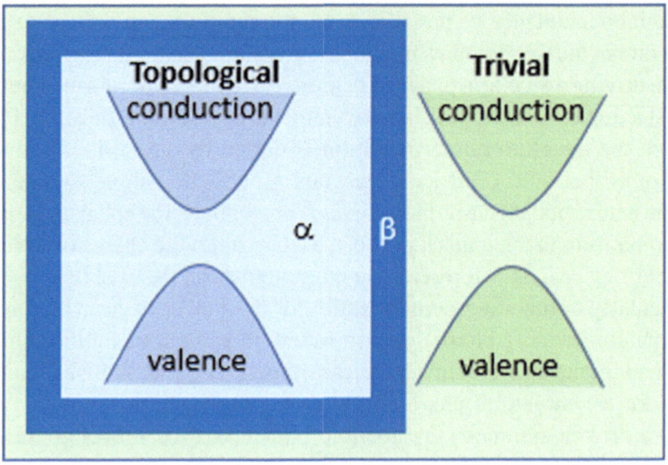

Fig. 12.12 A topologically interesting gapped 2D material **a** surrounded by a topologically trivial gapped material. The boundary **b** supports metallic in-gap states

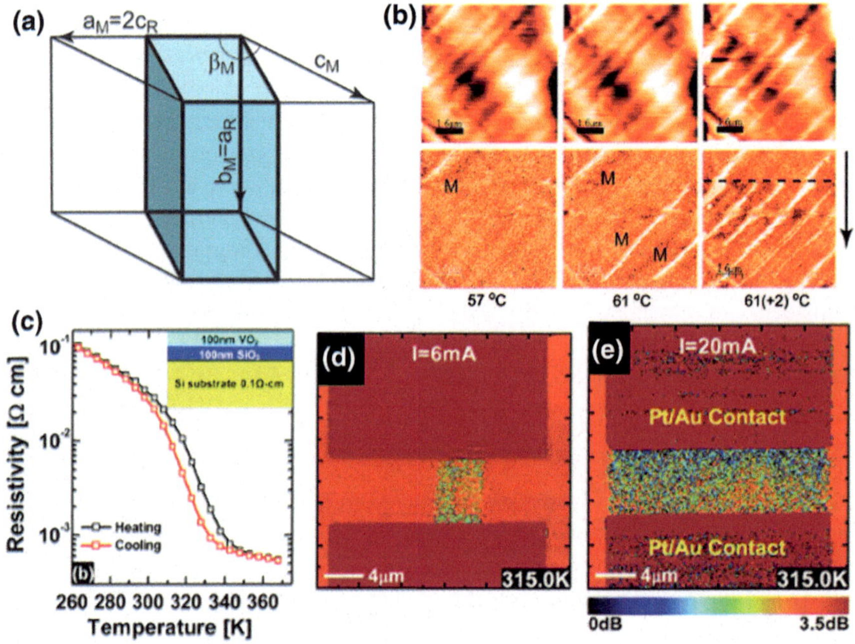

Fig. 12.13 **a** Low temperature monoclinic (light solid lines) and high-temperature tetragonal (thick solid lines) arrangement of V atoms in vanadium dioxide. **b** AFM (top) and reflected microwave power (bottom) images of vanadium dioxide platelet at 330 K, 334 K, and 334/336 K. **c** Temperature dependence of DC resistivity through the metal-insulator transition. **d**, **e** Reflected microwave power of a vanadium dioxide film between source and drain contacts, with the current passed through the contacts is indicated at the top of the panel. Reprinted with permission from [73] for (**a**, **b**) and [72] for (**c**, **d**, **e**). Copyright 2010 and 2013 American Chemical Society.

uses an electric field to drive the transition to the metallic phase at a temperature slightly on the insulating side of the transition (Fig. 12.13c). The simultaneously acquired microwave reflectance shows that the current is not carried uniformly by the vanadium dioxide channel. Rather, a metallic filament forms to carry the current, and the filament expands to occupy the entirety of the channel as the drive current is increased.

Scanning microwave techniques have also recently been used to visualize the 1D metallic states at the edge of topologically protected 2D systems. Conceptually, the simplest to understand are the results on the integer quantum Hall state of two-dimensional electron gases [74, 75]. Figure 12.14a shows data from an etch-defined island in a shallow GaAs/AlGaAs heterostructure at low temperature and a magnetic field where we expect both bulk and edge state conduction. The imaginary part of the reflectance shows both a moderately conducting bulk and a highly conductive 1D edge surrounding it. The real part of the conductance indicates that there is an insulating strip that separates the two. Comparing the magnetic field dependence of these with the resistivity reinforces this interpretation. When the resistivity goes to zero, the bulk states are gapped out, and conductivity is only through the metallic edge states. Indeed, concomitantly, the position-dependent microwave signal drops in the bulk, while the signal remains strong at the edge.

More broadly, local microwave techniques have been used to not only confirm expected characteristics of the quantum Hall state, but also reveal surprising details about the nature of the edge states in graphene [76] and 2D topological insulators [77]. Counterintuitively, spatial signatures of the edge state of graphene were not coincident with observation of quantum Hall resistivity plateaus as a function of applied magnetic field. In the 2D topological insulator, zero-field edge states were only seen in samples beyond a critical thickness, in agreement with electrical transport. Counter to intuition, these edge states persisted to an applied field of 9T, although the origin of this behavior remains unknown.

12.4 Summary and Conclusion

As time marches on, the variety of electrical devices continues to proliferate. Smaller feature size and devices, new materials, new process techniques, higher analog and digital frequencies, lower and higher power and lower and higher temperature all share a common need for improved electrical characterization capability and the attributes of sMIM are anticipated to support that trend. sMIM is a flexible electrical measurement technique that is being applied to an ever-broadening scope of applications. It directly senses electrical properties that are relevant to devices that operate at RF and microwave frequencies. Being a RF technique, it is used to measure both conductive and insulating samples in both bulk and thin film form and is not impeded by the presence of thin insulating layers on surfaces. It can examine local variations of electrical properties both in the quiescent state as well as when the material or device is excited such as by electrical or optical stimulation. It can characterize large sized

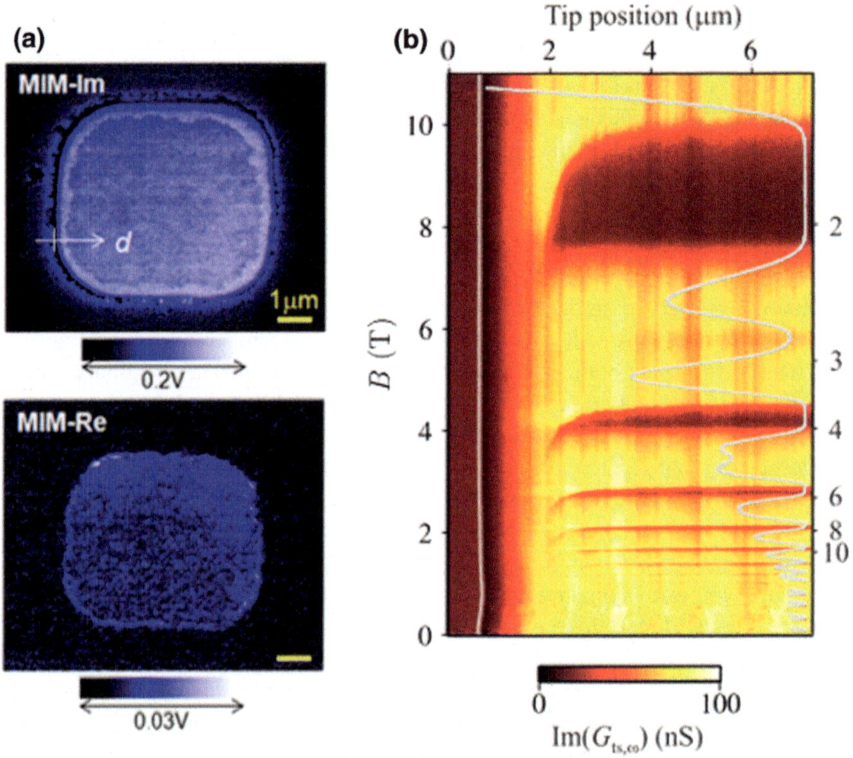

Fig. 12.14 **a** SMIM image of a shallow 2DEG at non-integer filling factor. **b** Field and linear distance dependence the imaginary part of the reflected signal, showing correspondence of the SMIM and resistivity signals. Reprinted (**a**) with permission from [74]. Copyright 2011 by the American Physical Society. http://dx.doi.org/10.1103/PhysRevLett.107.176809 (**b**) is reprinted from [75] under the Creative Commons Attribution-NonCommercial-ShareAlike 3.0 license

features as well as to reach down to nanoscale dimensions. Being probe-based, it can directly image position-dependent variations of electrical states. It directly measures device response as well as reach into buried devices through plugs or interconnects. Since it can operate at cryogenic temperatures, it is well suited for characterizing materials and devices that may be relevant for quantum computing.

sMIM continues to evolve and be applied to a growing range of materials and applications. An area of development where Near-Field Microwave Microscopy will be valuable is in the quantification of dielectrics and doping concentration. As has been discussed previously in this chapter sMIM has a monotonic response with dielectric value and doping concentration. With further development of the modeling and refinements in the experimental technology, there are promising developments to make such measurements routine. Some hurdles that will need to be overcome to achieve a routine implementation for quantitative measurements are better understanding of the effects of the sample surface conditions, a method of calibration that addresses the probe tip radius and changing probe conditions, and the maturing of models that can better predict a systems response and be used to guide the conversion of the sMIM response to the corresponding physical unit of interest.

Near-field microwave microscopy will be combined with other techniques utilizing different frequencies of light to further the understanding of material's behavior. By integrating different wavelengths of light a large dynamic range of signals is available to measure the system response. Overlaying a range of measurements will allow the study of local regions without the requirement to change setup and find the same region to measure with different techniques. By further combining higher frequency with low frequency or DC methods novel behavior may be discovered. Finally, all electronic and charge transport mechanisms in real materials can exhibit variable temporal response. Teasing out these dependencies with respect to time can help guide making improvements to materials. For example it could help guide improving ionic conductors and photovoltaic cells by modifying grain boundary properties which trap and hinder charge flow. Coupling sMIM and microwave imaging to electrical pump-probe type experiments and studying the temporal movement of charge and charge conductors through materials offer another powerful technique to understand nanoelectronic materials.

The application of scanning microwave techniques to the field of quantum materials holds promise. As researchers advance these materials to applications, which will certainly involve integrating the material into some mesoscopic heterogeneous device, scanning microwave techniques will be uniquely useful for characterization in integrated structures. Simply knowing that the electronic phase of the material is the same expected from bulk samples is a key piece of metrology. The only intrinsic limitations of the technique in this regard is the limitation identifying only metal-to-insulator transitions, and only 2D topological materials (not possible to see the bulk in a 3D topological material). Obviously, the broad application to a wider variety of quantum materials than has been studied to date is also likely.

Acknowledgements & Attributions PrimeNano authors KR, YY, and OA contributed Section 2, Sandia authors DS and SM contributed Section 3, and all authors contributed to Sections 1 and 4. Sandia National Laboratories is a multimission laboratory managed and operated by National Technology & Engineering Solutions of Sandia, LLC., a wholly owned subsidiary of Honeywell International, Inc., for the U.S. Department of Energy's National Nuclear Security Administration under contract DE-NA0003525.

References

1. K. Lai et al., Modeling and characterization of a cantilever-based near-field scanning microwave impedance microscope. Rev. Sci. Instrum. **79**(6) (2008)
2. Z. Y. Wang et al., Quantitative measurement of sheet resistance by evanescent microwave probe. Appl. Phys. Lett. **86**(15) (2005)
3. H. P. Huber et al., Calibrated nanoscale dopant profiling using a scanning microwave microscope. J. Appl. Phys. **111**(1), 014301 (2012)

4. O. Amster et al., Practical quantitative scanning microwave impedance microscopy of semiconductor devices. in *2017 IEEE 24th International Symposium on the Physical and Failure Analysis of Integrated Circuits* (2017)
5. T. Monti et al., High-resolution dielectric characterization of minerals: a step towards understanding the basic interactions between microwaves and rocks. Int. J. Miner. Process. **151**, 8–21 (2016)
6. Z. Y. Wang et al., Evanescent microwave probe measurement of low-*k* dielectric films. J. Appl. Phys. **92**(2), 808–811 (2002)
7. Y. T. Cui et al., Unconventional correlation between quantum hall transport quantization and bulk state filling in gated graphene devices. Phys. Rev. Lett. **117**(18) (2016)
8. K. J. Lai et al., Mesoscopic percolating resistance network in a strained manganite thin film. Science **329**(5988), 190–193 (2010)
9. E. Y. Ma et al., Charge-order domain walls with enhanced conductivity in a layered manganite. Nat. Commun. **6**, 7595 (2015)
10. S. R. Johnston et al., Measurement of surface acoustic wave resonances in ferroelectric domains by microwave microscopy. J. Appl. Phys. **122**(7) (2017)
11. S. M. Anlage et al., Near-Field Microwave Microscopy of Materials Properties, *in Microwave Superconductivity,* eds. by H. Weinstock, M. Nisenoff (Springer Netherlands, Dordrecht, 2001), pp. 239–269
12. A. Imtiaz, T. M. Wallace, P. Kabos, Near-field scanning microwave microscopy. IEEE Microwave Mag. **15**(1), 52–64 (2014)
13. B. T. Rosner, D. W. van der Weide, High-frequency near-field microscopy. Rev. Sci. Instrum. **73**(7), 2505–2525 (2002)
14. A. Imtiaz et al., Nanometer-scale material contrast imaging with a near-field microwave microscope. Appl. Phys. Lett. **90**(14), 143106 (2007)
15. Q. Zhang, P. J. McGinn, Imaging of oxide dielectrics by near-field microwave microscopy. J. Eur. Ceram. Soc. **25**(4), 407–416 (2005)
16. J. Smoliner et al., Scanning microwave microscopy/spectroscopy on metal-oxide-semiconductor systems. J. Appl. Phys. **108**(6), 064315 (2010)
17. K. Lai et al., Atomic-force-microscope-compatible near-field scanning microwave microscope with separated excitation and sensing probes. Rev. Sci. Instrum. **78**(6) (2007)
18. D. E. Steinhauer et al., Quantitative imaging of dielectric permittivity and tunability with a near-field scanning microwave microscope. Rev. Sci. Instrum. **71**(7), 2751–2758 (2000)
19. A. N. Reznik, V. V. Talanov, Quantitative model for near-field scanning microwave microscopy: application to metrology of thin film dielectrics. Rev. Sci. Instrum. **79**(11), 113708 (2008)
20. A. P. Gregory et al., Measurement of the permittivity and loss of high-loss a near-field scanning microwave microscope. Ultramicroscopy **161**, 137–145 (2016)
21. Y. Cho, Scanning nonlinear dielectric microscopy. J. Mater. Res. **26**(16), 2007–2016 (2011)
22. H.P. Huber et al., Calibrated nanoscale capacitance measurements using a scanning microwave microscope. Rev. Sci. Instrum. **81**(11) (2010)
23. C. Gao et al., Quantitative scanning evanescent microwave microscopy and its applications in characterization of functional materials libraries. Meas. Sci. Technol. **16**(1), 248–260 (2005)
24. Q. Zhang, P. J. McGinn, Characterization of Calcium Titanate-Magnesium Titanate Eutectic by Scanning Microwave Microscopy. J. Am. Ceram. Soc. **89**(12), 3817–3823 (2006)
25. H. F. Cheng, Y. C. Chen, L. N. Lin, Evanescent microwave probe study on dielectric properties of materials. J. Eur. Ceram. Soc. **26**(10–11), 1801–1805 (2006)
26. Y. Qi et al., Local dielectric measurements of $BaTiO_3$–$CoFe_2O_4$ nanocomposites through microwave microscopy. J. Mater. Res. **22**(5), 1193–1199 (2011)
27. K. Lai et al., Tapping mode microwave impedance microscopy. Rev. Sci. Instrum. **80**(4) (2009)
28. T. Dargent et al., An interferometric scanning microwave microscope and calibration method for sub-fF microwave measurements. Rev. Sci. Instrum. **84**(12), 123705 (2013)
29. J.S. McMurray, J. Kim, C. C. Williams, Quantitative measurement of two-dimensional dopant profile by cross-sectional scanning capacitance microscopy. J. Vac. Sci. Technol. B: Microelectron. Nanometer Struct. Process., Measur., and Phenom. **15**(4), 1011–1014 (1997)

30. J.J. Kopanski et al., Towards reproducible scanning capacitance microscope image interpretation. J. Vac. Sci. Technol. B: Microelectron. Nanometer Struct. **22**(1), 399 (2004)
31. R. Stephenson et al., Nonmonotonic behavior of the scanning capacitance microscope for large dynamic range samples. J. Vac. Sci. Technol. B: Microelectron. Nanometer Struct. **18**(1), 405 (2000)
32. A. Imtiaz, S. M. Anlage, A novel STM-assisted microwave microscope with capacitance and loss imaging capability. Ultramicroscopy **94**(3–4), 209–216 (2003)
33. W. Kundhikanjana et al., Unexpected surface implanted layer in static random access memory devices observed by microwave impedance microscope. Semicond. Sci. Technol. **28**(2) (2013)
34. G. Gramse et al., Calibrated complex impedance and permittivity measurements with scanning microwave microscopy. Nanotechnology. **25**(14) (2014)
35. E. Brinciotti et al., Probing resistivity and doping concentration of semiconductors at the nanoscale using scanning microwave microscopy. Nanoscale **7**(35), 14715–14722 (2015)
36. M. Kasper et al., Metal-oxide-semiconductor capacitors and Schottky diodes studied with scanning microwave microscopy at 18 GHz. J. Appl. Phys. **116**(18), 184301 (2014)
37. F. Wang et al., Quantitative impedance characterization of sub-10 nm scale capacitors and tunnel junctions with an interferometric scanning microwave microscope. Nanotechnology **25**(40), 405703 (2014)
38. A. Buchter et al., Scanning microwave microscopy applied to semiconducting GaAs structures. Rev. Sci. Instrum. **89**(2) (2018)
39. D. P. Adams, T. M. Mayer, B. S. Swartzentruber, Nanometer-scale lithography on Si(001) using adsorbed H as an atomic layer resist. J. Vac. Sci. Technol., B **14**(3), 1642–1649 (1996)
40. J. B. Ballard et al., Multimode hydrogen depassivation lithography A method for optimizing atomically precise write times. J. Vac. Sci. Technol. B, **31**(6) (2013)
41. M. Rudolph et al., Probing the limits of Si:P delta-doped devices patterned by a scanning tunneling microscope in a field-emission mode. Appl. Phys. Lett. **105**(16) (2014)
42. T. C. Shen et al., Ultradense phosphorous delta layers grown into silicon from PH_3 molecular precursors. Appl. Phys. Lett. **80**(9), 1580–1582 (2002)
43. G. Gramse et al., Nondestructive imaging of atomically thin nanostructures buried in silicon. Sci. Adv. **3**(6) (2017)
44. D. A. Scrymgeour et al., Determining the resolution of scanning microwave impedance microscopy using atomic-precision buried donor structures. Appl. Surf. Sci. **423**, 1097–1102 (2017)
45. A. Imtiaz et al., Imaging the p-n junction in a gallium nitride nanowire with a scanning microwave microscope. Appl. Phys. Lett. **104**(26), 263107 (2014)
46. S. Berweger et al., Near-field microwave microscopy of one-dimensional nanostructures, in *2016 IEEE Mtt-S International Microwave Symposium* (2016)
47. J. C. Weber et al., A near-field scanning microwave microscope for characterization of inhomogeneous photovoltaics. Rev. Sci. Instrum. **83**(8), 083702 (2012)
48. M. Tuteja et al., Direct observation of electrical properties of grain boundaries in sputter-deposited CdTe using scan-probe microwave reflectivity based capacitance measurements. Appl. Phys. Lett. **107**(14) (2015)
49. K. J. Lai et al., Nanoscale electronic inhomogeneity in In_2Se_3 nanoribbons revealed by microwave impedance microscopy. Nano Lett. **9**(3), 1265–1269 (2009)
50. S. Z. Butler et al., Progress, challenges, and opportunities in two-dimensional materials beyond graphene. ACS Nano **7**(4), 2898–2926 (2013)
51. C. Ashworth, 2D materials: the thick and the thin. Nat. Rev. Mater. **3**, 18019 (2018)
52. W. Kundhikanjana et al., Hierarchy of electronic properties of chemically derived and pristine graphene probed by microwave imaging. Nano Lett. **9**(11), 3762–3765 (2009)
53. S. S. Hong et al., Ultrathin topological insulator Bi_2Se_3 nanoribbons exfoliated by atomic force microscopy. Nano Lett. **10**(8), 3118–3122 (2010)
54. D. Wu et al., Thickness-dependent dielectric constant of few-layer In_2Se_3 nanoflakes. Nano Lett. **15**(12), 8136–8140 (2015)

55. Y. Liu et al., Mesoscale imperfections in MoS_2 atomic layers grown by a vapor transport technique. Nano Lett. **14**(8), 4682–4686 (2014)
56. S. Berweger et al., Microwave near-field imaging of two-dimensional semiconductors. Nano Lett. **15**(2), 1122–1127 (2015)
57. Y. Liu et al., Thermal oxidation of WSe_2 nanosheets adhered on SiO_2/Si substrates. Nano Lett. **15**(8), 4979–4984 (2015)
58. J.S. Kim et al., Toward air-stable multilayer phosphorene thin-films and transistors. Sci. Rep. **5**, 8989 (2015)
59. P. J de Visser et al., Spatial conductivity mapping of unprotected and capped black phosphorus using microwave microscopy. 2D Mater. **3**(2), 021002 (2016)
60. B. Keimer, J. E. Moore, The physics of quantum materials. Nat. Phys. **13**(11), 1045–1055 (2017)
61. Y. Tokura, M. Kawasaki, N. Nagaosa, Emergent functions of quantum materials (vol 13, pg 1056, 2017). Nat. Phys. **13**(11), 1069–1069 (2017)
62. The rise of quantum materials. Nat. Phys. **12**(2), 105–105 (2016)
63. L. D. Landau, E. M. Lifshitz, L. P. Pitaevskiĭ, *Statistical physics. Course of theoretical physics* (Pergamon Press, Oxford; New York, 1980)
64. S. L. Sondhi et al., Continuous quantum phase transitions. Rev. Mod. Phys. **69**(1), 315–333 (1997)
65. K. A. Moler, Imaging quantum materials. Nat. Mater. **16**(11), 1049–1052 (2017)
66. C. K. Chiu et al., Classification of topological quantum matter with symmetries. Rev. Mod. Phys. **88**(3) (2016)
67. Y. F. Ren, Z. H. Qiao, Q. Niu, Topological phases in two-dimensional materials: a review. Rep. Prog. Phys. **79**(6) (2016)
68. A. Stern, N. H. Lindner, Topological quantum computation-from basic concepts to first experiments. Science **339**(6124), 1179–1184 (2013)
69. S. Hyun et al., Coexistence of metallic and insulating phases in epitaxial $CaRuO_3$ thin films observed by scanning microwave microscopy. Appl. Phys. Lett. **80**(9), 1574–1576 (2002)
70. H. Takahashi, Y. Imai, A. Maeda, Observation of mesoscopic phase separation in Se_2 by scanning microwave microscopy. Physica C (Amsterdam, Neth.) **518**, 33–35 (2015)
71. W. Kundhikanjana et al., Direct imaging of dynamic glassy behavior in a strained manganite film. Phys. Rev. Lett. **115**(26) (2015)
72. H. Madan et al., Quantitative mapping of phase coexistence in mott-peierls insulator during electronic and thermally driven phase transition. ACS Nano **9**(2), 2009–2017 (2015)
73. A. Tselev et al., Mesoscopic metal-insulator transition at ferroelastic domain walls in VO_2. ACS Nano **4**(8), 4412–4419 (2010)
74. K. Lai et al., Imaging of coulomb-driven quantum Hall edge states. Phys. Rev. Lett. **107**(17), 176809 (2011)
75. M. E. Suddards et al., Scanning capacitance imaging of compressible and incompressible quantum Hall effect edge strips. New J. Phys. **14**(8), 083015 (2012)
76. Y. T. Cui et al., Unconventional correlation between quantum hall transport quantization and bulk state filling in gated graphene devices. Phys. Rev. Lett. **117**(18), 186601 (2016)
77. E. Y. Ma et al., Unexpected edge conduction in mercury telluride quantum wells under broken time-reversal symmetry. Nat. Commun. **6**, 7252 (2015)

Printed by Printforce, the Netherlands